T0139172

SPECTRAL
TECHNIQUES
IN
PROTEOMICS

SPECTRAL TECHNIQUES IN PROTEOMICS

Edited by

DANIEL S. SEM

CRC Press
Taylor & Francis Group
Boca Raton London New York

CRC Press is an imprint of the
Taylor & Francis Group, an **informa** business

CRC Press
Taylor & Francis Group
6000 Broken Sound Parkway NW, Suite 300
Boca Raton, FL 33487-2742

First issued in paperback 2019

© 2007 by Taylor & Francis Group, LLC
CRC Press is an imprint of Taylor & Francis Group, an Informa business

No claim to original U.S. Government works

ISBN-13: 978-1-57444-580-0 (hbk)
ISBN-13: 978-0-367-38928-4 (pbk)

This book contains information obtained from authentic and highly regarded sources. Reasonable efforts have been made to publish reliable data and information, but the author and publisher cannot assume responsibility for the validity of all materials or the consequences of their use. The authors and publishers have attempted to trace the copyright holders of all material reproduced in this publication and apologize to copyright holders if permission to publish in this form has not been obtained. If any copyright material has not been acknowledged please write and let us know so we may rectify in any future reprint.

Except as permitted under U.S. Copyright Law, no part of this book may be reprinted, reproduced, transmitted, or utilized in any form by any electronic, mechanical, or other means, now known or hereafter invented, including photocopying, microfilming, and recording, or in any information storage or retrieval system, without written permission from the publishers.

For permission to photocopy or use material electronically from this work, please access www.copyright.com (http://www.copyright.com/) or contact the Copyright Clearance Center, Inc. (CCC), 222 Rosewood Drive, Danvers, MA 01923, 978-750-8400. CCC is a not-for-profit organization that provides licenses and registration for a variety of users. For organizations that have been granted a photocopy license by the CCC, a separate system of payment has been arranged.

Trademark Notice: Product or corporate names may be trademarks or registered trademarks, and are used only for identification and explanation without intent to infringe.

Library of Congress Cataloging-in-Publication Data

Spectral techniques in proteomics / editor, Daniel S. Sem.
    p. ; cm.
  "A CRC title."
  Includes bibliographical references and index.
  ISBN-13: 978-1-57444-580-0 (alk. paper)
  ISBN-10: 1-57444-580-4 (alk. paper)
  1. Proteins--Spectra. 2. Proteomics--Methodology. 3. Mass spectrometry. I. Sem, Daniel S.
  [DNLM: 1. Proteomics--methods. 2. Mass Spectrometry--methods. 3. Spectrum Analysis--methods. QU 58.5 S741 2007]

QP551.S675 2007
572'.633--dc22
                             2006103310

Visit the Taylor & Francis Web site at
http://www.taylorandfrancis.com

and the CRC Press Web site at
http://www.crcpress.com

Dedication

In loving thanks to my wife, Teresa, and children, Lucas, Camille, and Isaac, for being a constant source of inspiration and support and for tolerating the countless hours I had to spend immersed in my laptop.

Table of Contents

PART III Protein–Protein (or Peptide) Interactions: Studies in Parallel and with Mixtures

PART IV Chemical Proteomics: Studies of Protein–Ligand Interactions in Pools and Pathways

Preface

A significant challenge in presenting an overview of *Spectral Techniques in Proteomics* is in defining the scope of the topic. Proteomics means different things to different people; for years, the dominant technique employed was 2D gel electrophoresis, followed by mass spectrometry (MS). While many exciting MS applications are presented (e.g., matrix-assisted laser desorption/ionization [MALDI], electrospray ionization [ESI], tandem MS, liquid chromatography [LC]-MS, surface-enhanced laser desorption/ionization [SELDI], isotope-coded affinity tag [ICAT]), a comprehensive survey of MS methods and applications in proteomics is certainly beyond the scope of this book. Since *Spectral Techniques in Proteomics* is intended for a broad audience of protein biochemists and biophysicists, topics such as *structural proteomics* and *chemical proteomics* will also be covered, along with fluorescence/array-based screening, SPR (surface plasmon resonance), and other "lab-on-a-chip" technologies.

Furthermore, a disproportionate amount of time will be spent on some less established spectroscopic methods in proteomics, with forward-looking speculation on future applications. The intention of this book is therefore to facilitate the innovation, development, and application of new spectroscopic methods in proteomics while giving a modest overview of existing and proven techniques. To this end, a broader view of proteomics is taken in order to include studies that go beyond the usual scope of 2D gels and MS, attempting to address function and mechanism at the level of protein–ligand interactions. After all, this is the realm in which protein spectroscopists have always excelled and felt most at home.

Proteomics is defined broadly as the "systems-based" study of proteins in the organelles, cells, tissues, or organs of an organism. It is the study of the protein complement of the genome in time and space. In practice, this definition sometimes limits proteomics to the study of proteins in 2D gels, since this is one of the few contexts in which so many proteins can be studied at once. It is also possible to simplify a proteome into a more manageable subset (a subproteome) by focusing on a smaller number of proteins related in a systems-based manner. Such systems of interrelated proteins can include: (1) regulatory cascades connected via protein–protein interactions; (2) metabolic pathways; (3) proteins with related modifications (acylation, phosphorylation, glycosylation, methylation, etc.); and (4) any collection of proteins associated with a biological effect, such as uncontrolled cell growth (cancer), an immune response, or drug metabolism. This approach to simplifying a proteome into systems-related subproteomes is described in chapter 1.

Thus, for purposes of this book, proteomic studies are extended to include the parallel study of subsets of related proteins, some of which were described previously. Such subsets might also include proteins that comprise a unique basis set of protein folds in an organism's proteome, as currently defines the scope of most *structural proteomic* initiatives. Another systems-related subset could comprise proteins with similar binding sites (chapter 2), as currently defines the scope of *chemical proteomic* studies. In this case, the subset of proteins can be considered to be part of a biologically relevant network in the sense that they represent all of the protein–ligand interactions that would occur when an organism is exposed to a given chemical perturbant (drug, pollutant, chemical genetic probe, etc.). One prominent example of such systems-related proteins is those that use the same cofactor or prosthetic group, such as kinases, which all bind ATP (chapter 18).

This broader definition of proteomics and systems-based studies is not only a convenience for framing proteome-wide questions, but also has biological relevance. This book aims to provide a broad overview of the spectroscopic toolbox that can be applied in such systems-based studies of proteins, whether they are studied in the context of proteome or subproteome mixtures (traditional proteomics) or as individual/purified proteins studied in parallel (the broader, systems-based view of proteomics), as in structural proteomics.

This book begins in part I by defining the scope of the field in order to give coherence to the chapters from the various expert contributors. Proteomics is defined in a way that is relevant to a spectroscopist (chapters 1 and 2) and then a very brief overview of commonly used spectroscopic methods is given (chapter 3). In part II, commonly used MS methods are presented, including separation techniques that typically precede ESI studies, as well as MALDI MS/MS-based protein identification. SELDI is presented as a tool that combines separation with MS analysis on the same chip. Part III focuses on studies of protein–protein interactions using a variety of techniques, including near-infrared (NIR) fluorescence, nuclear magnetic resonance (NMR), and MS. Part IV covers protein–ligand interactions with techniques ranging from MS to SPR to NMR. Recent developments in ICAT labeling strategies are covered, and the section ends with a discussion of metabonomics, since metabolites represent an important functional output of the proteome. Part V covers advances in structural proteomics using NMR, x-ray crystallography, and electron paramagnetic resonance (EPR). The book ends with chapter 20, a summary of current technology and future prospects extracted from the various contributors, again to give added coherence to the topic.

Spectral Techniques in Proteomics will be useful for graduate students and other scientists wanting to develop and apply spectroscopic methods in proteomics. It will also be of value to more experienced researchers thinking of moving into this field or those in proteomics looking to broaden the scope of their studies. In short, it is intended for anyone wanting to take a systems-based approach to studying proteins, their function, and their mechanisms using various spectroscopic tools.

Daniel Sem

Editor

Daniel Sem is an assistant professor in the chemistry department at Marquette University in Milwaukee, Wisconsin. He also serves as director of the Chemical Proteomics Facility at Marquette (CPFM) and is a member of the Marine and Freshwater Biomedical Sciences Center at the University of Wisconsin–Milwaukee in the endocrine disruptor core group. His current research is focused on the development and application of chemical proteomic probes for the study of protein–ligand interactions. Emphasis is on fluorescence and NMR-based assays, as well as on proteins that are drug targets and proteins that are antitargets, leading to the adverse and toxic side effects of drugs and pollutants.

Prior to joining Marquette, Dr. Sem cofounded Triad Therapeutics in San Diego, California, where he served as vice-president for biophysics. In that capacity, he was involved in NMR-based characterization of large protein–ligand complexes, cheminformatic characterization of combinatorial libraries, bioinformatic analysis of gene families, high-throughput screening, and enzymology/assay development. Triad was the first company founded around NMR-driven, structure-based drug design. It had a technology based on a systems-based approach to drug design, targeting gene families of proteins like kinases and dehydrogenases with focused combinatorial chemistry libraries.

Dr. Sem graduated from the University of Wisconsin–Milwaukee with a B.S. in chemistry (summa cum laude) and from University of Wisconsin–Madison with a Ph.D. in biochemistry, specializing in mechanistic enzymology. He then pursued postdoctoral studies at McArdle Laboratory for Cancer Research (Madison, Wisconsin), followed by the Scripps Research Institute (La Jolla, California), where he did NMR-based structural biology. He has 20 years of experience using spectral techniques to study protein–ligand interactions in basic and applied research settings.

Contributors

Alex Aronov
Vertex Pharmaceuticals Incorporated
Cambridge, Massachusetts

Peter E. Barker
Biotechnology Division
National Institute of Standards and
  Technology
Gaithersburg, Maryland

Guy Bemis
Vertex Pharmaceuticals Incorporated
Cambridge, Massachusetts

Richard Cammack
Department of Life Sciences
Pharmaceutical Sciences
  Research Division
King's College
London, United Kingdom

Lisa H. Cazares
The Center for Biomedical Proteomics
Eastern Virginia Medical School
Norfolk, Virginia

Volker Dötsch
Institute for Biophysical Chemistry
Center for Biomolecular Magnetic
  Resonance
University of Frankfurt
Frankfurt, Germany

Cynthia Enderwick
Ciphergen Biosystems Inc.
Fremont, California

Eric Fung
Ciphergen Biosystems Inc.
Fremont, California

Michael B. Goshe
Department of Molecular and
  Structural Biochemistry
North Carolina State University
Raleigh, North Carolina

Per Hägglund
Biochemistry and Nutrition Group
Technical University of Denmark
Lyngby, Denmark

Mark R. Hansen
Altoris, Inc.
San Diego, California

Brian Hare
Vertex Pharmaceuticals Incorporated
Cambridge, Massachusetts

Qiuhong He
Departments of Radiology and
  Bioengineering
University of Pittsburgh Cancer Institute
MR Research Center,
  University of Pittsburgh
Pittsburgh, Pennsylvania

Elaine Holmes
Department of Biomolecular Medicine
Faculty of Medicine
Imperial College London
South Kensington, London,
  United Kingdom

Marc Jacobs
Vertex Pharmaceuticals Incorporated
Cambridge, Massachusetts

Ole Nørregaard Jensen
Protein Research Group
Department of Biochemistry and
  Molecular Biology
University of Southern Denmark
Odense, Denmark

Richard Kho
Altoris, Inc.
San Diego, California

Uma Kota
Department of Molecular and
  Structural Biochemistry
North Carolina State University
Raleigh, North Carolina

Martin R. Larsen
Department of Biochemistry and
  Molecular Biology
University of Southern Denmark
Odense, Denmark

Kim Lau
Department of Biochemistry and
  Molecular Biology
Bio21 Molecular Science and
  Biotechnology Institute
University of Melbourne
Victoria, Australia

Walter S. Liggett
Statistical Engineering Division
National Institute of Standards and
  Technology
Gaithersburg, Maryland

John C. Lindon
Department of Biomolecular Medicine
Faculty of Medicine
Imperial College London
South Kensington, London,
  United Kingdom

Lee Lomas
Ciphergen Biosystems Inc.
Fremont, California

John L. Markley
Center for Eukaryotic Structural
  Genomics
National Magnetic Resonance Facility
  at Madison
Biochemistry Department
University of Wisconsin–Madison
Madison, Wisconsin

Amber L. Mosley
Stowers Institute for Medical Research
Kansas City, Missouri

David G. Myszka
Center for Biomolecular Interaction
  Analysis
School of Medicine
University of Utah
Salt Lake City, Utah

Christian Neusüß
Bruker Daltonik GmbH
Leipzig, Germany

Jeremy K. Nicholson
Department of Biomolecular Medicine
Faculty of Medicine
Imperial College London
South Kensington, London,
  United Kingdom

Matthias Pelzing
Bruker Daltonik GmbH
Leipzig, Germany

Al Pierce
Vertex Pharmaceuticals Incorporated
Cambridge, Massachusetts

Anthony W. Purcell
Department of Biochemistry and
  Molecular Biology
Bio21 Molecular Science and
  Biotechnology Institute
University of Melbourne
Victoria, Australia

Rebecca L. Rich
Center for Biomolecular Interaction
  Analysis
School of Medicine
University of Utah
Salt Lake City, Utah

Emmanuelle Sachon
Protein Research Group
Department of Biochemistry and
  Molecular Biology
University of Southern Denmark
Odense, Denmark

Daniel S. Sem
Chemical Proteomics Facility at
  Marquette
Department of Chemistry
Marquette University
Milwaukee, Wisconsin

O. John Semmes
The Center for Biomedical Proteomics
Eastern Virginia Medical School
Norfolk, Virginia

Jeremy Spater
Departments of Radiology and
  Bioengineering
University of Pittsburgh Cancer Institute
MR Research Center
University of Pittsburgh
Pittsburgh, Pennsylvania

Sunitha B. Thakur
Departments of Radiology and
  Bioengineering
University of Pittsburgh Cancer Institute
MR Research Center
University of Pittsburgh
Pittsburgh, Pennsylvania

Sakanyan Vehary
ProtNeteomix
Université de Nantes
Nantes, France

Hugo O. Villar
Altoris, Inc.
San Diego, California

Michael P. Washburn
Stowers Institute for Medical Research
Kansas City, Missouri

Andrew I. Webb
Department of Biochemistry and
  Molecular Biology
Bio21 Molecular Science and
  Biotechnology Institute
University of Melbourne
Victoria, Australia

Scot R. Weinberger
GenNext Technologies™
Montara, California

Nicholas A. Williamson
Department of Biochemistry and
  Molecular Biology
Bio21 Molecular Science and
  Biotechnology Institute
University of Melbourne
Victoria, Australia

Garabet Yeretssian
Biotechnologie, Biocatalyse,
  Biorégulation
Faculté des Sciences et des Techniques
Université de Nantes
Nantes, France

Harmon Zuccola
Vertex Pharmaceuticals Incorporated
Cambridge, Massachusetts

Abbreviations

2D two dimensional
AAs amino acids
ALICE acid-labile isotope-coded extractants
ANTS 8-aminonaphthalene-1,3,6-trisulfonate
APC antigen presenting cell
APCI atmospheric pressure chemical ionization
APTA (3-acrylamidopropyl)-trimethylammonium chloride
APPI atmospheric pressure photo ionization
AQUA absolute quantification
ATP adenosine triphosphate
BLAST basic local alignment search tool
BSA bovine serum albumin
CA-ENMR capillary-array ENMR
CBB Coomassie Brilliant Blue
CCA canonical correlation analysis
CC-ENMR convection-compensated ENMR
CD circular dichroism
CDK cyclin-dependent kinase
CE capillary electrophoresis
CEC capillary electrochromatography
CESG Center for Eukaryotic Structural Genomics
CGE capillary gel electrophoresis
CHD coronary heart disease
CID collision-induced dissociation
CIEF capillary isoelectric focusing
CLOUDS classification of unknowns by density superposition
COMET Consortium for Metabonomic Toxicology
CORES complexes restricted by experimental structures
COSY correlation spectroscopy
CRINEPT cross relaxation-enhanced polarization transfer
CSF cerebrospinal fluid
CT constant time
CTL cytotoxic T lymphocyte
CZE capillary zone electrophoresis
DA discriminant analysis
DCs dendritic cells
DC direct current
DHB 2,5-dihydroxybenzoic acid

DisProt Database of Protein Disorder
ECD electron capture dissociation
EGFR epidermal growth factor receptor
EI electron ionization
EIE extracted ion electropherogram
ELISA enzyme-linked immunosorption assay
ENMR electrophoretic NMR
EOF electro-osmotic flow
EPR electron paramagnetic resonance
ESI electrospray ionization
ESR electron spin resonance
FAB fast atom bombardment
FP fluorescence polarization
FRET fluorescence resonance energy transfer
FT Fourier transform
FTICR Fourier transform ion cyclotron resonance
GC-MS gas chromatography-mass spectrometry
GFP green fluorescent protein
GIST global internal standard technology
GlcNAc N-acetylglucosamine
GPCR G-protein coupled receptor
GPI glycosyl-phosphatidylinositol
GST glutathione-S-transferase
HCCA alpha-cyano-4-hydroxycinnamic acid
HILIC hydrophilic interaction liquid chromatography
HPLC high-performance liquid chromatography
HSQC heteronuclear single quantum coherence
HTS high-throughput screening
ICAT isotope-coded affinity tag
ICGs interchromatin granules
IEF isoelectric focusing
Ig immunoglobulin
IGOT isotope-coded glycosylation site-specific tagging
IMAC immobilized metal affinity chromatography
INEPT insensitive nuclei enhanced by polarisation transfer
IR infrared
Irk insulin receptor kinase
IRMPD infrared multiphoton photodissociation
IT ion trap
ITP isotachophoresis
LC liquid chromatography
LC/MS/MS liquid chromatography tandem mass spectrometry
LCM laser capture microdissection
LDI laser desorption/ionization
LDL low-density lipoprotein
LIF laser-induced fluorescence

LIMS laboratory information management systems
LLE liquid–liquid extraction
mAb monoclonal antibody
MALDI matrix-assisted laser desorption/ionization
MAS magic angle spinning
MCAT mass-coded abundance tagging
MEKC micellar electrokinetic chromatography
MEM maximum entropy method
MHC major histocompatibility complex
MS mass spectrometry
MS/MS tandem mass spectrometry
MudPIT multidimensional protein identification technology
MW molecular weight
NACE-MS nonaqueous CE-MS
NIR near infrared
NMR nuclear magnetic resonance
NMRFAM National Magnetic Resonance Facility at Madison
NOE nuclear Overhauser effect
NPC nuclear pore complex
PAGE polyacrylamide gel electrophoresis
PC phosphatidylcholine
PC principal component
PCA principal components analysis
PCR polymerase chain reaction
PDB Protein Data Bank
PECAN protein energetic conformational analysis from NMR chemical shifts
PhIAT phosphoprotein isotope-coded affinity tag
PhIST phosphoprotein isotope-coded solid-phase tag
PI-PLC phosphatidylinositol phospholipase C
PISTACHIO probabilistic identification of spin systems and their assignments
     including coil-helix inference as output
PKA c-AMP dependent kinase (protein kinase A)
PLS partial least squares
PML promyelocytic leukemia
PTM post-translational modification
Q quadrupolar (as in Q-TOF)
QC quality control
QD quantum dot
QqQ triple quadrupole
QUEST quantitation using enhanced signal tags
RCSB Research Collaboratory for Structural Biology
RDCs residual dipolar couplings
RMSD root mean square deviation
RP reverse phase
RPLC reverse phase liquid chromatography
RR resonance Raman

SA sinapinic acid
SAX strong anion exchange
SBDD structure-based drug design
SCX strong cation exchange
SDS sodium dodecyl sulfate
SEAC surface-enhanced affinity capture
SELDI surface-enhanced laser desorption/ionization
SEND surface-enhanced neat desorption
SEREX serological expression of cDNA expression libraries
SILAC stable isotope labeling by amino acids in cell culture
SMRS standard metabolic reporting structures
SPE solid phase extraction
SPITC 4-sulfophenyl isothiocynate
SPR surface plasmon resonance
STE stimulated echo
TAP tandem affinity purification
TcR T cell receptor
TLF time-lag focusing
TFA trifluoroacetic acid
TOF time of flight
TROSY transverse relaxation optimized spectroscopy
TSP 3-(trimethylsilyl) propionic 2,2,3,3-d_4 acid
VICAT visible isotope-coded affinity tag
VLDL very low density lipoprotein

Part I

The Scope of Proteomic and
Chemical Proteomic Studies

1 The Systems-Based Approach to Proteomics and Chemical Proteomics

Daniel S. Sem

CONTENTS

1.1 INTRODUCTION

Simply stated, proteomics is the study of the protein complement of a genome using the tools of protein biochemistry on a proteome-wide scale. It is devoted to monitoring changes in expression levels or post-translational modifications of all the proteins in an organism, organ, cell, or organelle as a function of time or biological state (e.g., diseased vs. healthy). Ideally, it should also address protein structure–function in terms of interactions with substrates, drugs, inhibitors, lipids, DNA, or other proteins.

It is possible to infer some information about protein expression levels based on changes in mRNA detected using microarray technology—an elegant coupling of microfluidics, "lab-on-a-chip," and detection (usually fluorescence-based) technologies. But, mRNA levels are not always correlated well with protein levels, and they reveal nothing about post-translational modification or protein interactions. As such, the field of proteomics serves an essential function, despite the additional technical challenges involved in analyzing proteins in comparison with polynucleotides [1–4]. Most significant is the challenge of dealing with sample complexity and dynamic range.

3

1.2 COMPLEXITY AND DYNAMIC RANGE CHALLENGES

The human genome comprises over 30,000 genes, which encode many more proteins; many are variants due to alternative splicing and PTMs (post-translational modifications). Over 400 PTMs are known to date, so there is tremendous complexity in the proteome as it is expressed in a given cell type. While it is a challenge to resolve the thousands of proteins expressed in a proteome, it is an even greater challenge because these proteins may be present at very different concentrations, ranging over six to nine orders of magnitude depending on the cell type. For example, serum contains albumin as the most abundant protein (at ~40 mg/mL and ~50% of blood protein), while other proteins of interest, such as interleukin-6, are present at <5 pg/mL [5]. The need to quantify so many proteins over such a wide concentration range is one of the greatest challenges in proteomics. Other challenges include the need to assess interactions between the proteins and their various ligands that define biological networks or systems. Furthermore, at least in some cases, these interactions should also be measured within a cell (chapter 15) to ensure their biological relevance in the context of potential accessory proteins or cofactors, as well as under physiologically relevant conditions (water activity, pH, ionic strength, lipids, etc.).

1.3 THE SYSTEMS-BASED APPROACH

1.3.1 SYSTEMS-BASED RELATIONSHIPS

Proteomics is considered a subdiscipline of systems biology. So what is systems biology? Weston and Hood define it as "the analysis of the relationships between elements in a system in response to genetic or environmental perturbations, with the goal of understanding the system or the emergent properties of the system" [6]. "System" is a broad term borrowed from other fields (e.g., engineering), but in a biological context the word usually refers to organelles, cells, organs, or organisms. Such a definition is therefore based largely on the physical location of the proteins studied, their network of interactions, and their collective role in defining a biological entity (such as a mitochondrion or a liver cell) or function (such as the immune response).

It is also possible to define systems at the level of molecules based on the network of interactions that occur between them, usually within the context of a single cell or organelle. This typically involves the measurement of all pairs of protein–protein interactions that can occur, using techniques such as the yeast two-hybrid system. In this manner it is possible to establish networks of proteins that participate in interactions with each other. These pairs of interactions, summed across a whole proteome, comprise a protein interaction map such as that shown in figure 1.1 and discussed in chapter 14 by Rich and Myszka.

In a broad sense, systems or networks of interacting proteins can be defined based on the presence of interactions between [7]:

(a) Two proteins directly, as in a regulatory cascade when a protein kinase phosphorylates another protein

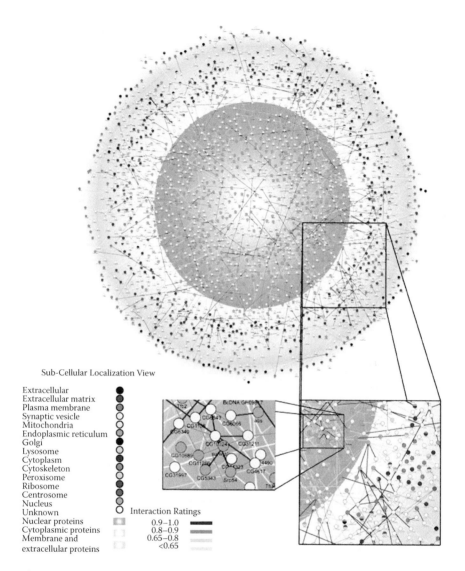

Sub-Cellular Localization View

Extracellular
Extracellular matrix
Plasma membrane
Synaptic vesicle
Mitochondria
Endoplasmic reticulum
Golgi
Lysosome
Cytoplasm
Cytoskeleton
Peroxisome
Ribosome
Centrosome
Nucleus
Unknown

Interaction Ratings
0.9–1.0
0.8–0.9
0.65–0.8
<0.65

Nuclear proteins
Cytoplasmic proteins
Membrane and
extracellular proteins

FIGURE 1.1 Systems relationships in a so-called "protein interaction map" for *Drosophilia melanogaster*. Proteins are coded by subcellular location as well as interactions (indicated with lines). Most probable interactions are indicated with darker lines. (Reprinted from Giot, L. et al., *Science*, 302, 1727–1736, 2003. With permission from the American Association for the Advancement of Science, 2003, and discussed further in chapter 14.)

(b) Two proteins that are sequential enzymes in a metabolic pathway, whereby they are related by binding to a common ligand that is a substrate for one enzyme and a product for another

(c) Two proteins that bind to a common ligand, such as a nonspecific drug that binds to multiple targets (as is common for protein kinase inhibitors) or even a cofactor (NAD(P)H binds to all dehydrogenases)

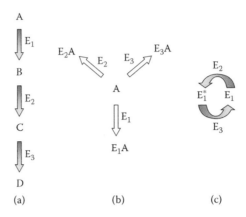

FIGURE 1.2 Systems relationships that are relevant in proteomics and *chemical proteomics*. Proteins can be related by: (a) *metabolic pathway*, with protein pairs related by binding to the same molecule, which is a substrate for one enzyme and a product for another; (b) *ligand binding*, where protein pairs are related by binding the same ligand (not necessarily enzyme and substrate); or (c) *regulatory cascade*, where proteins interact directly with each other, as when a protein kinase phosphorylates another protein. (Adapted from Sem, D.S., *Expert Rev. Proteomics*, 1, 165–178, 2004.)

These interactions are defined schematically in figure 1.2 and are central to defining the field of *chemical proteomics*.

1.3.2 SUBPROTEOMES

To simplify the complexity of a proteome in order to make proteomic studies more manageable, it is common to create subproteomes—that is, to study a subset of the whole proteome, where that subset is defined based on some systems-based relationship between proteins. Proteins can be grouped into subproteomes in the many ways mentioned in the previous sections. These include:

1. location within the cell (e.g., golgi, lysosome, nucleus)
2. participation in a large functionally defined protein complex (e.g., ribosome, transcription initiation complex, cytoskeleton)
3. shared post-translational processing (e.g., phosphorylation, glycosylation)
4. affinity for a ligand (e.g., ATP, NAD(P)H, drugs)
5. chemically reactive groups (e.g., cysteine thiols, lysine amines)
6. shared biological function (e.g., immunoproteome)

These groupings are based on physical location in the cell, chemical properties (ligand binding, PTM), or functional role. Classifications were made based on practical considerations in that each provides a means to isolate the subproteome, although it is often not possible to isolate category 6 subproteomes. Systems defined by networks of protein–protein interactions, as identified in figure 1.1, are often not easy to isolate.

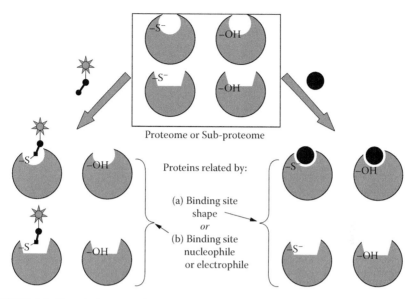

FIGURE 1.3 Chemical proteomic studies. Such studies employ probes to "profile" proteins. This can be done using: (a) activity-based profiling, with probes that covalently label proteins (left arrow); or (b) affinity-based profiling, with probes that noncovalently bind to proteins (right panel). (Adapted from Sem, D.S., *Expert Rev. Proteomics*, 1, 165–178, 2004.)

Subproteomes categorized based on physical location in the cell are the easiest to define since separation techniques permit isolation of organelles [8]. The most general breakdown of cellular location is into the nucleus, cytoplasm, or membrane/extracellular region. A more detailed breakdown would include the 14 categories of subcellular locations and organelles defined in figure 1.1 [9]. Subproteomes defined based on protein–ligand interactions are presented at greater length later in this book (part IV) and fall largely within the scope of chemical proteomics, described in the next section.

1.3.3 CHEMICAL PROTEOMICS

Chemical proteomics is a branch of proteomics [7,10–14] with a focus on directly detected protein–ligand interactions measured across a systems-related group of proteins. It is therefore a mechanistic complement to chemical genetics. In chemical genetics, the observable is a phenotypic change induced by a chemical knockout [15], but without any direct characterization of protein–ligand interactions. Bogyo defined chemical proteomics as being focused on the structure, function, and role of proteins in different biological systems using chemical probes [10]. Thus, the field relies heavily on chemical probes to define systems-related proteins. These probes can be activity based (covalently labeling all proteins that share a common electrophile or nucleophile), as illustrated in the left side of figure 1.3.

An example might be labeling of all thiol groups, as done with the ICAT (isotope-coded affinity tag) technology pioneered by Aebersold [16] and presented in chapter 13.

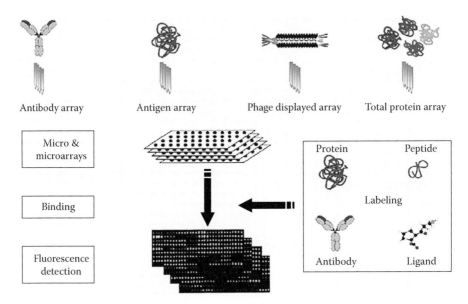

FIGURE 1.4 Various subproteomes quantified using fluorescence detection of microarrays. Different ways of capturing subproteomes are shown using antibodies, antigens, ligands, and other affinity–capture arrays. Discussed further in chapter 10.

Detection of probes can also take place by using fluorescence, as developed by Bogyo and others [10,17]. Probes can also be affinity based (right side of figure 1.3), where they bind noncovalently to families of proteins [7,12,18] with related binding sites (so-called pharmaco families [14,19,20]). Many subproteomes are isolated and characterized based on common binding sites, so such classification is extremely important for the fields of proteomics and chemical proteomics. It is also important in drug design, where binding site similarities can determine how specific a drug is for its intended protein target relative to antitargets, which include drug-metabolizing enzymes as well as proteins that produce undesired side effects. The science of classifying protein binding sites based on ligand-binding preferences is a growing field that has evolved independently of proteomics. It is discussed in detail in chapter 2 by Villar et al.

1.3.4 APPLICATIONS

Most proteomic studies now employ mass spectral detection, usually of proteins extracted from 2D gels (chapter 4) or fluorescence studies of microarrays (chapter 10 and fig. 1.4). An alternative to matrix-assisted laser desorption/ionization (MALDI) analysis of extracted proteins is to use in-line purification of proteins by capillary electrophoresis (CE), followed by electrospray ionization-mass spectrometry (ESI-MS) (chapter 4). Although these methods provide good resolution, to help address the complexity problem in proteomics, they usually also require further simplification of proteome samples. The systems-based approaches described earlier for simplifying proteomes are therefore crucial in most proteome studies. In particular, it is

Biochemical Surfaces for Specific Protein Interaction Studies

| Proactivated surface | Antibody–antigen | Receptor–ligand | DNA–protein |

FIGURE 1.5 Surfaces used on SELDI chips to select for various subproteomes. Discussed further in chapter 7.

routine to analyze subproteomes rather than entire proteomes. This book presents studies of many different subproteomes, including:

- immunoproteomes (chapters 9, 10)
- glycoproteomes (chapters 6, 13)
- phosphoproteomes (chapter 13)
- transcriptional regulatory pathways (chapter 11)
- ATP-binding proteins/protein kinases (chapter 18)
- the metabonomes (chapter 16)

Many of these studies require some technique to first purify the desired subproteome, usually based on an affinity purification step before MS analysis. As an alternative, surface-enhanced laser desorption/ionization-mass spectrometry (SELDI-MS; chapter 7) employs affinity purification on the chip itself, which is coated with an appropriate ligand that captures the desired subproteome (fig. 1.5). Related affinity-capture techniques are also possible using microarrays (fig. 1.4).

Finally, it should be noted that "systems" of proteins typically include pools of proteins that define a subproteome, as is common in proteomics and in most of the chapters of this book. But systems of proteins subjected to proteomic studies might also include purified/single proteins studied in parallel. That is, for a pool of N systems-related proteins, one can study: (1) one pool of N-proteins in a mixture; or (2) N individual proteins in parallel [7]. The latter approach is taken in the field of *structural proteomics*, using methods such as nuclear magnetic resonance (NMR; chapter 17), x-ray crystallography (chapter 18), or even electron paramagnetic resonance (EPR; chapter 19).

1.4 SUMMARY

The most significant challenge in proteomics is how to detect so many proteins that cover such wide concentration or dynamic ranges. One solution is to simplify proteomes into subproteome fractions and then to analyze these. Subproteomes are defined as proteins related in a systems-based manner so that they can be physically isolated for study. In this regard, some of the most practical subproteomes for spectral studies are those associated with organelles (isolated by centrifugation) or those defined by binding to certain classes of ligands (isolated by affinity chromatography).

The latter classification is central to the field of chemical proteomics and to most of the studies presented in this book.

1.5 FUTURE PROSPECTS

The complete systems-based characterization of a proteome will ultimately provide a description of how that system responds to a biological stimulus such as exposure to an environmental insult like a pollutant or a drug. Such a complete characterization of a proteome would also provide an explanation of the underlying biology that differentiates a disease state from a healthy state. To achieve this goal, the map of protein interactions in figure 1.1 should be expanded to include protein–ligand or protein–DNA interactions and should indicate relative levels of the various proteins as well as the subcellular localization of the proteins involved. This map should also indicate how the system changes over time after exposure to a biological stimulus. Ideker et al. [1] have made a significant step towards creating such a comprehensive proteome map that includes changes in expression levels along with all protein–protein and protein–DNA interactions associated with galactose use in yeast.

However, more is needed to describe a proteome fully. Such maps could be extended to include PTMs, levels of mRNA, and levels of metabolites. Improved spectral tools for analyzing proteomes will be needed to create such maps. Furthermore, the complexity of visualizing and analyzing all of this information has created a need for improved bioinformatic tools, which are rapidly evolving along with supporting databases. To help avoid data overload and to coordinate the growing volume of proteomic data, the Human Proteome Organization (HUPO) has a Web site to centralize this information: http://www.hupo.org/information/mission.htm.

The goal of monitoring changes in proteomes to identify biological states associated with pathology is also important in a practical sense because it permits the early diagnosis of disease. An exciting advance on the horizon in this regard is the use of SELDI-MS to profile proteomes (chapter 7) by comparing normal and disease states and then using these profiles to predict disease. Early successes in predicting ovarian [21] and prostate [22] cancer have been reported. But before such a technique can find wide clinical applications [6], certain issues need to be resolved, such as reproducibility of the profile data (chapter 8) as well as ascertaining the acceptability of diagnosing based on a profile that involves unknown proteins. It is anticipated that future advances will address these concerns, along with improvements in the SELDI separation technology to permit analysis of different subproteome fractions. The latter would also be important in permitting more comprehensive and faster chemical proteomic studies. Also, metabonomic data are increasingly used to diagnose diseases, with many successes reported in clinical settings (see chapter 16).

Finally, studies of protein–ligand interactions across a proteome are most relevant if done in the context of a living cell or even a multicellular organism (*in vivo*). To this end, recent developments in molecular imaging [23–25] will be an important complement to *in vitro* proteomic studies. One exciting advance in this regard is the use of NMR to provide structural information about protein–ligand interactions inside living cells [26], as presented in chapter 15.

REFERENCES

1. Ideker, T., Thorsson, V., Ranish, J.A., Christmas, R., Buhler, J., Eng, J.K., Bumgarner, R., Goodlett, D.R., Aebersold, R., and Hood, L., Integrated genomic and proteomic analyses of a systematically perturbed metabolic network, *Science*, 292, 929–934, 2001.
2. Gygi, S.P., Rist, B., Gerber, S.A., Turecek, F., Gelb, M.H., and Aebersold, R., Quantitative analysis of complex protein mixtures using isotope-coded affinity tags, *Nat. Biotechnol.*, 17, 994–999, 1999.
3. Griffin, T.J., Gygi, S.P., Ideker, T., Rist, B., Eng, J., Hood, L., and Aebersold, R., Complementary profiling of gene expression at the transcriptome and proteome levels in *Saccharomyces cerevisiae*, *Mol. Cell. Proteomics*, 1, 323–333, 2002.
4. Baliga, N.S., Pan, M., Goo, Y.A., Yi,E.C., Goodlett, D.R., Dimitrov, K., Shannon, P., Aebersold, R., Ng, W.V., and Hood, L., Coordinate regulation of energy transduction modules in *Halobacterium* sp. analyzed by a global systems approach, *Proc. Natl. Acad. Sci. U.S.A.*, 99, 14913–14918, 2002.
5. Anderson, N.L., and Anderson, N.G., The human plasma proteome: history, character, and diagnostic prospects, *Mol. Cell. Proteomics*, 1, 845–867, 2002.
6. Weston, A.D., and Hood, L.H., Systems biology, proteomics, and the future of health care: toward predictive, preventative, and personalized medicine, *J. Proteome Res.*, 3, 179–196, 2004.
7. Sem, D.S., Chemical proteomics from an NMR spectroscopy perspective, *Expert Rev. Proteomics*, 1, 165–178, 2004.
8. Wilson, K., Walker, J., and Wilson, J.M., *Principles and Techniques of Practical Biochemistry*, Cambridge University Press, New York, 2000.
9. Giot, L. et al., A protein interaction map of *Drosophila melanogaster*, *Science*, 302, 1727–1736, 2003.
10. Jeffery D., and Bogyo, M., Chemical proteomics and its application to drug discovery, *Curr. Opin. Biotechnol.*, 11, 602–609, 2000.
11. Adam, G.C., Sorensen, E.J., and Cravatt, B.F., Chemical strategies for functional proteomics, *Mol. Cell. Proteomics*, 1, 781–790, 2002.
12. Pullela, P.K., and Sem, D.S., NMR-driven chemical proteomics: The functional and mechanistic complement to proteomics. In *Separation Methods in Proteomics*, ed. Smejkal, G.B. and Lazarev, A., CRC Press, Boca Raton, FL, 467–487, 2006.
13. Sem, D.S., Villar, H., and Kelly, M., NMR on target, *Mod. Drug Discovery*, August 26–31, 2003.
14. Sem, D.S. et al., Systems-based design of bi-ligand inhibitors of oxidoreductases: Filling the chemical proteomic toolbox, *Chem. Biol.*, 11, 185–194, 2004.
15. Peterson, R.T., Link, B.A., Dowling, J.E., and Schreiber, S.L., Small molecule developmental screens reveal the logic and timing of vertebrate development, *Proc. Natl. Acad. Sci. USA*, 97, 12965–12969, 2000.
16. Gygi, S.P., Rist, B., Gerber, S.A., Turecek, F., Gelb, M.H., and Aebersold, R. Quantitative analysis of complex protein mixtures using isotope-coded affinity tags. *Nat. Biotechnol.*, 17, 994–999, 1999.
17. Greenbaum, D., Medzihradszky, K.F., Burlingame, A., and Bogyo, M., Epoxide electrophiles as activity-dependent cysteine protease profiling and discovery tools. *Chem. Biol.*, 7, 569–581, 2000.
18. Yao, H., and Sem, D.S., Cofactor fingerprinting with STD NMR to characterize proteins of unknown function: Identification of a rare cCMP cofactor preference, *FEBS Lett.*, 579, 661–666, 2005.

19. Kho, R. et al., A path from primary protein sequence to ligand recognition, *Proteins*, 50, 589–599, 2003.

20. Kho, R. et al., Genome-wide profile of oxidoreductases in viruses, prokaryotes, and eukaryotes, *J. Proteome Res.*, 2, 626–632, 2003.

21. Petricoin, E.F. et al., Use of proteomic patterns in serum to identify ovarian cancer, *Lancet*, 359, 572–577, 2002.

22. Petricoin, E.F., III, et al., Serum proteomic patterns for detection of prostate cancer, *J. Natl. Cancer Inst.*, 94, 1576–1578, 2002.

23. Meade, T.J., Taylor, A.K., and Bull, S.R., New magnetic resonance contrast agents as biochemical reporters, *Curr. Opin. Neurobiol.*, 13, 597–602, 2003.

24. Nivorozhkin, A.L. et al., Enzyme-activated Gd^{3+} magnetic resonance imaging contrast agents with a prominent receptor-induced magnetization enhancement, *Angew. Chem. Int. Ed. Engl.*, 40, 2903–2906, 2001.

25. Costache, A.D., Pullela, P.K., Kasha, P., Tomasiewicz, H., and Sem, D.S., Homology-modeled ligand-binding domains of zebrafish estrogen receptors α, β_1 and β_2: From *in silico* to *in vivo* studies of estrogen interactions in *Danio rerio* as a model system, *J. Molecular Endocrinol.*, 19, 2979–2990, 2005.

26. Serber, Z., Ledwidge, R., Miller, S.M., and Dotsch, V., Evaluation of parameters critical to observing proteins inside living *Escherichia coli* by in-cell NMR spectroscopy, *J. Am. Chem. Soc.*, 123, 8895–8901, 2001.

2 Similarities in Protein Binding Sites

Hugo O. Villar, Mark R. Hansen, and Richard Kho

CONTENTS

2.1 INTRODUCTION

Each protein is unique. Each has a unique primary sequence, folds into a unique tertiary structure, and carries out a unique set of functions. Despite each protein's uniqueness, there are commonalities that can be observed across even the most diverse set of proteins. The conserved aspects of protein structure and function have generated fundamental insights, some of which may have important implications for drug discovery. We are particularly interested in the features that are conserved in protein binding sites because they can have implications for the development of novel techniques in drug design.[1] The success of fragment-based[2] inhibitor design, popular in methods that use nuclear magnetic resonance (NMR), depends on the cross-reactivity of scaffolds that can span the diversity of protein binding sites.[3] If no similarities were found across proteins, then it stands to reason that the libraries would be much larger than if similarities were found. Conserved features are also important for proteomics because they allow the classification of proteins and consequently provide a framework for the organization of the proteome. In our case, the classification we seek is at the interface with chemistry, where patterns of ligand binding can be used to group or differentiate proteins, giving rise to the subfield of chemoproteomics.

The fact that conserved features exist in proteins can be deduced from a large number of indirect observations. For example, all small molecular weight drugs have side effects, most of which are mediated via interactions with proteins other than the intended target. Also, those molecules are metabolized by proteins other than the target and therefore proteins involved in transport or metabolism should also share some similarities with the protein target. For quite some time,[4] the literature has reported that small molecules can commonly bind to multiple proteins. The converse is true as well: A single protein can interact with a variety of chemicals, even though they may have little resemblance to each other.

The process of small-molecule recognition by a protein is complex and should not be oversimplified.[5] Most of our observations and insights gained through the years have been due to the data coming from x-ray crystallography. However, these data provide only a snapshot of the binding process and do not present the dynamic aspects, which may play a significant role. Other spectroscopic techniques are essential to provide complementary information to develop a more thorough understanding of the interaction process.

This chapter will provide a broad but not comprehensive overview of some of the current state of the art in protein binding site characterization, ligand recognition, and classification of proteins based on binding sites. The results are presented in the context of small-molecule drug design, with two major emphases. The first is with respect to the role of similarities in protein binding sites in modular approaches for drug discovery. The second is on how these similarities are incorporated into technologies and information mining strategies to take advantage of the ever increasing amount of data on protein–ligand interactions and binding site characterization.

2.2   THE PROCESS OF LIGAND RECOGNITION

The first theoretical models of small-molecule protein interaction were based on the lock and key concept. Despite its simplicity, Fischer's theory accurately describes the need to have complementarities between the protein and the ligand.[6] Unlike locks and keys, however, ligand and protein are dynamic entities. The number of cross-interactions between proteins and ligands are problematic for the lock and key model because small molecules and proteins are promiscuous; a key can fit multiple locks and a lock can accept a variety of keys.

The induced-fit model proposed by Koshland appears more reasonable in this respect[6] because it views the process as an adaptation of the structures of the ligand and the receptor to each other. Since the proteins are adapting to the ligand and vice versa, it is possible to see how promiscuous compounds and proteins may arise. The lock and key model can then be viewed as a snapshot of the induced fit that occurs when the protein and ligand conformations are the same in the unbound states as in the bound state. Proteins are not in a single static conformation, but rather in a statistical ensemble in thermodynamic equilibrium. NMR and other spectroscopic techniques[7-10] corroborate this idea, first introduced by Straub.[11] Because the proteins are in equilibrium among a number of preexisting conformations, displacement of the equilibrium can occur upon ligand binding, towards the protein conformer that has the most favorable interactions with the ligands.

The conformational energy space for the protein defines the accessibility of different protein conformers. If the native state of the protein is characterized by multiple minima, with low energy barriers separating the different conformations, the lock and key model would be far from a reasonable representation. A complete flexibility on the part of the protein corresponds to an induced fit mechanism and, in that case, the protein could adapt to the shape and requirements of the ligand. When looking at the protein as an ensemble of different conformations in thermo-dynamic equilibrium, the binding pocket is observed to be flexible because it can exist in the different states that coexist in equilibrium, from which the ligands can select the most favorable conformation. Therefore, in general, the structural confor-mation of the binding site is dictated by the ligand, as it optimizes its interactions with the ensemble of conformations that the protein can assume.[12]

The different models of ligand recognition may, however, coexist in a single process. For example, a stability study on 16 structurally diverse proteins was used to study flexibility in the binding site.[13] Binding sites appear to have regions of high structural stability and regions with low structural stability. Highly stable regions may in fact behave as a lock, while the more flexible regions may be able to undergo induced fit.

The ability of ligands to bind to proteins does not depend only on the types of residues available at the contact interface or the ability of the protein or ligand to accommodate each other via an accessible conformational change.[14] Other crucial elements are the solvation and desolvation processes of the interacting pair. The environment in which the interactions take place often dictates the solvation effects. Some binding pockets are enclosed in a deep, hydrophobic space within the protein, while others are more open and have greater exposure to solvent.[15] The energies involved in desolvation of ligand and protein are critical determinants of binding and therefore solvation can alter the ability of the ligands to interact with the target protein.

2.3 AMINO-ACID PREFERENCE IN BINDING SITES

Protein binding sites are characterized as having certain amino acids with properties that confer an ability to form binding interactions. The structural data available in protein crystallographic databases have shown that hydrophobic residues are over-utilized in the interior of proteins, and hydrophilic amino acids abound at the surfaces. It has long been known that antibodies have complementarity-determining regions with a distinct frequency of specific amino acids like Tyr and Trp.[16] Despite the limited types of amino acids present, the differential usage of these residues was proposed to account for the specificity observed in antibodies. A similar finding has been described for enzymes.[17,18] Large bulky amino acids, such as Trp and Tyr, and His and Arg, are overrepresented in binding sites compared to bulk protein. Even in the case of protein–protein interactions, where a small number of surface residues are responsible for most of the energy of interaction, the residues at the interface have a composition very different from that of the rest of the protein.

The preference for certain residues in antibodies, enzyme binding sites, or at protein–protein interaction sites limits the number of possible binding motifs, which in turn limits the specificity that can be achieved. At the same time, it suggests that certain features may repeat across different proteins, which could be useful for classification of binding sites.

The fact that commonalities exist in binding sites is reinforced by the observation that a bound ligand can stabilize proteins against thermal denaturation. When screening for ligands using this property, up to 10% of the hits were found to have biologically relevant activities and consequently reflect binding at the active site or a modulating site.[19] The high hit rate is clearly larger than what would be expected if the ligands were binding at random throughout the protein surface. It suggests that binding sites, as compared to other regions of the protein, have characteristics that make them hospitable to ligands. All of these studies support the idea of amino-acid preferences found in protein binding sites.

The prevalence of conserved residues has been used to predict the location of binding sites without the use of structural information. Neural network algorithms using protein sequence profiles were developed to successfully identify sites of protein–protein[20] and protein–DNA[21] interactions. Furthermore, a study by La et al.[22] showed that binding pockets in proteins could be identified from sequence information using phylogenetic motifs, which are sequence regions conserved in a protein family phylogeny.[22]

2.4 CONSERVED SEQUENCE AND STRUCTURAL MOTIFS AT BINDING SITES

Although sequence information is certainly useful, structural characterization of binding sites affords even greater utility. Similar spatial arrangements of particular residues in different proteins have been known for some time. The most familiar example is probably that of serine proteases, which share the same catalytic triad despite having diverse folding motifs and over 60 different phylogenetic families.[23] The triad of histidine, aspartate, and serine residues is arranged similarly in three-dimensional space. A fatty acid cleaving protein, lipase, is a key enzyme in the regulation of lipids and shares this same catalytic triad,[24] while acetylcholinesterases have a very similar triad, with a conservative substitution of glutamate for aspartate.[25]

Nucleotide cofactor recognition by proteins also shows remarkable similarities despite the considerable differences in primary sequence and chain folding. The nucleotide recognition domain in glycosyltransferases[26] is also found in multiple species. Recognition of adenylate by structurally diverse proteins has been shown to have a characteristic signature in overall energy calculations, despite variations in the utilization of different residues.[27] From the ligand point of view, most cofactors show some significant degree of conformational similarity when bound. The conservation was demonstrated for glutathione[28] and adenylate, which show substantial similarity in the ligand conformation when bound to the protein. As it was pointed out, the conservation of bound ligand conformation is a constraint that can be used even if the binding motifs on the complementary protein active site are not obviously homologous in other proteins.

The classification of proteins based on binding site characteristics does not necessarily agree with the classifications carried out by other means. Cappello et al.[29]

showed that characterization of the adenine binding sites in terms of the possible hydrogen bond patterns can be used as a protein classification scheme. The resulting classification does not correspond to other classifications based on sequence or structure. Differing architectures of the binding site can provide for similar patterns of binding. For adenine, hydrophobic residues involved in stacking interactions with the aromatic portions of the ligand were found to be especially variable. Nevertheless, proteins with a different fold or even belonging to a different protein class can have adenine binding sites with similar properties in terms of the interface composition and hydrogen bond interaction patterns.

In the case of the ubiquitous cofactor, nicotinamide adenine dinucleotide (NAD), we found that it was possible to make a connection between the sequences of proteins that utilize NAD and the NAD binding conformation.[30] Our laboratory employed sequence clustering algorithms to characterize all NAD utilizing enzymes. The Swiss–Prot Database was chosen as the source for most of the sequence information related to NAD(P)-dependent oxidoreductases due to the high level of annotation. There were 4,613 enzyme sequences that utilize NAD(P) to perform their enzymatic functions. These sequences were subjected to an all-against-all sequence comparison using the basic local alignment search tool (BLAST), and the sequence identities were used to populate a similarity matrix. Divisive hierarchical cluster analysis grouped the sequences into 94 distinct sequence families. These sequence families correlated strongly with protein fold classifications whenever structural information was available for multiple members of a sequence family.

Among the 94 sequence families, 53 were structurally characterized at the time. Each of the structurally characterized proteins in a sequence family correlated to a single protein fold and, remarkably, to a common bound conformation for NAD(P). Analysis of the crystal structures of oxidoreductases with bound NAD(P) cofactor revealed 16 different conformations and, in every case, a sequence family for which structural information was available corresponded to one and only one cofactor bioactive conformation.

The results of this study are interesting because the protein classifications were carried out without incorporating protein structure information—that is, from sequence information alone.[30] The correlation of sequence to NAD(P) bound conformation would suggest that protein sequences of certain gene families, when combined with the appropriate classification techniques, can be used to predict ligand binding conformation. Somehow, for the large oxidoreductase gene family, the sequences contain information about the ligand-preferred orientation and consequently the relative organization of the binding site. The method was used extensively[31] to mine the complete genomes of 25 organisms representing bacteria, protists, fungi, plants, and animals, and 811 viruses, to identify and classify NAD(P)-dependent enzymes.

In general, the distribution of these enzymes by oxidoreductase family was correlated to the number of different catalytic mechanisms in each family and suggests another important aspect of the studies. A better understanding of the ligands that proteins bind can provide us with insight into the potential mechanisms and functions of the enzymes. The sequence-based clustering methods group proteins according to the recognition motifs they use, even when they may have different overall fold. This is critical for modern techniques of drug discovery that rely on conserved binding site features to create chemical libraries for entire families of protein targets.[32]

2.5 THREE-DIMENSIONAL DESCRIPTORS FOR BINDING SITE CHARACTERIZATION

The use of three-dimensional pharmacophore descriptors derived from protein binding sites has recently been proposed to classify proteins.[33] The method relies on descriptors formed from molecular fragments that have been docked, minimized, filtered, and clustered in protein active sites. It builds upon a drug discovery technique—multiple copy simultaneous search (MCSS)—used for the buildup of ligands in binding sites.[34] The MCSS method is utilized to search for optimal positions and orientations of a set of functional groups. These fragments provide coordinates from their position, which in turn provide a summary of the shape, electrostatics, locations, and angles of entry into pockets of recognition sites. The descriptors can be used to correlate the active site pharmacophores with activity or function, although with a mixed degree of success.

A particular concern for such a technique is how protein flexibility or minor changes in structure might influence the results. The descriptors are robust with respect to small changes in protein structure, as shown by a number of compounds that were cocrystallized in a protein. While the classification technique works well with tight protein families that have small root mean square deviations among family members, protein families having larger variations in active site structure are not classified optimally. For example, nuclear receptors give a tightly correlated group despite the variety of ligands considered, whereas in metalloproteases, their overall shape is a less useful feature in classifying members of that family.[33] Nevertheless, the method is of particular interest for drug discovery because it demonstrates a correlation between binding site descriptors and biological classes, based on the characteristics of small molecules.

A related method was presented in the literature[35] in which an interaction fingerprint was defined such that a one-dimensional binary string was used to represent three-dimensional structural binding information from a protein–ligand complex. Each fingerprint represents the "structural interaction profile" of the complex that can be used to organize, analyze, and visualize information encoded in ligand–receptor complexes.[35] The method was used to analyze approximately 90 known x-ray crystal structures of protein kinase-inhibitor complexes obtained from the Protein Data Bank. The fingerprints allowed organization of the structures in terms of the similarities and diversity among their small-molecule binding interactions.

Knowledge from the exponentially growing body of structurally characterized protein–ligand complexes will increasingly be exploited in structure-based drug design.[32] These types of classifications based on the characteristics of protein–ligand interactions can be facilitated by receptor ligand databases. Relibase[36] was developed as a database system particularly designed to handle protein–ligand-related problems and tasks. Features of Relibase include the detailed analysis of superimposed ligand binding sites, ligand similarity and substructure searches, and three-dimensional searches for protein–ligand and protein–protein interaction patterns.

2.6 EXPERIMENTAL EVIDENCE OF BINDING SIMILARITIES

An alternative approach to the characterization of the properties of binding sites is to supplement computational parameters with experimental information. A large data collection comprising affinity estimates of small molecules for different proteins was generated. All measurable interaction strengths were recorded, including modest affinities that would be disregarded in most pharmacological screens. These affinity values can then be used as descriptors for the proteins and their ligands.[37–39] As the database grew, patterns started to emerge in the data that suggested the existence of statistical relationships among affinity values, even when the proteins were unrelated by structure or function. In many cases, the relationships adopt a linear form, where the affinity of a compound for a given protein could be expressed as a weighted sum of its affinity for other unrelated proteins.

This is a very important observation that suggests the presence of redundancies in the data that are accumulated in screening. In assembling a database of information intended for wide use in drug discovery, the redundancies should be minimized. Based on this idea, data for over 500 proteins was reduced to a significantly smaller set of less than 20 proteins that retained essentially all the information.[38] The set containing the most information was chosen based on orthogonalization procedures.[39] As a result, this subset of proteins, as a reference set, adequately represents all the data in the set studied to date. Because the original database included diverse proteins with few structural or functional similarities,[40] it is also possible that the smaller reference panel can represent as-of-yet untested or uncharacterized proteins. It is an interesting phenomenon that a small reference panel could potentially represent most of the protein capabilities of the proteins that can be found in a proteome.

Another study profiled a family of proteins using 20 kinase inhibitors, including 16 that are approved drugs or in clinical development, by analysis against a panel of 119 protein kinases.[41] Specificity was found to vary widely and is not strongly correlated with chemical structure or the identity of the intended target. The results represent a systematic, small molecule–protein interaction map for clinical compounds across a large number of related proteins.

There is a clear interest in exploiting the information contained in chemical databases of protein–ligand interactions and in the development of new tools centered on the use of such information. For the most part, the basic concept is simple in that compounds that share similarities in their binding to a set of proteins are expected to elicit similar pharmacological responses. The BioPrint database was constructed by systematic profiling of nearly all drugs available on the market, as well as numerous reference compounds.[42,43] The database is composed of several large datasets: compound structures and molecular descriptors; *in vitro* absorption, distribution, metabolism, and excretion (ADME) and pharmacology profiles; and complementary clinical data including therapeutic use information, pharmacokinetics profiles, and adverse drug reaction (ADR) profiles. The platform represents a systematic effort to enhance the use and reliability of *in silico* methods to predict potential clinical liabilities.[44]

The experimental activity profiles define an "activity space" in which drugs and reference compounds are positioned in coordinates that describe inhibitory propensities, thereby unambiguously characterizing a molecule in terms of its receptor binding properties. Even if their implementation is novel, conceptually similar approaches have been described in the literature. These approaches implicitly acknowledge the existence of similarities in binding sites and the transferability of information that can be obtained even among unrelated targets.[37–43]

The binding process can be simulated by computational means using docking procedures.[45] The redundancy in the data is also observed in docking scores. Even in that case, the scores for the interaction between a series of proteins and a ligand can be used to represent the interaction scores of an unrelated protein for the same set of compounds. As with the experimental results, transferability of binding information is found even when there is no obvious primary or tertiary homology among the structures. Some general characteristics of the binding site such as its shape and size may limit the types of ligands that can be favorably accommodated. As such, the correlations observed could be a reflection of the types of compounds a site is unable to recognize as much as what types of compounds it actually prefers to bind.

2.7 EXPLOITING THE SIMILARITIES IN PROTEIN BINDING SITES: MODULAR APPROACHES TO DRUG DESIGN

The major focus of applied work on protein–ligand interactions is for purposes of drug design. The existence of similarities in binding sites has multiple implications for those endeavors. The first is that a drug is not a key that can open only one lock. Rather, all drugs show varying degrees of interaction with a number of proteins. Successful drugs, therefore, are not ones that interact exclusively with a target of interest. Rather, they are compounds that display the highest affinity for protein targets attributed to the desired pharmacological outcome and lowest affinity to those detrimental to the desired effects. A second important lesson is that the effort to evaluate compounds against the target and the proteins related to it, based on sequence or structural family, may be misplaced. The results show that it is just as possible for the compounds to interact with completely unrelated proteins. As a practical matter, both conclusions reinforce the importance of accruing uniform datasets on compounds that bind to proteins. The resulting database can have significant utility when properly mined.

The increasing evidence for similarities in binding sites suggests that an effective way to create new ligands for proteins would be to anchor a promiscuous compound in the binding site and use flanking regions to gain specificity.[2] This provides the basis for modular approaches to drug design, such as the well known structure activity relationship by nuclear magnetic resonance (SAR by NMR) technique[46] or the SHAPES procedures[47] based on a parallel application of other spectroscopic techniques.[2,10,48]

The concept has been exploited in a systems-based approach for gene families that share a common cofactor.[49] The basic premise is to identify a mimic for the cofactor that has drug-like properties and can be used as the central scaffold for a parallel synthesis effort. The resulting libraries contain chemicals able to bind several different members of that gene family. Many challenges have to be overcome. First, as discussed earlier, the cofactor or other common ligands do not share the same conformation for all members of the gene family. Indeed, several different conformations are observed for the same cofactor in crystallographic studies. Once a protein family is subdivided into classes that bind the cofactor in a similar way and a mimic for the cofactor has been identified, the next challenge is to decide how to extend the scaffold into a pocket in the protein that would confer selectivity. Spectroscopic techniques such as NMR[10,48] are ideally suited for this purpose. The result is a ligand that spans more than one pocket in the binding site, with thermodynamic advantages for binding. An example using oxidoreductases illustrates the approach very clearly. The existence of similarities in binding sites supports this and other modular approaches to drug discovery.

2.8 FUTURE PROSPECTS

As new techniques are developed for the characterization of interactions among proteins and between proteins and small molecules, a better understanding of the individual processes is achieved. However, we need to develop more sophisticated tools to fully extract the information that these novel techniques provide, as well as to identify relationships across datasets and disciplines. The large datasets that are being compiled in genomics, structural biology, *in-vitro* testing, *in-vivo* testing, metabolism and toxicology, and clinical trials would be wasted without effective tools for mining them.[32] One of the most common queries is about similarities and differences. In the realm of molecular pharmacology, the questions of which molecules are similar and which are not can be reduced to the types of interactions that they make with the biological system. Those interactions occur with proteins, and the fact that the same chemicals are able to interact with a variety of proteins indicates some degree of similarity among the proteins, even though the resemblance may not be obvious.

A critical need in molecular pharmacology is the development of a better understanding of protein binding similarities and the dynamic process that occurs during the recognition process.[50] The arrangement of proteins and compounds into classes is a central problem in drug discovery, and the vast amount of data being accumulated requires new ways to classify proteins and ligands. The importance is not merely theoretical, but has great practical implications. Problems of selectivity are squarely in this realm. Adverse events, environmental challenges, drug–drug interactions, and toxicological risk assessment are all related to this central problem: Small molecules interact with multiple proteins in ways that are not necessarily anticipated. If we are to increase the efficiency of the drug discovery process, we need to understand how those interactions come about and apply our expanding knowledge base in drug design to reinforce desirable characteristics and avoid unwanted ones.

The interface between chemistry and the proteome, commonly referred to as chemoproteomics or chemical proteomics, will continue to provide a unique perspective of biological systems. We showed some initial tentative steps into that area, where the classification of small molecules can be done based on a proteome or the proteome classified according to its chemical preferences. Chemoproteomics is likely to continue to advance and its development will provide critical new tools to probe biological function and advance our knowledge of systems biology.

REFERENCES

1. Kauvar, L.M., Villar H.O., Deciphering cryptic similarities in protein binding sites. *Curr. Opin. Biotechol.* 9, 390, 1998.
2. Erlanson, D.A., McDowell, R.S., O'Brien, T., Fragment-based drug discovery. *J. Med. Chem.* 47, 3463, 2004.
3. Jacoby, E., Davies, J., Blommers, M.J., Design of small molecule libraries for NMR screening and other applications in drug discovery. *Curr. Top. Med. Chem.* 3, 11, 2003.
4. LaBella, F.S., Molecular basis for binding promiscuity of antagonist drugs. *Biochem. Pharmacol.* 42, Suppl:S1–8, 1991.
5. Lauffenburger, D.A., Linderman, J.J., *Receptors: Models for Binding Trafficking and Signaling*, Oxford University Press, New York, 1998.
6. Koshland, D.E., Jr., The key-lock theory and the induced fit theory. *Angew Chem-Int. Ed.* 33, 2375, 1995.
7. Busenlehner, L.S., Armstrong, R.N., Insights into enzyme structure and dynamics elucidated by amide H/D exchange mass spectrometry. *Arch. Biochem. Biophys.* 433, 34, 2005.
8. Eyles, S.J., Kaltashov, I.A., Methods to study protein dynamics and folding by mass spectrometry. *Methods* 34, 88, 2004.
9. Bruschweiler, R., New approaches to the dynamic interpretation and prediction of NMR relaxation data from proteins. *Curr. Opin. Struct. Biol.* 13, 175, 2003.
10. Pellecchia, M., Sem, D.S., Wuthrich, K., NMR in drug discovery. *Natl. Rev. Drug Discovery* 1, 211, 2002.
11. Straub, F.B., Formation of the secondary and tertiary structure of enzymes. *Adv. Enzymol. Related Areas Molecular Biol.* 26, 89, 1964.
12. Ma, B., Shatsky, M., Wolfson, H.J., Nussinov, R., Multiple diverse ligands binding at a single protein site: A matter of pre-existing populations. *Protein Sci.* 11, 184, 2002.
13. Petock, J.M., Torshin, I.Y., Weber, I.T., Harrison, R.W., Analysis of protein structures reveals regions of rare backbone conformation at functional sites. *Proteins* 53, 872, 2003.
14. Klebe, G., Bohm, H.J., Energetic and entropic factors determining binding affinity in protein–ligand complexes. *J. Recept Signal Transduct Res.* 17, 459, 1997.
15. Feig, M., Brooks, C.L., III, Recent advances in the development and application of implicit solvent models in biomolecule simulations. *Curr. Opin. Struct. Biol.* 14, 217, 2004.
16. Mian, I.S., Bradwell, A.R., Olson, A.J., Structure, function and properties of antibody binding sites. *J. Molecular Biol.* 217, 133, 1991.
17. Villar, H.O., Kauvar, L.M., Amino acid preferences at protein binding sites. *FEBS Lett.* 349, 125, 1994.
18. Villar, H.O., Koehler, R.T., Amino acid preferences of small, naturally occurring polypeptides. *Biopolymers* 53, 226, 2000.
19. Bowie, J.U., Pakula, A.A., Screening method for identifying ligands for target proteins. U.S. Patent 5585277, 1996.
20. Zhou, H.X., Shan, Y., Prediction of protein interaction sites from sequence profile and residue neighbor list. *Proteins* 44, 336, 2001.
21. Ahmad, S., Sarai, A., PSSM-based prediction of DNA binding sites in proteins. *Bioinformatics* 19, 33, 2005.
22. La, D., Sutch, B., Livesay, D.R., Predicting protein functional sites with phylogenetic motifs. *Proteins* 58, 309, 2005.

23. Rawlings N.D., O'Brien E., Barrett A.J., MEROPS: The protease database. *Nucleic Acids Res.* 30, 343, 2002.

24. Contreras, J.A., Karlsson, M., Osterlund, T., Laurell, H., Svensson, A., Holm, C., Hormone-sensitive lipase is structurally related to acetylcholinesterase, bile salt-stimulated lipase, and several fungal lipases. Building of a three-dimensional model for the catalytic domain of hormone-sensitive lipase. *J. Biol. Chem.* 271, 31426, 1996.

25. Sussman, J.L., Harel, M., Frolow, F., Oefner, C., Goldman, A., Toker, L., Silman, I., Atomic structure of acetylcholinesterase from *Torpedo californica*: A prototypic acetylcholine-binding protein. *Science* 253, 872, 1991.

26. Kapitonov, D., Yu, R.K., Conserved domains of glycosyltransferases. *Glycobiology* 9, 961, 1999.

27. Moodie, S.L., Mitchell, J.B., Thornton, J.M., Protein recognition of adenylate: An example of a fuzzy recognition template. *J. Molecular Biol.* 263, 486, 1996.

28. Koehler, R.T., Villar, H.O., Bauer, K.E., Higgins, D.L., Ligand-based protein alignment and isozyme specificity of glutathione S-transferase inhibitors. *Proteins* 28, 202, 1997.

29. Cappello, V., Tramontano, A., Koch, U., Classification of proteins based on the properties of the ligand-binding site: the case of adenine-binding proteins. *Proteins* 47, 106, 2002.

30. Kho, R., Baker, B.L., Newman, J.V., Jack, R.M., Sem, D.S., Villar, H.O., Hansen, M.R., A path from primary protein sequence to ligand recognition. *Proteins* 50, 589, 2003.

31. Kho, R., Newman, J.V., Jack, R.M., Villar, H.O., Hansen, M.R., Genome-wide profile of oxidoreductases in viruses, prokaryotes, and eukaryotes. *J. Proteome Res.* 2, 626, 2003.

32. Mestres, J., Computational chemogenomics approaches to systematic knowledge-based drug discovery. *Curr. Opin. Drug Discovery Dev.* 7, 304, 2004.

33. Arnold, J.R., Burdick, K.W., Pegg, S.C., Toba, S., Lamb, M.L., Kuntz, I.D., SitePrint: Three-dimensional pharmacophore descriptors derived from protein binding sites for family-based active site analysis, classification, and drug design. *J. Chem. Inf. Computer Sci.* 44, 2190, 2004.

34. Caflisch, A., Miranker, A., Karplus, M., Multiple copy simultaneous search and construction of ligands in binding sites: Application to inhibitors of HIV-1 aspartic proteinase. *J. Med. Chem.* 36, 2142, 1993.

35. Deng, Z., Chuaqui, C., Singh, J., Structural interaction fingerprint (SIFt): A novel method for analyzing three-dimensional protein-ligand binding interactions. *J. Med. Chem.* 47, 337, 2004.

36. Hendlich, M., Bergner, A., Gunther, J., Klebe, G., Relibase: Design and development of a database for comprehensive analysis of protein–ligand interactions. *J. Mol. Biol.* 326, 607, 2003.

37. Kauvar, L.M., Higgins, D.H., Villar, H.O., Sprotsman, J.R., Engqvist Goldstein, A., Bukar R., Bauer, K.E., Dilley, H., Rocke, D.M., Predicting ligand binding to proteins by affinity fingerprinting. *Chem. Biol.* 2, 107, 1995.

38. Kauvar, L.M., Villar, H.O., Sportsman, J.R., Higgins, D.L., Schmidt, D.E., Protein affinity map of chemical space. *J. Chromatogr. B* 715, 93, 1998.

39. Dixon, S.L., Villar, H.O., Bioactive diversity and screening library selection via affinity fingerprinting. *J. Chem. Inf. Computer Sci.* 38, 1192, 1998.

40. Hsu, N., Cai, D., Damodaran, K., Gomez, R.F., Keck, J.G., Laborde, E., Lum, R.T., Macke, T.J., Martin, G., Schow, S.R., Simon, R.J., Villar, H.O., Wick, M.M., Beroza, P., Novel cyclooxygenase-1 inhibitors discovered using affinity fingerprints. *J. Med. Chem.* 47, 4875, 2004.

41. Fabian, M.A., Biggs, W.H., Treiber, D.K., Atteridge, C.E., Azimioara, M.D., Benedetti, M.G., Carter, T.A., Ciceri, P., Edeen, P.T., Floyd, M., Ford, J.M., Galvin, M., Gerlach, J.L., Grotzfeld, R.M., Herrgard, S., Insko, D.E., Insko, M.A., Lai, A.G., Lelias, J.M., Mehta, S.A., Milanov, Z.V., Velasco, A.M., Wodicka, L.M., Patel, H.K., Zarrinkar, P.P., Lockhart, D.J., A small molecule–kinase interaction map for clinical kinase inhibitors. *Nat. Biotechnol.* 23, 329, 2005.

42. Horvath, D., Jeandenans, C., Neighborhood behavior of *in silico* structural spaces with respect to *in vitro* activity spaces—A novel understanding of the molecular similarity principle in the context of multiple receptor binding profiles. *J. Chem. Inf. Computer Sci.* 43, 680, 2003.

43. Krejsa, C.M., Horvath, D., Rogalski, S.L., Penzotti, J.E., Mao, B., Barbosa, F., Migeon, J.C., Predicting ADME properties and side effects: The BioPrint approach. *Curr. Opin. Drug Discovery Dev.* 6, 470, 2003.

44. Engelberg, A., Iconix Pharmaceuticals, Inc.—Removing barriers to efficient drug discovery through chemogenomics. *Pharmacogenomics* 5, 741, 2004.

45. Koehler, R.T., Villar, H.O., Statistical relationships among docking scores for different protein binding sites. *J. Computer Aided Molecular Design* 14, 23, 2000.

46. Shuker, S.B., Hajduk, P.J., Meadows, R.P., Fesik, S.W., Discovering high-affinity ligands for proteins: SAR by NMR. *Science* 274, 1531, 1996.

47. Fejzo, J., Lepre, C.A., Peng, J.W., Bemis, G.W., Ajay, Murcko, M.A., Moore, J.M., The SHAPES strategy: An NMR-based approach for lead generation in drug discovery. *Chem Biol.* 6, 755, 1999.

48. Villar, H.O., Yan, J., Hansen, M.R., Using NMR for ligand discovery and optimization. *Curr. Opin. Chem. Biol.* 8, 387, 2004.

49. Sem, D.S., Bertolaet, B., Baker, B., Chang, E., Costache, A.D., Coutts, S., Dong, Q., Hansen, M., Hong, V., Huang, X., Jack, R., Kho, R., Lang, H., Ma, C., Meininger, D., Pellecchia, M., Pierre, F., Villar, H.O., Yu L., Systems-based design of bi-ligand inhibitors of oxidoreductases: Filling the chemical proteomic toolbox. *Chem. Biol.* 11, 185, 2004.

50. Cavasotto, C.N., Abagyan, R.A., Protein flexibility in ligand docking and virtual screening to protein kinases. *J. Molecular Biol.* 337, 209, 2004.

3 Survey of Spectral Techniques Used to Study Proteins

Daniel S. Sem

CONTENTS

3.1 INTRODUCTION

Proteins can be studied with a wide range of spectral techniques, of which only a small subset are widely use in proteomics. Protein spectral techniques, for the purposes of this book, are categorized as:

- Mass spectrometry (MS) techniques
- Spectroscopic techniques

Mass spectrometry involves the measurement of protein mass-to-charge (m/z) ratios, from which molecular weights of intact proteins and their fragments can be calculated. Spectroscopic techniques all involve monitoring the interaction of electromagnetic radiation with matter (table 3.1). Since there are far too many mass spectrometry and spectroscopic techniques to discuss in a chapter as short as this, even in a cursory manner, emphasis will be placed on those currently being applied in proteomics. For more detailed discussion of these methods, the reader is referred to some of the many excellent books and articles that served as primary sources for this chapter [1–9].

3.2 MASS SPECTROMETRY [3]

3.2.1 BACKGROUND AND HISTORY

Mass spectrometry was applied to small molecules long before it was used to study proteins. The first mass spectrometer can be dated to 1912 (J. J. Thompson), with the first atomic weight measurement made in 1919. The 1950s saw the development of quadrupole analyzers and the first gas chromatography (GC)-MS. The 1960s saw the development of tandem MS and electrospray ionization (ESI), with subsequent development of liquid chromatography (LC)-MS in 1973 (McLafferty). But MS did not find broad use for protein studies until the 1980s—first with the development of FAB (fast atom bombardment) MS in 1981 and then application of ESI to macromolecules in 1984; these were followed by development of ion cyclotron resonance MS. These developments opened the door to the application of MS in proteomics in the 1990s, when protein sequencing with matrix-assisted laser desorption/ionization tandem mass spectrometry (MALDI MS/MS) was introduced.

Today, MS has evolved to be the most prominent spectral technique used in proteomic studies. Many technological developments have been and continue to be made, at an exciting pace. This chapter and this book attempt only to provide a snapshot in time of some of the more widely used MS techniques and make no attempt to provide a comprehensive overview of the field. A general description is provided for the more commonly used mass spectrometers in terms of ionization method and mass analyzers, since these are the components of greatest variability and importance in the purchase of an instrument.

3.2.2 IONIZATION METHOD

Proteins must be introduced into the mass spectrometer and ionized so that they can travel through the mass analyzer and to the detector (fig. 3.1). In the early days of

TABLE 3.1
Spectroscopic Techniques Used to Study Proteins

Technique	Measured (most common)
Electronic transitions	
Fluorescence,[a] including FP[a] (fluorescence polarization) and FRET[a] (fluorescence resonance energy transfer)	Binding; structure; dynamics
UV-visible (UV-vis) absorbance spectroscopy	Electronic structure; binding
CD (circular dichroism) and MCD (magnetic circular dichroism)	Secondary structure; bonding
Vibrational transitions	
IR[a] (infrared)	Structure; bonding
Raman,[a] resonance Raman,[a] and polarized Raman	Structure; bonding
VCD (vibrational CD)	Structure; bonding
Electron and/or nuclear spin transitions	
NMR[a] (nuclear magnetic resonance)	Structure and dynamics
CIDNP (chemically induced dynamic nuclear polarization)	Structure
EPR[b] (electron paramagnetic resonance)	Structure (paramagnetic)
ENDOR (electron nuclear double resonance)	Structure (paramagnetic) and bonding
ESEEM (electron spin-echo envelope modulation)	Structure (paramagnetic) and bonding
Other	
SPR[a] (surface plasmon resonance)	Binding
X-ray crystallography[a]	Structure
SLS/DLS (static light scattering/dynamic light scattering)	Size (mol. wt.; aggregation)
SAXS (small- [or low]-angle x-ray scattering)	Size (mol. wt.; aggregation)
Mössbauer and magnetic Mössbauer	Bonding (metal)
XAFS/EXAFS (x-ray absorption fine structure/extended x-ray absorption fine structure)	Bonding (metal)
XANES (x-ray absorption near-edge structure)	Bonding (metal)
XAS (x-ray absorption spectroscopy)	Bonding (metal)

[a] Techniques discussed in this chapter and elsewhere in this book.
[b] Also referred to as ESR (electron spin resonance).

mass spectrometry, EI (electron ionization) was the method of choice for ionization, but this would fragment the molecule. Such fragmentation provides useful structural information for small molecules, but makes spectra of proteins overly complex and uninterpretable. For this reason, softer ionization methods were developed. FAB was developed first, but had a limited molecular weight range (<7000 g/mol). Later, MALDI and ESI were developed for protein applications, and these are currently the most widely used ionization techniques.

3.2.2.1 MALDI

MALDI is described in detail in chapter 5, and elsewhere in this book. In MALDI, the protein sample is mixed with a solid "matrix" material, then introduced into the

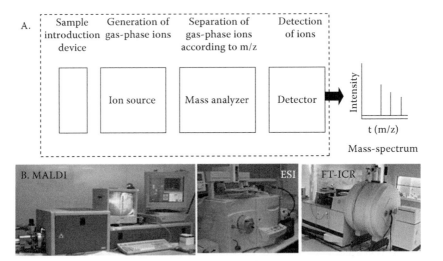

FIGURE 3.1 (A) Major components of a mass spectrometer. (Adapted from chapter 5.) (B) Typical mass spectrometer systems.

spectrometer on a probe. This matrix/sample mixture is then irradiated with a laser, which leads to two effects:

- Matrix ionizes and subsequently ionizes protein.
- Protein and matrix are desorbed from the solid matrix into the vapor phase.

From here, the ionized protein travels to the mass analyzer, which is commonly a TOF (time-of-flight) analyzer. MALDI is able to detect proteins in excess of 250,000 Da, can tolerate protein mixtures, and is tolerant of salts that can be present even in the millimolar range. Its main disadvantages are that resolution and accuracy are not as good as with ESI and that it requires a sample preparation in a solid matrix.

3.2.2.2 ESI

ESI and MALDI are briefly compared in chapter 13. In ESI, the sample is dissolved in aqueous solution and is introduced as a fine spray under very high voltage. The small droplets in the spray are drawn into the spectrometer inlet by electrostatic attraction and pass through a stream of dry gas and/or heat, which causes the droplets to decrease in size via evaporation. As the droplets get smaller, a point is reached at which charge–charge repulsions become significant and charged proteins are expelled through what is known as a "Taylor cone." Unlike with MALDI, ionization occurs at atmospheric pressure. The charged proteins then travel to the mass analyzer. The ionization conditions in ESI are such that multiply charged (protonated) proteins are obtained, so many m/z signals are detected for a single unfragmented protein. Various software tools are available to deconvolute and simplify these spectra. Since highly charged species will have lower m/z values, detectors with a lower mass range can be used.

Like MALDI, ESI is quite sensitive, even into the subpicomolar range. But ESI provides much better mass resolution than MALDI, although it has a more limited mass range (<75 kDa). An advantage of ESI is that the sample is introduced as a liquid; therefore, it can be coupled to the in-line purification (high-performance liquid chromatography [HPLC]; capillary electrophoresis [CE]) of complex protein mixtures (see chapter 4) where proteins, as they elute from a column, are injected directly into the mass spectrometer. These LC-MS applications are widely used in proteomics, where complex protein mixtures are the norm. Finally, since ESI is the softest ionization method, it even permits the detection of protein–ligand complexes.

3.2.3 MASS ANALYZERS

Mass analyzers (fig. 3.1) separate the charged protein ions based on m/z ratios. There have been many mass analyzers developed in the last 15 years, since the early days of MS when separation was based only on radial movement in a magnetic field. Most common now are TOF mass analyzers coupled to MALDI and quadrupole mass analyzers coupled to ESI instruments.

3.2.3.1 TOF

TOF-based separation of charged proteins is conceptually quite simple. Ions are accelerated into a tube of fixed length, all starting with the same energy. Since initial energy is given by $1/2 \; mv^2$ and energy is constant, it follows that ions of lower mass (m) will travel faster (v) and reach the detector first. Thus, mass is related to the time it takes to travel through this fixed-length, field-free tube. Although resolution is somewhat lower with TOF compared to other methods, its strength is that it has the highest mass detection range. Resolution can be increased somewhat by adding a reflectron.

3.2.3.2 Quadrupole

Quadrupole analyzers act as ion filters by trapping ions of specific m/z ratios in a quadrupole comprising four rods. Two of these rods have a direct current (DC) voltage with superimposed RF (radio frequency) voltage, and the other two have a DC voltage of opposite sign and RF voltage that is phase shifted 180°. Different m/z ratios are selected at specific RF values, so a mass spectrum is created by scanning through RF values. Quadrupole analyzers are used often with ESI because they are tolerant of the relatively higher pressures (poorer vacuums) associated with the ESI process. Also, their relatively low m/z detection range is not a problem with ESI, which produces multiply charged species and therefore low m/z ratios even for large proteins.

3.2.3.3 Ion Trap

Ion traps capture ions of specific m/z values in a process similar to that described in the previous section. They are used commonly in tandem MS (MS/MS) to capture ions for controlled fragmentation and subsequent mass analysis, described in section 3.2.4.

3.2.3.4 FT-ICR

The highest resolution and most expensive mass analyzer uses FT-ICR (Fourier transform-ion cyclotron resonance). While MALDI coupled to TOF and ESI coupled to a quadrupole mass analyzer have an accuracy of 0.1–0.01%, the accuracy with FT-ICR is as good as 0.001%. In FT-ICR, an ionized protein circles in a magnetic field at a given frequency and is excited with a corresponding radio frequency signal. This produces a time-dependent oscillation signal that can be Fourier transformed to a frequency domain spectrum, where frequencies can be converted to m/z values. Besides m/z resolution, FT-ICR resonance is able to perform tandem MS. But because of its high cost, tandem MS is usually performed using other methods, described next.

3.2.4 TANDEM MS

Currently used ionization methods (ESI or MALDI) are "soft" in that they do not produce any significant fragmentation of proteins. While this is desirable in that it greatly simplifies mass spectra, it should be noted that fragmentation patterns can be a rich source of structural information. To this end, tandem MS (aka MS/MS or MS^n) was developed to achieve the benefits of softer ionization, which yields simplified spectra, and controlled fragmentation, which yields structural information. Most commonly, tandem MS is used to sequence proteins extracted from 2D electrophoresis gels after proteolysis into peptides of manageable size. These peptides are then subjected to fragmentations by MS^n, which yields distinctive mass spectral patterns that can be used to search databases of spectra to obtain peptide sequences (see chapter 5).

In terms of instrumentation, the most commonly used method for peptide ionization in tandem MS is a soft method (usually MALDI) to generate a spectrum for unfragmented peptide. What distinguishes tandem MS is the next step, whereby a single protein or peptide ion is selected (e.g., in an ion trap) and then fragmented with CID (collision-induced dissociation) and an MS/MS (MS^2) spectrum is obtained that provides sequence data. The process can be repeated by selecting any one of these daughter ions in an ion trap and subjecting it to further fragmentation in a second CID process. This would generate granddaughter ions, which defines the MS^3 spectrum. Depending on hardware, the CID process can be repeated multiple times at increasingly high energy to generate MS^n spectra, where $(n - 1)$ is the number of CID steps.

3.3 SPECTROSCOPIC TECHNIQUES

3.3.1 BACKGROUND AND SURVEY

Spectroscopy has been defined [1] as "the study of the interaction of electromagnetic radiation with matter, excluding chemical effects" (i.e., photochemistry). The origin of the field has its roots in the detection of black body radiation by Planck in 1900, which was later recognized by Einstein (in 1905) and others as due to the release of photons ($E = h\nu$) by matter. This led ultimately to the first atomic and then molecular spectra and an understanding of how matter absorbs and releases energy in quantized packets of energy.

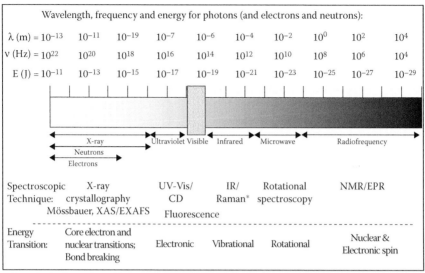

Wavelength, frequency and energy for photons (and electrons and neutrons):

λ (m) = 10^{-13}	10^{-11}	10^{-19}	10^{-7}	10^{-6}	10^{-4}	10^{-2}	10^{0}	10^{2}	10^{4}
ν (Hz) = 10^{22}	10^{20}	10^{18}	10^{16}	10^{14}	10^{12}	10^{10}	10^{8}	10^{6}	10^{4}
E (J) = 10^{-11}	10^{-13}	10^{-15}	10^{-17}	10^{-19}	10^{-21}	10^{-23}	10^{-25}	10^{-27}	10^{-29}

X-ray   Ultraviolet Visible Infrared   Microwave   Radiofrequency
Neutrons
Electrons

Spectroscopic Technique:	X-ray crystallography Mössbauer, XAS/EXAFS	UV-Vis/ CD Fluorescence	IR/ Raman*	Rotational spectroscopy	NMR/EPR
Energy Transition:	Core electron and nuclear transitions; Bond breaking	Electronic	Vibrational	Rotational	Nuclear & Electronic spin

*Excitation is often in the visible or near IR range, but what is monitored are vibrations with IR frequencies

FIGURE 3.2 Electromagnetic spectrum with corresponding energies, wavelengths, and frequencies. Spectroscopic techniques used to study proteins are indicated below the spectrum, along with the energy transitions that can be measured at the specified frequencies.

The interaction of light or other electromagnetic radiation (fig. 3.2) with matter produces three detectable effects:

- Scattering of photons
- Absorption of photons
- Emission of photons

The various spectroscopic techniques used to characterize proteins involve monitoring one or more of these effects, using electromagnetic radiation with wavelengths (and corresponding energies given by $E = h\nu$ and $\nu = c/\lambda$) that range in wavelength from $\lambda = 1$ m (radio frequency) to $\lambda = 10^{-11}$ m. The various spectroscopic techniques are associated with specific wavelength and energy ranges, as outlined in figure 3.2. All techniques involve the irradiation of a protein sample, followed by detection of photon (or electron) scattering, absorption, or emission. A range of spectroscopic techniques are used to study proteins, and it would be impossible to cover them in any detail in one chapter. Rather, emphasis is placed on methods that are currently finding (or show promise for) wide application in proteomics. A number of other techniques, not discussed here, are of significant value in characterizing individual proteins and some of these are listed in table 3.1.

3.3.2 UV-Visible [1,2]

UV-visible (UV-vis) spectroscopy measures transitions between electronic states corresponding to molecular orbitals that are occupied by electrons. These transitions occur upon absorption of photons with wavelengths in the ultraviolet (200–400 nm), visible (400–750 nm), or near-infrared (NIR) (>750 nm) ranges. In protein biochemistry, spectral absorbance is measured to determine a protein concentration or to monitor an enzymatic rate where a chromophore such as NAD(P)H is produced or consumed. Absorbance spectra of proteins can also be used to identify bound cofactors or prosthetic groups with characteristic spectral bands, such as flavins, hemes, pyridoxal phosphate, etc.

Absorbance occurs at a wavelength characteristic of a given chromophore, according to the Beer–Lambert law:

$$A = \varepsilon\, c\, l \tag{3.1}$$

where ε is the extinction coefficient, c is concentration of the chromophore, and l is path length of the sample cuvette.

Absorbance involves transition of electrons between molecular orbitals, typically π bonding, π^* antibonding, and n nonbonding (lone pair) orbitals. The peptide bond of proteins has forbidden n to π^* transitions at 210–220 nm. A more useful UV region for proteins is at higher wavelengths, where tryptophan (λ_{max} = 280 nm), tyrosine (λ_{max} = 274nm), phenylalanine (λ_{max} = 257 nm), and cystine (λ_{max} = 250 nm) absorb. Tryptophan is the strongest of these chromophores and is therefore frequently used to determine protein concentration.

In terms of relevance for proteomics, UV-vis absorbance spectroscopy is frequently used to monitor protein elution from a column (HPLC; CE) and to quantify protein concentration. Since protein absorptions are not in the visible range, they are of little help for visualizing protein bands in gels. Therefore, a number of protein stains have been developed over the years (fig. 3.3); one of the most widely used is Coomassie brilliant blue (CBB). The Coomassie stains bind to proteins through a combination of hydrophobic and electrostatic interactions—the latter between basic amino acids and the dye's sulfonate groups [10]. Staining intensity is proportional to concentration of a given protein, which has therefore led to the use of CBB in the Bradford assay for determining protein concentration [11]. A disadvantage of dyes like CBB is that they preferentially stain more basic proteins. To overcome this bias in staining, counter-ion stains were developed that include cationic and anionic dyes [12–14]. Silver staining provides a more sensitive alternative to CBB, but suffers the serious drawbacks of: (1) having a very limited linear range [15]; and (2) being incompatible with mass spectrometry, unless the silver can be removed.

3.3.3   FLUORESCENCE [4]

The previous section dealt with absorption of light photons, leading to the excitation of electrons to higher energy states. While an excited state electron can relax back to its ground state simply by releasing heat (radiationless decay), it can also release photons of lower energy (longer wavelength) in a process termed fluorescence. Initial excitation is rapid ($\sim 10^{-15}$ sec); then there is a lag period termed the fluorescence lifetime ($\sim 10^{-9}$ sec) before fluorescence occurs. Experimental fluorescence measurements are characterized by an intensity as well as a λ_{max}, the wavelength where

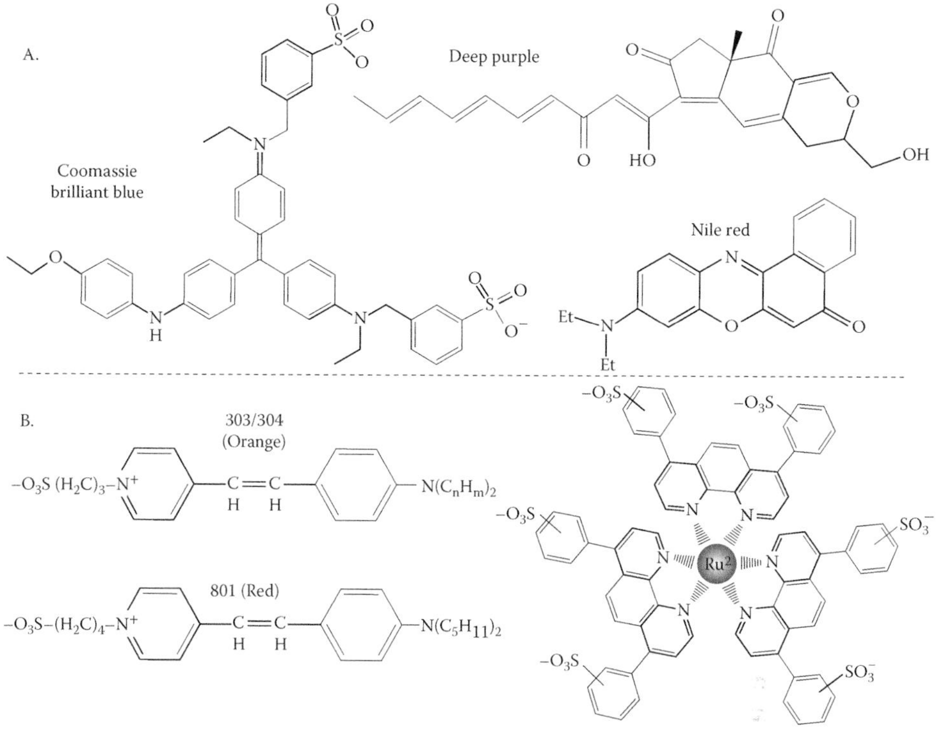

FIGURE 3.3 Chemical structures of commonly used visible, fluorescent, and luminescent protein stains. (A) Coomassie brilliant blue (CBB) and the newer deep purple and Nile red stains. (B) SYPRO stains. Some SYPRO fluorescent stains are based on merocyanine dyes (left two structures), such as SYPRO red and orange. (Steinberg, T.H. et al., *Anal. Biochem.*, 239, 223–237; 238–245, 1996.) Subscripts (n, m) are (6, 13) for compound #304 or (5, 11) for compound #303, both of which are orange-colored stains. (Haugland, R.P. et al. U.S. Patent #5,616,502, April, 1997.) Some SYPRO stains, such as SYPRO ruby (Berggren, K. et al., *Electrophoresis*, 21, 2509–2521, 2000; Steinberg, T.H. et al., *Electrophoresis*, 21, 486–496, 2000; Berggren, K. et al., *Anal. Biochem.*, 276, 129–143, 1999.), are luminescent and contain ruthenium (or other transition metal) chelates. A sample structure is shown in the right panel, but many variations are possible—especially with regard to sulfonate substitutions. The structure shown is for "compound 1" from Bhalgat, M.K. et al. PCT Patent #WO 00/25139, May, 2000.

maximum emission occurs. A fluorescence emission is also characterized by a lifetime (τ) and a quantum yield (Φ_F). Φ_F is the fraction of molecules that relax by fluorescence and is given by $\Phi_F = \tau/\tau_F$, where τ_F is the hypothetical lifetime if fluorescence were the only relaxation mechanism (i.e., no radiationless decay). An approximate value of τ_F in seconds is given by $10^{-4}/\varepsilon_{max}$, where ε_{max} is the extinction coefficient at the λ_{max} for absorbance.

As with UV-vis absorbance spectroscopy, the most useful amino acid is tryptophan, which fluoresces at $\lambda_{max} = 348$ nm. Somewhat less fluorescent amino acids are tyrosine ($\lambda_{max} = 303$ nm) and phenylalanine ($\lambda_{max} = 282$ nm). While these fluorescent residues are excellent probes for ligand binding and conformational changes, due to altered quenching effects, they are of little use for detecting proteins in 2D gels. Therefore, a number of fluorescent protein stains (fig. 3.3) have been developed, such as SYPRO Ruby, a ruthenium-based luminescent stain that interacts with basic groups of proteins through its sulfonate groups. Other fluorescent stains include Deep Purple, Nile Red, SYPRO Red, SYPRO Orange, and the lanthanide chelates [16,17].

One recent gel-imaging application employed CBB, with the gel scanned using a 680-nm laser for excitation, yielding a significant increase in sensitivity over visible CBB staining. Other exciting advances in fluorescence imaging are coming from the nanotechnology field, with the introduction of quantum dots (QDs) as highly fluorescent protein tags. QDs have tunable emission wavelengths based on particle size, yet excitation can be done with a single light source, thereby permitting multiplexed screening of protein–ligand interactions for high-throughput screening (HTS) [18,19] and even *in vivo* imaging [20].

In a chemical proteomic application of fluorescence (chapter 1 and Jeffery and Bogyo [21]), Bogyo and others reacted pools of proteins with a fluorescent electrophile, such as fluorescein tethered to an epoxide, which reacts with protein thiols. Such studies are done in the presence or absence of inhibitors, and the protein mixtures are separated on gels and then fluorescence-imaged. All proteins that bind the inhibitor are protected from fluorescence labeling, thereby identifying the proteins in a proteome that bind the inhibitor.

Fluorescence-based detection of proteins and protein–ligand interactions in proteomics extends beyond 2D gels. For example, Vehary and Garabet (chapter 10) present the use of array-based NIR fluorescence screening of antibody–antigen interactions. Such highly parallel approaches to screening protein–ligand interactions are reminiscent of the HTS approaches used to screen protein–ligand interactions. Prominent among the fluorescent techniques used in HTS are FP (fluorescence polarization) and FRET (fluorescence resonance energy transfer). Since HTS studies of systems-related proteins are relevant for chemical proteomic work, these methods will be discussed briefly here.

FP. When a fluorescent molecule is excited with plane polarized light, light can be emitted via fluorescence in the same plane as the incident light (emission intensity $= I_{\parallel}$), as long as there is no rotation of the molecule in the timeframe between excitation and emission (the fluorescence lifetime). But, this is of course not realistic, since molecules tumble in solution. The smaller a molecule is the faster it tumbles, leading to increasing fluorescence emission in the plane perpendicular to the incident light (intensity $= I_{\perp}$). The light is said to be highly polarized if $I_{\parallel} > I_{\perp}$ (little tumbling). The light becomes increasingly depolarized as $I_{\perp}$ increases, reaching a maximum depolarization at $I_{\perp} = I_{\parallel}$. The fluorescence polarization is given by:

$$P = \frac{I_{\parallel} - I_{\perp}}{I_{\parallel} + I_{\perp}} \tag{3.2}$$

If a trace amount of a small, fast tumbling ligand (usually fluorescently tagged) is titrated with a protein that binds to it, P will increase. From this titration curve, a K_d (dissociation constant) can be calculated for the fluorescent ligand. Likewise, a fluorescent ligand can be displaced by an unlabeled ligand and its K_d can be calculated as well. Such FP displacement assays are commonly used in HTS.

FRET. If a molecule contains a fluorescent donor and acceptor pair, such that the emission spectrum of the donor overlaps with the absorption spectrum of the acceptor, a FRET effect will be observed. That is, the emission from the donor will be decreased due to the presence of the acceptor. There will also be emission from the acceptor that occurs due to donor excitation and energy transfer to the acceptor. This effect can be quantified as the quenching caused by acceptor, given by:

$$\Phi_{DA}/\Phi_D = 1 - E_T \tag{3.3}$$

where E_T is the efficiency of the FRET-based depopulation of the excited state, Φ_{DA} is the quantum yield for donor in the presence of acceptor, and Φ_D is the quantum yield for the donor in the absence of acceptor. The efficiency of energy transfer (E_T) is dependent on distance, so it is possible to calculate donor–acceptor distances (R), based on experimentally determined E_T values, according to:

$$R = R_o \frac{\left(1 - E_T\right)}{E_T} \tag{3.4}$$

where R_o is a constant for a donor–acceptor pair, termed the Förster distance, which is the distance at which transfer efficiency is 50%. Thus, with appropriately labeled proteins or complexes, it is possible to measure conformational changes. It is also possible to monitor enzymatic reactions, such as the protease cleavage of a peptide bond in a linker joining a donor–acceptor pair. This provides an enzymatic rate measurement, which can be decreased in the presence of inhibitor. Many variations of FRET-based assays have been developed for HTS, largely because it is a robust assay with little background.

3.3.4 MAGNETIC RESONANCE

3.3.4.1 NMR [1,2,5]

NMR (nuclear magnetic resonance) spectroscopy is widely used to study the structure and dynamics of proteins. It is based on the measurement of the spin angular momentum properties of nuclei such as protons. In the presence of a static magnetic field (Bo), the nuclei of protons and other NMR active atoms (e.g., ^{13}C, ^{15}N, ^{31}P, ^{19}F) behave as if they are spinning and possess a spin angular momentum I. I is related to magnetic moment by $\mu = \gamma I$, where γ is the gyromagnetic ratio. As such, these nuclei have a magnetic moment (μ) that precesses around Bo at certain fixed angles dictated by the rules of quantum mechanics.

For nuclei with spin quantum number $I = \frac{1}{2}$, this precession occurs around the $+Bo$ axis ($E = -I_z \gamma Bo = -m\hbar\gamma Bo = -\frac{1}{2}\hbar\gamma Bo$, where I_z is the z-component of I) or around the $-Bo$ axis ($E = \frac{1}{2}\hbar\gamma Bo$), and transition between these states occurs upon excitation with photons of energy $\Delta E = \hbar\gamma Bo = \hbar\omega_o$. m is the magnetic quantum number, which can have values of $-I$, $-I + 1,\ldots I$, so $m = +\frac{1}{2}$ or $-\frac{1}{2}$ for a spin $\frac{1}{2}$ nucleus. The $m = +\frac{1}{2}$ state is called the α-state, while the $m = -\frac{1}{2}$ is called the β-state; for positive γ, the alpha state is lower in energy.

But, all nuclei experience slightly different magnetic environments and are therefore excited to undergo α to β spin-state transitions at slightly different energies. This excitation energy is given by $\Delta E = \hbar\gamma Bo(1 - \sigma)$, where σ is the shielding constant that adjusts for the environmental differences. An NMR spectrum therefore consists of absorption intensities plotted as a function of chemical shift, which is itself proportional to the transition energy ΔE. Because the chemical shift values are extremely sensitive indicators of environment, they are often used qualitatively to monitor ligand binding to proteins (see chapter 15). But, proteins can have thousands of ^{1}H NMR signals, which makes spectra too complicated to interpret. For this reason, 2D and 3D NMR spectra are usually measured, whereby absorption signals are spread into additional dimensions in order to remove spectral overlap.

Also, proteins can be isotopically labeled with ^{15}N and/or ^{13}C and simplified 2D or 3D spectra can be obtained. For example, a 2D ^{1}H-^{15}N HSQC (heteronuclear single quantum coherence) spectrum might show a cross-peak for each amide in a protein, such that coordinates for each cross-peak are the ^{1}H and ^{15}N chemical shifts for the two directly bonded amide atoms. One could simplify spectra further by labeling only specific amino acids so that only these are visible in a 2D spectrum. Volker Dötsch uses such HSQC spectra in chapter 15 to monitor ligand binding to proteins *in vitro* and inside living cells.

Protein NMR data are used most often to calculate 3D structures based on interproton distances, which are obtained from NOE (nuclear Overhauser effect) measurements. These distance constraints (as well as others) are used in molecular mechanics or distance geometry calculations to obtain structures of proteins. Since this process is too complicated a topic to discuss in the limited space of this chapter, see Cavanagh et al. [5] and Clore and Gronenborn [22] for a complete description, as well as chapter 17 for an overview of the role of NMR in structural proteomics efforts. Although NMR structure determinations are usually only practical for proteins smaller than 30 kDa, the method is very complementary to x-ray crystallography because it often can provide data for proteins that cannot be crystallized. NMR spectroscopy data can also provide a description of protein motion spanning pico-second to millisecond timescales.

3.3.4.2 EPR [1,2,6]

Just like nuclei, unpaired electrons behave as if they have spin in a magnetic field (Bo) and can orient either with or against that field. EPR (electron paramagnetic resonance; aka ESR, electron spin resonance) spectra comprise signals that have characteristic: (1) g-value; (2) intensity; (3) line width; and (4) splitting. These are analogous to the properties defining NMR spectra, with g-value comparable to the NMR chemical shift. EPR spectra are not absorbance spectra as in NMR, but are rather the first derivative of an absorbance spectrum. Electron spins can couple to nuclear spins, with the coupling referred to as "hyperfine splitting." This yields ($2nI + 1$) resonance lines, where n is the number of identical nuclei with spin quantum number I.

For example, an unpaired electron that is delocalized over an ^{14}N ($I = 1$) atom would give 3 lines, whereas delocalization over an Mn^{2+} ($I = 5/2$) atom would give 6 lines. Values of g and A (the hyperfine coupling constants), along with measurements of anisotropy, can provide detailed information about the ligand environment around metals bound to proteins. Such information is highly complementary to structural proteomic information obtained with NMR and x-ray crystallography and can often be obtained for proteins that cannot be structurally characterized by those methods (see chapter 19). Like NMR, EPR can be used to measure protein motion, quantified as an order parameters:

$$S = \text{(observed anisotropy)/(maximum anisotropy)} \qquad (3.5)$$

If specific spin labeling (usually nitroxide) is used on different surface regions, one can monitor changes in distances between unpaired electrons (for multiple labels) as a measure of conformational changes, along with the timescale of such motions. Finally, EPR also provides information about radical intermediates in enzymatic reactions.

3.3.5 IR AND RAMAN

Molecules vibrate with characteristic frequencies and modes that contain varying contributions from bond stretching and bending motions, and these frequencies can be measured with infrared (IR) and Raman spectroscopy. In a typical IR spectrum, absorption intensity is plotted as a function of $\overline{v}_{max}$ (= wave number = $1/\lambda$ in cm^{-1}), which is proportional to the energy of the vibrational transition. The relevant spectral region is 400–4000 cm^{-1}, and a given molecule can have as many as ($3N - 6$) vibration modes for N atoms. Although there can be a large number of bands for molecules as large as proteins, in practice one focuses on certain regions characteristic of functional groups like C=O and N–H.

IR is commonly used in protein studies to determine secondary structure—the relative contribution of α-helix and β-sheet. For example, characteristic IR bands for use in quantifying secondary structure are the so-called "Amide I" bands for C=O stretching in peptide bonds. These occur at 1652 cm^{-1} for α-helices, 1643 cm^{-1} for random coils, and 1632 cm^{-1} for β-sheets. To aid in the extraction of structural information from IR data in structural proteomics projects, a database of IR bands that have been correlated with structure has recently been created [23].

Additional information on structure and conformational changes can also be obtained using D_2O/H_2O exchange studies. One simply monitors the shift in the Amide II N–H band at 1546 cm^{-1} to that for N–D at 1430/55 cm^{-1}. Other changes can be monitored, such as in C=O stretching upon ligand binding or during pH titrations to obtain pK_a values. Selective isotopic labeling, especially of ligands, can also be used to identify bands of interest definitively by taking advantage of the change in vibration frequency induced by changes in mass, according to:

$$v_{vib} = (1/2\pi)\sqrt{K/\mu} \qquad (3.6)$$

where K is the force constant for the bond vibration and μ is the reduced mass for the two directly bonded atoms ($1/\mu = 1/mass_A + 1/mass_B$).

A disadvantage of IR for studies of proteins is that the O–H stretch of water is significant and obscures the protein bands. Even when subtraction of a water reference from a sample is done using a double beam instrument, this background can still be a problem. This is because of the high concentration of water ($55\,M$) compared to protein (low millimolar). Raman spectra, which provide similar information to that obtained from IR spectra, do not suffer from this large background water signal because water is a very weak Raman scatterer. Raman is also distinct from IR in that the selection rule that determines band intensity is that the vibrations must produce a change in polarizability, as opposed to in dipole moment for IR. Another distinguishing feature of Raman is that one measures changes in frequencies of scattered light, and what is plotted in the spectrum are these frequency changes (Δv). In contrast, IR spectra are plots of frequencies for the exciting photons.

Raman is used in many of the same applications as described previously for IR, but with the benefit of less interfering signal from water. Raman bands for highly polarizable groups will be stronger than the corresponding IR bands. For example, the cystine S–S-stretching mode gives an intense band at 675 cm^{-1} and a C–S band at 510 cm^{-1}. While Amide I bands (see preceding discussion) can be studied with IR and Raman, Amide II bands are much weaker in Raman. On the other hand, an additional Amide III band can be studied only with Raman. This band is from the N–H in-plane bending and the C–N stretching for peptide bonds. It can be used to distinguish α-helices (1235 cm^{-1}, intense), β-sheets (1260–1295 cm^{-1}, weak), or random coils (1245 cm^{-1}).

Like IR, Raman provides bond vibrational energy measurements. But a related technique, resonance Raman (RR), can provide significant improvement in signal-to-noise ratios and selectivity. In RR a chromophore such as a heme is excited in the UV-visible range, and vibrations associated with the chromophore (e.g., heme macrocycle stretching and deformation modes) are monitored. RR is more sensitive than traditional Raman by 2 to 3 orders of magnitude and also provides a means to probe groups of interest selectively. For example, if a given protein or mixture of proteins contains two different hemes with different λ_{max} values for their Soret bands, one can be selectively probed by exciting at its λ_{max} value.

A Raman advance of particular relevance for functional proteomics is the use of surface-enhanced resonance Raman (SERR) scattering to measure enzymatic rates with unprecedented sensitivity. The approach uses silver nanoparticles and masked enzyme substrates (initially invisible to SERR) and was able to assay 14 different enzymes with a sensitivity adequate to detect only 500 enzyme molecules—levels found in individual cells [24]. The technique relies on sample adsorption onto a colloidal metal surface, where surface plasmons (collective motions of conducting electrons) are produced. This amplifies the electromagnetic field near the surface so that subsequent scattering signals from protein are more intense.

3.3.6 SPR

Surface plasmon resonance (SPR) is increasingly used for HTS of protein–ligand interactions. Studies are typically done using purified proteins, but because of advances in chip and microfluidic technology, it is possible to assay large arrays of proteins in parallel. This lab-on-a-chip approach to parallel screening is a tremendous source of information in chemical proteomics, where large volumes of binding data are needed for families of proteins. Details of this approach to screening are presented in chapter 14 by Rich and Myszka, so only a brief discussion of the technique is given here.

SPR relies on thin metallic films to which a ligand is covalently attached. Binding of this tethered ligand to a receptor causes a change in refractive index, and this is monitored in an SPR experiment. This is achieved by shining a light on the metallic surface on the side opposite the ligand/receptor and measuring the reflected light. Changes in the reflected light occur due to changes in refractive index on the other side of the metallic thin film (typically gold on glass). This occurs because, upon interaction with light, plasmons at the surface produce "evanescent" waves that penetrate the film to around 300 nm and are sensitive to refractive index changes on the opposite surface of the film. Certain angles of incident light will match (resonate with) this plasmon frequency, which decreases the intensity of reflected light. Thus, it is possible to monitor changes in light intensity as one moves in and out of resonance as ligand/receptor interactions alter the refractive index.

3.3.7   X-Ray Crystallography [1,2,7,8]

The majority of protein structures are determined with x-ray crystallography, with a smaller fraction determined using NMR (15–20%). As such, most structural proteomics projects make central use of x-ray crystallography or a combination of crystallography and NMR [25,26]. The first step in determining a protein structure using x-ray crystallography is to obtain protein, usually by overexpressing in *Escherichia coli* and then to obtain diffraction-grade crystals. These tend to be the rate-limiting steps, although attempts have been made to lessen this limitation by use of automation. Companies such as Hampton Research (Alsio, California) and Emerald BioSystems (Bainbridge Island, Washington) provide kits to aid in the screening of an array of crystallization conditions, and the Hauptman–Woodward Medical Research Institute (http://www.hwi.buffalo.edu) offers a crystal screening service.

The process of determining a crystal structure is too complex to discuss at length here, so the reader is referred to an excellent book by McPherson for more details [7]. Briefly, protein crystals are irradiated with x-rays; this produces diffraction patterns that ultimately provide amplitudes and positions of scattered waves. But, the phases of these scattered x-ray waves must be determined before a structure can be calculated, and this is typically accomplished by comparison to a structurally similar protein (molecular replacement) or by comparison of the diffraction patterns for proteins soaked with heavy metals or with methionine residues substituted with selenomethionine residues (multiple isomorphous replacement). Once this "phase problem" is solved, an electron density map can be calculated from the diffraction patterns. Then, a covalent model of the protein is fitted into this electron density map. This is then subjected to rounds of refinement where small adjustments are made to the structure to improve the fit to the electron density, optimized by comparing calculated and experimental diffraction patterns. During this process, a molecular mechanics force field is used, as is also the case in NMR-based structure calculations.

Most structural proteomics projects strive to determine only structures that are likely to be significantly different from existing structures, based on a bioinformatic analysis. Another strategy, outlined in chapter 18, is to focus on families of related proteins such as kinases. This gene family-based (and systems-based) approach can be very efficient, since existing structures can be used to solve the phase problem. Studies of multiple protein-inhibitor complexes for multiple related proteins are needed in order to learn the rules governing substrate or inhibitor binding and selectivity (see chapter 2). Such information not only provides basic information about a gene family, but also has great utility in drug discovery and toxicoproteomics, where selective binding to a specific protein target or subset of targets determines efficacy of a drug as well as toxic side effects.

3.4 FUTURE PROSPECTS

Mass spectrometry has been and will likely remain for some time the most widely used spectral technique in proteomics. Many new innovations are on the horizon, some of which are discussed in the various chapters of this book. Therefore, further elaboration will not be made here.

The dominant role of 2D polyacrylamide gels in proteomic studies means that new tools for fluorescent staining and imaging of gels, especially with metal chelates and nanoparticles, will continue to be developed. Along these lines, chemical proteomic probes (often fluorescent) that explore protein–ligand interactions across proteomes will help define similarities and differences in binding sites (chapter 2). To complement studies of protein-ligand interactions in mixtures are lab-on-a-chip technologies, which allow parallel studies of many protein-ligand interactions using fluorescence (chapter 10), SPR (chapter 14) and other methods. When it is not possible to obtain isolated proteins, new approaches to more efficient in-line (chapter 4), on-chip (chapter 6), or even in-solution/*in-situ* (chapter 12) analysis of protein mixtures can be used.

In terms of structural proteomics, NMR and crystallography are the dominant methods, and there have been many recent reports on the value of using both methods synergistically [25,26]. For example, in one attempt to structurally characterize 263 proteins, only 8% could be characterized by both methods. But, 43 could be studied by only NMR and 43 by only x-ray crystallography. Thus, combined use of both methods made it possible to characterize 41% of all protein targets. Increasingly creative ways to use both techniques synergistically are currently under development [27,28].

Besides the previously mentioned techniques, which are widely used in proteomics, a wealth of spectroscopic tools (table 3.1) can be applied to study any individual protein identified in a proteomics effort and in need of functional characterization (i.e., basic biochemistry/biophysics). Some of these are even being used in reasonably high-throughput characterizations, such as EPR (chapter 19) and surface-enhanced resonance Raman (section 3.3.5). Other techniques with promise for high-throughput characterizations include the recent application of HTXAS (high-throughput x-ray absorption spectroscopy) and related methods in a metalloproteomics effort [29–31], and CD in a *Mycoplasma genitalium* proteomics project [32]. It is likely that many of the spectral techniques in table 3.1, which are presently somewhat underrepresented in proteomic efforts, will see increasing applications in multiplexed and/or high-throughput formats, especially as emphasis moves to the functional characterization of these newly identified proteins.

ACKNOWLEDGMENTS

Thanks to Dr. James Kincaid for helpful comments on the resonance Raman section of this chapter. Research supported by American Heart Association (AHA-05303072), Biomedical Technology Alliance (LEG FY06-12368 01-KMS) and NIH-NSF Instrumentation grants (S10 RR01901Z and CHE-0521323).

REFERENCES

1. Campbell, I.D., and Dwek, R.A., *Biological Spectroscopy*, Benjamin/Cummings Publishing, Menlo Park, CA, 1984.
2. Cooper, A., *Biophysical Chemistry*, Royal Society of Chemistry, Cambridge, UK, 2004.
3. Siuzdak, G., *Mass Spectrometry for Biotechnology*, Academic Press, San Diego, CA, 1996.
4. Lakowicz, J.R., *Principles of Fluorescence Spectroscopy*, Plenum, New York, 1983.
5. Cavanagh, J., Fairbrother, W.J., Palmer, A.G., III, and Skelton, N.J., *Protein NMR Spectroscopy*, Academic Press, San Diego, CA, 1996.
6. Ingram, D.J., *Biological and Biochemical Applications of ESR*, Adam Hilger, Ltd., London, 1969.
7. McPherson, A., *Introduction to Macromolecular Crystallography*, Wiley–Liss, Hoboken, NJ, 2003.
8. Petsko, G.A., and Ringe, D., *Protein Structure and Function*, New Science Press, London, 2004.
9. Fasman, G., *Circular Dichroism and the Conformational Analysis of Biomolecules*, Plenum, New York, 1996.
10. Fazekas, D.S., Groth, S., Webster, R.G., and Datyner, A., Two new staining procedures for quantitative estimation of proteins on electrophoretic strips. *Biochim. Biophys. Acta*, 71, 377–385, 1963.
11. Bradford, M.M., A rapid and sensitive method for quantification of microgram quantities of protein utilizing the principle of protein-dye binding. *Anal. Biochem.*, 72, 248–254, 1976.
12. Choi, J.K., Yoon, S.H., Hong, H.Y., Choi, D.K., and Yoo, G.S., A modified Coomassie blue staining of proteins in polyacrylamide gels with Bismark brown R. *Anal. Biochem.*, 236, 82–84, 1996.
13. Choi, J.K., and Yoo, G.S., Fast protein staining in sodium dodecyl sulfate polyacrylamide gel using counter-ion dyes, Coomassie brilliant blue R-250 and neutral red. *Arch. Pharm. Res.*, 25, 704–708, 2002.
14. Choi, J.K., Tak, K.H., Jin, L.T., Hwang, S.Y., Kwon, T.I., and Yoo, G.S., Background-free, fast protein staining in sodium dodecyl sulfate polyacrylamide gel using counter-ion dyes, zincon and ethyl violet. *Electrophoresis*, 23, 4053–4059, 2002.
15. Merril, C.R., and Goldman, D., *Two-Dimensional Gel Electrophoresis of Proteins*, Academic Press, Orlando, FL, 1984, 93–109.

16. Smejkal, G.B., proteins staining in polyacrylamide gels. In *Separation Methods in Proteomics*, ed. Smejkal, G.B. and Lazarev, A., CRC Press, New York, 2006.

17. Bell, P.J., and Karuso, P., Epicocconone, a novel fluorescent compound from the fungus epicoccumnigrum. *J. Am. Chem. Soc.*, 125, 9304–9305, 2003.

18. Chan, W.C., Maxwell, D.J., Gao, X., Bailey, R.E., Han, M., and Nie, S., Luminescent quantum dots for multiplexed biological detection and imaging. *Curr. Opin. Biotechnol.*, 13, 40–46, 2002.

19. Han, M., Gao, X., Su, J.Z., and Nie, S., Quantum-dot-tagged microbeads for multiplexed optical coding of biomolecules. *Nat. Biotechnol.*, 19, 631–635, 2001.

20. Gao, X., and Nie, S., Molecular profiling of single cells and tissue specimens with quantum dots. *Trends Biotechnol.* 21, 371–373, 2003.

21. Jeffery, D., and Bogyo, M., Chemical proteomics and its application to drug discovery. *Curr. Opin. Biotechnol.* 11, 602–609, 2000.

22. Clore, G.M., and Gronenborn, A.M., NMR structure determination of proteins and protein complexes larger than 20 kDa. *Curr. Opin. Chem. Biol.*, 2, 564–570, 1998.

23. Hering, J.A., Innocent, P.R., and Haris, P.I., Towards developing a protein infrared spectra databank (PISD) for proteomic research. *Proteomics*, 4, 2310–2319, 2004.

24. Moore, B.D., Stevenson, L., Watt, A., Flitsch, S., Turner, N.J., Cassidy, C., and Graham, D., Rapid and ultrasensitive determination of enzyme activities using surface-enhanced resonance Raman scattering. *Nat. Biotechnol.*, 22, 1133–1138, 2004.

25. Yee, A.A., Savchenko, A., Ignachenko, A., Lukin, J., Xu, X., Skarina, T., Evdokimova, E., Liu, C.S., Semesi, A., Guido, V., Edwards, A.M., and Arrowsmith, C.H., NMR and x-ray crystallography, complementary tools in structural proteomics of small proteins. *J. Am. Chem. Soc.*, 127, 16512–16517, 2005.

26. Snyder, D.A., Chen, Y., Denissova, N.G., Acton, T., Aramini, J.M., Ciano, M., Karlin, R., Liu, J., Manor, P., Rajan, P.A., Rossi, P., Swapna, G.V.T., Xiao, R., Rost, B., Hunt, J., and Montelione, G.T., Comparisons of NMR spectral quality and success in crystallization demonstrate that NMR and x-ray crystallography are complementary methods for small molecule protein structure determination. *J. Am. Chem. Soc.*, 127, 16505–16511, 2005.

27. Page, R., Peti, W., Wilson, I.A., Stevens, R.C., and Wuthrich, K., NMR screening and crystal quality of bacterially expressed prokaryotic and eukaryotic proteins in a structural genomics pipeline. *Proc. Nat. Acad. Sci., USA*, 102, 1901–1905, 2005.

28. Peti, W., Page, R., Moy, K., O'Neil-Johnson, M., Wilson, I.A., Stevens, R.C., and Wuthrich, K., Towards miniaturization of a structural genomics pipeline using micro-expression and microcoil NMR. *J. Struct. Funct. Genomics*, 9, 1–9, 2005.

29. Scott, R.A., Shokes, J.E., Cosper, N.J., Jenney, F.E., and Adams, M.W., Bottlenecks and roadblocks in a high-throughput XAS for structural genomics. *J. Synchotron Radiat.*, 12, 19–22, 2005.

30. Shi, W., Zhan, C., Ignatov, A., Manjasetty, B.A., Marinkovic, N., Sullivan, M., Huang, R., and Chance, M.R., Metalloproteomics: High-throughput structural and functional annotation of proteins in structural genomics. *Structure*, 13, 1473–1486, 2005.

31. Strange, R.W., and Hasnain, S.S., Combined use of XAFS and crystallography for studying protein–ligand interactions in metalloproteins. *Methods Mol. Biol.*, 305, 167–196, 2005.

32. Balasubramanian, S., Schneider, T., Gerstein, M., and Regan, L., Proteomics of *Mycoplasma genitalium*: Identification and characterization of unannotated and atypical proteins in a small genome. *Nucleic Acids Res.*, 28, 3075–3082, 2000.

33. Steinberg, T.H., Jones, I.J., Haugland, R.P., and Singer, V.L., SYPRO orange and SYPRO red protein gel stains: One-step fluorescent staining of denaturing gels for detection of nanogram levels of proteins. *Anal. Biochem.*, 239, 223–237, 1996.

34. Steinberg, T.H., Haugland, R.P., and Singer, V.L., Applications of SYPRO orange and SYPRO red protein gel stains. *Anal. Biochem.*, 239, 238–245, 1996.

35. Haugland, R.P., Singer, V.L., Jones, L.J., and Steinberg, T.H., Nonspecific protein staining using merocyanine dyes. U.S. Patent #5,616,502, April, 1997.

36. Berggren, K., Chernokalskaya, E., Steinberg, T.H., Kemper, C., Lopez, M.F., Diwu, Z., Haugland, R.P., and Patton, W.F., Background-free, high-sensitivity staining of proteins in one- and two-dimensional sodium dodecyl sulfate-polyacryalamide gels using a luminescent ruthenium complex. *Electrophoresis*, 21, 2509–2521, 2000.

37. Steinberg, T.H., Chernokalskaya, E., Berggren, K., Lopez, M.F., Diwu, Z., Haugland, R.P., and Patton, W.F., Ultrasensitive fluorescence protein detection in isoelectric focusing gels using a ruthenium chelate stain. *Electrophoresis*, 21, 486–496, 2000.

38. Berggren, K., Steinberg, T.H., Lauber, W.M., Carroll, J.A., Lopez, M.F., Chernokalskaya, E., Zieske, L., Diwu, Z., Haugland, R.P., and Patton, W.F., A luminescent ruthenium complex for ultrasensitive detection of protein immobilized on membrane supports. *Anal. Biochem.*, 276, 129–143, 1999.

39. Bhalgat, M.K., Diwu, Z., Haugland, R.P., and Patton, W.F., Luminescent protein stains containing transition metal complexes. PCT Patent # WO 00/25139, May, 2000.

Part II

Mass Spectral Studies of Proteome and Subproteome Mixtures

4 Capillary Electrophoresis— Mass Spectrometry for Characterization of Peptides and Proteins

Christian Neusüß and Matthias Pelzing

CONTENTS

4.1 INTRODUCTION

Bioanalytical tasks in the postgenome era increasingly require selective and sensitive methods for the characterization of peptides and proteins. There are many specific areas of interest, including protein identification, characterization of post-translational modifications (PTMs), biomarker discovery, clinically related applications, and quality control in the pharmaceutical biotechnological industries.

Mass spectrometry has become the technique of choice for the characterization of peptides and proteins as a result of its selectivity, identification power, and general applicability [1]. However, separation is often necessary for the characterization and quantification of the compounds of interest in these increasingly complex samples. For the analysis of peptides and proteins CE-MS (capillary electrophoresis-mass spectrometry) is becoming a popular alternative, or complementary technique to the widely used liquid chromatography (LC)-MS. CE-MS has the following beneficial characteristics:

- different selectivity compared to chromatographic approaches
- electrically driven open tubular approach
- fast separation
- low sample amount required

The key advantage of CE-MS is the ability for optimum separation with the injection of small sample amounts (typically tens to a few hundred nanoliters). If higher sample amounts are available, LC-MS becomes the technique of choice due to its improved concentration sensitivity. In combination with ease of use, this has resulted in LC-MS being the principle technique for the analysis of enzymatic (e.g., tryptic) protein digests in classical proteomics. There are, however, many interesting applications of CE-MS in the field of peptide and protein analysis. This chapter aims to highlight the most interesting examples. For a more detailed study of this subject, we recommend several recent reviews [2–4].

4.2 CAPILLARY ELECTROPHORESIS-MASS SPECTROMETRY (CE-MS)—INTERFACES AND MS-TECHNIQUES

4.2.1 INTERFACES

Electrospray ionization (ESI) is the obvious method of choice when coupling CE and MS for the analysis of peptides and proteins [5]. These analytes are already dissociated in liquid, enabling an efficient transfer into the gas phase by ESI. Furthermore, ESI enables the formation of multiply charged ions for larger polypeptide detection. This is important because most mass spectrometers have a limited mass range. In comparison to LC-ESI-MS, two additional key requirements have to be considered: (1) the electric CE circuit needs to be closed in the vicinity of the sprayer; and (2) low and BGE (background electrolyte)-dependent flows need to be handled.

In the sheath-flow interface as developed by Smith and coworkers [6], an additional sheath liquid, typically coaxially delivered by a capillary surrounding the metal needle, provides electrical contact and a constant (electrolyte-independent) flow. In the so-called sheathless design, the electrical contact is provided by means of an electrically conducting coated capillary tip [7]. Similarly, the electrical contact for the CE and ESI potential can be secured by applying a liquid junction between the CE capillary and an interface spray tip [8]. In both cases a constant flow is often achieved by the use of a coated capillary to produce a high and pH-independent EOF (electro-osomotic flow). The sheath–liquid interface is most widely applied due to its robustness, ease of use, and the absence of a coating [9]. However, if the ultimate in sensitivity is required, the sheathless design is preferred due to the higher ionization efficiency in this nanospray design [10].

Atmospheric pressure chemical ionization (APCI) and atmospheric pressure photo ionization (APPI) are alternative ionization methods for liquid phase separation. With respect to interfacing CE with MS, these ionization techniques are of particular interest when considering the potential use of nonvolatile electrolyte composition— for example, those used for chiral separation (cyclodextrins) or micellar electrokinetic chromatography. However, so far the benefits of APCI and APPI do not outweigh the disadvantage of a reduced sensitivity for peptide and protein analysis in comparison to ESI. No promising application has been published in this field to date.

CE fractions can also be used for subsequent analysis by matrix-assisted laser desorption/ionization time-of-flight mass spectrometry (MALDI-TOF MS). This off-line approach is gaining interest as a result of recent developments in tandem MS (MS/MS) capabilities of newly developed instrumentation. However, the time benefit of the fast CE separation is lost and method development is time consuming. This chapter will not deal with this emerging technology any further; however, the separation-related discussions about CE-ESI-MS can also be applied to CE-MALDI-MS.

4.2.2 MASS SPECTROMETERS

CE has been coupled to many different types of mass analyzers—that is, ion traps (IT), quadrupoles (Q), Fourier transform ion cyclotron resonance (FTICR), sector field (E, B), time of flight (TOF), and hybrid type instruments (e.g., Q-TOF, Q-FTICR). Quadrupoles have historically been the most commonly used mass analyzers for CE-MS, probably due to their presence in many labs. However, slow scan speeds, low resolutions, and relatively low sensitivity in full scan mode constitute important limitations of quadrupole analyzers. Therefore, triple quadrupole (QqQ) instruments especially are mainly used for quantification of target molecules. Time-of-flight mass analyzers, however, have a high duty cycle, good sensitivity, and high resolution. When considering future CE applications using high field strengths (e.g., chip CE-MS), where the expected peak width is in the range of milliseconds, only modern TOF analyzers with the ability to acquire more than 20 spectra per second present a viable alternative for MS detection. Ion traps are frequently used for the CE-MS analysis of peptides because they provide adequate scan speed and sensitivity and, more importantly, are able to carry out rapid MS^n experiments. The latter performance characteristic also makes IT-MS appealing for structure elucidation.

The combination of CE with Fourier transform-mass spectrometer (FTICR-MS) and other hybrid MS like Q-TOF have been reported [11]. FTICR-MS offers excellent sensitivity, high resolution (>100,000 with low parts per million mass measurement error), and the possibility to carry out MS^n experiments. The new generation of FTICR-MS, with magnetic field strengths of greater than 9 T, is capable of subattomole sensitivity at nearly 1 sec acquisition times [12]. However, due to the high costs involved, sophisticated MS equipment (e.g., Q-TOF, FTICR) can hardly be considered as a routine detector for CE at the present time.

4.3   BASIC SEPARATION PRINCIPLES

All electrophoretic techniques are based on applying high electrical fields to achieve separation of mostly charged species. The wide range of separation modes and the choice of the electrolyte (pH, additives, etc.) allow for optimization to tackle a wide variety of separation problems. Different techniques can be set up on the basis of the selection and arrangement of the buffer solutions in the capillary.

In *capillary zone electrophoresis* (CZE) the capillary is filled with a single background electrolyte of relatively high concentration, to provide uniform field strength when an electrical potential is applied to the capillary. The separation of ions in a background electrolyte (BGE) is based on distinct migration velocities as a direct result of differences in charge-to-size ratio and represents by far the most applied CE mode in combination with MS detection. Its standing as a technique of choice is not solely due to its selectivity and separation efficiency, but also because volatile electrolytes, as required for sensitive MS detection, are easily available for a broad pH range (<2 up to >12) as long as no special BGE additives like cyclodextrins (for enantiomeric separation) are used. Along with water, a broad range of solvents (alcohols, acetonitrile, etc.) is used in combination with MS, as can be seen in a large number of nonaqueous CE-MS (NACE-MS) studies. However, the use of NACE for biological samples is focused on specific applications like very hydrophobic (membrane) proteins or peptides. These applications are rare because most biological matrices are water based.

Capillary electrochromatography (CEC) is a hybrid of chromatography with an electroosmotic flow (EOF)-driven system. The association of these techniques attempts to combine the advantage of the flat flow profile of the EOF (i.e., high separation efficiency) with the versatility and loading capacity of chromatography. Packed columns and monoliths are principally used for separation, but open tubular CEC has gained popularity due to its simplicity and the possibility of using it in commercial CE instrumentation. For details of CEC-MS, see the review by Choudhary et al. [13].

Capillary isoelectric focusing (CIEF) provides the highest possible separation efficiency for the separation of proteins. Using a gel-free approach, it is possible to couple CIEF online to ESI-MS, offering a powerful tool for the analysis of very complex samples. The limitations for the coupling of CIEF to MS are the volatility of the ampholytes and the transfer of the very narrowly focused zones into the MS without peak broadening. The combination of the cathodic with gravimetric mobilization appears to be the most suitable manner to deal with the peak broadening issue [14,15]. To reduce the influence of the ampholyte on the ionization efficiency of the ESI, a dialysis interface prior to the ESI has been developed [16]. Furthermore, the introduction of a second dimension, like CZE or reverse phase liquid chromatography (RPLC), provides a tool to separate the ampholytes from the analytes (see section 4.5).

The combination of *capillary gel electrophoresis* (CGE) with mass spectrometry is highly desired—for example, when carrying out size-sieving, as is routinely applied in conjunction with laser-induced fluorescence (LIF) detection for the sequencing of DNA or the analysis of intact proteins. However, gels or monomeric impurities strongly decrease the ionization efficiency of ESI and cause contamination of the mass spectrometer. Therefore, few studies deal with this subject.

In *micellar electrokinetic chromatography* (MEKC), additional separation selectivity is achieved based on the pseudochromatographic system of the micelles. However, micelle-forming molecules, like sodium dodecyl sulfate, are among the strongest inhibitors of ESI efficiency. Therefore, there is a definite need to prevent the micelles (and their monomers) from entering the MS by using APCI, counter-migrating micelles, or by partial filling techniques. However, very few routine applications in bioanalysis have been reported, especially where sensitive MS detection is required.

The online coupling of *isotachophoresis* (ITP) with mass spectrometry has gained attention because of its higher loading capacity in comparison to CZE. However, the low separation efficiency and difficulties in finding appropriate spacers limit its applicability. Thus, ITP is mainly used as an online preconcentration step prior to CZE (transient ITP).

4.4 CZE-MS OF PEPTIDES

4.4.1 SELECTIVITY

Peptides are usually separated at low pH (<2–4), where they all migrate as cations. For coupling with MS, formic acid- and acetic acid-based BGEs are predominately used because of their volatility [2]. To avoid interactions of the peptides with the negatively charged silanol groups of the inner surface of fused silica, different approaches are used. The use of BGEs of very low or very high pH results in reduction of the charge density of the fused silica (pH < 2) or a repulsion of anions from the highly negatively charged wall, respectively. Separation in an acidic environment is by far the most widely used approach because this facilitates positive ESI detection, which is generally favorable for peptide detection. Coatings are frequently applied to the capillary to facilitate use at moderate pH or to ensure a strong EOF even at low pH, in order to obtain sufficient flow for a sheathless interface. Organic solvents are often added to reduce hydrophobic interactions with the capillary wall.

The mobility of a peptide strongly correlates with its charge in solution and the molecular mass. Several semiempirical models based on charge, size, and hydrophobicity are used to predict the mobility of peptides. In a recent study, Simó and Cifuentes demonstrated such an approach by the rapid realization of optimized CZE-MS conditions for the analysis of peptides obtained from an enzymatic protein hydrolysate in a single run, including information about "unexpected" fragments from protein digests [17].

Some studies have been performed to compare LC-MS and CZE-MS. Varesio et al. compared CZE-MS and nano-LC-MS for the detection of the Alzheimer's-related amyloid-β peptide [18]. Due to the higher loading capacity made possible with a column-switching setup, only nano-LC-MS was able to detect the peptide of interest in the biological fluids. Figeys and coworkers compared solid phase extraction (SPE)-CZE-MS/MS and nano-LC-MS/MS for the analysis of digests of gel-separated proteins [19]. The authors observed improved sensitivity for the sheathless SPE-CZE-MS/MS approach and applied this technique to the analysis of low-abundant proteins from a gel-separated yeast proteome.

A systematic comparison of CZE and LC, with subsequent online tandem mass spectrometry, for the analysis of protein digests has been presented recently [20]. In this study an SPE-CZE-micro-ESI-MS/MS setup was directly compared to nano-LC/nano-ESI using the same sample of a tryptic digest of bovine serum albumin as a reference standard. Measurements were made on a single ion trap mass spectrometer with identical acquisition parameters. Both systems showed the separation and detection of low levels of peptides in this mixture of moderate complexity, with most peptides identified using both techniques. However, specific differences were evident. Nano-LC-MS is about five times more sensitive than CZE-MS; this is, however, a smaller difference than expected. The CZE-MS technique showed a smaller loss of peptides, particularly with larger peptides (missed cleavages) and is about four times faster than the nano-LC-MS approach.

Because the separation is based on different underlying principles, LC and CZE are orthogonal with respect to separation selectivity. Depending on the separation problem, one or the other system will be more suited. Without a specific separation issue nano-LC will remain the routine tool for proteomics, with known and reproducible separation and high sensitivity. CZE does, however, have great potential, whenever small sample amounts, the analysis of larger proteins (e.g., beside small peptides), or short analysis times are required. A combination of both techniques will provide complementary information for complex samples.

4.4.2 SENSITIVITY

When considering sensitivity it is necessary to distinguish between the absolute sensitivity and the concentration sensitivity. Generally, the absolute sensitivity of CZE-MS is very good due to sharp peaks and effective ionization for both micro- and nanospray. This high absolute sensitivity (low fmol to amol on capillary) is an advantage if only small amounts of sample are available (<<10 μL). Detection limits of 7 amol of carbonic anhydrase in a crude blood extract were achieved in the early work of Valaskovic et al. by reducing the inner diameter of the capillary down to 5 μm and the spray tip to 2–5 μm [21]. However, lifetimes of these gold-coated ESI tips are short (±1 h), limiting this approach to research applications only. On a routine basis, high attomole to low femtomole amounts of peptides or proteins can be detected [22,23].

When applying a standard hydrodynamic injection, about 1% of the capillary (i.e., <10–50 nL) can be filled with sample without peak broadening. This leads to the disadvantage of low concentration sensitivity in comparison to techniques such as RPLC-MS, where 10 µL can easily be injected. However, various methods can be used to increase loading capacity in CE-MS. Stacking techniques, like the injection of a sample of low conductivity or the injection of a sample between plugs of different pH (pH-mediated stacking) [23] or different mobility (transient ITP) [24], can generally enhance the loading capacity by a factor of 10. In some cases, this enrichment might attain even higher levels. Using the principle of RPLC, the sample can be loaded on a plug of absorbent material placed at the beginning of the capillary. After elution with an appropriate solvent, the analytes are separated in the remainder of the capillary [25]. This online SPE approach results in a loading capacity similar to LC systems, even though the sample capacity is limited. In this way, limits of detection in the low nanomolar range have been achieved [26–28]. This approach has been further developed by increasing selectivity of the absorbing material (i.e., applying immobilized metal ions to selectively enrich phosphopeptides) [29]. Using immobilized antibodies, Guzman and coworkers demonstrated a promising approach for the analysis of target molecules in extremely complex mixtures [30].

4.4.3 SPEED

In contrast to high-performance liquid chromatography (HPLC), which produces a pneumatically driven (parabolic) flow, CE generates an electrically driven motion of fluid (electro-osmotic flow) with a flat flow profile, resulting in reduced band dispersion and eventually leading to narrower peaks. The observed migration of the analytes in CE is the sum of their mobility and the mobility resulting from the electro-osmotic flow. Both forces are related to the electric field strength, which is limited by the applicable voltage (in most instruments, ±30 kV) and the length of the capillary. Since, as a first approximation, the length of the capillary has no influence on the separation efficiency, very fast and yet efficient separation can be performed using short capillaries. However, the geometry of the CE instrument and the inlet of the mass spectrometer usually limit this approach to a minimum length of about 50 cm for commercial instrumentation.

Figure 4.1 shows a complete separation of a tryptic digest of bovine serum albumin (BSA) in less than 2.5 min. The separation was performed with a home-built CE system (Frank-Michael Matysik, University of Leipzig) in a 28 cm × 50 µm bare fused silica capillary. Due to the selectivity of the MS detector, a baseline separation of the peptides is not necessary. The fast migrating peptides show a peak width at half maximum (PWHM) of 1.5 secs, demanding a fast mass spectrometric detector. As discussed in section 4.2.2, modern TOF analyzers are able to perform up to 20 spectra per second. For the results shown, increased sensitivity was achieved by accumulating 5,000 spectra for each data point, resulting in a repetition rate of 3.6 spectra per second.

Recently, CE has benefited from developments in micromachined devices ("microfluidics," "lab-on-a-chip"; see later discussion). One of the advantages of these techniques is the reduced length of the separation channels and thus the increase in electrical field strength. When considering high-throughput analyses and two (or more) dimensional separation techniques, separation speed is an important issue.

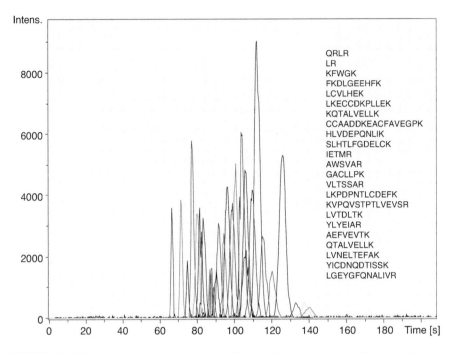

FIGURE 4.1 Electropherogram of a tryptic digest of bovine serum albumin, showing extracted ion traces of detected peptides listed in the order of their migration time. Conditions: 20 pmol/µL, fused silica capillary 50 µm (ID) × 28 cm (total length), electrolyte: 0.7 M HCOOH, 5 mM NH$_3$, 10% CH$_3$CN; separation voltage: 20 kV; injection: hydrodynamic 5 mbar 10 s (ca. 30 fmol), MS: ESI-oaTOF (micrOTOF-Bruker Daltonik).

4.4.4 SELECTED APPLICATIONS

4.4.4.1 Characterization of Post-Translational Modifications

Post-translational modifications (PTMs) play an important role in the functionality of biological systems. Thus, their characterization is crucial, though complicated by their low abundance, low ESI efficiency, and structural variability (i.e., glycosylation heterogeneity).

Two key aspects play a role in the use of CZE for the characterization of post-translational modified peptides (the characterization of PTMs based on intact protein analysis is discussed later): (1) phosphorylation, sulfation, and, in part, glycosylation (e.g., neuraminic acid) introduce a negative charge and therefore change the mobility of ions, enabling a clear separation of modified and nonmodified peptides; and (2) glycosylation significantly alters the effective size of the peptide. In contrast to RPLC, where no retention is often observed for glycopeptides with a high degree of glycosylation, these extremely hydrophilic compounds are readily accessible for CZE.

The selectivity of CZE-MS/MS for the characterization of phosphopeptides is illustrated in figure 4.2. Part A shows the analysis of the dephosphorylated and digested SIC 1 protein (*Saccharomyces cerevisiae*), including an MS/MS spectrum taken online of the potentially phosphorylated peptide (SQESEDEEDIIINPVR). In

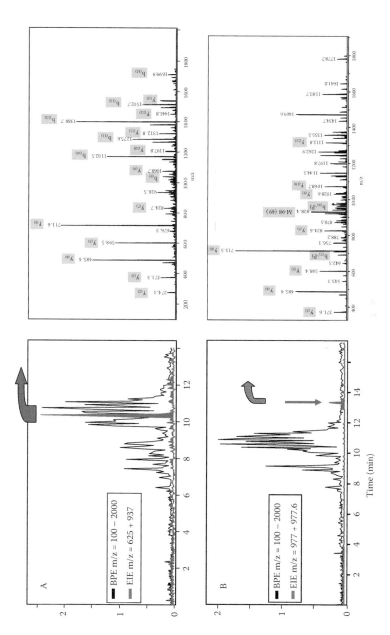

FIGURE 4.2 Separation of the tryptic digest of SIC1 protein (*S. cerevisiae*). The extracted ion electropherogram (EIE) and MS/MS spectra of the peptide SQESEDEEDIIINPVR is shown as nonphosphorylated in part A and phosphorylated in part B. Conditions: fused silica capillary, 50 μm (ID) × 80 cm (total length), electrolyte: 0.5 M HCOOH, 5 mM NH₃, 10% CH₃CN; separation voltage: 30 kV; injection: hydrodynamic 50 mbar 10 s, MS: three-dimensional ion trap (Esquire 3000plus-Bruker Daltonik).

Part B, characterization of the phosphorylated peptide is presented. The extracted ion electropherogram (EIE) of the same phosphorylated peptide shows a distinct difference between the mobility of the phosphorylated and nonphosphorylated peptide, resulting from the negative charge of the phosphate group. The MS/MS spectrum obtained online by an ion trap mass spectrometer clearly shows the identification of the phosphorylation, but does not locate the position in this case.

Further selectivity and sensitivity can be obtained using online enrichment of phosphopeptides by immobilized metal–ion affinity chromatography (IMAC).[28] This elegant method allows for the selective preconcentration of phosphopeptides and washing of matrix components and results in detection limits in the low nanomolar range (or below).

Several studies deal with the characterization of glycopeptides by CZE-MS/MS, and some of the most interesting are presented here. Demelbauer et al. showed MS/MS fragmentation of the glycosylation structure and subsequent peptide fragmentation (identification) by MS^3 for antithrombin III, applying an ion trap MS [31]. In another study, several glycopeptides from urine samples of patients suffering from N-acetylhexosaminidase deficiency were identified using a basic electrolyte and data-dependent MS/MS experiment in a QTOF MS [32]. Boss et al. achieved almost 100% sequence coverage for the digest of recombinant human erythropoietin by applying CZE-MS after sample prefractionation by HPLC [33].

Enzymatically released glycans can be analyzed by CZE-MS/MS if charge bearing groups (neuraminic acids) are present or after derivatization by, for example, 8-aminonaphtalene-1,3,6-trisulfonate (ANTS). Gennaro et al. used this method to study glycans released from the model protein, fetuin [34]. The ANTS-labeled method was superior to the analysis of the underivatized sample, not so much with respect to separation of (in this case) sialylated glycans, but rather concerning the ability to run extensive MS/MS experiments, since the ANTS-label (and thus a fixed negative charge) remains on the fragments and allows detailed structure analyses. Up to MS^5 could be performed within a CZE peak based on predefined masses.

4.4.4.2 Clinical Applications

In recent years CE-MS has been developed as a promising tool for biomarker discovery in the clinical field [35]. The technique involves a CZE separation and online ESI-TOF MS detection. As a separation principle, CZE has the advantage of the open-tubular approach (i.e., applicability over a wide range of hydrophobicity and size). This approach is used to characterize hundreds of polypeptides in the range of 1–10 kDa, allowing for a distinction to be made between groups of patients (i.e., healthy and diseased) based on pattern recognition algorithms [36]. In addition to cerebrospinal liquid [37,38] and serum [39], urine has generally been taken as a medium, primarily due to its reduced matrix complexity. A simple preconcentration, using off-line SPE, enriches the peptides and removes salts and other interfering compounds from the urine. In this way CE-MS was used to detect and distinguish several renal diseases [40]. The technique can even be used to successfully monitor the efficiency of therapy [38]. In a direct comparison of CZE-MS and surface-enhanced laser desorption ionization (SELDI), it turned out that CE-MS is

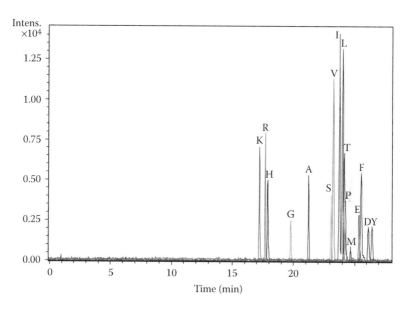

FIGURE 4.3 EIE (traces of different masses) of a standard containing 16 amino acids (Sigma); concentration 1 μM, formic acid based electrolyte, +30 kV, fused silica capillary, 50 μm (ID) × 115 cm (total length), injection (pH-mediated stacking—see text): hydrodynamically, 50 mbar 1000 s (250 nL injected); MS: ESI-oaTOF (micrOTOF-Bruker Daltonik).

advantageous with respect to a number of identified polypeptides, robustness, and reproducibility [41].

4.4.4.3 Amino Acids

Amino acids (AAs) were one of the first compound classes investigated by CE [42] because they are charged in solution. Their determination in various samples like body fluids, plants, and food has become an important field of analytical chemistry over the past 25 years. Generally, the determination of AAs using chromatographic or electrophoretic separation techniques, with UV absorbance or fluorescence detection, demands a pre- or postcolumn derivatization step to improve separation efficiency and detection sensitivity. Nevertheless, spectrophotometric detection methods of underivatized amino acids lack good sensitivity. Mass spectrometric detection in combination with CE separation achieves detection limits in the low parts per million down to parts per billion level using single quadrupole or triple quadrupole mass spectrometers [43,44]. In figure 4.3, the separation of an AA standard containing 16 amino acids is shown. Using a TOF mass spectrometer, detection limits in lower parts per billion levels are achievable and, in contrast to the target analysis with a triple quadrupole tandem MS, it is possible to identify unknown or rare amino acids and then calculate the elemental composition from the measured accurate mass. With detection limits in the nanomolar range, this CE-ESI-TOF MS method allows for the screening of AAs and related compounds in all body fluids without prior sample preparation or derivatization.

4.5   PROTEINS ANALYZED BY CE-MS

4.5.1   SEPARATION OF PROTEINS

CE is frequently used for the separation of intact proteins. The use of CE in protein and proteomic studies in recent years has been reviewed in great detail [45–47]. The reason for this widespread use of CZE is the open tubular principle, circumventing the problem that many proteins stick to surfaces and are thus difficult to handle in chromatography. The combination of CZE with ESI-MS is even more powerful due to the ability of ESI-MS to determine exact masses after charge deconvolution of the original mass spectrum.

In order to reduce interaction of the proteins with the capillary wall, coatings are usually applied for the analysis of intact proteins. Permanent coatings (like polyacrylamide) are available and can be used for CE-MS, as they are used for CE with optical detection. However, due to their high price and their pH instability in the basic solutions required for cleaning of the system, dynamic coatings are often preferable. To prevent interference of the coatings on the ESI-MS, the capillary is washed with the coating solution prior to the application of the electrolyte, which does not contain the coating reagent ("precoating" rather than "dynamic" coating). Thus, the coating has to be stable in the electrolyte without bleeding. Although several coating materials are available, the most widely applied coating is polybrene (hexadimethrinbromide), which provides a strong, reversed EOF.

Generally, similar background electrolytes are used for the analysis of peptides and proteins in CZE-MS (formic acid or acetic acid based). In this way small proteins are separated in a similar way to the peptides.

As an example of such studies, the analysis of protein components in single blood cells utilizing CZE-MS is mentioned. Cao and Moini reported the analysis of α- and β-chains of hemoglobin of a single intact red blood cell [48]. This application clearly illustrates the advantage of the low sample requirement of CE and strong detection of online ESI-MS. A total amount of about 450 amol hemoglobin was sufficient for detection by the ESI-TOF-MS. The same group characterized about 50 proteins in a cell lysate or the ribosomal subfraction of *Escherichia coli* by CZE-MS, applying a basic BGE in order to prevent protein precipitation in the capillary [49].

4.5.2   SEPARATION OF PROTEIN ISOFORMS

A key application of CE-MS in the area of intact proteins is the differentiation of modifications such as oxidation or glycosylation. This is illustrated in figure 4.4 for the characterization of glycoforms of bovine fetuin. This model protein contains three N-glycosylation and four O-glycosylation sites, resulting in broad glycosylation heterogeneity. Thus, fetuin cannot be analyzed by ESI-MS without prior separation. In an acetic acid-based electrolyte and coated capillary (reversed EOF), CZE sepa-rates isoforms sufficiently to obtain clear spectra in an ESI-TOF MS. On looking at the mass spectrum in a distinct time window (example shown in fig. 4.4B) and tracing the masses back, it is possible to observe clearly the different glycoforms. After charge deconvolution, these small mass differences can be attributed to certain structural differences (fig. 4.4C). The separation is shown in the electropherogram

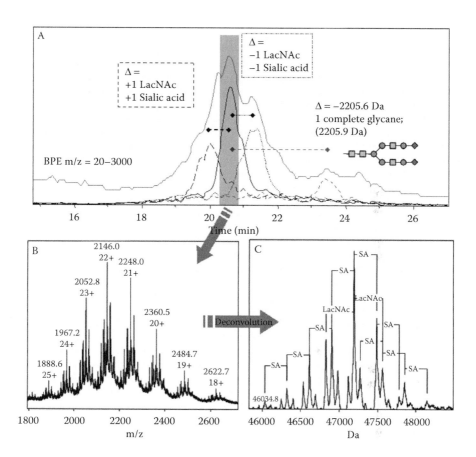

FIGURE 4.4 Electropherogram (A), typical mass spectrum (B), and typical deconvolution result (C) of intact bovine fetuin (SA = sialic acid, LacNAc = N-Acetyl-lactosamine). Conditions: 10 mg/mL, polybrene coated fused silica capillary 50 μm (ID) × 75 cm (total length), electrolyte: 1 M CH_3COOH, 20% CH_3OH; separation voltage: −30 kV; injection: hydrodynamic (50 mbar, 200 s); MS: ESI-oaTOF (micrOTOF-Bruker Daltonik). For experimental details, compare Neusüß, C. et al., *Electrophoresis*, 26, 1442–1450, 2005.

of figure 4.4 for differences in one N-acetyl-lactoseamine and one sialic acid and one complete N-glycan. This method strongly underlines the power of CE-MS in the field of intact protein characterization. A recent application of this approach resulted in the first in-depth characterization of recombinant human erythropoietin [50]. More than 135 isoforms could be distinguished for the reference material of the *European Pharmacopeia*, including heterogeneities differing in the content of sialic acids, hexose-N-acetyl-hexosamine, and acetylation.

4.5.3 CIEF-MS

The in-line coupling of CIEF with MS detection was mainly driven by the high-separation power of CIEF combined with the high-resolution power of an FTICR-type mass spectrometer. In this way a two-dimensional plot of pI versus mass, similar to classical two-dimensional gel electrophoresis, is obtained very rapidly and with much higher mass resolution and accuracy. The majority of the published data using CIEF-MS focuses on the analysis of complex biological samples, like physiological fluids containing a complex mix of proteins, or the analysis of whole cell lysates for proteomic studies. This technique was therefore evaluated and developed by different groups using cell lysates from *E. coli* as a model [14,51]. In figure 4.5, a two-dimensional display of the CIEF-FTICR analysis of lysates from (A) *Deinococcus radiodurans* and (B) *E. coli* is shown. In this way, Jensen et al. were able to detect 400–1,000 putative proteins in a mass range of 2–100 kDa from a total injection of 300 ng protein in a single CIEF-FTICR-MS analysis of cell lysates of *E. coli* and *D. radiodurans* [52].

To deal with the interference of the ampholytes used in CIEF, the 2D in-line coupling of CIEF to LC [53] and to CZE [54] have been reported. A microinjector interface and microdialysis junction have been applied, respectively.

Both approaches (CIEF-CZE and CIEF-RPLC) take advantage of several benefits of isoelectric focusing as the first dimension: Upon completion of analyte focusing, the self-sharpening effect greatly restricts analyte diffusion and contributes to analyte stacking in narrowly focused bands with a concentration factor of more than two orders of magnitude. Moreover, as the first separation dimension, CIEF resolves proteins/peptides on the basis of their differences in pI, therefore offering greater resolving power than that achieved in strong cation exchange chromatography. Furthermore, due to the flexibility in the choice of ampholyte systems, CIEF can be applied to acidic and very basic proteins/peptides in broad or narrow pH-bands.

The combination of the two highly resolving and orthogonal separation techniques of CIEF and RPLC or CZE, together with analyte focusing and concentration, significantly enhances the dynamic range and sensitivity of conventional mass spectrometry for the identification of low-abundance proteins. The CIEF-based multidimensional separation/concentration platform enables the identification of a greater number of soluble proteins while requiring 2–3 orders of magnitude lower protein loading when compared to classical methods like 2D gel electrophoresis.

4.6 TWO-DIMENSIONAL SEPARATIONS AND MINIATURIZATION

Beyond the previously discussed CIEF-LC-MS and CIEF-CZE-MS approaches, several other combinations of two separation techniques with independent (orthogonal) selectivity have been developed in order to handle biological samples of high complexity [55]. Different electrophoretic and chromatographic modes have been coupled to achieve good sensitivity (i.e., application of a system capable of high loadability as a first dimension) and high separation efficiency. CZE is attractive as a second dimension because of its speed and separation efficiency.

In some early work by Tomlinson et al., immunoaffinity enriched peptides were fractionated by RPLC and these fractions were then analyzed by SPE-CZE-MS [56]. Online multidimensional separation including CE-MS was pioneered by the group of Jorgenson [57]. They describe comprehensive RPLC-CZE-MS, where LC is

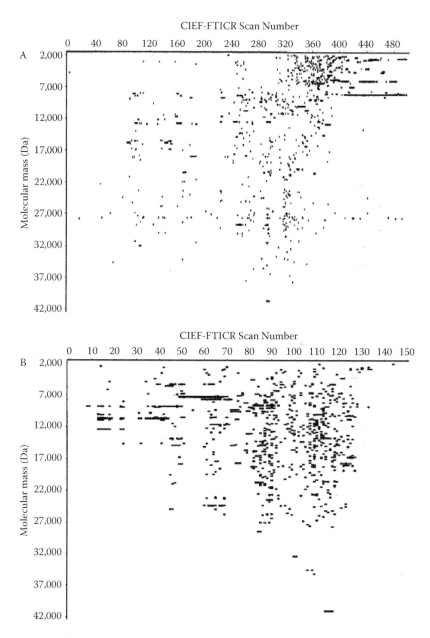

FIGURE 4.5 Two-dimensional display of the CIEF-FTICR analysis of lysates from (A) *D. radiodurans* and (B) *E. coli* grown in normal media. (Reprinted from Jensen, P. K. et al. *Electrophoresis*, 21, 1372–1380, 2000. With permission.)

coupled to the CZE by a flow-gating interface. Overlapping injections allowed for a CZE separation every 15 sec.

The emerging technology of microfluidic chip devices first enables the integration of two-dimensional (or even three-dimensional) separation systems (rather than "miniaturization" since electrophoresis in a capillary and on a chip is carried out in the same dimensions of the cross-section). CE is especially interesting since it requires only the implementation of electrodes, which can be controlled very rapidly. Applications involving mass spectrometry are still scarce since a robust and low-dead-volume coupling of the chip to the MS is difficult to achieve. Discussion of this topic is beyond the scope of this chapter. For further reading, see recent reviews (e.g., Mogenson et al. [58] and references therein).

Similarly, the integration of sample preparation into separation devices is highly attractive, since the idea of "lab-on-a-chip" has become popular. As an example, Wang and coworkers developed a miniaturized trypsin membrane reactor as part of an integrated CE-ESI MS chip.

4.7 FUTURE PROSPECTS

The application of CE-MS for the analysis of proteins will increase for those applications where separation speed, low sample quantities, or the separation selectivity is advantageous. This primarily includes the analysis of isoforms (glycoforms) of intact proteins, the analysis of post-translational modifications, and screening purposes (e.g., in the context of biomarker discovery or pattern recognition). If concentration sensitivity is not an issue, such as in impurity profiling, CE-MS is at least a complementary technique to LC-MS because it is an orthogonal separation technique and can easily separate low-abundant species from major species. CZE-MS will be an important option in two-dimensional separation and in conjunction with selective (online) enrichment techniques.

A simple and robust coupling, like the sheath-liquid interface with a grounded needle, strongly facilitates the use and will contribute to further distribution of this technology. As with LC-MS, CZE-MS will further benefit from ongoing improvements in mass spectrometric technology.

The general trend of miniaturization and multidimensional analyses will certainly push the analysis on microfluidic chip devices. In this area CE is highly attractive and may play a dominant role in separation technology. Whether MS detection is the choice in the near future will depend on the development of a robust chip–MS coupling applicable to CE separation without losing the excellent separation efficiency of electrophoresis in short microchannels.

ACKNOWLEDGMENTS

We thank Uwe Demelbauer for ideas, comments, and work on the glycoprotein analysis and Andrea di Fonzo (University of Milano) for providing samples of the SIC 1 protein digest. Anneke Lubben is acknowledged for the review of this manuscript.

REFERENCES

1. Aebersold, R. and Goodlett, D. R., Mass spectrometry in proteomics, *Chem. Rev.*, 101, 269–295, 2001.
2. Hernández-Borges, J. et al., Online capillary electrophoresis-mass spectrometry for the analysis of biomolecules, *Electrophoresis*, 25, 2257–2281, 2004.
3. Simpson, D. C. and Smith, R. D., Combining capillary electrophoresis with mass spectrometry for applications in proteomics, *Electrophoresis*, 26, 1291–1305, 2005.
4. Moini, M., Capillary electrophoresis mass spectrometry and its application to the analysis of biological samples, *Anal. Bioanal. Chem.*, 373(6), 466–480, 2002.
5. Severs, J. C. and Smith, R. D., Capillary electrophoresis-electrospray ionization mass spectrometry. In *Electrospray Ionization Mass Spectrometry, Fundamentals, Instrumentation and Applications*, ed. Cole, R. B., Wiley, New York, 1997, chap. 10.
6. Smith, R. D. et al., Capillary zone electrophoresis-mass spectrometry using an electrospray ionization interface, *Anal. Chem.*, 60(5), 436–441, 1988.
7. Olivares, J. A. et al., Online mass spectrometric detection for capillary zone electrophoresis, *Anal. Chem.*, 59(8), 1230–1232, 1987.
8. Lee, E. D. et al., Liquid junction coupling for capillary zone electrophoresis/ion spray mass spectrometry, *Biomed. Environ. Mass Spectrom.*, 18, 844, 1989.
9. Schmitt-Kopplin, P. and Frommberger, M., Capillary electrophoresis-mass spectrometry: 15 years of developments and applications, *Electrophoresis*, 24, 3837–3867, 2003.
10. Wilm, M. and Mann, M., Analytical properties of the nano-electrospray ion source, *Anal. Chem.*, 68(1), 1–8, 1996.
11. Tomer, K. B., Separations combined with mass spectrometry, *Chem. Rev.*, 101, 297–328, 2001.
12. Shen, Y. et al., High-throughput proteomics using high-efficiency multiple-capillary liquid chromatography with online high-performance ESI FTICR mass spectrometry, *Anal. Chem.*, 73, 3011–3021, 2001.
13. Choudhary, G. et al., Use of online mass spectrometric detection in capillary electrochromatography, *J. Chromatogr. A*, 887, 85–101, 2000.
14. Tang, Q., Harrata, K. A., and Lee, C. S., Two-dimensional analysis of recombinant *E. coli* proteins using capillary isoelectric focusing electrospray ionization mass spectrometry, *Anal. Chem.*, 69(14), 3177–3182, 1997.
15. Yang, L. et al., Capillary isoelectric focusing-electrospray ionization Fourier transform ion cyclotron resonance mass spectrometry for protein characterization, *Anal. Chem.*, 70(15), 3235–3241, 1998.
16. Lamoree, M. H., Tjaden, U. R., and van der Greef, J., Use of microdialysis for the online coupling of capillary isoelectric focusing with electrospray mass spectrometry, *J. Chromatogr. A*, 777, 31–39, 1997.
17. Simó, C. and Cifuentes, A., Capillary electrophoresis-mass spectrometry of peptides from enzymatic protein hydrolysis: Simulation and optimization, *Electrophoresis*, 24(5), 834–842, 2003.
18. Varesio, E. et al., Nanoscale liquid chromatography and capillary electrophoresis coupled to electrospray mass spectrometry for the detection of amyloid-β peptide related to Alzheimer's disease, *J. Chromatogr. A*, 974(1–2), 135–142, 2002.
19. Figeys, D. et al., Electrophoresis combined with novel mass spectrometry techniques: Powerful tools for the analysis of proteins and proteomes, *Electrophoresis*, 19(10), 1811–1818, 1998.

20. Pelzing, M. and Neusüß, C., Separation techniques hyphenated to electrospray tandem mass spectrometry in proteomics—Capillary electrophoresis versus nanoliquid chromatography, *Electrophoresis*, 26, 2717–2728, 2005.

21. Valaskovic, G. A., Kelleher, N. L., and Mc Lafferty, F. W., Attomole protein characterization by capillary electrophoresis-mass spectrometry, *Science*, 273, 1199, 1996.

22. Kelly, J. F., Ramaley, L., and Thibault, P., Capillary zone electrophoresis-electrospray mass spectrometry at submicroliter flow rates: Practical considerations and analytical performance, *Anal. Chem.*, 69(1), 51–60, 1997.

23. Neusüß, C., Pelzing, M., and Macht, M., A robust approach for the analysis of peptides in the low femtomole range by capillary electrophoresis-tandem mass spectrometry, *Electrophoresis*, 23(18), 3149–3159, 2002.

24. Shihabi, Z. K., Transient pseudo-isotachophoresis for sample concentration in capillary electrophoresis, *Electrophoresis*, 23(11), 1612–1617, 2002.

25. Tomlinson, A. J. et al., Preconcentration and microreaction technology online with capillary electrophoresis, *J. Chromatogr. A*, 744, 3–15, 1996.

26. Figeys, D., Ducret, A., and Aebersold, R., Identification of proteins by capillary electrophoresis-tandem mass spectrometry—Evaluation of an online solid-phase extraction device, *J. Chromatogr. A*, 763(1–2), 295–306, 1997.

27. Naylor, S. et al., Enhanced sensitivity for sequence determination of major histocombatibility complex class I peptides by membrane preconcentration—capillary electrophoresis-microspray-tandem mass spectrometry, *Electrophoresis*, 19(12), 2207–2212, 1998.

28. Bateman, K. P., White, R. L., and Thibault, P., Evaluation of adsorption preconcentration/capillary zone electrophoresis/nanoelectrospray mass spectrometry for peptide and glycoprotein analyses, *J. Mass Spectrom.*, 33(11), 1109–1123, 1998.

29. Cao, P. and Stults, J. T., Phosphopeptide analysis by online immobilized metal–ion affinity chromatography—capillary electrophoresis-electrospray ionization mass spectrometry, *J. Chromatogr. A*, 853(1–2), 225–235, 1999.

30. Guzman, N. A. and Stubbs, R. J., The use of selective adsorbents in capillary electrophoresis-mass spectrometry for analyte preconcentration and microreactions: A powerful three-dimensional tool for multiple chemical and biological applications, *Electrophoresis*, 22(17), 3602–3628, 2001.

31. Demelbauer, U. M. et al., Determination of glycopeptide structures by multistage mass spectrometry with low-energy collision-induced dissociation: Comparison of electrospray ionization quadrupole ion trap and matrix-assisted laser desorption/ionization quadrupole ion trap reflectron time-of-flight approaches, *Rap. Commun. Mass Spectrom.*, 18(14), 1575–1582, 2004.

32. Zamfir, A. and Peter-Katalinic, J., Glycoscreening by online sheathless capillary electrophoresis/electrospray ionization-quadrupole time of flight-tandem mass spectrometry, *Electrophoresis*, 22(12), 2448–2457, 2001.

33. Boss, H. J., Watson, D. B., and Rush, R. S., Peptide capillary zone electrophoresis mass spectrometry of recombinant human erythropoietin: An evaluation of the analytical method, *Electrophoresis*, 19(15), 2654–2664, 1998.

34. Gennaro, L. A. et al., Capillary electrophoresis/electrospray ion trap mass spectrometry for the analysis of negatively charged derivatized and underivatized glycans, *Rap. Commun. Mass Spectrom.*, 16(3), 192–200, 2002.

35. Kolch, W. et al., Capillary electrophoresis-mass spectrometry as a powerful tool in clinical diagnosis and biomarker discovery, *Mass Spectrom. Rev.*, 24, 959–977, 2005.

36. Kaiser, T. et al., Capillary electrophoresis coupled to mass spectrometer for automated and robust polypeptide determination in body fluids for clinical use, *Electrophoresis*, 25, 2044–2055, 2004.

37. Wetterhall, M. et al., Rapid analysis of tryptically digested cerebrospinal fluid using capillary electrophoresis-electrospray ionization-Fourier transform ion cyclotron resonance-mass spectrometry, *J. Prot. Res.*, 1(4), 361–366, 2002.
38. Wittke, S. et al., Discovery of biomarkers in human urine and cerebrospinal fluid by capillary electrophoresis coupled to mass spectrometry: Towards new diagnostic and therapeutic approaches, *Electrophoresis*, 26, 1476–1487, 2005.
39. Sassi, A. P. et al., An automated, sheathless capillary electrophoresis-mass spectrometry platform for discovery of biomarkers in human urine, *Electrophoresis*, 26, 1500–1512, 2005.
40. Weissinger, E. M. et al., Proteomic patterns established with capillary electrophoresis and mass spectrometry for diagnostic purposes, *Kidney Int.*, 65, 2426–2434, 2004.
41. Neuhoff, N. et al., Mass spectrometry for the detection of differentially expressed proteins: A comparison of surface-enhanced laser desorption/ionization and capillary electrophoresis/mass spectrometry, *Rapid Commun. Mass Spectrom.*, 18, 149–156, 2004.
42. Jorgenson, J. W. and Lukacs, K. D., Free-zone electrophoresis in glass capillaries, *Clin. Chem.*, 27(9), 1551–1553, 1981.
43. Klampfl, C. W. and Ahrer, W., Determination of free amino acids in infant food by capillary zone electrophoresis with mass spectrometric detection, *Electrophoresis*, 22(8), 1579–1584, 2001.
44. Soga, T. et al., Qualitative and quantitative analysis of amino acids by capillary electrophoresis-electrospray ionization-tandem mass spectrometry, *Electrophoresis*, 25(13), 1964–1972, 2004.
45. Manabe, T., Capillary electrophoresis of proteins for proteomic studies, *Electrophoresis*, 20, 3116–3121, 1999.
46. Dolník, V. and Hutterer, K. M., Capillary electrophoresis of proteins 1999–2001, *Electrophoresis*, 22(19), 4163–4178, 2001.
47. Patrick, J. S. and Lagu, A. L., Review applications of capillary electrophoresis to the analysis of biotechnology-derived therapeutic proteins, *Electrophoresis*, 22(19), 4179–4196, 2001.
48. Cao, P. and Moini, M., Separation and detection of the α- and β-chains of hemoglobin of a single intact red blood cell using capillary electrophoresis/electrospray ionization time-of-flight mass spectrometry, *J. Am. Soc. Mass Spectrom.*, 10(2), 184–186, 1999.
49. Moini, M. and Huang, H. Application of capillary electrophoresis/electrospray ionization-mass spectrometry to subcellular proteomics of *Escherichia coli* ribosomal proteins, *Electrophoresis*, 25(13), 1981–1987, 2004.
50. Neusüß, C., Demelbauer, U., and Pelzing, M., Glycoform characterization of intact erythropoietin by capillary electrophoresis-electrospray-time-of-flight mass spectrometry, *Electrophoresis*, 26, 1442–1450, 2005.
51. Jensen, P. K. et al., Probing proteomes using capillary isoelectric focusing-electrospray ionization Fourier transform ion cyclotron resonance mass spectrometry, *Anal. Chem.*, 71(11), 2076–2084, 1999.
52. Jensen, P. K. et al., Mass spectrometric detection for capillary isoelectric focusing separations of complex protein mixtures, *Electrophoresis*, 21, 1372–1380, 2000.
53. Chen, J. et al., Integration of capillary isoelectric focusing with capillary reversed-phase liquid chromatography for two-dimensional proteomics separation, *Electrophoresis*, 23(18), 3143–3148, 2002.
54. Mohan, D. and Lee, C. S., Online coupling of capillary isoelectric focusing with transient isotachophoresis-zone electrophoresis: A two-dimensional separation system for proteomics, *Electrophoresis*, 23(18), 3160–3167, 2002.

55. Wang, H. and Hanash, S., Multidimensional liquid phase based separations in proteomics, *J. Chromatogr. B*, 787, 11–18, 2003.
56. Tomlinson, A. J., Jameson, S., and Naylor, S., Strategy for isolating and sequencing biologically derived MHC class I peptides, *J. Chromatogr. A*, 744(1–2), 273–278, 1996.
57. Lewis, K. C. et al., Comprehensive online RPLC-CZE-MS of peptides, *J. Am. Soc. Mass Spectrom.*, 8, 495–500, 1997.
58. Mogensen, K. B., Klank, H., and Kutter, J. P., Recent developments in detection for microfluidic systems, *Electrophoresis*, 25, 3498–3512, 2004.

5 Protein and Peptide Analysis by Matrix-Assisted Laser Desorption/Ionization Tandem Mass Spectrometry (MALDI MS/MS)

Emmanuelle Sachon and Ole Nørregaard Jensen

CONTENTS

5.1 INTRODUCTION

Tracing its origins back to the beginning of the 20th century, mass spectrometry (MS) holds a special place among analytical techniques. Mass spectrometry measures an intrinsic property of a molecule—the molecular mass—with very high sensitivity, mass resolution, and mass accuracy. Therefore, mass spectrometry is used in an amazingly wide range of scientific and analytical applications. For large analyte species, such as intact proteins and strands of nucleic acids, molecular weights can be measured with an accuracy of 0.01%—that is, to within 4 Daltons (Da) or atomic mass units (amu) for a sample of 40,000 Da. This is usually sufficient to allow minor mass changes to be detected (e.g., the substitution of one amino acid for another or the presence of post-translational modifications).

Detailed information on protein primary structure is usually achieved by analysis of peptides that are derived from the protein by proteolytic digestion. In the case of peptides (relative molecular weight (Mr) 800–5,000), the mass accuracy obtained by modern mass spectrometers is well within 30 ppm. Using tandem mass spectrometry, structural information can be generated from proteins and peptides. This is achieved by fragmenting peptide ions inside the mass spectrometer and analyzing the product ions that are generated. Tandem mass spectrometry technology is extremely useful for *de novo* amino acid sequencing of peptides, for protein identification *via* partial sequence information, and for the identification of post-translational modifications. In the fields of protein chemistry and proteomics, mass spectrometry is used for

- *accurate molecular weight measurements* (e.g., determination of sample integrity and purity)
- *reaction monitoring* (e.g., enzyme reactions, chemical modification and protein digestion)
- *amino acid sequencing* (e.g., sequence validation, *de novo* sequencing, and protein identification by database searching with a peptide mass map [peptide mass fingerprint] or a peptide sequence "tag" from a proteolytic fragment)
- *protein structure determination* (e.g., protein folding monitored by H/D exchange, protein–ligand complex formation under physiological conditions, and macromolecular structure determination)

Protein mass spectrometry is usually performed using matrix-assisted laser desorption/ionization mass spectrometry (MALDI MS) or electrospray ionization mass spectrometry (ESI MS). In MALDI MS the sample is prepared as a solid crystalline deposit from which the ions are generated by laser irradiation. In ESI MS the sample is solubilized in an aqueous/organic solvent mixture and ionized in an electrostatic spray interface [1].

The aim of this chapter is to provide an overview of tandem mass spectrometry of peptides using MALDI-TOF-TOF (TOF: time-of-flight mass analyzer) or MALDI-Q-TOF (Q: quadrupole mass analyzer) mass spectrometers, including a brief overview of sample preparation methods. Other instrument configurations, like MALDI-iontrap [2] and MALDI-Fourier transform ion cyclotron resonance MS [3,4], will not be discussed in this chapter.

5.2 THE MASS SPECTROMETER

The five basic parts of any mass spectrometer are: a vacuum system; a sample introduction device; an ionization source; a mass analyzer; and an ion detector. The

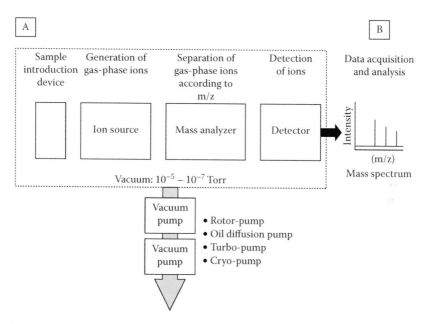

FIGURE 5.1 (A) Schematic representation of the five basic parts of a mass spectrometer: a vacuum system; a sample introduction device; an ionization source; a mass analyzer; and an ion detector. (B) Schematic representation of a mass spectrum. The relative intensity of the peaks is represented on the y-axis and the mass-to-charge ratio (m/z) of the different ions is represented on the x-axis.

mass spectrometer determines the molecular weight of chemical compounds by generating, separating, and detecting molecular ions according to their mass-to-charge ratio (m/z). Gas-phase ions are produced from a solid (MALDI) or liquid (ESI) sample in the ionization source by inducing the loss or the gain of a charge by neutral molecules (e.g., protonation, or deprotonation). Once the ions are formed in the gas phase they can be electrostatically directed into a mass analyzer, separated according to their m/z ratio, and finally detected. The result is a mass spectrum that can provide molecular weight or even structural information (fig. 5.1).

Matrix-assisted laser desorption/ionization was developed in the mid-1980s in Münster, Germany, by Karas and Hillenkamp [5]. MALDI provides an ideal ionization method for mass spectrometry of biomolecules and is extensively used for protein identification by peptide mass mapping (aka mass fingerprinting) in proteomics [6]. In MALDI analysis, the analyte is first cocrystallized with a large molar excess of a matrix compound, usually a UV-absorbing weak organic acid, to generate a solid sample. This "solid solution" is then irradiated by a pulsed UV laser, leading to sublimation of the matrix that in the process carries the analyte with it into the gas phase (fig. 5.2). The matrix therefore plays a key role by strongly absorbing the laser energy to softly "lift" the analyte species into the gas phase without destroying them. The matrix also serves as a proton donor and acceptor in the plasma (ionized gas) to ionize the analyte molecules. Intact proteins ($M_r > 8,000$) may generate singly protonated $[M+H]^+$ and multiply protonated $[M+nH]^{n+}$ ion species, whereas

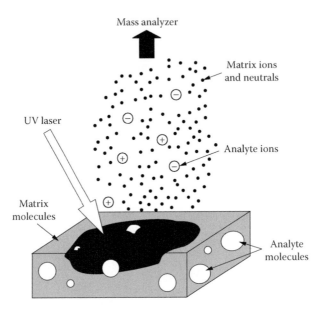

FIGURE 5.2 Schematic representation of the MALDI (matrix-assisted laser desorption/ ionization) process.

peptides in the Mr range of 600–5,000 predominantly generate singly protonated ion species [M+H]⁺. Following ionization, the gas phase ions are guided from the ion source into the mass analyzer that separates them by their m/z ratio.

The performance of the ion optics and the mass analyzer is critical because these components determine the mass accuracy, mass resolution, and mass range of the instrument. In the simplest mass spectrometers, ions are typically separated according to their m/z values by alternating electric fields (Q mass analyzers) or by their drift time in a field free region (TOF mass analyzers).

Quadrupoles are four precisely parallel rods with a direct current (DC) voltage and a superimposed radio-frequency (RF) potential. By altering the RF field (amplitude scan) one effectively scans over the m/z range. Quadrupole mass analyzers are the most common mass spectrometers because they are simple, robust, and easy to operate. They operate at an intermediate vacuum ($\sim 5 \times 10^{-5}$ torr), which reduces the demands and price of the vacuum system, and they are capable of analyzing ions up to m/z 4,000 or even more in specialized applications [7,8].

In time-of-flight mass analysis, ion m/z is determined by accurate measurements of ion drift time in a high vacuum. Ions travel from the ion source to the detector with a given amount of kinetic energy. Because all the different ion species have the same kinetic energy (E), yet a different mass (m), the ions reach the detector at different times because they have different velocities (v)—that is, $E = zeV_s = \frac{1}{2} m v^2$, in which V_s is the ion acceleration potential and ze the total ion charge. Low molecular weight ions reach the detector first because of their greater velocity. Larger ions take longer to travel through the TOF instrument because they move at a lower velocity. Thus, in the TOF mass analyzer, the m/z is determined as a function of the time of

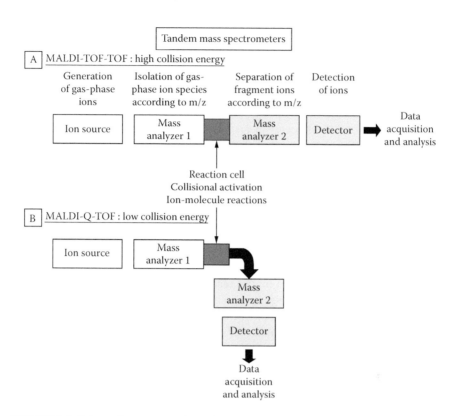

FIGURE 5.3 Schematic representation of two tandem mass spectrometers with MALDI ionization source. (A) The MALDI-TOF-TOF instrument; (B) the MALDI-Q-TOF instrument. Collision-induced dissociation is accomplished by selecting an ion of interest with the first mass analyzer (e.g., Q or TOF) and introducing that ion into a collision cell. The selected ion then collides with a collision gas, resulting in fragmentation. The fragments are then analyzed by the second analyzer (e.g., TOF) to obtain a fragment ion spectrum.

arrival of the ion (*equation $t^2 = m/z \; (d^2/(2 \; V_s e))$*), where the distance from the ion source to the detector is *d*).

5.2.1 MALDI Tandem Mass Spectrometry (MS/MS)

Amino acid sequencing of peptide ions by MALDI MS/MS has become a practical option only within the last decade. Tandem mass spectrometry enables peptide backbone fragmentation *via* collision-induced dissociation (CID) of selected peptide ions (precursor ions) by impact with an inert gas, such as argon. Subsequent mass analysis of the resultant peptide fragment ions reveals the complete or partial amino acid sequence of the peptide. Several MALDI MS/MS instrument geometries exist (fig. 5.3). In the case of MALDI-TOF-TOF instruments, the two TOF mass analyzers are collinear and separated by a collision cell. In contrast, in MALDI-Q-TOF instruments, the two mass analyzers, a quadrupole and a TOF analyzer, are orthogonal to each other. Since quadrupoles work at a relatively low vacuum (~5.10^{-5} torr), the extraction of the ions in the MALDI source can be performed at atmospheric pressure or at intermediate vacuum in MALDI-Q-TOF instruments; however, a high-vacuum MALDI source is needed for the MALDI-TOF-TOF instrument to optimize ion transmission efficiency.

The main difference between the two instrument geometries is that the MALDI-TOF-TOF mass spectrometer operates with high-energy ion fragmentation (kilo-electronvolt range) in MS/MS experiments, whereas the MALDI-Q-TOF operates with low energy (10–100 eV) collisions. The MALDI-Q-TOF usually exhibits better performance for sequencing of post-translationally modified peptides [9–12] and it seems to be the most versatile instrument for MS/MS of peptides, although it is often slightly less sensitive than a MALDI-TOF-TOF instrument (see following discussion).

5.2.2 SAMPLE PREPARATION FOR MALDI-MS ANALYSIS

With the development of robust and sensitive MALDI tandem mass spectrometers, the real challenge for the analysis of complex peptide or protein mixtures is the sample preparation step prior to mass spectrometric analysis. A range of sample preparation methods and strategies has been developed to obtain the best possible spectra from peptide mixtures, intact proteins, serum samples, phosphopeptides, etc. [9,13–16]. In the case of complex samples such as those encountered in many proteomics applications, peptide or protein fractionation by affinity purification methods is a useful approach. Affinity-based enrichment of post-translationally modified proteins and peptides serves to increase the relative abundance of a select class of modified polypeptide species in a sample. It is applied at the intact protein level or at the peptide level (i.e., after enzymatic or chemical degradation of proteins) [17]. Next, we briefly describe some of the classical and also recent sample preparation techniques for MALDI MS and MS/MS analysis of peptide and protein samples [18–22].

5.2.2.1 Dried Droplet Method

The dried droplet method for MALDI MS, as originally introduced by Karas and Hillenkamp (1988), is commonly used for MALDI MS of simple peptide or protein samples. In this method, mixing of an equal volume of analyte and of matrix solution is performed on the MALDI target and the mixture is allowed to dry in ambient air [5] before the sample is inserted into the mass spectrometer for analysis. Common matrices include sinapinic acid (SA) [23], alpha-cyano-4-hydroxycinnamic acid (HCCA) [24], and 2,5-dihydroxybenzoic acid (DHB) [25].

5.2.2.2 Thin Layer Method

This technique decouples matrix deposition from sample deposition. First, a matrix solution prepared by using a low-viscosity, volatile solvent (HCCA dissolved in an acetone/isopropanol mixture) is deposited on the MALDI plate. Fast evaporation of the solvent results in a thin, homogeneous layer of matrix crystals. A small volume of acidified sample solution is placed on top of the thin matrix layer and allowed to

TABLE 5.1
Summary of Some Techniques Using Microcolumns (GELoader Tips) Packed with Different Affinity Material for Rapid Sample Preparation Prior to MS

GELoader tip (Eppendorf)	Column Material	Affinity of the Column
	Reversed phase material [19]	Proteins
	Poros R1 (similar to C4)	Peptides
	Poros R2 (similar to C8-C18)	More hydrophilic peptides
	Oligo R3 (similar to C18)	
Column material	Immobilized metal affinity Chromatography (IMAC) material [16]	Phospho-peptides, his-tagged- and negative peptides
	HILIC [29]	Hydrophilic peptides
	Graphite powder [11]	Very hydrophilic peptides/phospho-peptides

dry [26]. The sample is then quickly rinsed by adding a droplet of 0.1% trifluoroacetic acid (TFA) to the sample deposit and then incubated for a few seconds followed by removal of the solvent. Inclusion of nitrocellulose into the fast evaporation matrix solution provides a more robust surface for sample deposition and is very well suited for peptide mass mapping applications in proteomics [18].

5.2.2.3 Microcolumn Affinity Purification

A significant improvement in sample preparation came with the introduction of custom-made disposable microcolumns as a fast cleanup step prior to MS. Some of the techniques using microcolumns (GELoader Tips) packed with different affinity material are summarized in table 5.1. Briefly, the material is packed into a GELoader Tip with air pressure using a syringe; the resulting microcolumn is then equilibrated before loading the sample. After loading, the sample is desalted by performing a washing step. Then, the desalted and concentrated sample is eluted from the micro-column in a single step, using a solution of MALDI matrix. All these techniques have been used successfully in the case of purification of different post-translationally modified peptides such as phosphopeptides [9,11,16,27] or glycopeptides [17,28,29]. A similar but less sensitive implementation of this technique is commercially available as ZipTips from Waters/Millipore.

5.2.2.4 In-Situ Liquid–Liquid Extraction (LLE) in MALDI MS

Liquid–liquid extraction takes advantage of the hydrophobic properties of certain peptides or proteins; it was recently introduced as a sample preparation technique

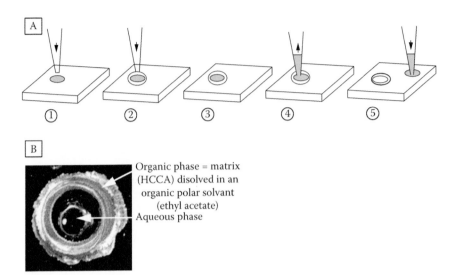

FIGURE 5.4 Schematic of the *in-situ* liquid–liquid extraction method. (A) The method proceeds in four steps: (1) An aqueous droplet (0.5 μL) containing peptides is placed on the MALDI target;(2) an equal volume of a saturated solution of a MALDI matrix (HCCA) in a water-immiscible solvent (ethyl acetate) is carefully added on the top of the analyte droplet; (3) during the incubation, the hydrophobic peptides are extracted from the aqueous sample and crystallized with the MALDI matrix; (4) the remaining hydrophilic fraction is transferred to a new position on the target and matrix solution is added. (B) The hydrophobic and hydrophilic MALDI deposits after solvent evaporation.

for MALDI MS [30]. LLE can be used as an enrichment step for lipid-modified proteins/peptides, which are very hydrophobic. The technique is simple and straightforward and affords a minimal loss of material because the two-phase separation is performed *in situ* (i.e., directly on the MALDI target). The *in-situ* LLE procedure is outlined in figure 5.4. The two liquid phases of the LLE method consist of (1) the aqueous sample solution (peptide or protein mixture); and (2) the water-immiscible organic phase containing a saturated solution of matrix (HCCA) in ethyl acetate.

First, a 0.5-μL aliquot of the aqueous sample is deposited on the MALDI plate. An equal volume of organic phase (matrix in ethyl acetate) is added to the aqueous sample, leading to a two-phase system consisting of a central aqueous phase surrounded by the organic phase (fig. 5.4). The ethyl acetate rapidly evaporates within 30 sec, leading to the generation of a circular deposit of matrix and analyte, which surrounds the remaining aqueous droplet. The aqueous droplet is then transferred to a new position on the target, matrix is added (dried droplet method), and the "organic deposit" and the "aqueous deposit" can be analyzed separately by MALDI MS (fig. 5.4), leading to more efficient detection and mass analysis of hydrophobic species.

5.2.3 PEPTIDE SEQUENCING BY MALDI MS/MS

Unambiguous identification of a protein may require amino acid sequencing in order to be able to search DNA, EST, or protein sequence databases. Locating modified

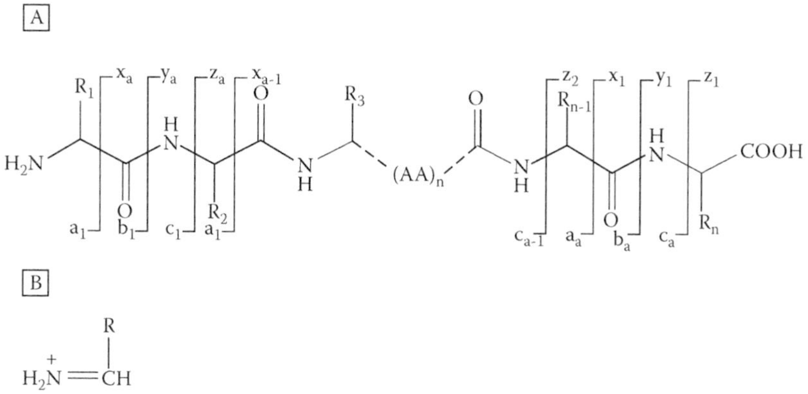

FIGURE 5.5 (A) Peptide fragmentation ions according to the Roepstorff et al. and Biemann nomenclature. (Biemann, K. *Biomed. Environ. Mass Spectrom.*, 16, 99, 1988; Roepstorff, P. and Fohlman, J. *Biomed. Mass. Spectrom.*, 11, 601, 1984). (B) Representation of an immonium ion of an amino acid: ion resulting from the multiple cleavage of the peptidic chain.

amino acid residues (e.g., phosphorylated, glycosylated, or oxidized residues) also calls for sequencing of peptides. Amino acid sequence information can be obtained by MALDI MS/MS experiments using MALDI-TOF-TOF instruments, MALDI-Q-TOF instruments, or other techniques, as mentioned in a previous section. Briefly, a particular peptide of interest, from the protein, is selected and dissociated in the mass spectrometer to produce a fragmentation mass spectrum indicative of the sequence of the peptide [31].

A drawback of peptide sequencing by MALDI MS/MS is the sometimes complex and labor-intensive interpretation of fragment ion spectra. Tandem mass spectra can be very complicated owing to the many types of fragment ions that arise during ion dissociation in MS/MS. These can be immonium ions, N-terminal fragments (a-, b-, and c-type ions), or C-terminal fragment ions (x-, y-, and z-type ions) (fig. 5.5). The fragmentation nomenclature has its origin from two publications: Roepstorff and Fohlman in 1984 and Biemann in 1988, who proposed a nomenclature system for fragment ions based on their origin along the backbone [32,33]. Furthermore, some of these fragment ions may display satellite peaks at −17 Da, due to loss of ammonia from lysine or arginine residues, or at −18 Da, due to loss of water from serine or threonine residues, or from acidic residues. Fragmentation also depends on amino acid sequence: Some peptides fragment excessively, resulting in extremely complex patterns that are difficult to interpret, whereas other peptides produce too few fragment ions with insufficient information to determine a sequence. Nevertheless, a range of software tools has been developed that can aid in the interpretation of tandem mass spectra of peptides (e.g., by performing sequence database searches to match known protein sequences to the mass spectra) [34]. Thus, in practical protein chemistry and proteomics, MALDI MS/MS is a highly useful technology.

Methods to simplify MS/MS spectra by limiting the types of fragments generated upon CID MS/MS analysis of peptides have been reported (for a review, see Roth

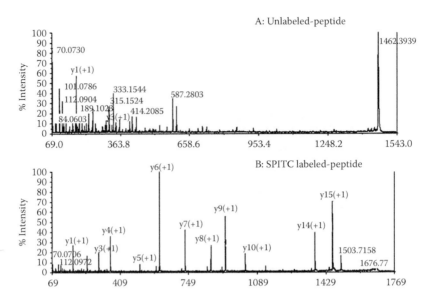

FIGURE 5.6 Tandem mass spectra obtained on a MALDI-TOF-TOF instrument (AB 4700 Proteomics Analyzer, Applied Biosystems). (A) Fragmentation spectrum of unmodified peptide STGSVVAQQPFGGAR at m/z = 1462.39; (B) fragmentation spectrum of modified peptide SPITC-STGSVVAQQPFGGAR at m/z = 1676.77. The y-type ions are indicated. The unmodified parent peptide appears at m/z = 1462.39.

et al. [35]). This is typically achieved by chemical derivatization of the N-terminal or the C-terminal residues of peptides in order to direct the fragment generation toward one type of peptide ion fragment. Keough et al. developed a strategy *de novo* peptide sequencing by introducing sulfonic acid derivatives to the N-termini of peptides [36,37]. In a MALDI MS/MS experiment, fragmentation of the derivatized peptides produced exclusively C-terminal fragments (y-series ions) in the resulting MS/MS spectra. As an example, MALDI MS/MS experiments have been performed using 4-sulfophenyl isothiocynate (SPITC) as a sulfonating reagent. The resultant sulfonate suppresses the generation of N-terminal fragments (b-series ions) by neutralizing them.

Thus, the pattern of fragment ions is significantly simplified since the presence of primarily only one type of ion (C-terminal y-ions) creates a "sequence ladder" in the mass spectrum, which facilitates interpretation. The tandem mass spectra obtained before and after derivatization are shown in figure 5.6. The facile peptide fragmentation has been explained in the framework of a mobile proton model wherein the mobile proton required to induce backbone fragmentation is provided by the labile proton of the sulfonic acid [38].

5.3   FUTURE PROSPECTS

MALDI mass spectrometry has developed into a very robust, sensitive, and relatively simple technology for mass determination of peptides and proteins. The advent of efficient and reproducible sample preparation methods for a range of applications in protein chemistry and proteomics has established MALDI MS as a versatile analytical technology in these fields. Recent advances in MALDI tandem mass spectrometry have provided new capabilities for large-scale MALDI MS/MS experiments and for amino acid sequencing of peptides in proteomics applications. The combination of capillary liquid chromatography and automated MALDI MS/MS is particularly promising for qualitative and quantitative analysis of complex peptide mixtures. Further developments of sample preparation methods, MALDI MS hardware, and the concomitant advances in bioinformatics for MALDI MS spectral analysis and data interpretation will no doubt lead to new and more widespread applications of this technique in the life sciences.

ACKNOWLEDGMENTS

Research in the Protein Research Group is financed by the Danish Natural Sciences Research Council and the Danish Technical Sciences Research Council. E. Sachon was the recipient of an EMBO long-term postdoctoral fellowship. O. N. Jensen is a Lundbeck Foundation research professor.

REFERENCES

1. Griffiths, W.J. et al., Electrospray and tandem mass spectrometry in biochemistry, *Biochem. J.*, 355, 545, 2001.
2. Krutchinsky, A.N. et al., Automatic identification of proteins with a MALDI-quadrupole ion trap mass spectrometer, *Anal. Chem.*, 73, 5066, 2001.
3. Solouki, T. et al., High-resolution multistage MS, MS2, and MS3 matrix-assisted laser desorption/ionization FT-ICR mass spectra of peptides from a single laser shot, *Anal. Chem.*, 68, 3718, 1996.
4. Solouki, T. et al., Electrospray ionization and matrix-assisted laser desorption/ionization Fourier transform ion cyclotron resonance mass spectrometry of permethylated oligosaccharides, *Anal. Chem.*, 70, 857, 1998.
5. Karas, M. and Hillenkamp, F., Laser desorption ionization of proteins with molecular masses exceeding 10,000 daltons, *Anal. Chem.*, 60, 2299, 1988.
6. Jensen, O.N. et al., Sample preparation methods for mass spectrometric peptide mapping directly from 2-DE gels, *Methods Molecular Biol.*, 112, 513, 1999.
7. Yost, R.A., Boyd, R.K., Tandem mass spectrometry: Quadrupole and hybrid instruments, *Methods Enzymol.*, 193, 154, 1990.
8. Finnigan, R.E., Quadrupole mass spectrometers. From development to commercialization, *Anal. Chem.*, 66, 969A, 1994.
9. Stensballe, A. and Jensen, O.N., Phosphoric acid enhances the performance of Fe(III) affinity chromatography and matrix-assisted laser desorption/ionization tandem mass spectrometry for recovery, detection and sequencing of phosphopeptides, *Rapid Commun. Mass Spectrom.*, 18, 1721, 2004.
10. Larsen, M.R. et al., Highly selective enrichment of phosphorylated peptides from peptide mixtures using titanium dioxide microcolumns, *Mol. Cell Proteomics*, 4, 873, 2005.

11. Larsen, M.R. et al., Improved detection of hydrophilic phosphopeptides using graphite powder microcolumns and mass spectrometry: Evidence for *in vivo* doubly phosphorylated dynamin I and dynamin III, *Mol. Cell Proteomics*, 3, 456, 2004.

12. Bennett, K.L. et al., Phosphopeptide detection and sequencing by matrix-assisted laser desorption/ionization quadrupole time-of-flight tandem mass spectrometry, *J. Mass Spectrom.*, 37, 179, 2002.

13. Meng, F. et al., Processing complex mixtures of intact proteins for direct analysis by mass spectrometry, *Anal. Chem.*, 74, 2923, 2002.

14. Gomez, S.M. et al., The chloroplast grana proteome defined by intact mass measurements from liquid chromatography mass spectrometry, *Mol. Cell Proteomics*, 1, 46, 2002.

15. Bunkenborg, J. et al., Screening for N-glycosylated proteins by liquid chromatography mass spectrometry, *Proteomics*, 4, 454, 2004.

16. Stensballe, A. et al., Characterization of phosphoproteins from electrophoretic gels by nanoscale Fe(III) affinity chromatography with off-line mass spectrometry analysis, *Proteomics*, 1, 207, 2001.

17. Jensen, O.N., Modification-specific proteomics: Characterization of post-translational modifications by mass spectrometry, *Curr. Opin. Chem. Biol.*, 8, 33, 2004.

18. Jensen, O.N., Podtelejnikov, A., Mann, M., Delayed extraction improves specificity in database searches by matrix-assisted laser desorption/ionization peptide maps, *Rapid Commun. Mass Spectrom.*, 10, 1371, 1996.

19. Gobom, J. et al., Sample purification and preparation technique based on nanoscale reversed-phase columns for the sensitive analysis of complex peptide mixtures by matrix-assisted laser desorption/ionization mass spectrometry, *J. Mass. Spectrom.*, 34, 105, 1999.

20. Laugesen, S. and Roepstorff, P., Combination of two matrices results in improved performance of MALDI MS for peptide mass mapping and protein analysis, *J. Am. Soc. Mass Spectrom.*, 14, 992, 2003.

21. Rappsilber, J. et al., Stop and go extraction tips for matrix-assisted laser desorption/ionization, nanoelectrospray, and LC/MS sample pretreatment in proteomics, *Anal. Chem.*, 75, 663, 2003.

22. Callesen, A.K. et al., Serum protein profiling by miniaturized solid-phase extraction and matrix-assisted laser desorption/ionization mass spectrometry, *Rapid Commun. Mass Spectrom.*, 19, 1578, 2005.

23. Beavis, R.C. and Chait, B.T., Cinnamic acid derivatives as matrices for ultraviolet laser desorption mass spectrometry of proteins, *Rapid Commun. Mass Spectrom.*, 3, 432, 1989.

24. Beavis, R. et al., Alpha-cyano-4-hydroxycinnamic acid as a matrix for matrix-assisted laser desorption mass spectrometry, *Org. Mass. Spectrom.*, 27, 156, 1992.

25. Strupat, K. et al., 2,5-Dihydroxybenzoic acid—A new matrix for laser desorption ionization mass spectrometry, *Int. J. Mass Spectrom. Ion Processes*, 111, 89, 1991.

26. Vorm, O. et al., Improved resolution and very high sensitivity in MALDI TOF of matrix surfaces made by fast evaporation, *Anal. Chem.*, 66, 3281, 1994.

27. Posewitz, M.C. and Tempst, P., Immobilized gallium(III) affinity chromatography of phosphopeptides, *Anal. Chem.*, 71, 2883, 1999.

28. Larsen, M.R. et al., Characterization of gel-separated glycoproteins using two-step proteolytic digestion combined with sequential microcolumns and mass spectrometry, *Mol. Cell. Proteomics*, 4, 107, 2005.

29. Hagglund, P. et al., A new strategy for identification of N-glycosylated proteins and unambiguous assignment of their glycosylation sites using HILIC enrichment and partial deglycosylation, *J. Proteome Res.*, 3, 556, 2004.

30. Kjellstrom, S., Jensen, O.N., In situ liquid-liquid extraction as a sample preparation method for matrix-assisted laser desorption/ionization MS analysis of polypeptide mixtures. *Anal Chem.*, 75, 2362, 2003.

31. Standing, K.G., Peptide and protein *de novo* sequencing by mass spectrometry, *Curr. Opin. Struct. Biol.*, 13, 595, 2003.

32. Biemann, K., Contributions of mass spectrometry to peptide and protein structure, *Biomed. Environ. Mass Spectrom.*, 16, 99, 1988.

33. Roepstorff, P. and Fohlman, J., Proposal for a common nomenclature for sequence ions in mass spectra of peptides, *Biomed. Mass. Spectrom.*, 11, 601, 1984.

34. Yates, J.R., III, Database searching using mass spectrometry data, *Electrophoresis*, 19, 893, 1998.

35. Roth, K.D. et al., Charge derivatization of peptides for analysis by mass spectrometry, *Mass Spectrom. Rev.*, 17, 255, 1998.

36. Keough, T. et al., A method for high-sensitivity peptide sequencing using postsource decay matrix-assisted laser desorption ionization mass spectrometry, *Proc. Natl. Acad. Sci. U.S.A.*, 96, 7131, 1999.

37. Keough, T. et al., Sulfonic acid derivatives for peptide sequencing by MALDI MS, *Anal. Chem.*, 75, 156A, 2003.

38. Wysocki, V.H. et al., Mobile and localized protons: A framework for understanding peptide dissociation, *J. Mass Spectrom.*, 35, 1399, 2000.

6 Characterization of Glycosylated Proteins by Mass Spectrometry Using Microcolumns and Enzymatic Digestion

Per Hägglund and Martin R. Larsen

CONTENTS

6.1 INTRODUCTION

Protein glycosylation is one of the most common types of post-translational modifications and plays numerous roles in cellular functions as diverse as immunity, cell signaling, and cell adhesion [1]. Moreover, glycosylation affects physiocochemical properties of proteins, such as folding, thermal stability, and solubility [2].

Mass spectrometry (MS) is an established tool for analysis of glycosylation in isolated proteins. Advances in the field of proteomics in conjunction with new generations of mass spectrometers with improved performance have opened up new possibilities for the detailed characterization of protein-linked glycans in complex mixtures (glycomics). Such characterization involves several steps—that is, identification of the glycosylated protein, the occupied sites, the nature and number of glycans attached to each site, and the linkages connecting the individual monosaccharides. Even though modern MS instruments may provide detailed information in these steps, there are still several analytical limitations, partly due to complexities caused by heterogeneous glycosylation. Thus, a comprehensive determination of glycosylation sites and glycan structures in a proteome remains a daunting task. Therefore, there is a strong need to develop new methods for enrichment and selective detection of glycoproteins and glycopeptides, methods should be applicable to the small sample quantities encountered in proteomics studies—in contrast to characterization of purified glycoproteins, where the sample amount often is not a limitation.

This chapter will give a short overview of the role of MS in the characterization of glycosylated proteins, with a focus on some of the more recent developments in sample preparation techniques. The use of chromatographic resins packed into microcolumns has been combined with novel approaches in enzymatic digestion for elucidation of structures and attachment points for protein-linked glycans. For a comprehensive overview of the role of MS in the characterization of glycans and glycosylated proteins, see Zaia [3] and Harvey [4].

6.2 GLYCAN STRUCTURES

6.2.1 N-Linked Glycans

The biosynthesis of protein N-glycosylation has been carefully investigated and the individual steps have been described in great detail [5]. Briefly, lipid-linked dolichol phosphate oligosaccharides in the endoplasmatic reticulum are attached to asparagine in the consensus sequence NXS/T/C, where X denotes any amino acid except proline. These oligosaccharides are subsequently processed by a series of glycosidases and glycosyltransferases, and a wide variety of glycan structures (glycoforms) may thus be generated. Common to most N-glycans is a core structure of two N-acetylglucosamine (GlcNAc) residues and three mannose residues. N-glycans are classified into three main classes: high mannose, complex, and hybrid types of glycosylation. The high mannose type of glycosylation contains branches of one or several mannose residues attached to the core structure. In the complex type, the branches may contain GlcNAc, galactose, and sialic acid residues. In the hybrid type of glycosylation, features of both high mannose and complex types of glycosylation may be present. In some hybrid and complex types of glycans, a fucose residue is linked to the proximal GlcNAc residue in the core [6]. In plants, the core structure may be modified by a 1-2 linked xylose residue or a 1-3 linked fucose residue [7].

6.2.2 O-Linked Glycans

In contrast to the well-defined consensus sequence for N-linked glycans described above, no common amino acid sequence defines the attachment point for O-linked glycans. The synthesis of O-glycans is typically initiated by the transfer of a single activated monosaccharide to a threonine or serine residue. However, in some cases, O-glycans can also be attached to tyrosine, hydroxyproline, and hydroxylysine [8]. An abundant class of O-glycans in mammals are the so-called "mucin type" O-glycans, which contain an N-acetylgalactosamine (GalNAc) residue attached to the peptide chain. However, no defined core structure is common to O-glycans, and a series of different motifs have been described (for a review on O-glycosylation, see Van den Steen et al [9]). Proteins modified by a single O-linked GlcNAc residue have attracted much interest since they have been implicated in cell signaling [10].

6.2.3 Glycosyl-Phosphatidylinositol (GPI) Anchors and C-Glycans

The attachment of a GPI anchor to a protein is not a glycosylation event in a strict sense, since no direct peptide–glycan bond is formed. The GPI anchor is attached to the protein via an amide bond between the C-terminal carboxyl group and an ethanolamine moiety from the GPI anchor [11]. The attachment is carried out by a transamidase that cleaves a short C-terminal peptide (15–30 amino acids) from the protein and transfers a preassembled GPI anchor to the newly formed carboxy-terminus [12]. Most GPI anchors contain a core structure composed of ethanolamine, phosphate, mannose, glucosamine, and inositol. But, several additional residues may also be attached. This core structure is attached to a lipid molecule, which anchors the GPI protein to the cell surface or other membranes.

C-glycosylation is a relatively newly discovered type of glycosylation that differs from the more common ones because the sugar is linked to the protein through a carbon–carbon bond to tryptophan residues [13]. This type of glycosylation has been found attached to tryptophan residues in certain membrane-associated and secreted proteins [14].

6.3 MASS SPECTROMETRY IN GLYCOMICS

Mass spectrometry has been applied to the characterization of glycosylated proteins and oligosaccharides since the 1980s, when the analysis was performed using fast atom bombardment (FAB) MS (e.g., Dell [15]). However, several limitations were observed with FAB MS, including substantial fragmentation, limited mass range, and poor sensitivity. Today, FAB MS has been replaced with softer ionization methods (i.e., matrix-assisted laser desorption/ionization [MALDI] [16,17] and electrospray ionization [ESI] [18]), which provide high sensitivity, less fragmentation, and the ability to analyze relatively complex mixtures.

6.3.1 MALDI-MS

MALDI-MS has evolved as a powerful technique for the structural characterization of glycans and glycopeptides. Several important parameters need to be considered prior to MALDI analysis, depending on the composition of the glycan in question. These include the sample preparation method, the choice of matrix, and the use of positive or negative ion mode. A wide range of matrices has been developed for glycan analysis (see Harvey [19] for a comprehensive review). The most frequently used matrix for analysis of free glycans and glycopeptides is 2,5-dihydroxybenzoic acid (DHB). DHB forms long, needle-like crystals that can be redissolved on the target to produce a more homogeneous crystal layer [20]. When embedded in a DHB matrix, neutral oligosaccharides and glycopeptides are normally ionized as alkali metal adducts in positive ion mode. Glycans containing sialic acid and other negatively charged residues may in some cases preferentially be analyzed in negative ion mode to avoid complex spectra due to salt formation [21].

When MALDI-MS is used in combination with time-of-flight (TOF) mass analyzers, it is also important to consider the use of the TOF analyzer in linear or reflector mode. Metastable fragmentation, which occurs in the TOF analyzer between the ion source and the reflectron, is a common problem observed in MALDI-TOF analysis of some types of glycans. This results in unfocused and consequently low-resolution signals in reflector ion mode. It does not contribute to any false signals in linear ion mode because the fragments will hit the detector at the same time as the parent ion. In-source decay (prompt fragmentation) has been shown to occur for very labile compounds, where the fragmentation takes place before the acceleration and consequently contributes to false signals in reflector and linear ion mode. This has mainly been reported for sialic acid containing glycans and glycopeptides [22].

One of the advantages of using MALDI-MS for analysis of glycans is that underivatized as well as derivatized glycans can be analyzed with apparent comparable sensitivity. Another advantage is that the signal intensity of neutral glycans does not decrease with increasing size [20]. As a consequence, MALDI-MS has been used extensively for profiling mixtures of free or derivatized glycans (e.g., for quantitation of permethylated carbohydrate mixtures [23]). This approach has proven reproducible and reflects the results obtained by using various labeling strategies. In addition, it is our experience that neutral glycosylated peptides give higher signal intensities in MALDI-MS when compared to ESI-MS (M. R. Larsen, unpublished results). In a recent comparison, we found that the new generation of tandem MALDI-MS instruments show some variation in their ability to analyze glycosylated peptides and glycans [24]. In our hands, a MALDI quadrupole orthogonal time-of-flight (QTOF) mass spectrometer showed superior performance compared with TOF-TOF instruments in fragmentation analysis of glycosylated peptides and glycans. One possible reason for this difference in performance could be that the low collision energy in the MALDI-QTOF is more suitable for glycan fragmentation compared with the high collision energy of TOF-TOF instruments. Another possible explanation is that the lower laser frequency of the MALDI-QTOF is more suitable for analysis of glycans and glycopeptides since these often are embedded in DHB and other soft matrices that generate large inhomogeneous crystals, which are more rapidly degraded by the high-frequency lasers found in high-throughput TOF-TOF instruments. It can also be speculated that lower laser frequencies prevent overheating of the local area on the MALDI target and therefore prevent temperature-dependent degradation of the analyte molecule.

6.3.2 ESI-MS

Only recently has ESI-MS been routinely applied to proteomic analysis of glycans and glycopeptides. These developments have been promoted by a change in the ionization interface towards spraying with significantly smaller droplets, which improves the ionization efficiency of hydrophilic glycans and glycopeptides (e.g., Juraschek et al. [25]). The hydrophilicity of oligosaccharides reduces the sensitivity of their analysis by ESI-MS because the surface activity is reduced when larger droplets are used [26]. Under these conditions, the highest ionization efficiencies are observed for hydrophobic peptides and other analyte ions with high surface activities. Thus, permethylation of free glycans, which increases their hydrophobicity, can improve the sensitivity of their analysis in ESI-MS [27].

Coupling of liquid chromatography (LC) to ESI-MS is preferred when complex mixtures are analyzed. This has been shown for free glycans using graphitized carbon cartridges for the analysis of reduced O-linked oligosaccharides released from bovine submaxillary mucin, bovine fetuin, and porcine gastric mucin [28,29]. Reverse-phase LC has not proven suitable for separation of free glycans [28], but can be applied to glycopeptides [30]. In LC-ESI MS, parent ion scanning can be used to localize glycopeptides by tracing the oxonium ions (m/z [mass-to-charge ratio] 163 [Hexose], 204 [N-acetylhexosamine (HexNAc)], 274 [sialic acid], 292 [sialic acid], and 366 [Hexose-HexNAc]) generated by mild fragmentation of the glycan structures attached to the peptide [31]. It has also been shown that the specificity of the precursor ion scanning can be markedly increased using optimized collision energies and high-resolution mass spectrometers [32].

6.4 CHARACTERIZATION OF GLYCAN STRUCTURES BY MS

One of the most common approaches for obtaining information about the structures of glycans is to use different exoglycosidases in combination with MS [33]. Here, the glycosylated protein is enzymatically digested with a specific protease, followed by chromatographic separation of the resulting peptide mixture using HPLC (high-performance liquid chromatography) or similar methods. Fractions containing glycosylated peptides can often be localized in MALDI mass spectra as a series of peaks with characteristic m/z spacing representing different glycoforms of the same peptide. Alternatively, glycopeptides can be localized by analyzing fractions with ESI-MS. The appearance of glycan-specific oxonium ions at high orifice potentials indicates the presence of glycopeptides. Fractions containing glycopeptides are then sequentially treated with exoglycosidases to cleave off monosaccharides from the nonreducing end of the glycan and the reaction is monitored by MALDI-MS.

This methodology provides information about the types of monosaccharides present in the glycan and their linkages, but it is limited by the availability of exoglycosidases with suitable specificities, and previous knowledge about the glycan composition is advantageous for the design of the sequential digestion. Another limitation of this method is that a large amount of starting material is needed for the sequential exoglycosidase treatment since sample is lost in the individual steps. Immobilized exoglycosidases on magnetic beads have been used to increase sensitivity [34].

Recent developments in tandem mass spectrometry have made it possible to elucidate glycan structures by MS alone at a sensitivity compatible with standard proteomic applications. Using tandem MS fragmentation to elucidate the structure of glycans has several advantages compared to exoglycosidase strategies including, for example, no restriction in specificity and higher sensitivity. However, a disadvantage of tandem MS fragmentation is that information on linkages and monosaccharide composition often cannot be obtained due to isobaric masses of monosaccharide isomers such as galactose and glucose. Ion trap instruments capable of performing MS^n (i.e., subsequent fragmentation of internal fragment ions from the parent ion), have been used to resolve linkage positions and branching patterns in permethylated oligosaccharides [35].

In 1988, Domon and Costello established the nomenclature for glycan fragmentation, as shown in figure 6.1A [36]. Fragment ions containing the intact nonreducing end of the glycan are labeled A, B, and C, where A denotes cross-ring cleavage and B and C glycosidic bond cleavages. Fragment ions containing the reducing end are labeled X, Y, and Z, where X denotes cross-ring cleavage and Y and Z glycosidic bond cleavages. Glycans are normally detected in MS as $(M+H)^+$, $(M+Na)^+$, or $(M+K)^+$. The fragmentation of those ions is very different; that is, $(M+H)^+$ ions fragment easily, whereas much more energy is needed to fragment the alkali metal adducts. The latter ions produce more cross-ring cleavages [37].

Fragmentation of glycosylated peptides is more complicated because the fragment ion spectrum contains a mixed population of fragment ions originating from the glycan and the attached peptide. The ions observed in fragmentation of glycosylated peptides are (1) loss of single monosaccharides from the glycosylated peptide, indicating that the charge is predominantly retained on the peptide (Y-type fragments); (2) B-type oxonium ions corresponding to fragmentation of the glycan structure with the charge retained on the glycan; and (3) y and b ions obtained from peptide backbone fragmentation (see Roepstorff and Fohlman [38] and fig. 6.1B for more details on peptide fragmentation). The degree of peptide backbone fragmentation depends on the size and the sequence of the peptide, the size of the attached glycan, and the collision energy. Glycosylated peptides with a long peptide chain tend to generate more peptide backbone fragmentation than the ones with small peptide chains. In addition, the smaller the glycan is, the more peptide backbone fragmentation is observed (M. R. Larsen, unpublished results). As for free glycans, glycopeptides can be detected in MS as $(M+H)^+$, $(M+Na)^+$, or $(M+K)^+$ depending on the purity of the sample; also, here the $(M+H)^+$ species fragment is more easily compared to the alkali metal adducts.

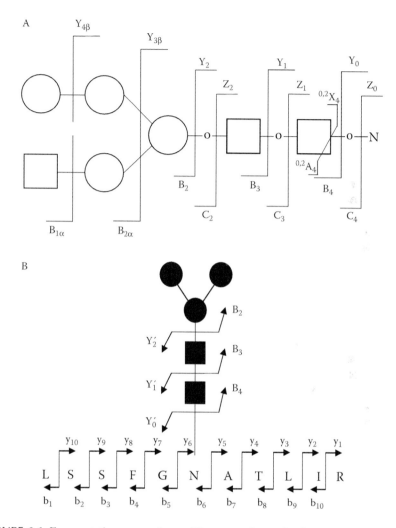

FIGURE 6.1 Fragmentation nomenclature. The nomenclature for fragmentation of glycans suggested by Domon and Costello (Domon, B. and Costello, C. E. *Glycoconjugate J.* 5(4), 397–409, 1988) is illustrated in A. B: Putative fragmentation of a glycosylated peptide including peptide fragmentation according to Roepstorff and Fohlman. (Roepstorff, P. and Fohlman, J. *Biomed. Mass Spectrom.* 11(11), 601, 1984.)

6.5 IDENTIFICATION OF GLYCOSYLATION SITES

Determination of glycosylation sites directly from intact glycopeptides can be problematic. This is particularly the case when the glycosylation is heterogeneous and the signal therefore is distributed over a population of different glycopeptides, which confounds the interpretation of the mass spectrum. The heterogeneity may also limit the detection of low-abundance peptides, particularly if poorly ionizable and/or large

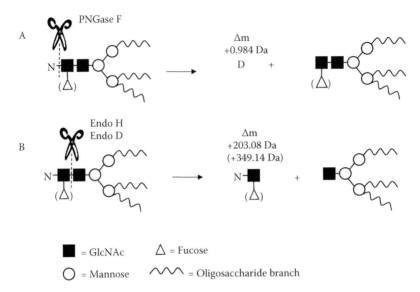

FIGURE 6.2 Enzymatic digestion of glycopeptides by endoglycosidases. This schematic picture only shows the conserved core structure of N-glycosylation with a putative fucose residue attached to the innermost GlcNAc residue. (A) Specificity of peptide-N-glycosidases, like PNGase F. (B) Specificity of endo-β-N-acetylglucosaminidases, like Endo D and Endo H. (Reprinted with permission from Hagglund, P. et al. *J. Proteome Res.* 3(3), 556–566, 2004. Copyright 2004, American Chemical Society.)

glycans are present. Determination of glycan attachment points in a peptide chain can be achieved by complete or partial removal of the glycan chain by fragmentation in the mass spectrometer or by enzymatic digestion.

In the "mild" fragmentation technique of electron capture dissociation (ECD), labile glycan bonds and peptide bonds are fragmented to a similar extent and ECD has thus successfully been used for determination of glycosylation sites using Fourier transform (FT) MS [39]. However, this technique is not yet fully applicable for analysis of complex protein mixtures. In the more common fragmentation technique of collision-induced dissociation (CID), glycans are generally fragmented at lower fragmentation energies than peptides and the peptide fragment ions are therefore often masked by high-abundance glycan fragment ions. Therefore, it may be convenient to remove the glycans by enzymatic cleavage, as described below. Obviously, a drawback with this method is that the information about the intact glycan structures attached to a given site will be lost.

6.5.1 N-Linked Glycans

The most commonly used *endo*-acting enzyme for deglycosylation of N-linked glycans is peptide-N-glycosidase F (EC 3.5.1.52). This enzyme cleaves the bond between asparagine and the innermost GlcNAc residue of the glycan (fig. 6.2A). In the enzymatic process, asparagine is deamidated to aspartate, thus yielding a mass increase of 1 Da that can be used to identify a glycosylation site in a peptide sequence.

However, deamidation is a ubiquitous protein modification *in vivo*, and it is therefore possible to confuse a "true" deamidation site for a glycosylation site. In order to decrease this problem, it is possible to use PNGase F in the presence of ^{18}O water. In this case, an ^{18}O-labeled deglycosylated peptide will gain a mass increase of 3 Da [40,41].

Another group of *endo*-acting enzymes active on N-linked glycans are the endo-β-N-acetylglucosaminidases (EC 3.2.1.96) that hydrolyze the bond between the two innermost GlcNAc residues attached to asparagine (fig. 6.2B). After this enzymatic reaction, one GlcNAc residue will be retained on the peptide that will give a mass increase of 203 Da relative to the unmodified peptide chain. If a fucose residue is attached to the innermost GlcNAc residue, a mass increase of 349 Da will result instead. In MS/MS experiments, the oxonium ion of the retained GlcNAc residue (m/z 204) can be used as a diagnostic marker for glycopeptides (e.g., in precursor ion scanning). Moreover, the signal from the GlcNAc oxonium ion can potentially be used to extract spectra of glycosylated peptides from large datasets, prior to in-depth analysis [42].

One drawback with these enzymes in a comprehensive analysis of glycosylation sites is that no single enzyme is active on all of the major classes of N-glycans (i.e., high mannose, complex, and hybrid). In order to circumvent this problem a combination of enzymes can be used that, when used in concert, are active on all the major classes. For example, Endoglycosidase H is active on high mannose type glycosylation and Endoglycosidase D is active on complex and hybrid type glycosylation when used together with a set of exoglycosidases [43].

6.5.2 O-Linked Glycans

No single enzyme can be used generally for removal of O-glycan structures. The only commercially available *endo*-acting enzyme is O-glycanase (endo-α-N-acetyl-galactosaminidase), which exclusively acts on the specific disaccharide galactosyl-β-1-3-GalNAc attached to serine or threonine [44]. As an alternative to enzymatic digestion, O-glycan attachment sites are often determined by beta-elimation reactions [45]. Beta-elimination of glycosylated serine and threonine residues yields dehydro-alanine and dehydrobutyric acid, respectively, and thus results in a mass decrease of 18 Da that can be used for site identification. Beta elimination can also be combined with a Michael addition reaction, in which a nucleophile attacks the double bond formed after beta elimination. In this manner, different nucleophiles can be used for identification and/or selective enrichment of modified peptides [46,47]. One major drawback with the beta-elimination reaction is that it is not specific for glycosylated residues; that is, phosphorylated serine and threonine residues may also be modified in beta-elimination reactions [46].

6.5.3 GPI Anchors

GPI anchors can be released from the C-terminus of proteins using phosphatidyli-nositol phospholipase C (PI-PLC). PI-PLC hydrolyzes the bond between the lipid and the phospatidylinositol moiety and thus retains the glycan part of the anchor on

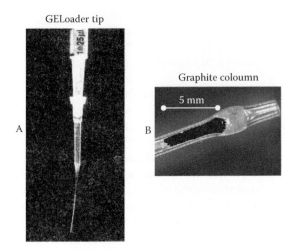

FIGURE 6.3 Pictures of custom-made microcolumns. (A) Microcolumn packed in a GELoader tip. (B) Close-up of graphite powder packed in a microcolumn.

the protein. PI-PLC digestion has been used together with hydrophilic interaction chromatography (HILIC) and MS to identify sites of GPI-anchor attachment [48].

6.6 MICROCOLUMN PURIFICATION STRATEGIES

Several instruments in the new generation of mass spectrometers provide similar results with respect to sensitivity of protein identification. As an example, we have tested several MALDI instruments and found that these show similar performance for identification of in-gel and in-solution digested proteins by peptide mass finger-printing [49]. Thus, the limitation in MS protein identification often lies in the sample preparation steps. Consequently, efforts have been made to develop new methods for sample preparation and improve existing techniques. For example, different matrices and additives have been evaluated for sample preparation in MALDI-MS (e.g., references 50 through 52). Another strategy has been to develop disposable microcolumns for desalting of samples prior to MS analysis.

A number of different microcolumns are commercially available (Ziptip™ [Millipore], NuTip™ [Glygen Corp.], and StageTips [Proxeon]). As an alternative to these, custom-made disposable microcolumns have been developed [53–56]. These columns offer several advantages compared to the commercial ones. First, they are easily prepared in disposable tips (fig. 6.3) and the minute amount of column material used for each column makes them very inexpensive. Second, the small bed volume and large surface area make them more sensitive than the commercial ones because it is possible to achieve a more efficient elution in a much smaller volume, as recently demonstrated [56]. Finally, custom-made microcolumns can be prepared using any kind of material for chromatographic separation available in the appropriate particle size. We and other groups have used several chromatographic materials for proteomics applications that can be used for characterization of glycosylated peptides or free glycan structures:

Reversed-phase (RP) resins (e.g., Poros R1, R2, and R3) are often used in proteomics to desalt and concentrate peptides or proteins prior to mass spectrometry. In the analysis of glycosylated peptides, the efficiency of RP purification is dependent almost exclusively on the peptide sequence, since the hydrophilic glycan contributes very little to the retention on such columns [28].

Graphite powder can be used as an alternative to RP materials, in particular for cleanup of small hydrophilic peptides, phosphorylated peptides, or glycosylated peptides [56–59]. In addition, graphitized carbon has been used extensively for purification of free or derivatized glycans [28]. Below is an example of the use of graphite microcolumns for the analysis of small glycosylated peptides.

Anion (e.g., Poros HQ) and cation (e.g., Poros S) exchange resins are frequently used for protein and peptide separation in proteomics [60] and can be used, for example, to separate charged glycosylated peptides from noncharged species.

Hydrophilic interaction liquid chromatography (HILIC) material can also be used for the purification of hydrophilic compounds (e.g., glycosylated peptides [61]), as described in more detail next.

6.7 TWO EXAMPLES OF USE OF MICROCOLUMNS AND ENZYMATIC DIGESTION FOR MS CHARACTERIZATION OF GLYCOPEPTIDES

6.7.1 Determination of Glycosylation Sites Using HILIC Microcolumns, Endoglycosidases, and MS

In a recent study, we investigated the use of HILIC for enrichment of glycopeptides prior to MS analysis [61]. Hydrophilic interaction liquid chromatography—also referred to as normal phase chromatography—is a well established technique for separation of polar molecules [62]. The technique is based on interactions between polar solutes and hydrophilic groups immobilized in the stationary phase. Typically, different types of chemically modified silica-based stationary phases are used. Samples are applied in a mobile phase with a high content of organic solvent (e.g., acetonitrile), and elution is achieved by lowering the organic content of the mobile phase. HILIC has been used successfully for separation of amino acids, peptides, oligonucleotides, glycans, and various small, polar compounds (e.g., see references 63 through 66).

We tested whether HILIC could potentially be used under optimized conditions to separate glycosylated peptides from nonglycosylated peptides prior to mass spectrometric analysis, using HILIC resin microcolumns. Initially, the method was developed by MS analysis of fetuin, a well-characterized glycoprotein containing three N-glycosylation sites. A tryptic digest of the protein was applied to an HILIC microcolumn in 80% acetonitrile and peptides were eluted in 0.5% formic acid. As seen in figure 6.4C, mainly high molecular weight signals were observed in the mass spectrum of the sample eluted from the HILIC microcolumn. To verify the presence of the peptides containing the three glycosylation sites, the sample was digested by

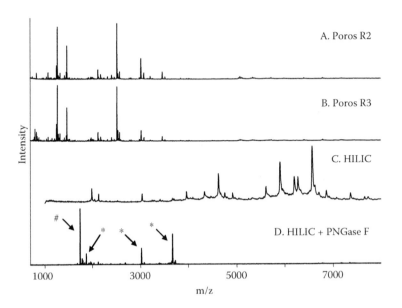

FIGURE 6.4 MALDI-TOF mass spectra of peptides from fetuin after tryptic digestion. (A) Spectrum of peptides retained on an R2 reversed-phase microcolumn. (B) Spectrum of peptides retained on an R3 reverse-phase microcolumn. (C) Spectrum of peptides retained on a HILIC microcolumn. (D) Spectrum of peptides after PNGase F digestion of the sample displayed in 2C. Peaks corresponding to deglycosylated peptides are marked with asterisks (*). The peak marked with a hash sign (#) corresponds to a glycopeptide containing an internal S–S bond. All spectra were acquired in positive reflector mode except 2C, which was acquired in negative linear mode. (Adapted with permission from Hagglund, P. et al. *J. Proteome Res.* 3(3), 556–566, 2004. Copyright 2004, American Chemical Society.)

an endoglycosidase (PNGase F). Indeed, as seen in figure 6.4D, the main signals in the mass spectrum of the sample after endoglycosidase digestion correspond to the three deglycosylated peptides containing the glycosylation sites in fetuin. In contrast, mass spectra of the same tryptic digest purified on microcolumns packed with standard reverse phase resins were dominated by signals from nonglycosylated peptides (fig. 6.4A and B)

In the same study, we combined the use of HILIC microcolumns and endo-glycosidase digestion for analysis of N-glycosylated peptides in human plasma [61]. Briefly, proteins from human plasma were separated by SDS-PAGE followed by in-gel tryptic digestion. The digested samples were applied to HILIC microcolumns as described previously, and the retained peptides and the flow-through peptides were digested with Endoglycosidase H and Endoglycosidase D, thus leaving one GlcNAc residue retained on the peptide (fig. 6.2B). Finally, the samples were analyzed by LC-MS/MS and 62 glycosylation sites were unambiguously identified. All the identified glycopeptides were retained on the HILIC columns. Only one peptide was identified among the flow-through fractions, and this peptide was also identified among the retained fractions.

6.7.2 PROTEINASE K DIGESTION COMBINED WITH MICROCOLUMN PURIFICATION AND MS

The current methods for glycoprotein characterization (i.e., localization of glyco-sylation site and characterization of the glycan composition/structure) cannot easily be applied to proteins in most proteomics studies due to the low sensitivity of those methods. Recently, we described the use of specific/nonspecific protease treatment in combination with sequential microcolumn purification and MALDI-MS for char-acterization of glycosylated proteins present in low amounts (low picomole level) [58]. In this method, the glycosylated protein is enzymatically digested into pep-tides/glycopeptides using a specific protease (e.g., trypsin); a small amount of the peptides (usually <2%) is used to identify the protein based on peptide mass mapping or MS/MS sequencing. The remaining sample is treated with the nonspecific protease proteinase K, which cleaves the peptides into small fragments (i.e., mono-, di-, or tripeptides). The glycan structure on the glycopeptides will act as steric hindrance for the proteinase K, thereby leaving a small "peptide tag" attached to the glycan. The digested glycopeptides can be specifically purified from the remaining peptides that are resistant to proteinase K or peptides originating from autoproteolysis of proteinase K by using sequential microcolumn purification employing reverse phase and graphite materials. The sample after proteinase K treatment is applied to a Poros R2 reverse-phase microcolumn. The flow-through is collected on a microcolumn packed with graphite powder and, after washing the column, the glycopeptides are eluted directly onto a MALDI target using 30% acetonitrile, where it is mixed with matrix (DHB). A similar strategy employing the nonspecific protease pronase was described by Juhasz and Martin [67] and has been used in a number of studies (e.g., An et al. [68]). However, a significant decrease in sensitivity is observed when this method is used. Pronase is composed of many endo- and exoproteinases that cleave each other in an unspecific manner, resulting in a large number of peptides of various lengths. This significantly reduces the sensitivity of the method due to ion suppres-sion effects.

Using the method described above, the analysis of 2 pmol of HPLC purified ovalbumin separated by SDS-PAGE yielded a glycopeptide profile in MALDI-MS negative ion mode as shown in figure 6.5A. Here, a total of 11 different signals are detected in the mass area > 1200 Da, which is the area where we expect the cleaved glycopeptides to appear. Each signal represents a different glycan structure attached to the same peptide (YNLT) as indicated in the figure. The mass of the attached peptide and a putative monosaccharide composition can be obtained by performing MALDI tandem MS or by using the GlycoMod software (http://www.expasy.com).

A MALDI tandem MS spectrum of the signal at 1886.9 Da from figure 6.5A in positive ion mode is shown in figure 6.5B. Here, the sequential loss of single monosaccharides from the glycopeptide indicates that the charge is predominantly retained on the peptide (Y-type fragments [36]). Additional signals originate from B-type oxonium ions, corresponding to fragmentation of the glycan structure with the charge retained on the glycan. The described method is fast and sensitive enough for investigation of the glycome of different types of cells or tissues in diseases or

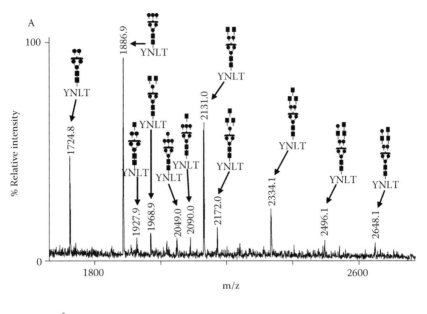

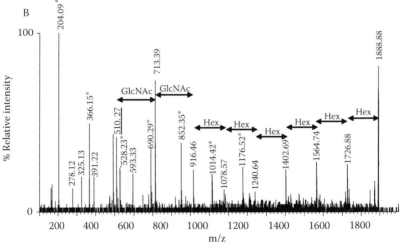

FIGURE 6.5 Analysis of gel-separated ovalbumin using sequential proteases and micro-columns. (A) MALDI-MS spectrum of glycosylated peptides from ovalbumin after selective purification using proteolytic enzymes in combination with reversed-phase and graphite microcolumns as described previously. (Larsen, M. R. et al. *Mol. Cell. Proteomics* 4(2), 107–119, 2005.) The spectrum was obtained in negative reflector ion mode and the peptide sequence and the putative monosaccharide compositions are shown. (B) Tandem MALDI-MS fragment ion spectrum of the glycopeptide at m/z 1888.8 in positive ion mode. The sequential loss of single monosaccharides from the glycopeptide is shown and the oxonium ions are marked with asterisks.

during different physiological stimuli where the proteins are present in low levels, especially if combined with liquid chromatography to reduce the complexity.

6.8   FUTURE PROSPECTS

A few examples of developments in MS analysis of glycoproteins have been presented here. As outlined in this chapter, an array of mass spectrometric and chromatographic techniques can now be used for analysis of glycoproteins. However, further developments in this field are needed to improve the detection of low-abundance glycoproteins in complex samples. This is particularly important for many clinical applications since the immediate problem associated with characterization of biomarkers in body fluids is the dynamic range of protein concentration [69]. This results in difficulties in detecting low-abundance biomarkers leaking from diseased cells/tissues due to the limited dynamic range of the current techniques.

Abnormal glycosylation is observed in many diseases (e.g., human cancers [70,71]), making glycosylated proteins ideal candidates for biomarkers. It is well known that several glycosylated proteins change in terms of expression and in terms of the type of glycan structures attached and glycan attachment site—for example, upon tumor invasion and metastasis [70,71]. Since glycosylated proteins are mainly membrane attached or secreted, they will eventually end up in body fluids. By focusing the biomarker detection on glycosylated proteins, the problem of large dynamic range in protein concentrations can be reduced, since glycosylated proteins can be affinity purified from plasma using carbohydrate-binding lectins. Several lectins are commercially available and many more are on their way, holding promise for future biomarker detection.

For a comprehensive understanding of the glycosylation pattern in a particular proteome, it would be desirable to obtain quantitative information about the glycosylation status of its individual proteins. Quantification of glycosylated proteins in a proteomics context has been shown previously [72]. Glycosylated proteins were conjugated to a solid support using hydrazide chemistry followed by proteolytic digestion, isotope labeling of the glycopeptides, and final release of the N-linked glycopeptides by PNGase F digestion. The released peptides were analyzed and quantified using LC-MS/MS. Even though this method provided quantitative information about protein expression, it is likely that many quantitative changes are related to differences in the relative quantities of different glycans attached at a particular glycosylation site. It is therefore important that these types of studies are further developed to provide quantitative information also on the glycan level.

Development of new, sensitive hybrid types of MS instruments capable of performing fragmentation by ECD, electron transfer dissociation and infrared multiphoton photodissociation (IRMPD) appears to be promising for future analysis strategies in glycomics. In ECD, precursor ions are irradiated with low-energy electrons yielding predominantly peptide backbone fragmentation (mainly c and z ions). In IRMPD, precursor ions are subjected to infrared irradiation and slowly heated until they begin to dissociate, giving rise to amide bond cleavage or fragmentation of the most labile bonds (e.g., glycosidic bonds). The combination of the two techniques will undoubtedly contribute to the field of glycomics in the future, as peptide sequence information is obtained using the ECD fragmentation and information about the glycan structure is obtained using the IRMPD fragmentation option. The limiting step is the slow fragmentation compared to CID; however, we believe that this hurdle will be overcome in the future.

In contrast to protein expression analysis and simple protein identifications, information about post-translational modifications can only be determined on the protein level and not by complementary approaches on the RNA transcript level (e.g., microarrays). Therefore, we believe that the area of glycomics in particular and modification-specific proteomics in general will continue to develop in the field of proteomic research because information about glycosylations and other post-translational modifications is important for a complete understanding of many of the mechanisms behind the dynamics and regulations of living cells.

ACKNOWLEDGMENTS

The present work was made possible through a grant to the Danish Biotechnology Instrument Center (DABIC) from the Danish Research Agency. MRL was financed through a Steno scholarship from The Danish Natural Science Research Council. PH was financed through a long-term fellowship from the Federation of European Biochemical Societies.

REFERENCES

1. Varki, A., Biological roles of oligosaccharides—All of the theories are correct, *Glycobiology* 3(2), 97–130, 1993.
2. Lis, H. and Sharon, N., Protein glycosylation—Structural and functional aspects, *Eur. J. Biochem.* 218(1), 1–27, 1993.
3. Zaia, J., Mass spectrometry of oligosaccharides, *Mass Spectrom. Rev.* 23(3), 161–227, 2004.
4. Harvey, D. J., Identification of protein-bound carbohydrates by mass spectrometry, *Proteomics* 1(2), 311–328, 2001.
5. Kornfeld, R. and Kornfeld, S., Assembly of asparagine-linked oligosaccharides, *Annu. Rev. Biochem.* 54, 631–664, 1985.
6. Taniguchi, T. et al., The structures of the asparagine-linked sugar chains of bovine interphotoreceptor retinol-binding protein—Occurrence of fucosylated hybrid-type oligosaccharides, *J. Biol. Chem.* 261(4), 1730–1736, 1986.
7. Rayon, C., Lerouge, P., and Faye, L., The protein N-glycosylation in plants, *J. Exp. Bot.* 49(326), 1463–1472, 1998.
8. Spiro, R. G., Protein glycosylation: Nature, distribution, enzymatic formation, and disease implications of glycopeptide bonds, *Glycobiology* 12(4), 43R–56R, 2002.
9. Van den Steen, P. et al., Concepts and principles of O-linked glycosylation, *Crit. Rev. Biochem. Mol. Biol.* 33(3), 151–208, 1998.
10. Wells, L., Vosseller, K., and Hart, G. W., Glycosylation of nucleocytoplasmic proteins: Signal transduction and O-GlcNAc, *Science* 291(5512), 2376–2378, 2001.
11. Ferguson, M. A. J. et al., Glycosyl-phosphatidylinositol moiety that anchors *Trypanosoma brucei* variant surface glycoprotein to the membrane, *Science* 239(4841), 753–759, 1988.

12. Gerber, L. D., Kodukula, K., and Udenfriend, S., Phosphatidylinositol glycan (Pi-G) anchored membrane-proteins—Amino-acid-requirements adjacent to the site of cleavage and Pi-G attachment in the cooh-terminal signal peptide, *J. Biol. Chem.* 267(17), 12168–12173, 1992.

13. Hofsteenge, J. et al., New-type of linkage between a carbohydrate and a protein—C-glycosylation of a specific tryptophan residue in human Rnase U-S, *Biochemistry* 33(46), 13524–13530, 1994.

14. Furmanek, A. and Hofsteenge, J., Protein C-mannosylation: Facts and questions, *Acta Biochim. Pol.* 47(3), 781–789, 2000.

15. Dell, A., Fab-mass spectrometry of carbohydrates, *Adv. Carbohydr. Chem. Biochem.* 45, 19–72, 1987.

16. Karas, M. and Hillenkamp, F., Laser desorption ionization of proteins with molecular masses exceeding 10,000 daltons, *Anal. Chem.* 60(20), 2299–2301, 1988.

17. Hillenkamp, F. et al., Matrix-assisted laser desorption ionization mass-spectrometry of biopolymers, *Anal. Chem.* 63(24), A1193–A1202, 1991.

18. Fenn, J. B. et al., Electrospray ionization for mass-spectrometry of large biomolecules, *Science* 246(4926), 64–71, 1989.

19. Harvey, D. J., Matrix-assisted laser desorption/ionization mass spectrometry of carbohydrates, *Mass Spectrom. Rev.* 18, 349–451, 1999.

20. Harvey, D. J., Quantitative aspects of the matrix-assisted laser-desorption mass-spectrometry of complex oligosaccharides, *Rapid Commun. Mass Spectrom.* 7(7), 614–619, 1993.

21. Harvey, D. J., Identification of protein-bound carbohydrates by mass spectrometry, *Proteomics* 1(2), 311–328, 2001.

22. Mortz, E. et al., Does matrix-assisted laser desorption ionization mass spectrometry allow analysis of carbohydrate heterogeneity in glycoproteins? A study of natural human interferon-gamma, *J. Mass Spectrom.* 31(10), 1109–1118, 1996.

23. Viseux, N. et al., Qualitative and quantitative analysis of the glycosylation pattern of recombinant proteins, *Anal. Chem.* 73(20), 4755–4762, 2001.

24. Bache, N. et al., Comparison of MALDI tandem mass spectrometers. Part 2: Modified peptides and oligonucleotides. In 52nd ASMS Conference on Mass Spectrometry and Allied Topics, Nashville, Tennessee, 2004, abstract A042122.

25. Juraschek, R., Dulcks, T., and Karas, M., Nanoelectrospray— More than just a minimized-flow electrospray ionization source, *J. Am. Soc. Mass Spectrom.* 10(4), 300–308, 1999.

26. Cech, N. B. and Enke, C. G., Practical implications of some recent studies in electrospray ionization fundamentals, *Mass. Spectrom. Rev.* 20(6), 362–387, 2001.

27. Viseux, N., deHoffmann, E., and Domon, B., Structural analysis of permethylated oligosaccharides by electrospray tandem mass spectrometry, *Anal. Chem.* 69(16), 3193–3198, 1997.

28. Packer, N. H. et al., A general approach to desalting oligosaccharides released from glycoproteins, *Glycoconjugate J.* 15(8), 737–747, 1998.

29. Karlsson, N. G. and Packer, N. H., Analysis of O-linked reducing oligosaccharides released by an in-line flow system, *Anal. Biochem.* 305(2), 173–185, 2002.

30. Carr, S. A., Huddleston, M. J., and Bean, M. F., Selective identification and differentiation of N-linked and O-linked oligosaccharides in glycoproteins by liquid-chromatography mass-spectrometry, *Protein Sci.* 2(2), 183–196, 1993.

31. Huddleston, M. J., Bean, M. F., and Carr, S. A., Collisional fragmentation of glycopeptides by electrospray ionization Lc Ms and Lc Ms Ms—Methods for selective detection of glycopeptides in protein digests, *Anal. Chem.* 65(7), 877–884, 1993.

32. Jebanathirajah, J., Steen, H., and Roepstorff, P., Using optimized collision energies and high-resolution, high-accuracy fragment ion selection to improve glycopeptide detection by precursor ion scanning, *J. Am. Soc. Mass Spectrom.* 14(7), 777–784, 2003.

33. Stahl, B. et al., The oligosaccharides of the Fe(Iii)-Zn(Ii) purple acid-phosphatase of the red kidney bean—Determination of the structure by a combination of matrix-assisted laser-desorption ionization mass-spectrometry and selective enzymatic degradation, *Eur. J. Biochem.* 220(2), 321–330, 1994.

34. Krogh, T. N., Berg, T., and Hojrup, P., Protein analysis using enzymes immobilized to paramagnetic beads, *Anal. Biochem.* 274(2), 153–162, 1999.

35. Sandra, K., Devreese, B., Van Beeumen, J., Stals, I., and Claeyssens M. The Q-trap mass spectrometer, a novel tool in the study of protein glycosylation, *J. Am. Soc. Mass Spectrom.* 15(3), 413–423, 2004.

36. Domon, B. and Costello, C. E., A systematic nomenclature for carbohydrate fragmentations in Fab-Ms Ms spectra of glycoconjugates, *Glycoconjugate J.* 5(4), 397–409, 1988.

37. Harvey, D. J., Collision-induced fragmentation of underivatized N-linked carbohydrates ionized by electrospray, *J. Mass Spectrom.* 35(10), 1178–1190, 2000.

38. Roepstorff, P. and Fohlman, J., Proposal for a common nomenclature for sequence ions in mass-spectra of peptides, *Biomed. Mass Spectrom.* 11(11), 601, 1984.

39. Mirgorodskaya, E., Roepstorff, P., and Zubarev, R. A., Localization of O-glycosylation sites in peptides by electron capture dissociation in a Fourier transform mass spectrometer, *Anal. Chem.* 71(20), 4431–4436, 1999.

40. Gonzalez, J. et al., A method for determination of N-glycosylation sites in glycoproteins by collision-induced dissociation analysis in fast-atom-bombardment mass-spectrometry—Identification of the positions of carbohydrate-linked asparagine in recombinant alpha-amylase by treatment with peptide-N-glycosidase-F in O-18-labeled water, *Anal. Biochem.* 205(1), 151–158, 1992.

41. Kuster, B. and Mann, M., O-18-labeling of N-glycosylation sites to improve the identification of gel-separated glycoproteins using peptide mass mapping and database searching, *Anal. Chem.* 71(7), 1431–1440, 1999.

42. Hagglund, P. et al., Analysis of N-glycosylated peptides using online LC-ESI MS and off-line MALDI MS. In 52nd ASMS Conference on Mass Spectrometry and Allied Topics, Nashville, Tennessee, 2004, abstract A040389.

43. Koide, N. and Muramats.T., Endo-beta-N-acetylglucosaminidase acting on carbohydrate moieties of glycoproteins—Purification and properties of enzyme from *Diplococcus pneumoniae*, *J. Biol. Chem.* 249(15), 4897–4904, 1974.

44. Iwase, H. and Hotta, K., Release of O-linked glycoprotein glycans by endo-alpha-N-acetyl-D-galactosaminidase, *Methods Molecular Biol.* 14, 151–159, 1993.

45. Greis, K. D. et al., Selective detection and site-analysis of O-GlcNAc-modified glycopeptides by beta-elimination and tandem electrospray mass spectrometry, *Anal. Biochem.* 234(1), 38–49, 1996.

46. Wells, L. et al., Mapping sites of O-GlcNAc modification using affinity tags for serine and threonine post-translational modifications, *Mol. Cell. Proteomics* 1(10), 791–804, 2002.

47. Mirgorodskaya, E. et al., Mass spectrometric determination of O-glycosylation sites using beta-elimination and partial acid hydrolysis, *Anal. Chem.* 73(6), 1263–1269, 2001.

48. Omaetxebarria, M.J. et al., Isolation and characterization of glycosylphosphatidylinositol-anchored peptides by hydrophilic interaction chromatography and MALDI tandem mass spectrometry, *Anal. Chem.* 78(10), 3335–3341, 2006.

49. Horning, O. et al., Comparison of MALDI tandem mass spectrometers. Part 1: Protein identification. In 52nd ASMS Conference on Mass Spectrometry and Allied Topics, Nashville, Tennessee, 2004, abstract A042128.

50. Karas, M. et al., Matrix-assisted laser-desorption ionization mass-spectrometry with additives to 2,5-dihydroxybenzoic acid, *Org. Mass Spectrom.* 28(12), 1476–1481, 1993.

51. Mechref, Y. and Novotny, M. V., Matrix-assisted laser desorption ionization mass spectrometry of acidic glycoconjugates facilitated by the use of spermine as a co-matrix, *J. Am. Soc. Mass Spectrom.* 9(12), 1293–1302, 1998.

52. Papac, D. I., Wong, A., and Jones, A. J. S., Analysis of acidic oligosaccharides and glycopeptides by matrix assisted laser desorption ionization time-of-flight mass spectrometry, *Anal. Chem.* 68(18), 3215–3223, 1996.

53. Erdjument-Bromage, H. et al., Examination of micro-tip reversed-phase liquid chromatographic extraction of peptide pools for mass spectrometric analysis, *J. Chromatogr. A* 826(2), 167–181, 1998.

54. Gobom, J. et al., Sample purification and preparation technique based on nanoscale reversed-phase columns for the sensitive analysis of complex peptide mixtures by matrix-assisted laser desorption/ionization mass spectrometry, *J. Mass Spectrom.* 34(2), 105–116, 1999.

55. Wilm, M. and Mann, M., Analytical properties of the nanoelectrospray ion source, *Anal. Chem.* 68(1), 1–8, 1996.

56. Larsen, M. R., Cordwell, S. J., and Roepstorff, P., Graphite powder as an alternative or supplement to reversed-phase material for desalting and concentration of peptide mixtures prior to matrix-assisted laser desorption/ionization-mass spectrometry, *Proteomics* 2(9), 1277–1287, 2002.

57. Larsen, M. R. et al., Improved detection of hydrophilic phosphopeptides using graphite powder microcolumns and mass spectrometry—Evidence for *in vivo* doubly phosphorylated dynamin I and dynamin III, *Mol. Cell. Proteomics* 3(5), 456–465, 2004.

58. Larsen, M. R., Hojrup, P., and Roepstorff, P., Characterization of gel-separated glycoproteins using two-step proteolytic digestion combined with sequential microcolumns and mass spectrometry, *Mol. Cell. Proteomics* 4(2), 107–119, 2005.

59. Chin, E. T. and Papac, D. I., The use of a porous graphitic carbon column for desalting hydrophilic peptides prior to matrix-assisted laser desorption/ionization time-of-flight mass spectrometry, *Anal. Biochem.* 273(2), 179–185, 1999.

60. Nuhse, T. S. et al., Large-scale analysis of *in vivo* phosphorylated membrane proteins by immobilized metal ion affinity chromatography and mass spectrometry, *Mol. Cell. Proteomics* 2(11), 1234–1243, 2003.

61. Hagglund, P. et al., A new strategy for identification of N-glycosylated proteins and unambiguous assignment of their glycosylation sites using HILIC enrichment and partial deglycosylation, *J. Proteome Res.* 3(3), 556–566, 2004.

62. Alpert, A. J., Hydrophilic-interaction chromatography for the separation of peptides, nucleic-acids and other polar compounds, *J. Chromatogr.* 499, 177–196, 1990.

63. Wuhrer, M. et al., Normal-phase nanoscale liquid chromatography—Mass spectrometry of underivatized oligosaccharides at low-femtomole sensitivity, *Anal. Chem.* 76(3), 833–838, 2004.

64. Yoshida, T., Peptide separation in normal phase liquid chromatography, *Anal. Chem.* 69(15), 3038–3043, 1997.

65. Schlichtherle-Cerny, H., Affolter, M., and Cerny, C., Hydrophilic interaction liquid chromatography coupled to electrospray mass spectrometry of small polar compounds in food analysis, *Anal. Chem.* 75(10), 2349–2354, 2003.

66. Paek, I. B. et al., Hydrophilic interaction liquid chromatography-tandem mass spectrometry for the determination of levosulpiride in human plasma, *J. Chromatogr. B* 809(2), 345–350, 2004.

67. Juhasz, P. and Martin, S. A., The utility of nonspecific proteases in the characterization of glycoproteins by high-resolution time-of-flight mass spectrometry, *Int. J. Mass Spectrom.* 169, 217–230, 1997.

68. An, H. J. et al., Determination of N-glycosylation sites and site heterogeneity in glycoproteins, *Anal. Chem.* 75(20), 5628–5637, 2003.

69. Anderson, N. L. and Anderson, N. G., The human plasma proteome—History, character, and diagnostic prospects, *Mol. Cell. Proteomics* 1(11), 845–867, 2002.

70. Granovsky, M. et al., Suppression of tumor growth and metastasis in Mgat5-deficient mice, *Nat. Med.* 6(3), 306–312, 2000.

71. Kannagi, R. et al., Carbohydrate-mediated cell adhesion in cancer metastasis and angiogenesis, *Cancer Sci.* 95(5), 377–384, 2004.

72. Zhang, H. et al., Identification and quantification of N-linked glycoproteins using hydrazide chemistry, stable isotope labeling and mass spectrometry, *Nat. Biotechnol.* 21(6), 660–666, 2003.

7 Surface-Enhanced Laser Desorption/Ionization Protein Biochip Technology for Proteomics Research and Assay Development

Scot R. Weinberger, Lee Lomas, Eric Fung, and Cynthia Enderwick

CONTENTS

7.1   INTRODUCTION

Surface-enhanced laser desorption/ionization (SELDI) protein array technology represents a collection of analytical tools and protocols that address the challenges of protein separation, protein purification, and protein detection by mass spectrometry. Commercially, SELDI has been embodied within Ciphergen Biosystems Inc. Protein-Chip® array products (Fremont, California). SELDI array surfaces function as solid phase extraction media that support on-probe isolation and cleanup of analytes prior to mass spectrometric investigation. Often, a complex mixture of proteins and other biological compounds can be divided into a series of manageable populations by employing specific or quasispecific interaction motifs that exist between the analyte and the array's surface. Analytes with physical and chemical properties complementary to array surface functional groups are adsorbed, while others are washed away during the sample preparation process. Further, contaminants such as salts, detergents, elutropic agents, lipids, and chemical denaturants that squelch the creation of ions during mass spectrometric proteomic studies are, for the most part, also washed away, producing a "purified" protein pool for facile mass spectrometry (MS) analysis.

After adsorption and purification upon the array surface, retained proteins are subsequently desorbed and ionized using matrix-assisted, surface-enhanced laser desorption/ionization and detected typically using a time-of-flight (TOF) mass spectrometer. For the most part, SELDI array technology has been used for the discovery and characterization of many biomarkers of organic disease and cancer as well as extended to routine toxicological studies employing samples of various origins. Most recently, SELDI technology has demonstrated the ability to translate the fundamental discovery of protein biomarkers into predictive assays for the purpose of diagnosing the presence of metabolic disease and cancer. Furthermore, this platform has been used to investigate biomolecular interactions and protein signaling pathways, as well as being used as a convenient tool to accelerate the development of preparative scale protein chromatography.

This chapter serves to introduce the reader to the fundamentals of SELDI biochip array technology while providing a recent review of new technologies and achievements in basic proteomic and clinical proteomic research. A general overview of SELDI array and MS technology is first given, followed by a discussion of combining bead-based purification and enrichment schemes along with SELDI analysis. Finally, a brief discussion of future technological developments and applications is provided.

7.2   OVERVIEW OF SELDI PROTEIN ARRAY TECHNOLOGY

7.2.1   SELDI, Laser Desorption/Ionization, and MALDI

Three widely used photoinduced ionization techniques are employed for the analysis of solid state samples by mass spectrometry: laser desorption/ionization (LDI);

matrix-assisted laser desorption/ionization (MALDI), and surface-enhanced laser desorption/ionization (SELDI). Each of these techniques relies upon the energy inherent in a focused laser beam to promote the creation of gaseous ions from solid-state matter. Samples are most often presented as crystals or thin films upon a sample support, which is typically referred to as a probe. LDI occurs when the sample directly absorbs energy from the laser and heats up via direct or secondary thermal changes, thus producing desorbed ions. In MALDI, thermal energy is transferred to the analyte from energy-absorbing compounds that are typically referred to as the "matrix." Analyte is first embedded as impurities within matrix crystals or thin films. Laser energy is then applied to the sample, inducing ionization and transformation from the solid to the gas state.

LDI analyses date back to the early 1960s, when LDI was mostly used to study a number of small, inorganic salts.[1-3] Laser-based analysis of large biopolymers was first facilitated by the development of MALDI.[4-7] For LDI and MALDI applications, the sample probe surface plays a passive role in the analytical scheme; the probe merely presents the sample to the mass spectrometer for analysis. In order to produce usable MS signal, crude samples must first be fractionated and purified of any ionization suppressants, such as salts, chaotropic agents, detergents, and so on. Furthermore, in the case of MALDI, biopolymer analysis is only possible when analytes are cocrystallized with a solution of matrix.

SELDI, as originally defined by Hutchens and Yip, consists of two subsets of technology: surface-enhanced affinity capture (SEAC) and surface-enhanced neat desorption (SEND).[8] By far, the SELDI array technology showing the most utility to date is SEAC. Because of the latter, SEAC is often generally referred to as SELDI in the published domain. In SEAC, the probe surface plays an active role in the extraction, presentation, structural modification, and/or amplification of the sample. Figure 7.1 depicts a number of chemical and biochemical SELDI array surfaces used in SEAC applications. Chemical surface arrays are derivatized with classic chromatographic separation moieties, such as reverse phase (RP), ion exchange, immobilized metal affinity capture (IMAC), and normal phase media. Surfaces such as these, with broad binding properties, are typically used for protein profiling and *de novo* biomarker discovery, where large populations of proteins are compared (e.g., from diseased vs. normal samples) with the goal of elucidating differentially expressed elements. Biomolecules bind to these surfaces through hydrophobic, electrostatic, coordinate covalent bond, or Lewis-acid/base interaction.

Biochemical arrays are created by immobilizing bait molecules upon the surface of preactivated SELDI surfaces. Linkage proceeds via covalent attachment using primary amines or alcohols. In this way, specific protein interaction arrays of virtually any content may be created, including antibodies, receptors, enzymes, DNA, small molecules, ligands, and lectins. In contrast to standard chromatographic media, these surfaces provide much more enrichment of captured analytes, due to the high specificity of biomolecular interactions. Because specific biochemical interaction motifs demonstrate high-affinity and low-equilibrium dissociation constants, biochemical surfaces facilitate a vast array of on-chip, microscale experiments, including SELDI immunoassay, targeted protein identification and/or purification, ligand binding domain analysis, epitope mapping experiments, and post-translational modification

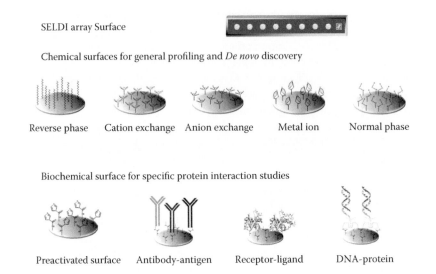

SELDI array Surface

Chemical surfaces for general profiling and *De novo* discovery

Reverse phase   Cation exchange   Anion exchange   Metal ion   Normal phase

Biochemical surface for specific protein interaction studies

Preactivated surface   Antibody-antigen   Receptor-ligand   DNA-protein

FIGURE 7.1 SELDI protein biochip array surfaces.

detection, as well as reliable quantitative studies, even when fishing for target proteins within a complex biological milieu.

SEND is a process by which analytes, even those of large molecular weight, may be desorbed and ionized without the need for the addition of matrix. SEND is accomplished by the attachment of an energy-absorbing compound to the probe surface via covalent modification or physical adsorption.[8–10] Recently, a series of SEND-based arrays have been created using a novel polymer approach in which cinnamic acid-based MALDI matrices are converted to reactive monomers via conjugation with methacryloyl chloride and allowed to polymerize to form a polymeric film capable of supporting laser induced desorption/ionization of compounds deposited upon its surface (see fig. 7.2).

In addition to these pure "matrix" polymers, polymer blends have been created by mixing preactivated matrices with other preactivated monomers intended to impart specific array surface properties such as proton donor groups for the purpose of facilitating ionization and film stabilizers to ensure strong polymer adhesion to the underlying array substrate, as well as solid phase extraction groups intended to create selective protein interaction properties, as is the case for SEAC (see fig. 7.3).[10] When compared to MALDI or SEAC, SEND arrays have demonstrated not only a simplified workflow, but also improved analytical sensitivity and associated mass detection limit. The latter is attributed to the diminished levels of chemical noise produced when using these arrays as well as to a marked reduction in unwanted sample loss to the array surface. Compared to SEAC, this technology remains in an early stage of adoption.

7.2.2 SELDI Sample Preparation: Retentate Chromatography

The process of SELDI-based sample preparation using four different chemical surfaces is demonstrated in figure 7.4. A series of orthogonal SELDI surfaces—reverse

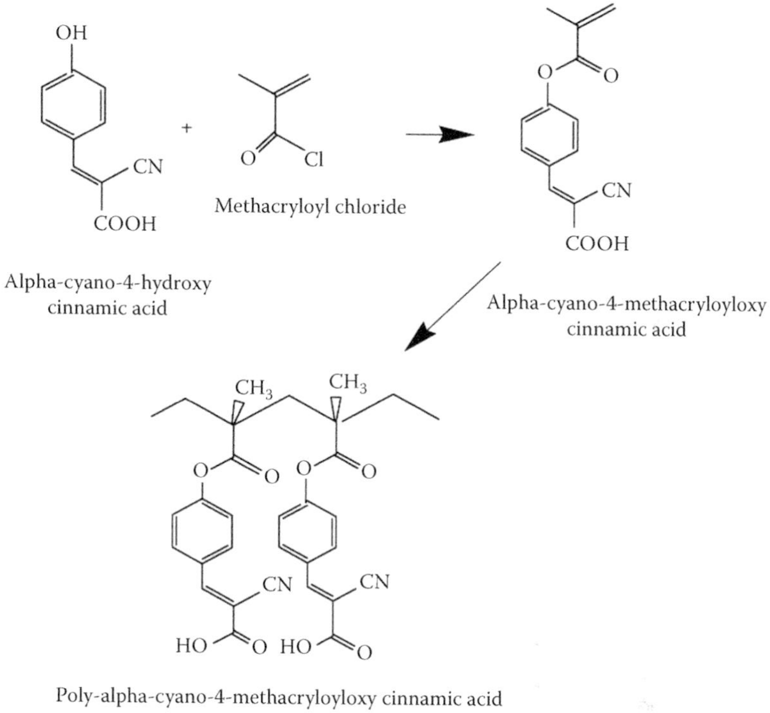

OH

+

Methacryloyl chloride

Alpha-cyano-4-hydroxy
cinnamic acid

CN

COOH

CN

COOH

Alpha-cyano-4-methacryloyloxy
cinnamic acid

CH₃   CH₃

CN   CN

HO   O HO   O

Poly-alpha-cyano-4-methacryloyloxy cinnamic acid

FIGURE 7.2 Shotten–Bauman reaction scheme for the creation of activated cinnamic acid monomers and the ultimate creation of a matrix polymeric film. The creation of poly-alpha-cyano-4-methacryloxycinnamic acid is depicted.

phase, anionic (−), cationic (+), and IMAC—are arranged in parallel. A complex, heterogeneous sample solution is deposited upon every chemically active "spot" of each array. After binding, the active surfaces on each array are washed with appropriate buffers in gradient manner. Within a given array, subsequent spots experience a greater degree of stringency, removing analytes with weak surface interaction potential and enriching those of strong surface affinity. The variety of retained proteins is depicted schematically for each spot on each array. In some cases, particularly when using specific biomolecular interactions, purification to almost complete homogeneity is possible, without substantial loss of analyte.

When compared to traditional sample cleanup methods such as microcolumn chromatography or purification using pipette tips packed with chromatographic resin, on-array cleanup diminishes sample loss by eliminating the opportunity for analytes to bind nonspecifically to the walls of sample handling vessels as well as avoiding the dilution inherent in eluting samples from a stationary phase. For example, figure 7.5 illustrates the benefit of extracting a protein denaturant, urea, from the resulting peptide pool of an enzymatic digest of bovine serum albumin (BSA). In this case, the total starting protein amount was 0.5 pmol. Note that removal of urea was required for generating ion signal. At the 0.5-pmol level, on-chip cleanup and packed pipette tip purification appear to work equally well in terms of identified

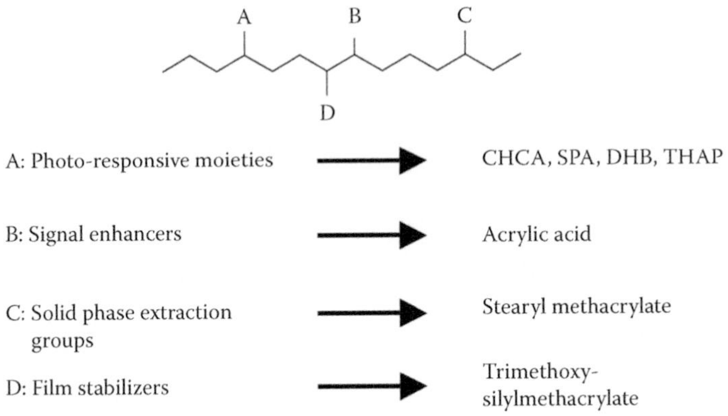

A   B   C

D

A: Photo-responsive moieties   ⟶   CHCA, SPA, DHB, THAP

B: Signal enhancers   ⟶   Acrylic acid

C: Solid phase extraction   ⟶   Stearyl methacrylate
   groups

D: Film stabilizers   ⟶   Trimethoxy-
                                             silylmethacrylate

FIGURE 7.3 SEND polymeric blend arrays are created by the mixed polymerization of activated matrix and other selective monomers. The resultant heterogeneous polymer blend demonstrates a plurality of surface interaction and laser interaction properties. "A" represents the photoresponsive pendant groups (matrices): CHCA is alpha-cyano-4-hydroxycinnamic acid, SPA is 3,5-dimethoxy-4-hydroxycinnamic acid (sinapinic acid), DHB is 2,5-dihyroxy-benzoic acid (gentisic acid), and THAP is 2,4,6-trihydroxyacetophenone. "B" represents monomers with signal-enhancing capabilities such as organic acids or selective cationization groups. "C" represents groups intended to impart specific surface interaction potentials, as is the case for SEAC. "D" signifies a number of film-stabilizing groups that augment film adhesive and cohesive properties.

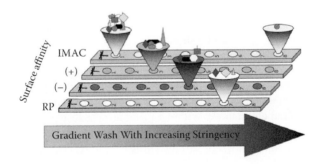

FIGURE 7.4 (See color insert.) The principles of retentate chromatography.

peptides. However, when total protein starting material is significantly limited, as is the case in figure 7.6, significant sample loss and dilution during pipette tip purification make detection of the resultant peptides impossible, while on-array cleanup clearly provides a number of useful signals for protein identification by peptide fingerprinting or tandem MS analysis.

In addition to the benefits of diminished work flow and improved sample recovery, protein extraction upon SELDI arrays also reveals physical–chemical properties of the analyte, such as hydrophobicity, total charge, pI, phosphorylation, glycosylation, and primary composition. Such *de facto* knowledge may be further exploited during directed protein discovery or purification experiments by using specific arrays and/or

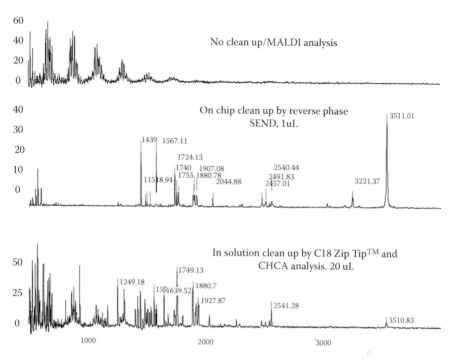

FIGURE 7.5 Direct MALDI analysis, SEND on-chip, and ZipTip cleanup of 0.5 pmol of a BSA tryptic digest in the presence of 2 *M* urea.

1 Femtomole / uL BSA Tryptic Digest

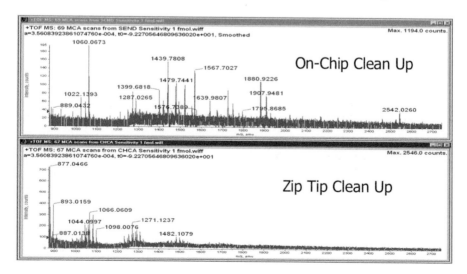

FIGURE 7.6 On-chip and ZipTip cleanup of 1 fmol of a BSA digest in the presence of 2 *M* urea.

TABLE 7.1
Commonly Used Matrices in SELDI Biochip Analysis

Matrix	Application	Solvent
Sinapinic acid[11]	Proteins to large peptides	ACN, H_2O, MeOH, TFA, FA
α-Cyano-4-hydroxycinnamic acid (HCCA)[12]	Amino acids, peptides, small proteins	ACN, H_2O, MeOH, TFA
2,5-Dihyroxy benzoic acid (DBA)[13]	Small peptides and glycoproteins	MeOH, H_2O
Ferulic acid[14]	Large proteins	IPA, H_2O
2,6,Dihyroxy-acetophenone[15]	Peptides, proteins, oligonucleotides	ACN, MeOH, DAHC, H_2O

Notes: ACN: acetonitrile; MeOH: methanol; TFA: trifluoroacetic acid; FA: formic acid; IPA: isopropanol; DAHC: aqueous diammonium hydrogen citrate (~100 mM).

matched chromatographic beads to further enrich for target proteins from nascent complex mixtures. As can be surmised, the SELDI process detects extracted or retained analytes and not those analytes eluted from its stationary phase. This is in stark contrast to classical elution-based chromatography. This unique form of solid phase extraction has been coined retentate chromatography, where retained analytes are ultimately detected using mass spectrometry.

7.2.3 SAMPLE DETECTION IN SELDI MASS SPECTROMETRY: ROLE OF THE MATRIX AND MATRIX SOLVENT SYSTEM

After proteins are captured on the surface of SELDI arrays, detection using matrix-assisted, surface-enhanced laser desorption/ionization (SELDI) mass spectrometry then ensues. For SEAC arrays, adsorbed proteins are first eluted and then cocrystallized using a matrix solution. For the most part, the classical matrix compounds and solutions employed for ultraviolet MALDI analysis are used in SELDI. Table 7.1 illustrates five of these most commonly used matrices along with their companion solvent systems.[11–15] As previously described, SEND arrays contain matrix polymers capable of supporting desorption and ionization of adsorbed species and as such do not require the addition of matrix solution prior to MS analysis.

While seemingly simple, proper addition of matrix during SELDI analysis is key to obtaining sensitive and reproducible results. During the initial sample adsorption process, proteins diffuse their way into the polymer network of the SELDI array and become adsorbed according to their surface interaction potentials. The rate of penetration and the final adsorptive distribution of these proteins within the array's polymer are dependent upon the sample incubation time, polymer thickness and porosity, protein hydrodynamic radius, viscosity of the incubation buffer, and the temperature of the incubation system. Once applied to the SELDI array, the matrix solvent system not only delivers the required matrix, it also functions as an elutropic agent to elute adsorbed proteins, which have partitioned into the array's stationary phase.

As one might suspect, this elution solution creates a new distribution of proteins between the array's stationary phase and the matrix solvent system. As was the case for adsorption, the final concentration of released protein within the matrix solution is dependent upon the previously noted parameters, as well as the elutropic strength of the matrix solvent system. Only those proteins released from the array polymer and suspended within the matrix solution are able to cocrystallize with the matrix; as such, they are the only members of the adsorbed population that can be detected. Under these circumstances, it becomes apparent that reproducible results are best achieved when incubation and elution periods and temperatures are maintained constant. Furthermore, because the drying times associated with matrix crystallization are also dependent upon gases, such as water vapor found in the head-space above the array, the relative humidity of the laboratory as well as its temperature must be prudently controlled.

7.2.4 SELDI TOF MS DETECTION

After elution and cocrystallization with matrix, proteins captured on SELDI arrays are read by using a laser desorption ionization source typically coupled to a TOF mass spectrometer. Almost universally, ionization is achieved using a pulsed nitrogen laser (λ = 337 nm, 3- to 4-nsec pulse duration). Depending upon the matrix used and the ion source employed, typical applied laser fluence ranges from about 10 to 200 $\mu J/mm^2$ for low-pressure sources and from about 50 to 400 $\mu J/mm^2$ for elevated-pressure sources. These devices are often simply referred to as protein array or ProteinChip® readers.

Two basic TOF MS geometries are used in SELDI protein array readers: linear TOF systems and orthogonal TOF MS systems. A detailed description of a linear TOF MS protein array reader is given in Merchant and Weinberger[16] and Weinberger et al.[17] Briefly, ions are created as previously noted within an ion optic assembly that is intimate with the TOF analyzer. Typical ion source pressures range between 2 and 0.5 μtorr. After ion creation, a brief lag period then ensues to allow formed ions to occupy positions correlated with their nascent velocities. At the conclusion of this waiting phase, an ion extraction field is applied to accelerate the ions to constant kinetic energy. These ions then "fly" across a fixed distance at constant velocity prior to striking an ion detector. The time between ion extraction and ion detection is inversely related to the square root of its mass-to-charge ratio. Pulsed ion extraction is employed to allow correction of initial ion velocity differences. This practice is typically referred to as time-lag focusing (TLF). For more details regarding TLF, see references 16 through 18. Taken collectively, this array reader is technically referred to as a time-lag focusing, laser desorption/ionization, linear time-of-flight mass spectrometer, hence the preferred term SELDI ProteinChip reader.

Historically, two principal series of linear SELDI array readers have been offered commercially by Ciphergen Biosystems. They are the Protein Biology System (PBS) series and the new ProteinChip System (PCS). For further details regarding the PBS series of ProteinChip array readers, see Dalmasso.[19] A detailed discussion of the PCS protein array reader is given in Cannon et al.[20]

In addition to the standard ProteinChip reader, a SELDI interface for orthogonal TOF mass spectrometers has been developed. This technology, originally described in Merchant and Weinberger[16] and Weinberger et al.,[17] is embodied in Ciphergen's Protein

Chip Interface (PCI-1000). The PCI-1000 converts the Applied Biosystems/MDS Sciex QStar™ series of qQ-TOF liquid chromatography MS systems (Concord, Ontario) into ProteinChip readers. Orthogonal qQ-TOF MS technology for electrospray ionization was initially described in Standing et al.,[21] while subsequent work in creating a MALDI-based qQ-TOF system was reported in Shevchenko et al.[22] and Laboda et al.[23]

In contrast to the linear approach, orthogonal SELDI chip readers form ions at elevated pressure (typically 20–50 mtorr). In this manner, ions are cooled and stabilized so that they can be further analyzed using the tandem MS system without the introduction of unwanted ion decay products. The inherent instability of MALDI-generated ions to undergo unimolecular decay is universally accepted.[24] While unimolecular decay has been exploited to provide primary peptide sequence,[25–30] it has also been a major limiting factor in the hyphenation of MALDI ion sources to complex mass analyzers such as reflectron TOF and ion trap MS, as well as ion cyclotron resonance (ICR) MS systems.

In addition to stabilizing formed ions, a collisional cooling orthogonal TOF MS system provides several benefits directly attributable to the uncoupling of ion formation from the TOF MS measurement process.[17] When compared to classical parallel ion extraction as used in the linear geometry, these advantages include: improved mass accuracy, the ability to use electrically nonconductive arrays, simplified m/z (mass-to-charge ratio) calibration, and simplified control of applied laser energy.

While the SELDI qQ-TOF has been reported to be useful for *de novo* biomarker discovery,[31,32] the overall detection sensitivity for ions in excess of 10,000 m/z is markedly compromised when compared to standard linear chip readers. For the most part, SELDI qQ-TOF technology has been used to provide a convenient means to perform SELDI-based, low-energy collision-induced dissociation (CID) MS[2] experiments for the purpose of protein and/or peptide ID, as well as epitope mapping and ligand binding domain experiments during the study of protein interactions and biomolecular pathways.

Over the course of the last decade, the primary utility for MALDI-TOF MS technology has moved away from its roots—the detection of large biopolymers—and has, for the most part, been focused upon the improved detection of protein digest products (peptides and peptide bioconjugates). This trend has been driven by the growing requirement to identify and characterize proteins initially isolated and detected using gel electrophoresis or liquid chromatography. As such, MALDI-TOF MS systems have been progressively improved in terms of mass resolving power and mass accuracy—unfortunately, at the expense of analytical sensitivity, particularly for the study of nascent proteins.

While seemingly similar to conventional MALDI-TOF MS systems, the design of SELDI array readers has evolved to address the challenging requirements of discovering and characterizing intact proteins. Corresponding improvements for sensitive detection of peptides and proteins have been made, while great strides have been taken to extend the linear dynamic range of detection and quantitative response. In the case of the linear array readers, improved protein utility has come at the expense of peptide mass resolving power. However, with the advent of collisional cooling, it is expected that future SELDI array readers will maintain utility in protein analysis without sacrificing the ability to routinely supply monoisotopic mass resolution in the classical peptide ion domain (m/z < 5,000).

7.3 SELDI PROTEOMICS RESEARCH

7.3.1 DIFFERENTIAL PROTEIN DISPLAY AND BIOMARKER DISCOVERY

Over the course of the last decade, SELDI technology has mostly been applied to challenges of proteomics research. Among today's most popular proteomic research activities is differential protein display, or expression monitoring. Differential protein display is a comparative technique that contrasts protein profiles between different organisms, individuals, pathogenic and/or metabolic conditions, and phenotypic response to environmental or chemical challenges. In this manner, differential protein studies require approaches that isolate and enrich major as well as minor protein constituents from a complex biological mixture.

Today, differential protein display studies are routinely performed in research proteomics, clinical proteomics, and toxicological arenas, with differing overall requirements and goals for each. For example, basic research activities often involve studies performed on pooled biological samples such as lysates, secretions from harvested cells, lysates, or products of cultivated bacteria, or combined pools of biological fluids and tissues from many laboratory animals or human subjects. Under these conditions, typical research proteomic studies are not hampered by sample limitations, and protein purification often relies upon tried and true techniques such as low- and high-pressure liquid chromatography, serial chromatography, and various electrophoretic approaches. Often, the goal in research proteomic studies is to catalogue and identify every protein associated with a particular sample. In this case, there are massive automation requirements, and detected proteins are universally reduced to peptides for the purpose of protein identification and characterization—typically by electrospray ionization mass spectrometry or MALDI MS.

In contrast, clinical proteomic studies endeavor to follow the progress of disease within an individual or small population with the ultimate goal of finding biomarkers potentially useful as diagnostic agents or new drug targets. Under such circumstances, sample or tissue availability is limited and the dependence on highly efficient, small-scale techniques is high. Typically, protein populations between groups are compared using univariate or multivariate analysis schemes with the ultimate goal of identifying a protein or groups of proteins whose expression levels correlate with a given clinical condition.[33] Automation requirements are primarily focused on running many samples in a massively parallel manner and only proteins of interest are further studied to provide insight into identification and post-translational modifications. Similarly, toxicological studies for the purpose of monitoring a physiological response to a pharmaceutical challenge, or stratifying laboratory animal profiles, are best performed using an individual subject. In this manner, sample requirements and analytical approaches for toxicological differential protein studies are similar to those of clinical proteomics research. Because of its reproducibility, throughput, and starting material requirements, SELDI MS has gained major acceptance as a tool of choice for clinical proteomics studies.

7.3.2 SELDI CLINICAL PROTEOMIC STUDIES

Clinical proteomics aims to scan the realm of expressed proteins to identify biomarkers that can answer specific clinical questions. The most obvious are markers that can be used for diagnosis or prognosis. Another important issue that clinical proteomics promises to help resolve is the ability to predict a patient's response to a specific drug. Clinical proteomics also has utility outside the doctor's office. For example, diagnostic markers can be candidates for drug targets and pharmaceutical companies use clinical proteomics to identify markers that predict the toxicity of candidate drugs.

The most straightforward approach to clinical proteomics begins with protein profiling or protein differential display studies, in which protein expression in control individuals is compared with that of patients. Physiological fluids, tissue homogenates, or cell lysates from control and disease samples are processed on the same types of SELDI array surfaces and the arrays read under the same data collection conditions. The premise of this approach is to establish composite fingerprint profiles of disease and nondisease states from a series of training samples and then to use these profiles to make a diagnosis on actual unknown patient samples. Biomarker identification is not strictly required for diagnostic purposes, but is required for insight into the underlying biological process as well as to assess the use of the biomarker as a therapeutic target.

In general, it is expected that markers specific to a given medical condition are most likely to be found in the afflicted organ or tissue and even the cellular population. Consequently, protein profiling of selected tissues or cells is often the first attempt at biomarker discovery. Laser capture microdissection (LCM) is a powerful tool that allows researchers to look at histologic sections of a tissue and specifically select the cells of interest.[34] Initially, most studies used laser capture microdissection to obtain samples from which mRNA could be obtained for cDNA microarray analysis. With innovative sample preparation techniques and the use of frozen sections, researchers have been able to maintain an environment friendly to proteins and thus allowed laser capture microdissection to be coupled to proteomics techniques. Because of the relatively low sample requirements of SELDI technology, researchers have been eager to use this technology to analyze LCM-procured samples.[35–37]

While examination of tissue is useful for finding biomarkers, more clinically convenient biomarkers are those found in body fluids such as serum and urine, since these can be obtained using noninvasive techniques and are therefore less stressful to the patient and less costly to the health care system. During the course of the last two years, researchers have used SELDI array technology to perform biomarker discovery research in a variety of diseases, including infectious disease,[38–46] Alzheimer's disease,[47–53] and cancer.[54–70] Further information regarding SELDI array technology and clinical proteomics research is found in the recent reviews found in references 71 through 97.

7.3.3 FRACTIONATION TO DECREASE SAMPLE COMPLEXITY AND INCREASE DYNAMIC RANGE OF DETECTION

While body fluids such as serum, plasma, or urine represent the easiest samples to obtain, the concentration of putative biomarkers for early detection may be low in abundance and the analysis is further complicated by the presence of high-abundance proteins such as albumin, transferin, or immunoglobulins (in serum, for example). Thus, body fluids such as plasma and/or serum generally require processing, such as fractionation, prior to analysis to: (1) partition away abundant proteins that may generally interfere with protein detection; (2) simplify the complex mixture into fractions containing a lower number of proteins; and (3) enrich for rare protein species. Fractionation of serum or plasma samples prior to analysis is particularly relevant because a decrease in the sample complexity and a reduction in the dynamic range of protein abundance can be achieved.[98–100] These two important features enhance the resolving power of detection and increase the sensitivity of analytical techniques such as 2-dimensional electrophoresis 2-DE and SELDI MS.

The presence of very high abundance proteins (e.g., albumin from serum) produces large signals with consequent signal overlap (2-dimensional electrophoresis) or signal suppression (mass spectrometry) for other proteins. Classical methodologies of serum fractionation consist of depleting the proteins that are very abundant (e.g., albumin, transferrin, IgG fraction) from the sample to be analyzed or fractionating the entire mixture by means of sequential elution steps off a single resin to segregate the abundant proteins in unique partitions. Other methods of fractionating a proteome are well established and include subcellular fractionation,[7,98,101–105] isoelectric separation,[99,101,106–110] one-dimensional electrophoresis,[99,103,111] molecular sieving,[100,101] and liquid chromatography.[99,101] With respect to liquid chromatography, examples of fractionation methods have been reported using ion exchange chromatography,[98,112] IMAC for calcium-binding proteins[113] or separation of phospho-proteins,[114] hydrophobic interaction chromatography,[113] affinity for immobilized heparin,[101] or immobilized lectins.[101,113,115,116]

Two-dimensional liquid chromatography methods have also been proposed for the fractionation of proteomes after global digestion using endoproteases such as trypsin. These methods generally combine reverse phase with ion exchange chromatography[117,118] or, to a lesser extent, with chromatofocusing,[119] size exclusion,[120] affinity capture,[116] or even a second reverse phase step.[121] However, multidimensional chromatography as used in proteomics fractionation generally does not exceed two dimensions due to the high number of fractions to manage (pH adjustment, desalting, re-injection in second dimension) and to analyze.

In SELDI analysis, complex biological fluids are often initially fractionated using strong anion exchange (SAX) beads and a variety of elutropic solvents followed by secondary fractionation on the array. Low-throughput applications rely upon the use of spin columns packed with quaternary ammonium beads, such as Ciphergen's Q-Hyper D resin. Figure 7.7 depicts a typical SAX–SELDI array fractionation scheme. Although this process is more complicated and time consuming when compared to array profiling alone, it generates the largest amount of information from a single sample. Once a particular protocol is defined, it is often desirable to scale the process to allow for higher throughput, utilizing a 96-well plate format. Although SELDI arrays are supplied in eight spots per array, they are designed to be multiplexed into current microwell formats for automation. All sample preparation can be achieved using standard robotic platforms (see fig. 7.8). This automated format not only yields a marked improvement in analytical throughput, but also

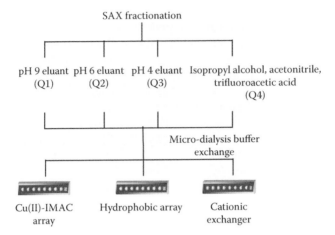

FIGURE 7.7 Strong anion exchange (SAX) fractionation strategy combined with SELDI array analysis. Samples are incubated with SAX beads. Adsorbed proteins are selectively eluted using four different elution conditions: Q1, pH 9 (typically 50 mM Tris HCl); Q2, pH 6 (50 mM sodium phosphate); Q3, pH 4 (50 mM sodium acetate); and Q4, an organic solution as indicated. In some cases, buffer exchange is achieved using microdialysis prior to performing retentate chromatography.

demonstrates superior qualitative and quantitative reproducibility when compared to manual processing. For serum profiling experiments, typical protein quantitative profiles demonstrate CVs of less than 20%.

7.3.4 FRACTIONATION USING AFFINITY HEXAPEPTIDE BEADS

The ability to detect low-abundance proteins remains a critical challenge in deciphering proteome and correlating proteome changes with metabolic events for diagnostic purposes. Depletion methodologies, as discussed earlier, are frequently used to remove the most abundant species. However, this removal not only fails to enrich trace proteins significantly, it may also nonspecifically deplete them due to their interactions with the high-abundance proteins being removed. Recently, we have reported a simple-to-use methodology that reduces the protein concentration range of a complex mixture, like whole serum, through the simultaneous dilution of high-abundance proteins and the concentration of low-abundance proteins. The principle of this novel strategy is based on the selective adsorption of proteins on a solid phase combinatorial ligand library under capacity-limited binding conditions.

The spatial arrangement of amino acids within a protein defines its physicochemical properties—for example, isoelectric point, charge density, hydrophobicity index, and even conformation. The latter determines the ability of a protein to interact *in vivo* with other molecules with complementary structures and forms the basis of protein separation by affinity chromatography, where the interacting molecule (ligand) is chemically attached to a solid carrier.[113] The proteins complementary to the immobilized ligands are captured from complex mixtures, up to the saturation of the available ligand.

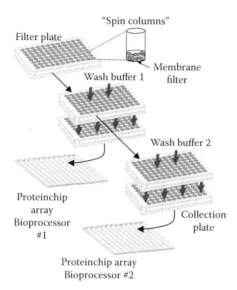

FIGURE 7.8 Automated high-throughput sample fractionation using 96 well plates in conjunction with a Ciphergen Bioprocessor. Stackable filter plates filled with chromatographic resin are used for parallel and serial scaling of sample fractionation. Resins are suspended in each well using a porous membrane. Gravity- or vacuum-assisted flow proceeds as samples are loaded upon the resins in each well. Eluted protein solutions are transferred to a bioprocessor for on-array extraction. The bioprocessor is a device that allows the juxtaposition of up to 12 different SELDI arrays creating a 96-well format. Arrays are faceted to a supporting base plate. A seal between the arrays and overlying well assembly is created by an intervening gasket that is intimate with the well assembly. The well assembly accepts up to 400 μL of solution and is directly compatible with most robotic systems.

With sufficient diversity of ligands, it is theoretically possible to have a ligand to every protein in a complex mixture, ensuring that each is adsorbed. When a biological extract like serum is exposed to such a ligand library under specific capacity-constrained conditions, an abundant protein will quickly saturate all of its available high-affinity ligands and the vast majority of the same protein will remain unbound. In marked contrast, a trace protein will not saturate all its high-affinity ligands and the majority of the same protein will be bound. Thus, based upon the saturation-overloading principle, a combinatorial solid phase library will enrich for trace proteins relative to their abundant counterparts. Following washing to remove non- or weakly bound proteins, elution of the adsorbed proteins from the beads will result in a solution with a narrower dynamic range of protein concentrations while still representing all proteins present in the original material.

To be effective, the library must meet three criteria:

1) A sufficient reproducible diversity of ligands must be present to reliably bind each protein in the mixture.
2) Dissociation constants of the ligands and proteins must be compatible with the protein concentration.

3) The ligand and its support must be compatible with the unfractionated test sample and have a binding capacity high enough to capture sufficient protein to be detected by current methods.

The technology is founded upon libraries of peptide ligands on which proteins can be adsorbed. Based on the pioneering work of Merrifield[114] on solid-phase synthesis using the "split, couple, recombine" method, libraries of peptide ligands are synthesized on resin beads.[115–117] Each bead has millions of copies of a single, unique ligand, and each bead potentially has a different ligand. Using just the 20 natural amino acids, a library of linear hexapeptides contains 20^6, or 64 million, different ligands. The addition of unnatural amino acids and D-enantiomers into branched, linear, or circular ligands generates a potential library diversity that is, for all practical purposes, unlimited and may contain a ligand to every protein present in a biological sample. To date, using libraries of moderate diversity, ligands have been identified to a number of plasma proteins.[118–120]

In the current format, a given volume of affinity beads, typically 100 µL to 1 mL, is incubated with at least a 10-fold excess of biological sample for about 1 to 3 h. Once ligands bind their corresponding proteins, the beads are washed to eliminate unbound or weakly bound proteins. Adsorbed proteins are subsequently released by means of classical elution methods used in chromatography. The eluted protein mixture can then be analyzed by standard methods such as 1-DE and 2-DE and/or SELDI MS.

Figure 7.9 compares SELDI serum profiles generated for unfractionated and affinity bead-processed human sera. As can be readily seen, the SELDI profile for serum processed with these affinity beads is markedly different from that of the unprocessed samples, demonstrating a substantially larger population of signals detected in low, intermediate, and high molecular weight ranges. These low molecular weight species may be of diagnostic relevance should they result from proteolytic breakdown of proteins due to perturbed biological pathways. Such enrichment was evidenced in a variety of very different biological samples, including human sera, low-concentration cell culture supernatant, and chicken egg white. Thus, the method largely addresses the problem of the dynamic range in clinical proteomic analyses.

7.3.5 SELDI CLINICAL PROTEOMIC RESEARCH EXPERIMENTAL DESIGN

A typical SELDI clinical proteomics study begins with a discovery phase, in which assay conditions are tested on a relatively small number of samples. The number of samples and types of samples are perhaps the most important parameters that determine the success of the study. Usually, profiling proceeds with at least 30 samples in each classification group (e.g., diseased vs. healthy or treated vs. untreated). The samples are another critical parameter. Because the initial sample set size is relatively small, inherent biological variability always threatens the ability to conclude that differences seen are consequences specific to the disturbance under study. Therefore, it is imperative that the study includes well-chosen samples (e.g., patients of the same age group or all of a single sex) and, equally important, appropriately chosen controls. Naturally, *in vitro* studies show less variability than do animal studies,

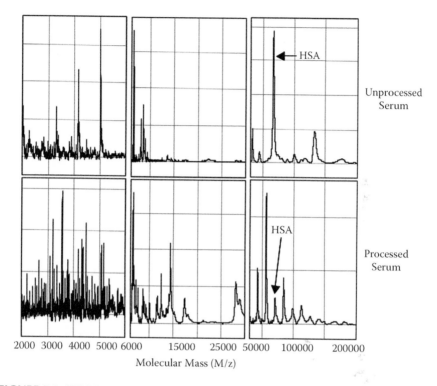

FIGURE 7.9 SELDI-TOF profiles for unprocessed human serum (upper panel) and human serum processed with combinatorial hexapeptide affinity beads (lower panel). For each sample, three different m/z ranges are depicted: low molecular weight (left panels), intermediate molecular weight (middle panels), and high molecular weight (right panels). For each molecular weight range, SELDI analysis of processed sera reveals a greater number of detected species. Furthermore, abundant species in the unprocessed sera, such as human serum albumin (HSA), are detected at more "equalized" levels when compared to minor serum constituents.

which in turn show less variability than human studies. No matter the source of samples, all should have been handled identically, and care should have been taken to minimize the number of freeze–thaw cycles.

Once the samples are procured, they can be processed directly on the arrays with minimal preparation. However, when there is adequate material (for serum, this is defined as 20 µL), it is recommended to perform some form of fractionation prior to any array binding procedure. As was previously explained, fractionation can significantly increase the number of peaks visualized and therefore increases the likelihood that biomarkers will be found. For serum, anion exchange fractionation is typically employed, while for cells or tissues, subcellular fractionation is often used. Following fractionation, each fraction is profiled under a series of SELDI array assay conditions, which can include different permutations of array surface chemistries, choice of matrix solutions, and laser energies. Consequently, each sample in the study generates multiple spectra and therefore generates a significant amount of data.

7.4 DATA ANALYSIS IN SELDI CLINICAL PROTEOMICS RESEARCH

Analysis of SELDI array data consists of several preprocessing and postprocessing steps. Data preprocessing involves the manipulations required to organize the data, such as TOF to mass calibration, baseline subtraction, and signal intensity normalization. Postprocessing consists of using analytical tools so that one can draw conclusions from the empirical results. These steps are similar in many ways to those used to process DNA microarray images. For example, SELDI mass spectra data must be processed to allow the application of downstream multidimensional methods such as clustering and classification. With this goal in mind, the following paragraphs outline the analysis procedure generally employed during the course of SELDI differential expression profiling.

The spectrum acquired from a unique sample run on a single array "spot" is analogous to the image acquired from the hybridization of a unique sample on a DNA microarray. Just as multiple DNA microarrays may be necessary to profile the entire genome, multiple combinations of sample fractions and SELDI array surface types may be necessary to profile the proteome sufficiently. There are two basic types of DNA arrays: the Affymetrix GeneChip® and the class of spotted microarrays including cDNA, *in situ* synthesized, or whole oligonucleotide arrays. SELDI arrays are most similar to the GeneChip platform and abundance data are based upon peak intensities—not ratios or log ratios thereof. Perhaps the most obvious difference between SELDI and GeneChip arrays is the form of the raw data. While DNA microarrays are intended to provide sequence-specific binding, adsorption to a chromatographic SELDI array is far less specific and typically produces a mixture of proteins that share physical or chemical properties complementary to the array's surface interaction potential. Consequently, mass spectra can contain hundreds of protein expression levels encoded in their peaks.

7.4.1 PROCESSING RAW SPECTRA: PREPROCESSING

A few steps in processing mass spectra can be done in isolation or on a per-spectrum basis. The first is mass calibration, or the conversion of the raw TOF data to molecular weight (MW). Typically, this process involves acquiring a spectrum from a standard with at least five proteins or peptides of various molecular weights, spanning the molecular weight range of interest, if possible. A quadratic equation relating ion TOF to mass-to-charge ratio is then fit to the TOF values of the standard peaks in this spectrum. The equation generated by this process can then be used on mass spectra that are collected under the same instrument conditions, including laser intensity, approximate date, mass range, and TLF conditions.

In addition to TOF-MW calibration, baseline subtraction or normalization is also performed. Baseline processing is directed towards eliminating signals associated with unwanted chemical noise. Finally, peak detection is a key component in identifying and quantifying protein peaks in the mass spectra. Several peak detection algorithms exist; see Fung and Enderwick[122] for further details. Typically, peak intensity is used to represent the quantity of protein expressed in the sample.

7.4.2 Multispectra Processing

In order to draw meaningful conclusions about protein abundance from a series of different SELDI measurements, some level of quantitative normalization is classically employed. Normalization is essential to eliminate any systematic effects between samples due to varying amounts of protein or degradation over time in the sample or variations in the analytical regimen. For the most part, quantitative normalization in SELDI measurements is achieved by referring to measured total ion current. In employing this method, it is assumed that, on average, the number of proteins overexpressed is approximately equal to the number of proteins under-expressed and that the number of proteins whose expression levels are changing are few relative to the total number of proteins bound to the array surface. The total ion current is calculated and divided by the number of points over which it was calculated, thus resulting in an average ion current. After an overall average of the ion current is calculated across all the spectra in the study, each spectrum is simply multiplied by a constant factor equal to the overall average ion current divided by the average ion current for that spectrum.

7.4.3 Clustering SELDI Array Data

Extracting marker abundance information from SELDI array data is an involved process, since there is no *a priori* knowledge of which proteins will be captured upon a particular surface from a given sample. However, for a group of spectra comprising a profiling study, one can leverage the information across all the spectra to find the union of all possible protein peaks in the study. This process involves a number of different steps. The first is to perform an initial pass of peak detection for each spectrum. For studies containing spectra from different samples, this step inevitably will find a different set of protein peaks per spectrum. The second step groups peaks of similar molecular weight across spectra together into peak "clusters" while allowing for slight variations in determined mass-to-charge ration. Each cluster therefore represents a particular protein.

At this stage, the data are in theory ready to be mined for structure and biomarker discovery. In practice, this is a good time to reassess the quality of the data. This includes scanning for aberrant spectra and looking for systematic flaws in the data. Aberrant spectra can result from problems in the analytical protocol as well as problems inherent to a given clinical sample. Many criteria can be used to define an aberrant spectrum. For example, algorithms contained within Ciphergen's data-reduction software program (CiphergenExpress™) records the total ion current value as well as the normalization factor for each spectrum. One could, for example, define an aberrant spectrum as one that has a normalization factor of greater than three standard deviations from the median. Note that this approach will always lead to some number of spectra being aberrant.

Another approach to quality control is to choose some number of reference peaks and define acceptance criteria for intensity (or signal to noise) and resolution for these peaks; this is the approach taken by the Early Detection Research Network.[123]

Any spectrum whose reference peaks fail to meet these criteria is rejected. More sophisticated methods to perform quality control have been devised. For example, Coombes et al. perform principal component analysis of detected peaks, and they reject chips whose reference spectra are mapped greater than a defined distance from the center of a principal component map.[124] Visualization of spectra using unsupervised methodologies can be particularly useful in revealing underlying structure. For example, it may reveal a systematic bias in a given array or on a given spot location. From here, one proceeds with postprocessing.

7.4.4 POSTPROCESSING I: FINDING SINGLE BIOMARKERS

Finding single genes or proteins responsible for differentiating disease versus normal or treated versus untreated is a natural first step in analyzing expression data.[125] SELDI analysis generally employs nonparametric statistical methods because one cannot uniformly assume that peak intensity data conform to a normal distribution and often SELDI studies have a small sample lot size. These methods include the Mann–Whitney, the nonparametric equivalent of the student's t-test, and the Kruskal–Wallis, the nonparametric equivalent of ANOVA, thus eliminating any assumption about the distribution of the peak intensity data.[126] Essentially, these nonparametric tests sort the peak intensities and their corresponding ranks are used in the p-value calculation. The p-value results of these tests, in conjunction with data visualization tools, help to identify potential markers.

These statistical tests simply give an indication of group mean differences, which may not always be helpful if there is a very large spread in the distribution of data. Simply increasing the sample size can improve p-values, while discrimination between groups may remain poor.[127] This is especially important when attempting to use the biomarker in a diagnostic assay. Furthermore, one may not find single biomarkers with acceptable p-values. At this point, it is prudent to turn to other analytical methods that are multivariate in nature and lend themselves to developing a clinical assay.

7.4.5 POSTPROCESSING II: MULTIVARIANT ANALYSIS

A number of different analysis tools identify and use multivariate patterns in the expression profile data for the purposes of identifying groupings and/or classifying groupings. These methods can generally be lumped into one of two categories: unsupervised learning in the form of cluster analysis or supervised learning in the form of classification methods. A number of these (only a fraction of existing literature is cited here) have been applied to gene expression data, including clustering and visualization,[128,129] self-organizing maps,[130] and support vector machines.[131]

Most SELDI clinical proteomic studies have relied upon regression tree-based methods[132,133] and, for the most part, use the algorithms embodied in Ciphergen's Biomarker Patterns™ software. Similar to the way in which a clinician makes his or her diagnosis by correlating and integrating various findings from a patient's physical examination with laboratory test results, the classification tree creates rules based on peak intensity, such as "peak intensity at 13,979 Da < 5.169 and at 16,760 Da > 12.283;

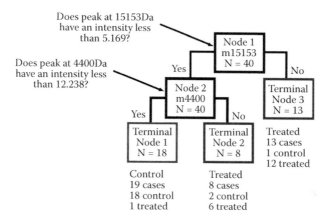

FIGURE 7.10 Classification tree example. Each node (square) is a decision point. Each sample is sifted down the tree based on how it answers the question in each node. For example, the first node asks the question, "Does peak at molecular weight 15,153 Da have a peak intensity of <5.169?" If the answer is yes, the sample goes to the left to node 2; otherwise, it goes to the right to terminal node 3. Terminal nodes are stopping points and the majority of samples determine the classification of each terminal node. In terminal node 3, there are one control and 12 untreated samples; the control is misclassified and the treated samples are correctly classified. See text for further details.

therefore, this sample is treated." In addition, for each peak cluster that is determined to be a good classifier, Biomarker Patterns software determines the intensity value that serves as a threshold above or below which a given classification is assigned (see fig. 7.10).

A number of characteristics of classification trees make them an attractive tool for protein expression studies. The model is easy to interpret compared to "black-box" classifiers such as neural networks and nearest neighbor classifiers. The protein peaks used in the model are easily attainable by examination of the rules, and these rules are easily validated by examination of the spectra. This openness of the tree-based model is an attractive feature for researchers wanting a diagnostic assay as well as potential therapeutic targets. In addition, a classification tree can sift through all the input variables and select the subset to use in the tree. As such, it alleviates some of the burden of performing feature selection up front.

7.5   PROTEIN IDENTIFICATION IN SELDI ARRAY RESEARCH

Regardless of study design, once a protein of interest has been detected, protein characterization efforts often ensue. Proteins are characterized by identifying post-translational modifications, providing primary sequence information, and, ultimately, by elucidating protein identity. As a typical starting point, proteins are chemically reduced and then exposed to endoprotease digestion. The resultant peptides may also be treated with dephosphorylating, deglycosylating, or exoproteolytic enzymes. At each step along the way, MS detection is performed to monitor changes in peptide

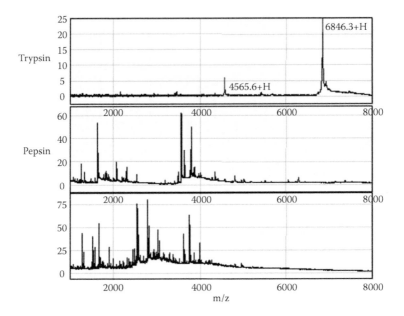

FIGURE 7.11 On-chip proteolysis of ribonuclease A. The top panel represents results when using trypsin at 37°C, pH = 8.0; the middle panel depicts the results for the use of pepsin at 37°C, pH = 1; and the lower panel illustrates the results for the use of thermolysin at pH = 8, 65°C.

m/z profile. Observed m/z changes are reconciled with known enzymatic or chemical modes of action to provide structural and identification insight.

In some instances, particularly when biochemical arrays have been employed, sufficient protein purification can be achieved on the array. Under these circumstances, captured proteins are often directly digested using chemical or enzymatic means with minimal denaturation or disruption of disulfide bonds. For on-chip identification, it is essential that protein denaturation strategies be directly compatible with MS detection, for while SELDI arrays may be used to capture and purify nascent proteins, their digest products will demonstrate measurable differences in surface affinity and could easily be lost during subsequent wash steps. Accordingly, proteases that operate under denaturing conditions such as extreme pH or elevated temperature have demonstrated utility for on-chip protein identification.

For example, figure 7.11 compares three on-array proteolytic strategies for bovine pancreas ribonuclease A (~13.7 kDa). When trypsin was used, essentially no proteolytic products were detected, and the acquired spectra only demonstrated the doubly and triply charged pseudomolecular ions of ribonuclease A (see upper panel). However, when enzymes such as pepsin (incubation at pH = 2) or thermolysin (incubation at 60°C) were employed, significant proteolysis was observed and many peptides useful for tandem MS analysis were provided. A thorough review of on-chip proteolytic methods is given in references 134 through 136.

When on-array proteolysis is not practical, protein identification generally proceeds by combining the complementary nature of SELDI arrays and bead-based chemistries. In this manner authentic samples containing the protein of interest are further purified by using chromatographic beads with identical surface interaction chemistries as found on the arrays originally used in protein discovery. *A priori* knowledge gained during refinement of the discovery protocol is used to establish optimal protein–bead binding and elution guidelines. This approach has also been extended to the convenient creation of large-scale protein preparative schemes, as discussed in Weinberger et al.[137] If additional purification is required, eluted proteins may be further purified using denaturing polyacrylamide gel electrophoresis and in-gel digestion, as described in Wang et al.[138]

Should the target protein be sufficiently purified to greater than 90% relative abundance, protein identification can often be achieved using single MS peptide mass fingerprinting studies with mass accuracy determinations on the order of about 0.02% absolute error. However, in the final analysis, the mass accuracy requirement of peptide-guided database searches is heavily dependent upon the purity of the protein pool prior to digestion, the number of peptides submitted for searching, and the genomic complexity of the studied organism. Should multiple proteins be simultaneously digested, a heterogeneous peptide pool is created, and successful database mining requires not only extreme accuracy, but also, in many instances, primary sequence information.

Tandem MS/MS approaches have demonstrated significant utility in providing primary sequence information.[24,139–143] Until recently, the only MS/MS approach available for laser desorption-based analyses was postsource decay analysis (PSD). While PSD is capable of providing reasonable sequence information for picomole levels of peptides, the overall efficiency of this fragmentation process is low and, when combined with the poor mass accuracy and sensitivity often demonstrated during this approach, its applicability to analysis of low-abundance proteins often found on ProteinChip arrays has been greatly limited. Recent advances in low-energy collision-induced dissociation (CID) MS/MS for SELDI have demonstrated enabling protein identification capability, even for complex protein mixtures directly analyzed upon SELDI arrays.[16,89] Of course, should in-gel digestion be performed, protein identification can proceed using any liquid chromatography (LC)- or MALDI-based tandem MS approach.

7.6 FUTURE PROSPECTS AND CONCLUSIONS

Over the course of the last 12 years, SELDI protein array technology has demonstrated utility in the analysis of protein complexes, the discovery of relevant biomarkers, and the facile creation of assays for clinical research, drug discovery, or basic biological research. To do so, significant advances in the areas of surface chemistry, protein purification, bioinformatics, and laser desorption-based mass spectrometry were achieved. Future improvements are expected on all of these fronts.

In terms of surface chemistry, ongoing efforts towards the creation of new SEND and SEAC surfaces with enhanced, sample-adsorptive properties and capacities is expected to create marked improvements in the analytical sensitivity and quantitative reproducibility of SELDI-based differential protein display studies. Moreover, microscale, multidimensional fractionation schemes are being developed based upon combinatorial ligand libraries, as well as other unique bead-based approaches, such as isoelectric focusing, in order to address the analytical challenges of protein abundance and species complexity in genuine biological samples. Surface chemistry advances will also lead to the routine extension of SELDI array analysis to the study of protein post-translational medications such as glycosylation, sulfonation, and phosphorylation, providing an additional dimension to chip-based proteomic analysis.

Improvements in chip reader technology are expected as work towards the creation of MS devices with simultaneous high detection sensitivity for proteins and mass resolving power for peptides is enabled by implementing new laser optic, ion optic, and mass analyzer approaches. Additionally, it is expected that the next generation of chip readers will not only demonstrate improved analytical performance when compared to their predecessors, but also provide a more robust, easier to use, and quantitative platform for the purpose of facilitating translational proteomic studies as well as the performance of *in vitro* diagnostic tests. Taken collectively, these improvements should enable the growth and utility of SELDI protein array technology in related biological research, drug discovery/development, and clinical research.

REFERENCES

1. Isenor, N. R., High energy ions from a Q-switched laser, *Canadian Journal of Physics* 42 (July, 1964), 1413–1416, 1964.
2. Fenner, N. C. and Daly, N. R., Laser used for mass analysis, *The Review of Scientific Instruments* 37(8), 1068–1070, 1966.
3. Vastola, F. J. and Pirone, R. H., Ionization of organic solids by laser irradiation, *Advances in Mass Spectrometry* 4, 107–111, 1968.
4. Tanaka, K., Waki, H., Ido, Y., Akita, S., Yoshida, Y., and Yoshida, T., Protein and polymer analyses up to m/z 100,000 by laser ionization time-of-flight mass spectrometry. In Proceedings of the Second Japan–China Joint Symposium on Mass Spectrometry, Osaka, Japan, 1987, 185–187.
5. Tanaka, K., Waki, H., Ido, Y., Akita, S., Yoshida, Y., and Yoshida, T., Protein and polymer analyses up to m/z 100,000 by laser ionization time-of-flight mass spectrometry, *Rapid Communications in Mass Spectrometry* 2(1988), 151–153, 1988.
6. Karas, M. and Hillenkamp, F., Laser desorption ionization of proteins with molecular masses exceeding 10,000 daltons, *Analytical Chemistry* 60(20), 2299–2301, 1988.
7. Karas, M. and Hillenkamp, F., UV laser desorption of ions above 10,000 daltons. In International Mass Spectrometry Conference, Bordeaux, France, 1988.
8. Hutchens, T. W. and Yip, T.-T., New desorption strategies for the mass spectrometric analysis of macromolecules, *Rapid communications in mass spectrometry* 7, 576–580, 1993.
9. Voivodov, K., Ching, J., and Hutchens, T. W., Surface arrays of energy absorbing polymers enabling covalent attachment of biomolecules for subsequent laser-induced uncoupling/desorption, *Tetrahedron Letters* 37(32), 5669–5672, 1996.
10. Weinberger, S. R., Lin, S., Viner Rosa, I., Naotaka, K., Chang, D., and Tang, N., New surface enhanced neat desorption SELDI protein biochip arrays for evaluating the low molecular weight proteome. In 51st ASMS Conference on Mass Spectrometry and Allied Topics, ASMS ASMS, Montreal, Canada, 2003, TOB 320.

11. Beavis, R. C. and Chait, B. T., Cinnamic acid derivatives as matrices for ultraviolet laser desorption mass spectrometry of proteins, *Rapid Communications in Mass Spectrometry* 3(12), 432–435, 1989.

12. Beavis, R. C., Chait, B. T., and T, C., Alpha-cyano-4-hydroxycinnamic acid as a matrix for matrix- assisted laser desorption mass spectrometry, *Org. Mass Spectrometry* 27(2), 156–158, 1992.

13. Strupat, K., Karas, M., and Hillenkamp, F., 2,5-Dihydroxybenzoic acid: a new matrix for laser desorption-ionization mass spectrometry, *International Journal of Mass Spectrometry Ion Processes* 111, 89–102, 1991.

14. Tang, K., Allman, S. L., and Chen, C. H., Matrix-assisted laser desorption ioization of oligonucleotides with various matrices, *Rapid Communications in Mass Spectrometry* 7, 943–948, 1993.

15. Schlunegger, U. P., Studies on the selection of new matrices for ultraviolet matrix-assisted laser desorption/ionization time-of-flight mass spectrometry, *Rapid Communications in Mass Spectrometry* 10, 1927–1933, 1996.

16. Merchant, M. and Weinberger, S. R., Recent advancements in surface-enhanced laser desorption/ionization time-of-flight mass spectrometry, *Electrophoresis* 21, 1164–1177, 2000.

17. Weinberger, S. R., Davis, S., Makarov, A. A., Thompson, S., Purves, R., and Whittal, R. M., Time-of-flight mass spectrometry. In *Encyclopedia of Analytical Chemistry*, ed. Meyers, R. A. John Wiley & Sons Ltd, Chichester, 2000, 11915–11984.

18. Wiley, W. C. and McLaren, I. H., Time-of-flight mass spectrometer with improved resolution, *The Review of Scientific Instruments* 26(12), 1150–1157, 1955.

19. Dalmasso Enrique, A., Discovery of protein biomarkers and phenomic fingerprints using SELDI, *High Throughput Screening: Supplement to Biomedical Products* July, 1999, 4–8, 1999.

20. Cannon, J. S., Gilbert, K., Zheng, W., and Yip, C., From biomarker discovery to assay using SELDI TOF-MS technology, *American Biotechnology Laboratory* 22(12), 18–20, 2004.

21. Standing, K. G., Ens, W., Krutchinsky, A. N., Chrnushevich, I. V., and Spicer, V. L., Collisional damping interface for an electrospray ionization time-of-flight mass spectrometer, *Journal of the American Society for Mass Spectrometry* 9, 569–579, 1998.

22. Shevchenko, A., Loboda, A., Shevchenko, A., Ens, W., and Standing, K. G., MALDI quadrupole time-of-flight mass spectrometry: A powerful tool for proteomic research, *Analytical Chemistry* 72(9), 2132–2141, 2000.

23. Laboda, A. V., Krutchinsky, A. N., Bromirski, M., Ens, W., and Standing, K. G., A tandem quadrupole/time-of-flight mass spectrometer (QqTOF) with a MALDI source: Design and performance, *Rapid Communications in Mass Spectrometry*, 14, 12, 1047–1057, 2000.

24. Spengler, B., Kirsch, D., and Kaufmann, R., Metastable decay of peptides and proteins in matrix-assisted laser desorption mass spectrometry, *Rapid Communications in Mass Spectrometry* 5, 198–202, 1991.

25. Chaurand, P., Luetzenkirchen, F., and Spengler, B., Peptide and protein identification by matrix-assisted laser desorption ionization (MALDI) and MALDI-post-source decay time-of-flight mass spectrometry, *Journal of the American Society of Mass Spectrometers* 10, 91–103, 1999.

26. Hoffmann, R., Metzger, S., Spengler, B., and Otvos, L., Jr., Sequencing of peptides phosphorylated on serines and threonines by post-source decay in matrix-assisted laser desorption/ionization time-of-flight mass spectrometry, *Journal of Mass Spectrometry* 34(11), 1195–1204, 1999.

27. Chaurand, P., Luetzenkirchen, F., Spengler, B., "Peptide and protein identification by MALDI and MALDI post-source decoy time-of-flight mass spectrometry, *J. Am. Soc. Mass Spectrom.*, 1012, 91–103.

28. Brown, R. S. and Lennon, J. J., Sequence specific fragmentation of matrix-assisted laser desorbed protein/peptide ions, *Analytical Chemistry* 67(21), 3990–3999, 1995.

29. Reiber, D. C., Brown, R. S., Weinberger, S., Kenny, J., and Bailey, J., Unknown peptide sequencing using matrix-assisted laser desorption/ionization and in-source decay, *Analytical Chemistry* 70(6), 1214–1222, 1998.

30. Reiber, D. C., Grover, T. A., and Brown, R. S., Identifying proteins using matrix-assisted laser desorption/ionization in-source fragmentation data combined with database searching, *Analytical Chemistry* 70(4), 673–683, 1998.

31. Conrads, T. P., Fusaro, V. A., Ross, S., Johann, D., Rajapakse, V., Hitt, B. A., Steinberg, S. M., Kohn, E. C., Fishman, D. A., Whitely, G., Barrett, J. C., Liotta, L. A., Petricoin, E. F., 3rd, and Veenstra, T. D., High-resolution serum proteomic features for ovarian cancer detection, *Endocrine-Related Cancer* 11(2), 163–178, 2004.

32. Keay, S., Conrads, T. P., and Veenstra, T. D., Mass spectral characterization of urine samples from patients with interstitial cystitis, Poster Presentation MPG 241, in the Proceedings of the 50th American Society of Mass Spectrometry and Allied Topics, Orlando, FL, June 2–6, 2002.

33. Weinberger, S. R., Dalmasso, E. A., and Fung, E. T., Current achievements using ProteinChip array technology, *Current Opinion in Chemical Biology* 6(1), 86–91, 2002.

34. Emmert-Buck, M. R., Bonner, R. F., Smith, P. D., Chuaqui, R. F., Zhuang, Z., Goldstein, S. R., Weiss, R. A., and Liotta, L. A., Laser capture microdissection, *Science* 274(5289), 998–1001, 1996.

35. Banks, R. E., Dunn, M. J., Forbes, M. A., Stanley, A., Pappin, D., Naven, T., Gough, M., Harnden, P., and Selby, P. J., The potential use of laser capture microdissection to selectively obtain distinct populations of cells for proteomic analysis—Preliminary findings, *Electrophoresis* 20(4–5), 689–700, 1999.

36. Jones, M. B., Krutzsch, H., Shu, H., Zhao, Y., Liotta, L. A., Kohn, E. C., and Petricoin, E. F., III, Proteomic analysis and identification of new biomarkers and therapeutic targets for invasive ovarian cancer, *Proteomics* 2(1), 76–84, 2002.

37. Wright, G. L., Jr., Cazares, L. H., Leung, S. M., Nasim, S., Adam, B. L., Yip, T. T., Schellhammer, P. F., Gong, L., and Vlahou, A., Proteinchip surface enhanced laser desorption/ionization (SELDI) mass spectrometry: A novel protein biochip technology for detection of prostate cancer biomarkers in complex protein mixtures, *Prostate Cancer Prostatic Disease* 2(5/6), 264–276, 2000.

38. Seo, G.-M., Kim, S.-J., and Chai Young, G., Rapid profiling of the infection of Bacillus anthracis on human macrophages using SELDI-TOF mass spectroscopy, *Biochemical and Biophysical Research Communications* 325(4), 1236–1239, 2004.

39. Lundquist, M., Caspersen, M. B., Wikstroem, P., and Forsman, M., Discrimination of *Francisella tularensis* subspecies using surface enhanced laser desorption ionization mass spectrometry and multivariate data analysis, *FEMS Microbiology Letters* 243(1), 303–310, 2005.

40. Hess, J. L., Blazer, L., Romer, T., Faber, L., Buller, R. M., and Boyle, M. D. P., Immunoproteomics, *Journal of Chromatography, B: Analytical Technologies in the Biomedical and Life Sciences* 815(1–2), 65–75, 2005.

41. Yuan, M. and Carmichael, W. W., Detection and analysis of the cyanobacterial peptide hepatotoxins microcystin and nodularin using SELDI-TOF mass spectrometry, *Toxicon* 44(5), 561–570, 2004.

42. Lai, R., Lomas, L. O., Jonczy, J., Turner, P. C., and Rees, H. H., Two novel noncationic defensin-like antimicrobial peptides from hemolymph of the female tick, *Amblyomma hebraeum*, *Biochemical Journal* 379(3), 681–685, 2004.

43. Hodgetts, A., Bosse J. T., Kroll, J. S., and Langford P. R., Analysis of differential protein expression in *Actinobacillus pleuropneumoniae* by surface enhanced laser desorption ionisation—ProteinChip (SELDI) technology, *Veterinary Microbiology* 99(3–4), 215–225, 2004.

44. Barzaghi, D., Isbister, J. D., Lauer, K. P., and Born, T. L., Use of surface-enhanced laser desorption/ionization-time of flight to explore bacterial proteomes, *Proteomics* 4(9), 2624–2628, 2004.

45. Zhu, X.-D., Zhang, W.-H., Li, C.-L., Xu, Y., Liang, W.-J., and Tien, P., New serum biomarkers for detection of HBV-induced liver cirrhosis using SELDI protein chip technology, *World Journal of Gastroenterology* 10(16), 2327–2329, 2004.

46. Schwegler, E. E., Cazares, L., Steel, L. F., Adam, B.-L., Johnson, D. A., Semmes, O. J., Block, T. M., Marrero, J. A., and Drake, R. R., SELDI-TOF MS profiling of serum for detection of the progression of chronic hepatitis C to hepato-cellular carcinoma, *Hepatology* (Baltimore, MD) 41(3), 634–642, 2005.

47. Anon, Ciphergen announces discovery of a potential multi-biomarker test for Alzheimer's disease using SELDI proteinchip technology, *Proteomics* 4(1), 280–281, 2004.

48. Head, E., Moffat, K., Das, P., Sarsoza, F., Poon, W. W., Landsberg, G., Cotman C. W., and Murphy, M. P., Beta-amyloid deposition and tau phosphorylation in clinically characterized aged cats, *Neurobiology of Aging* 26(5), 749–763, 2005.

49. Yalkinoglu, O., Koenig, G., Hochstrasser, D. F., Sanchez, J.-C., and Carrette, O., U.S. Patent Application: WO 2003-EP8879, 2004, SELDI-TOF MS detection and identification of protein biomarkers for diagnosing Alzheimer's disease.

50. Maddalena, A. S., Papassotiropoulos, A., Gonzalez-Agosti, C., Signorell, A., Hegi, T., Pasch, T., Nitsch, R. M., and Hock, C., Cerebrospinal fluid profile of amyloid beta peptides in patients with Alzheimer's disease determined by protein biochip technology, *Neurodegenerative Diseases* 1(4–5), 231–235, 2004.

51. Lewis, H. D., Beher, D., Smith, D., Hewson, L., Cookson, N., Reynolds, D. S., Dawson, G. R., Jiang, M., Van der Ploeg, L. H. T., Qian, S., Rosahl, T. W., Kalaria, R. N., and Shearman, M. S., Novel aspects of accumulation dynamics and Ab composition in transgenic models of AD, *Neurobiology of Aging* 25(9), 1175–1185, 2004.

52. Lewczuk, P., Esselmann, H., Groemer Teja, W., Bibl, M., Maler Juan, M., Steinacker, P., Otto, M., Kornhuber, J., and Wiltfang, J., Amyloid beta peptides in cerebrospinal fluid as profiled with surface enhanced laser desorption/ionization time-of-flight mass spectrometry: Evidence of novel biomarkers in Alzheimer's disease, *Biological Psychiatry* 55(5), 524–530, 2004.

53. Bradbury, L. E., LeBlanc, J. F., and McCarthy, D. B., ProteinChip array-based amyloid b assays, *Methods in Molecular Biology* (Totowa, NJ) 264 (Protein Arrays), 245–257, 2004.

54. Yu, J.-K., Zheng, S., Tang, Y., and Li, L., An integrated approach utilizing proteomics and bioinformatics to detect ovarian cancer, *Journal of Zhejiang University Science B* 6(4), 227–231, 2005.

55. Woong-Shick, A., Sung-Pil, P., Su-Mi, B., Joon-Mo, L., Sung-Eun, N., Gye-Hyun, N., Young-Lae, C., Ho-Sun, C., Heung-Jae, J., Chong-Kook, K., Young-Wan, K., Byoung-Don, H., and Hyun-Sun, J., Identification of hemoglobin-alpha and -beta subunits as potential serum biomarkers for the diagnosis and prognosis of ovarian cancer, *Cancer Science* 96(3), 197–201, 2005.

56. Sauter, E. R., Shan, S., Hewett, J. E., Speckman, P., and Du Bois, G. C., Proteomic analysis of nipple aspirate fluid using SELDI-TOF-MS, *International Journal of Cancer* 114(5), 791–796, 2005.

57. Pawlik, T. M., Fritsche, H., Coombes, K. R., Xiao, L., Krishnamurthy, S., Hunt, K. K., Pusztai, L., Chen, J.-N., Clarke, C. H., Arun, B., Hung, M.-C., and Kuerer, H. M., Significant differences in nipple aspirate fluid protein expression between healthy women and those with breast cancer demonstrated by time-of-flight mass spectrometry, *Breast Cancer Research and Treatment* 89(2), 149–157, 2005.

58. Malik, G., Ward, M. D., Gupta, S. K., Trosset, M. W., Grizzle, W. E., Adam, B.-L., Diaz, J. I., and Semmes, O. J., Serum levels of an isoform of apolipoprotein A-II as a potential marker for prostate cancer, *Clinical Cancer Research: An Official Journal of the American Association for Cancer Research* 11(3), 1073–1085, 2005.

59. Le, L., Chi, K., Tyldesley, S., Flibotte, S., Diamond, D. L., Kuzyk, M. A., and Sadar, M. D., Identification of serum amyloid a as a biomarker to distinguish prostate cancer patients with bone lesions, *Clinical Chemistry* 51(4), 695–707, 2005.

60. Junker, K., Gneist, J., Melle, C., Driesch, D., Schubert, J., Claussen, U., and von Eggeling, F., Identification of protein pattern in kidney cancer using ProteinChip arrays and bioinformatics, *International Journal of Molecular Medicine* 15(2), 285–290, 2005.

61. Albrethsen, J., Bogebo, R., Gammeltoft, S., Olsen, J., Winther, B., and Raskov, H., Upregulated expression of human neutrophil peptides 1, 2 and 3 (HNP 1-3) in colon cancer serum and tumours: a biomarker study, *BMC Cancer* 5, no pp. given, 2005.

62. Zhao, G., Gao, C.-F., Song, G.-Y., Li, D.-H., and Wang, X.-L., Identification of colorectal cancer using proteomic patterns in serum, *Ai zheng = Aizheng = Chinese Journal of Cancer* 23(6), 614–618, 2004.

63. Xiao, X., Zhao, X., Liu, J., Guo, F., Liu, D., and He, D., Discovery of laryngeal carcinoma by serum proteomic pattern analysis, *Science in China, Series C: Life Sciences* 47(3), 219–223, 2004.

64. Wong, Y. F., Cheung, T. H., Lo, K. W. K., Wang, V. W., Chan, C. S., Ng, T. B., Chung, T. K. H., and Mok, S. C., Protein profiling of cervical cancer by protein-biochips: Proteomic scoring to discriminate cervical cancer from normal cervix, *Cancer Letters* (Amsterdam, Netherlands) 211(2), 227–234, 2004.

65. Wadsworth, J. T., Somers, K. D., Stack, B. C., Jr., Cazares, L., Malik, G., Adam, B.-L., Wright, G. L., Jr., and Semmes, O. J., Identification of patients with head and neck cancer using serum protein profiles, *Archives of Otolaryngology—Head & Neck Surgery* 130(1), 98–104, 2004.

66. Vlahou, A., Giannopoulos, A., Gregory, B. W., Manousakas, T., Kondylis, F. I., Wilson, L. L., Schellhammer, P. F., Wright, G. L., Jr., and Semmes, O. J., Protein profiling in urine for the diagnosis of bladder cancer, *Clinical Chemistry* (Washington, DC) 50(8), 1438–1441, 2004.

67. Moody, T. W., Dudek, J., Zakowicz, H., Walters, J., Jensen, R. T., Petricoin, E., Couldrey, C., and Green, J. E., VIP receptor antagonists inhibit mammary carcinogenesis in C3(1)SV40T antigen mice, *Life Sciences* 74(11), 1345–1357, 2004.

68. Melle, C., Kaufmann, R., Hommann, M., Bleul, A., Driesch, D., Ernst, G., and Von Eggeling, F., Proteomic profiling in microdissected hepatocellular carcinoma tissue using protein chip technology, *International Journal of Oncology* 24(4), 885–891, 2004.

69. Koopmann, J., Zhang, Z., White, N., Rosenzweig, J., Fedarko, N., Jagannath, S., Canto, M. I., Yeo, C. J., Chan, D. W., and Goggins, M., Serum diagnosis of pancreatic adenocarcinoma using surface-enhanced laser desorption and ionization mass spectrometry, *Clinical Cancer Research* 10(3), 860–868, 2004.

70. Fowler, L. J., Lovell, M. O., and Izbicka, E., Fine-needle aspiration in PreservCyt: A novel and reproducible method for possible ancillary proteomic pattern expression of breast neoplasms by SELDI-TOF, *Modern Pathology* 17(8), 1012–1020, 2004.

71. Mazzulli, T., Low, D. E., and Poutanen, S. M., Proteomics and severe acute respiratory syndrome (SARS): Emerging technology meets emerging pathogen, *Clinical Chemistry* (Washington, DC) 51(1), 6–7, 2005.

72. Hortin, G. L., Can mass spectrometric protein profiling meet desired standards of clinical laboratory practice? *Clinical Chemistry* (Washington, DC) 51(1), 3–5, 2005.

73. Coombes, K. R., Analysis of mass spectrometry profiles of the serum proteome, *Clinical Chemistry* (Washington, DC) 51(1), 1–2, 2005.

74. Zhang, J., Zheng, Y., Feng, K., and Zou, D., Application of SELDI protein fingerprinting in early detection of cancer, *Shijie Huaren Xiaohua Zazhi* 12(12), 2773–2777, 2004.

75. Yang, T., Yang, J., and Wei, Y., Research advances in SELDI protein chip and its application, *Zhongliu Fangzhi Zazhi* 11(10), 1101–1104, 2004.

76. Walker, S. J. and Xu, A., Biomarker discovery using molecular profiling approaches, *International Review of Neurobiology* 61 (Human Brain Proteome), 3–30, 2004.

77. Tomosugi, N., Discovery of disease biomarkers by protein chip system; clinical proteomics as noninvasive diagnostic tool, *Rinsho Byori* 52(12), 973–979, 2004.

78. Tolson, J., Bogumil, R., Flad, T., and Wiesner, A., Detection of protein variants in SELDI-based multi-marker diagnostics, *BIOspektrum* 10 (Spec. Issue), 572–573, 2004.

79. Seibert, V., Wiesner, A., Buschmann, T., and Meuer, J., Surface-enhanced laser desorption ionization time-of-flight mass spectrometry (SELDI TOF-MS) and Protein-Chip technology in proteomics research, *Pathology, Research and Practice* 200(2), 83–94, 2004.

80. Saito, K., Analysis of human disease markers by mass spectrometry-linked protein chip system, *Bunshi Kokyukibyo* 8(2), 158–161, 2004.

81. Nomura, F. and Tomonaga, T., Current topics in clinical proteomics, *Chiba Igaku Zasshi* 80(1), 27–31, 2004.

82. Catimel, B., Rothhacker, J., and Nice, E., The use of biosensors for microaffinity purification: An integrated approach to proteomics, *J. Biochem. Biophys. Methods* 2001, 49:289–312.

83. Kroll, J., Rohn, S., and Rawel, H. M., Application of MALDI and SELDI mass spectrometry for the characterization of food proteins and posttranslational protein modifications, *Ernaehrung* (Vienna, Austria) 28(3), 102–109, 2004.

84. Huhalov, A., Spencer, D. I. R., and Chester, K. A., Mapping antibody: Antigen interactions by mass spectrometry and bioinformatics, *Methods in Molecular Biology* (Totowa, NJ) 248 (Antibody Engineering), 465–480, 2004.

85. Fiedler, G. M., Ceglarek, U., Lembcke, J., Baumann, S., Leichtle, A., and Thiery, J., Application of mass spectrometry in clinical chemistry and laboratory medicine, *Laboratoriumsmedizin* 28(3), 185–194, 2004.

86. von Eggeling, F., Melle, C., and Ernst, G., Biomarker discovery by tissue microdissection and ProteinChip array analysis, *Laboratoriumsmedizin* 27(3–4), 79–84, 2003.

87. Veenstra, T. D., Yu, L.-R., Zhou, M., and Conrads, T. P., Diagnostic proteomics: Serum proteomic patterns for the detection of early stage cancers, *Disease Markers* 19(4–5), 209–218, 2003.

88. Tomosugi, N., Screening of biomarkers in renal diseases by ProteinChip system, *Seibutsu Butsuri Kagaku* 47(1–2), 17–22, 2003.

89. Tang, N., Tornatore, P., and Weinberger, S. R., Current developments in SELDI affinity technology, *Mass Spectrometry Reviews* 23(1), 34–44, 2003.

90. Schweigert, F. J., Gericke, B., and Mothes, R., Food analysis with SELDI-TOF-MS, *Bioforum* 26(9), 534–536, 2003.

91. Sasaki, K., Tumor markers search by SELDI-TOF-MS and cancer diagnostic tests, *Kagaku Furontia* 10 (Posuto Genomu, Masu Supekutorometori), 181–190, 2003.

92. Saito, K., Seldi-Tof-Ms, *Kagaku Furontia* 10 (Posuto Genomu, Masu Supekutorometori), 37–47, 2003.

93. Reddy, G. and Dalmasso, E. A., SELDI ProteinChip array technology: Protein-based predictive medicine and drug discovery applications, *Journal of Biomedicine & Biotechnology* 2003(4), 237–241, 2003.

94. Petricoin, E. F. and Liotta, L. A., Clinical applications of proteomics, *Journal of Nutrition* 133(7S), 2476S–2484S, 2003.

95. Lotze, M. T., The critical need for cancer biometrics: Quantitative, reproducible measures of cancer to define response to therapy, *Current Opinion in Investigational Drugs* (Thomson Current Drugs) 4(6), 649–651, 2003.

96. Issaq, H. J., Conrads, T. P., Prieto, D. A., Tirumalai, R., and Veenstra, T. D., SELDI-TOF MS for diagnostic proteomics, *Analytical Chemistry* 75(7), 148A–155A, 2003.

97. Clarke, C. H., Weinberger, S. R., and Clarke, M. S. F., Application of ProteinChip array technology for detection of protein biomarkers of oxidative stress, *Critical Reviews of Oxidative Stress and Aging* 1, 366–379, 2003.

98. Anderson, L. N. and Anderson, N. G., The human plasma proteome: History, character and diagnostics prospects, *Mol. Cell. Proteomics* 21, 845–867, 2002.

99. Issaq, H., J., Conrads Thomas, P., Janini, G. M., and Veenstra Timothy, D., Methods for fractionation, separation, and profiling of proteins and peptides, *Electrophoresis* 21, 1082–1093, 2002.

100. Lopez, M. F., Better approaches to finding the needle in a haystack: Optimizing proteome analysis through automation, *Electrophoresis* 21(6), 1082–1093, 2000.

101. Herbert, B., Advances in protein solubilisation for two-dimensional electrophoresis, *Electrophoresis* 20(4–5), 660–663, 1999.

102. Quadroni, M. and James, P., Proteomics and automation, *Electrophoresis* 20(4-5), 664-77, 1999.

103. Leatherbarrow, R. J. and Dean, P. D. G., Studies on the mechanisms of binding of serum albumins to immobilized Cibacron blue F3G, *Journal of Analytical Biochemistry* 189, 27–34, 1980.

104. Birch, R. M., O'Byrne, C., Booth, I. R., and Cash, P., Enrichment of *Escherichia coli* proteins by column chromatography on reactive dye columns, *Proteomics* 3, 764–776, 2003.

105. Maccarrone, G. et al., Mining the human cerebrospinal fluid proteome by immuno-depletion and shotgun mass spectrometry, *Electrophoresis* 25, 2402–2412, 2004.

106. Mehta, A. I., Ross, S., Lowenthal, M. S., Fusaro, V., Fishman, D. A., Petricoin, E. F., III, and Liotta, L. A., Biomarker amplification by serum carrier protein binding, *Disease Markers* 19(1), 1–10, 2003.

107. Bjorhall, K., Miliotis, T., and Davidsson, P., Comparison of different depletion strategies for improved resolution in proteomics analysis of human serum samples, *Proteomics* 5, 307–317, 2005.

108. Gygi, S. P., Corthals, G. L., Zhang, Y., Rochon, Y., and Aebersold, R., Evaluation of two-dimensional gel electrophoresis-based proteome analysis technology, *Proceedings of the National Academy of Science U.S.A.* 97(17), 9390–9395, 2000.

109. Yamada, M., Murakami, K., Wallingford, J. C., and Yuki, Y., Identification of low-abundance of bovine colostral and mature milk using two-dimensional electrophoresis followed by microsequencing and mass spectrometry, *Electrophoresis* 23, 1153–1160, 2002.

110. Bjellqvist, B. et al., Micropreparative two-dimensional electrophoresis allowing the separation of samples containing milligram amounds of protein, *Electrophoresis* 14, 1375–1378, 1993.

111. Westbrook, J. A., Yan, J. X., Wait, R., Welson, S. Y., and Dunn, M. J., Zooming in on the proteome: Very narrow-range immobilised pH gradients reveal more protein species and isoforms, *Electrophoresis* 22, 2865–2871, 2001.

112. Cho, M. J. et al., Identifying the major proteome components of *Helicobacter pylori* strain 26695, *Electrophoresis* 23, 1161–1173, 2002.

113. Wilchek, M., Miron, T., and Kohn, J., Affinity chromatography. In *Methods in Enzymology*, vol. 104, ed. William B. Jacoby, 1984, 3–55. Burlington, MA, Elsevier U.S.A.

114. Merrifield, R. B., Automated synthesis of peptides, *Science* 150, 178–185, 1965.

115. Lam, K. S. et al., A new type of synthetic peptide library for identifying ligand-binding activity, *Nature* 354, 82–84, 1991.

116. Furka, A. and Sebetyen, F., General methods for rapid synthesis of multipcomponent peptide mixtures, *International Journal of Peptide and Protein Research* 37, 487–493, 1991.

117. Watts, A. D., Hunt, N. H., Hanbly, B. D., and Chaudhri, G., Separation of tumor necrosis factor alpha isoforms by two-dimensional polyacrylamide gel electrophoresis, *Electrophoresis* 18, 1086–1091, 1997.

118. Buettner, J. A., Dadd, C. A., Baumbach, G. A., Masecar, B. L., and Hammond, D. J., Chemically derived peptide libraries: a new resin and methodology for lead identification, *International Journal of Peptide and Protein Research* 47, 70–83, 1996.

119. Huang, P. Y. et al., Affinity purification of von Willebrand factor using ligands derived from peptide libraries, *Bioorganic Medicine Chemistry* 4, 699–708, 1966.

120. Kaufman, D. B. et al., Affinity purification of fibrinogen using a ligand from a peptide library, *Biotechnology Bioengineering* 77, 278–289, 2002.

121. Bastek, P. D., Land, J. M., Baumbach, G. A., Hammond, D. H., and Carbonnel, R. G., Discovery of alpha-1-proteinase inhibitor binding peptide from the screening of a solid phase combinatorial peptide library, *Sep. Science and Technolology* 35, 278–289, 2000.

122. Fung, E. T. and Enderwick, C., ProteinChip clinical proteomics: computational challenges and solutions, *Biotechniques* Suppl, 34–38, 40–41, 2002.

123. Semmes, O. J., Feng, Z., Adam, B.-L., Banez, L. L., Bigbee, W. L., Campos, D., Cazares, L. H., Chan, D. W., Grizzle, W. E., Izbicka, E., Kagan, J., Malik, G., McLerran, D., Moul, J. W., Partin, A., Prasanna, P., Rosenzweig, J., Sokoll, L. J., Srivastava, S., Thompson, I., Welsh, M. J., White, N., Winget, M., Yasui, Y., Zhang, Z., and Zhu, L., Evaluation of serum protein profiling by surface-enhanced laser desorption/ ionization time-of-flight mass spectrometry for the detection of prostate cancer: I. Assessment of platform reproducibility, *Clinical Chemistry* 51(1), 102–112, 2005.

124. Coombes, K. R., Fritsche, H. A., Jr., Clarke, C., Chen, J.-N., Baggerly, K. A., Morris, J. S., Xiao, L.-C., Hung, M.-C., and Kuerer, H. M., Quality control and peak finding for proteomics data collected from nipple aspirate fluid by surface-enhanced laser desorption and ionization, *Clinical Chemistry* 49(10), 1615–1623, 2003.

125. Long, A. D., Mangalam, H. J., Chan, B. Y., Tolleri, L., Hatfield, G. W., and Baldi, P., Improved statistical inference from DNA microarray data using analysis of variance and a Bayesian statistical framework. Analysis of global gene expression in *Escherichia coli* K12, *Journal of Biology Chemistry* 276, 19937–19944, 2001.

126. Sprinthall, R., *Basic Statistical Analysis.* Prentice-Hall, Englewood Cliffs, NJ, 1997.

127. Pajak, T. F., Clark, G. M., Sargent, D. J., McShane, L. M., and Hammond, M. E., Statistical issues in tumor marker studies, *Archives of Pathology Laboratory Medicine* 124(7), 1011–1015, 2000.

128. Ben-Dor, A., Shamir, R., and Yakhini, Z., Clustering gene expression patterns, *Journal of Computer Biology* 6, 281–297, 1999.
129. Eisen, M. B., Spellman, P. T., Brown, P. O., and Botstein, D., Cluster analysis and display of genome-wide expression patterns, *Proceedings of the National Academy of Science U.S.A.* 95(25), 14863–14868, 1998.
130. Golub, T. R., Slonim, D. K., Tamayo, P., Hurard, C., Gaasenbeek, M., Mesirov, J. P., Coller, H. A., Loh, M. L., Downing, J. R., Caligiuri, M. A., Bloomfield, C. D., and Lander, E. S., Molecular classification of cancer: class discovery and class prediction by gene expression monitoring, *Science* 286, 531–537, 1999.
131. Moler, E. J., Chow, M. L., and Mian, L. S., Analysis of molecular profile data using gnerative and discriminative methods, *Physiology Genomics* 4, 109–126, 2000.
132. Breiman, L., Friedman, J., Olshen, R., and Stone, C., *Classification and Regression Trees*, Wadsworth, Pacific Grove, CA, 1984.
133. Zhang, H., Yu, C. Y., Singer, B., and Xiong, M., Recursive partitioning for tumor classification with gene expression microarray data, *Proceedings of the National Academy of Science U.S.A.* 98, 6730–6735, 2001.
134. Lin, S., Tang, N., and Weinberger, S. R., Means of hydrolyzing proteins isolated upon ProteinChip array surfaces: Chemical and enzymatic approaches. In *Handbook of Proteomic Methods*, ed. Michael P. Conn, 59–72, 2003. Totowa, NJ, Humana Press.
135. Lin, S., Tornatore, P., King, D., Orlando, R., and Weinberger, S. R., Limited acid hydrolysis as a means of fragmenting proteins isolated upon ProteinChip array surfaces, *Proteomics* 1(9), 1172–1184, 2001.
136. Lin, S., Tornatore, P., Weinberger, S. R., King, D., and Orlando, R., Limited acid hydrolysis as a means of fragmenting proteins isolated upon ProteinChip array surfaces, *European Journal of Mass Spectrometry* 7(2), 131–141, 2001.
137. Weinberger, S. R., Boschetti, E., Santambien, P., and Brenac, V., Surface-enhanced laser desorption-ionization retentate chromatography mass spectrometry (SELDI-RC-MS): A new method for rapid development of process chromatography conditions, *Journal of Chromatography, B: Analytical Technologies in the Biomedical and Life Sciences* 782(1–2), 307–316, 2002.
138. Wang, S., Diamond, D. L., Hass, G. M., Sokoloff, R., and Vessella, R. L., Identification of prostate specific membrane antigen (PSMA) as the target of monoclonal antibody 107-1A4 by proteinchip; array, surface- enhanced laser desorption/ionization (SELDI) technology, *International Journal of Cancer* 92(6), 871–876, 2001.
139. Biemann, K. and Papayannopoulos, I. A., Amino acid sequencing of proteins, *Accounts of Chemistry Research* 27, 370–378, 1994.
140. Spengler, B., Kirsch, D., and Kaufmann, R., Peptide sequencing by matrix-assisted laser-desorption mass spectrometry, *Rapid Communications in Mass Spectrometry* 6, 105–108, 1992.
141. Yates, J. R., Eng, J. K., McCormack, A. L., and Schieltz, D., Method to correlate tandem mass spectra of modified peptides to amino acid sequences in the Protein Data Base, *Analytical Chemistry* 67(8), 1426–1436, 1995.
142. Kaufman, R., Spengler, B., and Lutzenkirchen, F., Mass spectrometric sequencing of linear peptides by product-ion analysis in a relfectron time-of-flight mass spectrometer using matrix-assisted laser desorption ionization, *Rapid Communications in Mass Spectrometry* 7, 902–910, 1993.
143. Kaufman, R., Kirsch, D., and Spengler, B., Sequencing peptides in a time-of-flight mass spectrometer: Evaluation of postsource decay following matrix assisted laser desorption ionization (MALDI), *International Journal of Mass Spectrometry and Ion Processes* 131, 355–385, 1994.

8 An Approach to the Reproducibility of SELDI Profiling

Walter S. Liggett, Peter E. Barker, Lisa H. Cazares, and O. John Semmes

CONTENTS

8.1 INTRODUCTION

The reproducibility of SELDI (surface-enhanced laser desorption/ionization) profiling is an issue in any research that envisions SELDI profiling as the basis for development of a biomarker. How this issue is to be approached is a question with many facets. One facet is dependence of the approach on the conception of the biomarker. A second facet is adaptation of the approach to the high dimensionality of SELDI profiles. A third facet is inclusion in the approach of concepts familiar from the univariate case. A fourth facet is consideration of physical sources of variation in the measurement system.

To some extent, the approach can depend on the form of the biomarker. A biomarker is usually a classifier that takes profiles—entire profiles or peak values computed from profiles—as inputs. Although knowledge of the protein species evident in the profiles is obviously helpful, the development can be entirely empirical in that it does not depend on any knowledge of protein function [1]. Moreover, restricting the classifier input to known proteins may not lead to the best performing classifier. Thus, when thought of as a measurement system, SELDI profiling is most generally regarded as having entire profiles as output, and the reproducibility issue is most generally framed in terms of entire profiles.

Semmes et al. [2] demonstrate the value of an approach to reproducibility that is by design less than fully general. Their approach consists of two experimental phases, one based on three m/z (mass-to-charge ratio) peaks in a standard pooled serum sample and the other based on a particular biomarker applied to 14 cases and 14 controls. In the context of this biomarker, their results demonstrate an encouraging degree of interlaboratory reproducibility. However, their framing of the reproducibility issue refers to a particular biomarker whereas the framing in this chapter extends to all biomarkers. Because of this, their results have a character different from those in this chapter.

A general approach to the reproducibility of SELDI profiles must address the high dimensionality of profiles—in other words, the fact that profiles are actually continuous functions of m/z. For the problem of biomarker specification with data from a case-control study, various methods apply to high-dimensional data. Often, a biomarker is a classifier that distinguishes cases from controls on the basis of the profile as a predictor. Derivation of such a classifier involves modeling the dependence of the case-control outcomes on the profiles. Tibshirani et al. [3], Morris et al. [4], and Carlson et al. [5] present accounts of classifier derivation based on extracting and quantifying peaks. Vannuchi et al. [6] discuss classifier derivation based on wavelets. Ramsay and Silverman [7] discuss classifier derivation based on penalized smoothing. Characterization of reproducibility is, however, a different problem and requires reconsideration of the approach to high dimension. Reproducibility involves modeling profile variation. Thus, in reproducibility studies, the profiles are not predictors but rather the dependent variable in the modeling.

Basic to reproducibility studies are replicate measurements. The particular experiment discussed in this chapter involves repeated measurements of a human serum standard, specifically, 88 spectra determined for mass-to-charge values between 3,300 and 30,700. Researchers have characterized the replicate-to-replicate variation in SELDI profiles in various ways familiar from univariate reproducibility studies. Bischoff and Luider [8] simply displayed spectra in separate panels to support their conclusion that there is more variability when the spectra are obtained with different protein chips than when they are all obtained with the same protein chip. Cordingley et al. [9] performed a ruggedness experiment in which aspects of the measurement protocol were intentionally changed to see the effects on the spectrum. Coombes et al. [10] used a batch of spectra all observed at about the same time to establish a multivariate control chart for judging other spectra observed later in their study.

This chapter offers a characterization of reproducibility in terms of sources of technical variation—that is, sources of variation in the measurement system. The underlying goal is association of the sources with physical processes that are part of the measurement system. This approach is general in that assumption of a particular biomarker or class of biomarkers does not constrain such a characterization.

As one advantage, this approach provides results that may help identify ways to improve the measurement system.

Most discussion of sources of variation in SELDI profiling has been linked to preprocessing steps such as baseline removal, profile alignment through registration of the m/z scales, and normalization. Malyarenko et al. [11] discuss the origin of the baseline in terms of the physics of the detector and the origin of the jitter in the m/z scale in terms of variation in ion path lengths associated with the spot of desorption on the chip surface. They also discuss spreading of the peaks, detector overload, and chemical adducts. Another generally recognized source of variation is the total amount of protein desorbed in forming a SELDI profile. This variation may be due to the heterogeneity of the sample in the matrix on the chip surface. Morris et al. [4] also discuss sources of variation.

Identification of unanticipated sources of variation is the focus of this chapter. As illustrated by Baggerly et al. [12,13], simple graphical comparisons sometimes reveal unanticipated sources. Such comparisons are the first step. This chapter presents a more sensitive approach to searching replicate profiles for unanticipated sources of variation.

8.2 IDENTIFYING SOURCES OF VARIATION

Consider correlation in replicate spectra to which preprocessing has been applied. A property of such spectra with interesting ramifications is high correlation of the spectral intensities at widely separated values of m/z. Such correlation implies sources of variation that affect more than one point in the spectrum, and such sources can often be associated with physical mechanisms in the measurement system. The approach to reproducibility presented in this chapter is based on such long-distance correlation and its implications.

The contribution of a source of variation to a batch of replicate spectra can be modeled as follows: Let the contribution of the source to a particular spectrum be proportional to $\eta(u)$, where u denotes the mass-to-charge ratio. Say that there are N spectra indexed by i, $i = 1,\dots, N$. Then the contribution to spectrum i is given by $f_i\eta(u)$. To be of interest, the f_i must be large enough that $f_i\eta(u)$ explains an appreciable part of the spectrum-to-spectrum variation remaining after preprocessing. The shape of $\eta(u)$ provides help in finding the physical location of the source in the measurement system. Of particular interest in this chapter are cases in which $\eta(u)$ is appreciably different from zero at widely separated values of u. In such cases, the source of variation can be seen as affecting proteins with dissimilar m/z.

As illustrated in this chapter, a statistical approach to finding long-distance correlation is functional canonical correlation analysis (CCA) [7], which, for example, has been applied to gait measurements [14] and to kidney degeneration [15]. In our illustration, functional CCA is applied to disjoint intervals of SELDI profiles.

In previous papers [16,17], we have used functional principal components analysis (PCA). Functional PCA characterizes a batch of functions, spectra in the present case, without any assumption about the functions being made up of peaks. Functional PCA involves computation of data-driven features, in particular, those given by the PCA weight functions. We showed how VARIMAX rotation [7] can be used to interpret these features [16,17]. This chapter demonstrates a data-driven approach to characterizing spectral variation that is applicable to m/z intervals much larger than the intervals we considered previously.

In mass spectrometry, there are reasons why characterizing spectral variation over a large m/z interval may be interesting. One reason is that strong correlation in the replicate-to-replicate spectral variation that occurs between widely separated values of m/z may be caused by sources of variation that lie in the sample preparation step rather than in subsequent use of the mass spectrometer. The reproducibility of this step has received little attention. In the case of SELDI-TOF (time of flight) mass spectrometry, the sample preparation step consists of applying the specimen as a coating on a protein chip with a surface that selectively binds to some proteins in the specimen but not to others. Fung and Enderwick [18] call this retentate chromatography. The unbound proteins are washed off, and the remaining proteins are cocrystallized with a matrix and introduced into the mass spectrometer. This sample preparation procedure contains at least one source of variation that can cause correlation between proteins with widely separated m/z values. This source is the nonuniformity of the crystallization, which has a scaling effect that is routinely eliminated from the spectra by normalization. Because of the possibility of other sources, characterization of long-distance correlation is important in assessing replicate spectral measurements.

This chapter proposes a method based on functional CCA [7] applied to spectral segments from disjoint m/z intervals. CCA differs from inspection of point-by-point correlation maps [10,19] in that CCA determines for each interval in a pair the combination of spectral intensities that gives the highest correlation for the pair. The method we have adopted is data driven and peak based. Our method is peak based in that it makes use of the assumption that except for differences in charge, the response to a protein falls within a limited m/z interval. This assumption allows us to divide the m/z domain into intervals and to account for the joint variation only two intervals at a time. Our method is data driven in that CCA constructs a weight function that defines a feature for each interval in the pair. These features are determined from the data rather than on the basis of predetermined peak shapes. Thus, CCA takes into account variation in peak shape due, for example, to instrument overload or chemical adducts to sample proteins. In this way, CCA exposes the full complexity of the long-distance correlation.

8.3 MATERIALS AND METHODS

8.3.1 DATA COLLECTION

The experiment we consider involves measurement of identical subsamples of a human serum standard. This quality control (QC) sample consists of pooled serum obtained from 360 healthy individuals—197 women and 163 men—resulting in almost 2 L of serum. Serum from each individual was collected by venipuncture into a 10cc SST vacutainer tube. Blood was allowed to clot at room temperature for 30 min, and the tubes were centrifuged at 3000 rpm for 10 min. Each individual serum sample was then decanted and pooled into a 3-L beaker on ice. The pooled serum was separated into 0.4-mL aliquots and stored at –80°C. Each aliquot of the QC sample had therefore only undergone one freeze–thaw cycle.

All serum processing steps were performed robotically as follows: Serum samples (20 µL) were mixed with 30 µL of 8 mol/L urea containing 1% CHAPS in phosphate-buffered saline (PBS). This was performed in a 96-well plate and incubated for 10 min at 4°C on a MicroMix shaker (DPC, Randolf, New Jersey) with settings of form = 20 and amplitude = 5. A volume of 100 µL of 1 mol/L urea containing 0.125% CHAPS was then added to each sample and mixed. A final dilution (1:5) was made in PBS and this diluted sample was then applied to each well of a bioprocessor (Ciphergen Biosystems) containing 11 IMAC-3 chips previously activated with $CuSO_4$. The bioprocessor was then sealed and agitated on the MicroMix shaker at the previous settings for 30 min. The excess serum mixture was discarded, and the chips were washed three times with PBS, followed by two washes with high-performance liquid chromatography (HPLC) water. The chips were then air-dried, and stored in the dark until subjected to SELDI analysis.

Before SELDI analysis, 1.0 µL of a solution of 12.5 mg/mL sinapinic acid in 500 mL/L acetonitrile containing 5 mL/L trifluoroacetic acid was applied onto each chip robotically. The sinapinic acid was applied twice, and the array surface was allowed to air dry between each application. Chips were placed in the PBS-II mass spectrometer (Ciphergen), and time-of-flight spectra were generated by averaging 192 laser shots in positive mode with a laser intensity of 220, detector voltage of 1,700, and a focus lag time of 900 ns. The laser intensity setting of 220 corresponds to laser energy of approximately 14 mJ. Mass accuracy was calibrated externally using the All-in-1 peptide molecular weight standard (Ciphergen). Each spectrum was subjected to baseline correction and normalization as provided by the Ciphergen software.

8.3.2 PREPROCESSING

The spectra obtained from the Ciphergen system required further preprocessing despite the preprocessing already applied. In ideal terms, one might imagine an observed spectrum to be a superposition of peaks of various sizes. Different peaks would correspond to different proteins or a protein with different charges. Each peak would be centered at the m/z for the protein and charge, and the area under the peak would be proportional to the concentration of the protein in the spot from which the proteins were desorbed. That a spectrum obtained from the Ciphergen system did not conform to this model exactly was remedied in part with further preprocessing. We registered the spectra, corrected the spectral baselines, and normalized the spectra. Also, we removed an outlier, thus reducing the number of spectra to 87. A preprocessing step that we did not apply was a square root or cube root transformation of the spectral intensities.

Lack of horizontal alignment among spectra is an issue because functional CCA is based on scores, each defined by the integral of a weight function times the spectrum. The weight function is intended to emphasize or de-emphasize certain portions of each spectrum—for instance, the portion containing a spectral peak. Emphasis will not be applied to proper portions of a spectrum if the spectrum is not properly aligned. Spectral registration in this chapter is the same as the registration discussed in Liggett et al. [17]. We represented the spectra after registration by a spline composed of fifth-order polynomials between the knots and having a continuous fourth derivative at every point. We denote registered spectra by $y_i(u)$, $i = 1,\ldots,$ N, where the independent variable u is the mass-to-charge ratio. To obtain $y_i(u)$, we interpolated spectrum i as originally observed with a cubic spline and evaluated this spline at $\delta_i + \gamma_i u$. Liggett et al. [17] discuss estimation of δ_i, γ_i, $i = 1,\ldots, N$. We checked to see if this shift and scale form for the registration was valid beyond the interval considered in Liggett et al. [17] and found no contradictory evidence.

We would like to apply functional CCA to functions consisting of the superposition of contributions from individual proteins as discussed previously. Let these functions be denoted by $\varphi_i(u)$. We assume that we in fact observe

$$y_i(u) = \beta_i \varphi_i(u) + \alpha_i(u)$$

where $\alpha_i(u)$ is the slowly varying baseline and β_i is a scale factor that does not depend on u. The purpose of our preprocessing is to reduce the influence of $\alpha_i(u)$ and β_i. Our approach to preprocessing does not provide estimates of $\varphi_i(u)$ but does provide modified spectra appropriate for functional CCA.

We did baseline correction separately for each of the intervals we used in our functional CCA. We denote these intervals by $[L_j, U_j]$. The baseline-corrected spectra are given by

$$y_{Bi}(u) = y_i(u) - \frac{1}{U_j - L_j} \int_{L_j}^{U_j} y_i(s)\,ds$$

Underlying this correction is the assumption that $\alpha_i(u)$ is essentially constant over $[L_j, U_j]$, which implies that $y_{Bi}(u)$ does not depend on $\alpha_i(u)$. It is easy to show that $y_{Bi}(u)$ is proportional to β_i, a fact that is important in the normalization step described later.

The mean of $y_{Bi}(u)$ over i is shown in figure 8.1. The 17 intervals $[L_j, U_j]$ are indicated. We note that the baseline-corrected spectra $y_{Bi}(u)$ are not always positive and are generally discontinuous from interval to interval. Although these characteristics of $y_{Bi}(u)$ are not always acceptable, they are acceptable in our application of functional CCA. We also note that the y-axis scales for the three panels differ by a factor of 100. The three largest peaks in the left panel of figure 8.1 are the basis of the first experimental phase described by Semmes et al. [2].

Our normalization is based on $\bar{y}_B(u)$, the mean of $y_{Bi}(u)$ over i. If the deviations of the individual spectra are small relative to their mean, then instead of dividing by an estimate of β_i, we can normalize by subtracting a quantity proportional to $\bar{y}_B(u)$. We normalized by computing the spectral deviations

$$x_i(u) = y_{Bi}(u) - b_i \bar{y}_B(u)$$

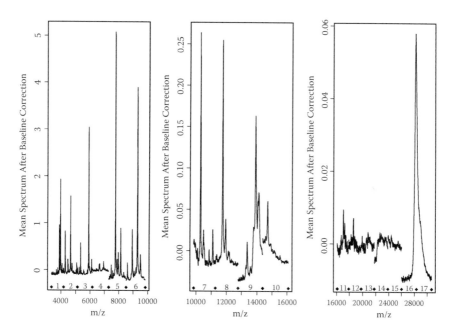

FIGURE 8.1 Mean intensity spectrum after baseline correction for each of the 17 intervals that are compared by functional CCA.

where the estimate of b_i is given by

$$b_i = \int y_{Bi}(s)\bar{y}_B(s)ds \Big/ \int \bar{y}_B^2(s)ds$$

The integrals that define b_i are over the union of all the intervals $[L_j, U_j]$. Billheimer [19] also follows this approach to normalization. Note that the spectral deviations $x_i(u)$ are not only normalized but also centered at the mean $\bar{y}_B(u)$. Normalized spectra without centering are given by $x_i(u) + \bar{y}_B(u)$.

An elementary look at the spectra after preprocessing is given in figure 8.2. This figure shows relations among three peaks; the largest peaks are in intervals 7, 8, and 9. Horizontally, this figure shows spectral deviations $x_i(u)$ at the interval 8 peak and, vertically, it shows the spectral deviations at the interval 7 peak and the interval 9 peak. Loosely speaking, it seems that the deviations are uncorrelated for intervals 7 and 8, but correlated for intervals 8 and 9. Generalizing, we see that after normalization, some pairs of peaks are uncorrelated and others are correlated. This suggests the complexity of the correlation structure in which we are interested. We do not pursue these observations further because figure 8.2 is only meant as an introduction.

8.3.3 FUNCTIONAL CCA

Generally speaking, we are interested in correlation between the spectra in two disjoint intervals. Out of the 17 intervals shown in figure 8.1, consider two intervals,

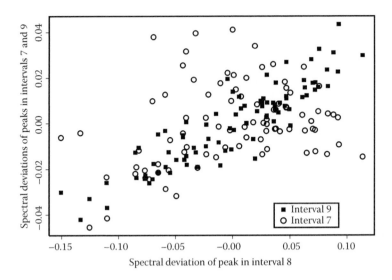

FIGURE 8.2 For the largest peaks in intervals 7–9, peak height scatter plots showing differing degrees of correlation.

j and k, where $1 \le j < k \le 17$. Letting u_j be an m/z point in one interval, and u_k be a point in the other, we could compute the correlation coefficient

$$\frac{\sum_i x_i(u_j)x_i(u_k)}{\sqrt{\left[\sum_i x_i^2(u_j)\right]\left[\sum_i x_i^2(u_k)\right]}}$$

Plotting this versus u_j and u_k gives a correlation map [10,19] that can be examined. Alternatively, we could compute two linear combinations, one of the spectral values in $[L_j, U_j]$ and the other of the spectral values in $[L_k, U_k]$ and examine the correlation between the two. The linear combinations are defined by the weight functions $\xi_j(s)$ and $\xi_k(s)$. Corresponding to these weight functions are two sets of variates: z_{ij}, $i = 1,\ldots, N$ and z_{ik}, $i = 1,\ldots, N$, where

$$z_{ij} = \int_{L_j}^{U_j} \xi_j(s)x_i(s)ds$$

and z_{ik} is defined similarly. These variates are also called scores. The correlation is given by

$$\frac{\sum_i z_{ij}z_{ik}}{\sqrt{\left[\sum_i z_{ij}^2\right]\left[\sum_i z_{ik}^2\right]}}$$

The leading canonical correlation in functional CCA can be specified in terms of two weight functions, $\xi_j(s)$ and $\xi_k(s)$, with variates (called canonical variates) that have maximum correlation subject to penalties on the smoothness of the weight functions. If these weight functions were given, then moments could be computed:

$$s_{jk} = \frac{1}{N} \sum_i \left[\int_{L_j}^{U_j} \xi_j(s) x_i(s) ds \right] \left[\int_{L_k}^{U_k} \xi_k(s) x_i(s) ds \right]$$

The leading canonical correlation coefficient for intervals j and k is the value of s_{jk} obtained by maximizing over the weight functions subject to two constraints:

$$s_{jj} + \lambda_j \int_{L_j}^{U_j} \left\{ D^2 \xi_j \right\}^2 ds = 1$$

$$s_{kk} + \lambda_k \int_{L_k}^{U_k} \left\{ D^2 \xi_k \right\}^2 ds = 1$$

where the symbol D^2 denotes second derivative. The two constraints not only limit the sizes of the weight functions but also impose smoothness on them. The second terms in the two constraints limit the sizes of the second derivatives of the weight functions. Positive values for these terms—that is, for λ_j and λ_k—are necessary if functional CCA is to give reasonable results [7]. Increasing the sizes of λ_j and λ_k increases the smoothness of the weight functions.

The baseline correction causes a problem in the use of this definition to compute the leading canonical correlation. The problem is that the constraints on the weight functions are not satisfied for the constant weight function. The reason is that the $x_i(u)$ are orthogonal to the constant weight function and consequently $s_{jj} = s_{kk} = 0$ for this weight function. We solved this problem by adding a small, randomly chosen constant to each $x_i(u)$ in each interval. Thus, for the constant weight function, s_{jk} remains 0 while s_{jj} and s_{kk} do not, and the weight functions for the leading canonical correlation have mean close to 0. We performed functional CCA computations with Ramsay's software [7].

A different pair of weight functions is derived for every pair of intervals. Denoting the weight functions for intervals j and k by $\xi_j^{\{j,k\}}(s)$ and $\xi_k^{\{j,k\}}(s)$ makes explicit the fact that an interval has a weight function that depends on the pairing. We denote the sets of scores similarly: $z_{ij}^{\{j,k\}}$ and $z_{ik}^{\{j,k\}}$. In the next section, we will consider the R^2 values for the spectral values $x_i(u)$ explained by the scores

$$R^2 = \frac{\left(\sum_i x_i(u) z_{ij}^{\{j,k\}} \right)^2}{\sum_i \left(x_i(u) \right)^2 \sum_i \left(z_{ij}^{\{j,k\}} \right)^2}$$

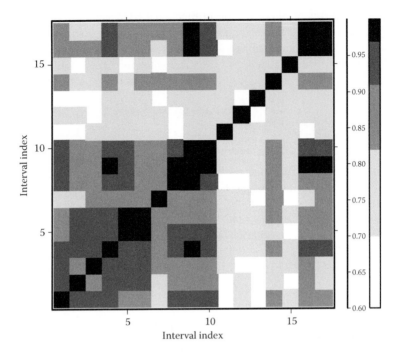

FIGURE 8.3 Leading canonical correlation coefficient for all pairs of intervals. Pairs with highest correlation are of greatest interest.

These R^2 values depend on the particular interval of the pair, the particular pair, and the mass-to-charge ratio u.

8.4 RESULTS

8.4.1 OVERVIEW

To gain insight into variation in the measurement procedure, we examined pairs of intervals with high canonical correlation. The leading canonical correlation coefficient for each interval pair is shown in figure 8.3. Note that this figure is symmetric because the correlation does not depend on the order of the intervals. Moreover, the values on the diagonal are 1. This figure directs our attention to seven interval pairs: (4–9), (5–6), (8–9), (9–10), (9–16), (9–17), and (16–17). Although other pairs also provide interesting insights, we will not discuss them.

One might ask how the choice of interval pairs depends on the choice of λ_j and λ_k. If λ_j and λ_k are too large, the canonical correlation will be less sensitive to relations among specific peaks, which is one aspect of a distinctive relation between intervals. To check this, we reduced λ_j and λ_k to 1/10 of the values on which figure 8.3 is based. With these values, we were still directed to the same seven interval pairs. In addition to the choice of λ_j and λ_k, we also questioned our choice of interval size, $U_j - L_j$.

In trying to understand the overall pattern in figure 8.3, one might start with the notion that high canonical correlation generally does not occur without distinct spectral peaks in both intervals. Figure 8.3 shows that the correlation is low for intervals 11 to 15. Figure 8.1 shows that these intervals contain only minor spectral peaks. It is interesting that interval 7 shows only low correlation despite its spectral peak. This corresponds to the lack of peak-height correlation shown in figure 8.2. Interval 7 provides evidence that our normalization is effective in removing variation that affects all peaks proportionally. We note that the smallest canonical correlation coefficient in figure 8.3 is larger than 0.6. The reason is that CCA always gives a positive coefficient because the algorithm is based on maximization.

8.4.2 PAIRS OF INTERVALS

To answer the question of why a pair of intervals has a high leading canonical correlation coefficient, we have attempted to identify the parts of the intervals primarily responsible. As shown in figures 8.4 through 8.9, for each interval, we display the mean spectrum, the weight function, and the R^2 values for the spectra regressed on the CCA scores. The R^2 value at a particular m/z is the fraction of the spectral variation explained by the scores. The mean spectrum shows locations of spectral peaks in the interval. The weight function shows how the parts of the interval contribute to the scores for that interval. The R^2 values show how the spectra in various parts of the interval are related to the scores, which underlie the canonical correlation coefficient that we want to explain. Of initial interest in figures 8.4 through 8.9 are m/z values for which the mean spectrum and the R^2 values are both high. High mean spectrum indicates high protein concentration, and high R^2 indicates a close relation to the scores that lead to the high canonical correlation coefficient.

Figure 8.4 provides the basis for examining the canonical correlation between intervals 4 (bottom) and 9 (top). High mean spectrum and high R^2 occur for interval 4 at m/z = 6,950 and for interval 9 at m/z = 13,900. Apparently, these points correspond to the doubly and singly charged versions of the same protein. In mass spectrometry, evidence of peaks corresponding to differing amounts of charge is not unusual. More than this, figure 8.4 indicates that the measurement-to-measurement spectral variations at these two m/z values are positively correlated. Positive correlation can be inferred from the values of the weight functions at the two m/z values. Identification of the underlying source of variation is an interesting question. That these variations occur despite the normalization suggests that this source of variation affects at least one but not all of the proteins in the specimen.

The R^2 curve for interval 9 (top) deserves further discussion. Variation of the spectral deviations $x_i(u)$ in this interval is dominated by a single component with the form $f_{i1}\eta_1(u)$. Because of the baseline correction applied, the integral of $\eta_1(u)$ over the interval must be close to 0. For this reason, the baseline correction spreads the variation of the spectral peaks near m/z = 14,000 over the entire interval. Thus, the R^2 curve is high throughout the interval except where $\eta_1(u)$ is nearly 0. This accounts for the two regions above m/z = 13,500 where the R^2 curve dips nearly to 0. The dip below m/z = 13,500 may be due in part to variation in the small peak evident in the mean spectrum. In contrast, the R^2 curve for interval 4 (bottom) seems to be affected by the only partially correlated variation in several peaks. Because of this effect of baseline correction, the R^2 curves for interval 9 in

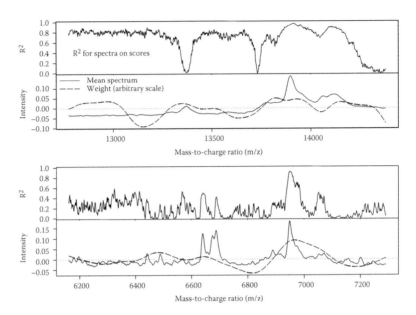

FIGURE 8.4 CCA results for intervals 4 (bottom) and 9 (top). High R^2 and high mean spectrum indicate highly correlated spectral peaks (e.g., m/z = 6,950 and m/z = 13,900). R^2 for spectra regressed on scores shows m/z values responsible for high canonical correlation. The weight functions convert spectra into scores.

figures 8.6 and 8.7 look similar to the R^2 curve for interval 9 in figure 8.4. They are not the same, however.

The weight functions shown in figure 8.4 are not easily interpreted. The main reason is that the measurement-to-measurement variations in the spectral deviations $x_i(u)$ are correlated from one m/z value to another. We force the weight functions to integrate to 0 over the interval because the spectral deviations $x_i(u)$ do. Moreover, the weight functions are computed subject to a smoothness penalty. It is hard to understand how the algorithm forms the trade-off between maximization and these requirements when the spectrum-to-spectrum correlation is high. Maybe the most that can be said is that the weight functions should be relatively large at the spectral peaks responsible for the high canonical correlation coefficient.

An interpretation similar to that of figure 8.4 seems applicable to the canonical correlation coefficients for intervals 9, 16, and 17. It turns out that the interval pairs (9–16), (9–17), and (16–17) have high leading canonical correlations because of the same two peaks, one of which, with m/z = 28,120, is rather broad and split between intervals 16 and 17. This split is shown in figure 8.1. For this reason, we created figure 8.5 by combining parts of intervals 9 and 10 and by combining parts of intervals 16 and 17. To adjust the baseline correction for intervals 9 and 10, we first added a constant to each value of $x_i(u)$ in interval 10 to attain continuity at the boundary between the two intervals. We then applied the baseline correction discussed earlier to the combined interval. We proceeded similarly with intervals 16 and 17. The results of functional CCA for the two combined intervals are shown in

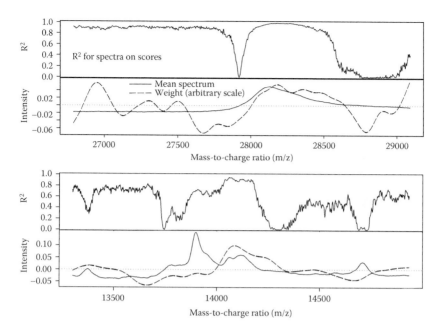

FIGURE 8.5 CCA results for parts of intervals 9 and 10 (bottom) and parts of intervals 16 and 17 (top). High R^2 and high mean spectrum indicate highly correlated spectral peaks (e.g., m/z = 14,060 and m/z = 28,120).

figure 8.5. Apparently, the leading canonical correlation coefficient is due to a protein that, when singly charged, has m/z value near 28,120 and when doubly charged has value near 14,060. The mean spectrum between 14,000 and 14,200 shows two peaks, one at 14,061 and the other at 14,118. The evidence that the first is the one primarily responsible for the high canonical correlation coefficient is somewhat uncertain.

A comparison of figures 8.4 and 8.5 has two interesting aspects. First, comparing interval 9 in the two figures, we see that the R^2 values and the weight functions suggest that the protein responsible for the canonical correlation in figure 8.4 is different from the one responsible for the canonical correlation in figure 8.5. Nonetheless, there is clearly substantial correlation between the variations in the intensities of these two proteins. Second, our figure 8.4 observation that the variation of the spectral deviations $x_i(u)$ is dominated by a single component applies to both combined intervals in figure 8.5. The single component seems even stronger in the combined interval made up of parts of intervals 16 and 17.

Figure 8.6 shows the results for intervals 8 and 9. On the basis of high R^2 value and high mean spectrum, one would point to a single peak in each interval, one with m/z value near 11,740 in interval 8 and one with value near 13,900 in interval 9. The scatter plot of the heights of these two peaks in figure 8.2 shows distinct correlation. In contrast to figures 8.4 and 8.5, the peak locations do not suggest a single protein with different charges. Rather, there appear to be two proteins with concentrations that vary together from spot to spot. The weight function for interval 9 is puzzling, however, because it is nearly 0 near the peak at 13,900. Maybe all we

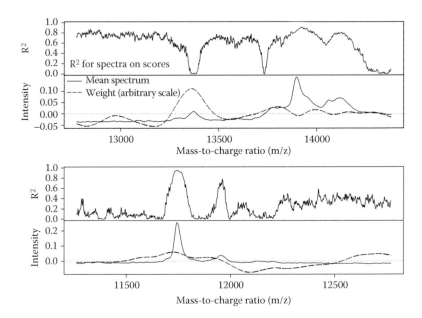

FIGURE 8.6 CCA results for intervals 8 (bottom) and 9 (top). High R^2 and high mean spectrum indicate highly correlated spectral peaks (e.g., m/z = 11,740 and m/z = 13,900).

can say is that the group of proteins surrounding m/z = 14,000 is correlated with the protein underlying the peak in interval 8. Part of the reason why the weight function is high just below 13,500 is the baseline correction.

Another sort of behavior is shown in figure 8.7, which displays the results for intervals 9 and 10. In interval 10, there does not seem to be a single peak responsible for the correlation but rather a collection of proteins with m/z values between 14,400 and 14,700. These seem to be correlated with the group of proteins surrounding m/z = 14,000 in interval 9. That the two groups are contiguous might suggest an explanation.

The results for intervals 5 and 6 are shown in figure 8.8 for the leading canonical correlation coefficient and figure 8.9 for the second canonical correlation coefficient. The second canonical correlation is obtained by maximization as in the case of the first but with the added constraint that for each interval, the second set of scores and the first set of scores must be orthogonal in a special sense [7]. Note that as shown in figure 8.1, intervals 5 and 6 contain the largest peaks in all the 17 intervals considered. One would expect that, to a large extent, these peaks determine the normalization and thus that their variations are small after normalization.

Figure 8.8 shows that the leading canonical correlation coefficient is due to a pair of peaks, one with m/z value near 8,140 and the other with value near 8,940. Neither of these peaks is the largest in the interval. This pair of peaks is captured in a single rotated component in the functional PCA reported by Liggett et al. [17]. That both peaks are in the same rotated component suggests that they are correlated. We note that these two peaks have m/z values close to the m/z values of two peaks identified by Li et al. [20] for use in their biomarkers.

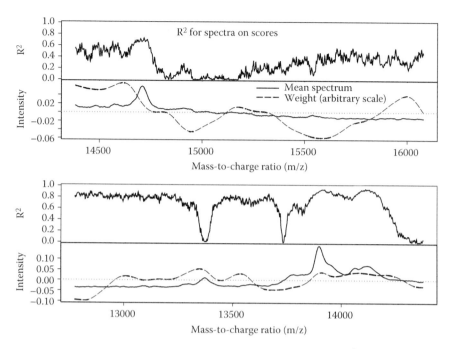

FIGURE 8.7 CCA results for intervals 9 (bottom) and 10 (top). High R^2 and high mean spectrum indicate highly correlated spectral peaks (e.g., m/z in vicinities of 14,000 and 14,500).

Figure 8.9 shows that the second canonical correlation coefficient involves an interesting relation between the two largest peaks and the R^2 curves. The R^2 curves and the weight functions show that regions with m/z values somewhat higher than the largest peaks are responsible for the second coefficient. That higher m/z values correspond to longer flight times might be part of the explanation for the second coefficient. In other words, figure 8.9 shows a correlated distortion in the largest peaks that consists of changes in the trailing part of the peak.

8.5   DISCUSSION

It is of interest to attempt further explanation of the instances of high canonical correlation noted in figure 8.3 and expanded on in figures 8.4 through 8.9. Of particular interest are explanations that suggest follow-on investigations of sources of variation in the measurement system. For follow-on investigation, explanations must be plausible but not necessarily conclusive. Further explanations of figures 8.4 through 8.9 follow.

Figures 8.4 and 8.5 each suggest a source of variation that affects the replicate spectra at two values of m/z, one of which is a multiple of the other. Such values of m/z seem to correspond to a single protein that is both singly and doubly charged during ionization in the mass spectrometer. Thus, these figures suggest a source that affects the concentration of one protein on the chip introduced into the spectrometer. Constant proportions of this protein are then singly and doubly charged. One might

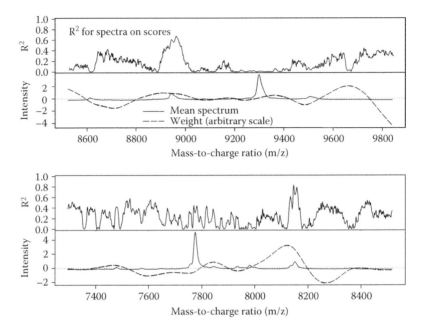

FIGURE 8.8 Leading CCA results for intervals 5 (bottom) and 6 (top). High R^2 and high mean spectrum indicate highly correlated spectral peaks (e.g., m/z = 8,140 and m/z = 8,940). Note that the highest values of R^2 do not coincide with the highest peaks in the mean spectrum.

suppose that such a source is located in the sample preparation part of the measurement procedure. Perhaps the source is spot-to-spot variation in surface retention due to variation in the protein chip or in the effectiveness of washing. Note that the source cannot be the nonuniform crystallization on the surface of the protein chip because the spectral preprocessing removed the effects of this nonuniformity.

Figures 8.6 and 8.8 suggest a source of variation that affects the replicate spectra at two values of m/z with no special relation to each other. Thus, these figures suggest a source that affects the concentrations of two different proteins. The source might affect both proteins similarly because they have similar chemical properties. Because of the amount of separation between the m/z values, one might suppose that such a source is located in the sample preparation part of the measurement procedure.

Figure 8.7 suggests a source of variation that affects the replicate spectra at values of m/z that extend over parts of intervals 9 and 10. The proteins corresponding to these values of m/z may differ from each other only in minor ways. Thus, mechanisms that cause minor modifications of proteins play a role in what is observed in figure 8.7. It is hard to tell whether these mechanisms precede or follow the source of variation responsible for the correlation.

SELDI-TOF mass spectrometry is based on modification of the relative concentrations of the proteins in the specimen through retentate chromatography [18]. Figures 8.4 through 8.8 suggest that the relative concentrations achieved by this process vary from spot to spot on the protein chips. One might suppose that the protein chips do not capture any noticeable amount of some proteins, that they capture

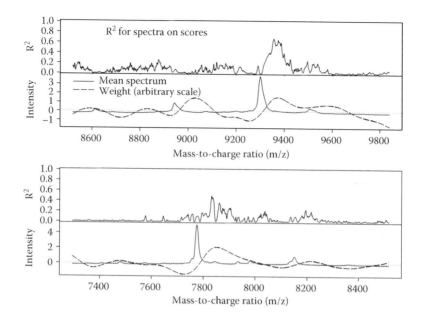

FIGURE 8.9 Second CCA results for intervals 5 (bottom) and 6 (top). The high R^2 values at m/z values above the two largest peaks in figure 8.1 suggest that the high second canonical correlation is caused by instrument-related peak distortion.

almost the entire amount of other proteins, and that they capture some but not all of still other proteins. Proteins in this third category, which show only weak binding to the chip, might exhibit the most spot-to-spot variation in mass-spectral intensity.

Figure 8.9 suggests a source of variation with a different mechanism. Apparently, there are two proteins with spectral-peak distortion that is correlated from replicate to replicate. These two peaks are the largest in the m/z interval on which we base our normalization. This suggests that these peaks largely determine the normalization and therefore that the heights of these peaks vary little after normalization. However, there may be a source of variation (perhaps the nonuniform crystallization) that sometimes causes the amount of material in the spectrometer to be so large that the detector with its associated analog-to-digital converter overloads and distorts the trailing edges of these peaks. This explanation seems consistent with what is reported by Malyarenko et al. [11]. Such a source would cause the correlated peak distortion observed in figure 8.9.

Figures 8.7 and 8.9 illustrate a point about methodology. These figures show the importance of the fact that functional CCA is data driven rather than dependent on assumptions about peak shape. It seems that a peak-based analysis would miss the correlations shown by these two figures.

The sources of variation hypothesized to explain figures 8.4 through 8.9 provide a basis for reduction of the effects of these sources. Is an effort to achieve such reduction necessary, or is characterization of the measurement variation sufficient? Such characterization can be made in terms of variances in the case of univariate measurements but also requires correlations in the case of spectral measurements.

In particular, it is clear that the measurement variation in SELDI-TOF mass spectrometry has complicated correlation properties. Even greater complication can be expected in characterizing measurements made over substantial periods of time or in different laboratories. Thus, characterizing the measurement variation in mass spectrometry is difficult. Consider the application of mass spectrometry in a case-control study. In such a study, derivation of a biomarker may require some type of dimension reduction technique. Performance of such a technique will be affected by the complicated correlation structure, but how it will be affected is a difficult question. Moreover, because of the complexity of the variation, the detection of measurement-system problems—the purpose of control charts in the univariate case—is difficult. These difficulties suggest that the most fruitful approach to measurement variation in SELDI-TOF mass spectrometry might be an effort to reduce the effects of sources of variation.

8.6 FUTURE PROSPECTS

The approach to reproducibility presented in this chapter has limitations to be overcome in future applications. The purpose our approach is identification of sources of variation, although foremost in many applications is the question of whether the measurement system has sufficient reproducibility for the purpose. The specifics of our approach are ones that seem to give reasonable results for one data set, although for another data set—with a different protein chip, for example—these specifics might not be appropriate. The statistical method behind our approach is functional CCA, which may not always be the best way of identifying sources of variation. Finally, the experimental design underlying our approach incorporates only one batch of replicates rather than different batches of replicates obtained under different measurement conditions. Overcoming these limitations might require investment in creative data analysis.

This chapter characterizes reproducibility to gain a scientific understanding of sources of variation. Such an approach is consistent with the fundamentals of metrology. Sometimes, however, scientists ask whether SELDI profiling is sufficiently reproducible for a particular purpose. Answering this question on the basis of an understanding of sources of variation is not easy. In the univariate case, the uncertainty associated with a determination by some measurement method can usually be assessed, and this assessment used to answer the question of the adequacy of the reproducibility. In the case of SELDI profiling, no single number portrays the uncertainty. One can imagine a collection of variances and correlations as portraying the uncertainty in SELDI profiling, but this may not provide an easy answer to adequacy for a particular purpose. It seems that methods for the adequacy question are just now being developed.

The results in this chapter show what can be done with SELDI profiles from a particular reference material obtained with a particular measurement system. Based on these profiles, we made choices as part of our data analysis. One group of choices determined the preprocessing—for example, the lengths of the intervals used as the basis for functional CCA. Another group of choices determined the details of the functional CCA—for example, the choice of the coefficient that determines the size of the smoothness penalty. In another application with another reference material or another protein chip, these choices would have to be revisited.

The issue of sources of variation in the sample preparation step has considerable breadth because there are so many alternative protein chips. The results obtained here apply to the IMAC-3(Cu) chip. These results provide little insight into sources of variation in sample preparation with other protein chips. Those whose research requires other chips should be prepared to do studies such as the one described here.

As discussed in section 8.2, one can think of the data analysis in this chapter as representation of normalized replicate spectra $x_i(u) + \bar{y}_B(u)$ in terms of sources of variation, each corresponding to a component given by $f_i \eta(u)$, where i indexes the spectra and u denotes the mass-to-charge ratio. Our approach is to use functional CCA to look for evidence of a source of variation with high values of $\eta(u)$ at well-separated values of m/z. Alternative correlation approaches include interpretation of correlation maps computed from the spectra or from peaks detected in the spectra [10,19]. A different alternative is principal components analysis, which represents the replicate spectra as the sum of terms of the form $f_i \eta(u)$ [16,17]. The most important aspect of fitting sources of variation is lack of identifiability: The spectra cannot be represented uniquely in terms of sources of variation. In the case of principal components analysis, restrictions are imposed to make the representation unique. However, these restrictions may not lead to the most useful representation. From the possible representations, the one with sources of variation that suggest physical mechanisms is most desirable. For the spectra analyzed in this chapter, our functional CCA approach leads to plausible physical mechanisms. For other data sets, another approach might be required.

An approach to reproducibility based on sources of variation leads to ideas on reducing the effects of the sources of variation on the profiles. As illustrated in this chapter, functional CCA helps the user understand sources of variation in the sample preparation step, the cornerstone of the SELDI methodology. Functional CCA might be used to compare the effects of different sample preparation protocols on the profiles. One could use the approach illustrated to see which protocol has smaller long-distance correlation.

Two batches of replicate profiles, one for each sample preparation protocol, comprise a more complicated design than the simple one-batch design discussed here. One can think of extensions of the statistical analysis presented here to even more complicated designs that involve several alternative measurement system configurations, for example. Of particular importance is the related comparison on batches of replicate profiles from different laboratories. The sources of variation with the biggest effects in one laboratory might differ from the sources of variation with the biggest effects in another laboratory. A statistical analysis that attributed sources of variation to each laboratory would be useful.

DISCLAIMER

Certain commercial entities, equipment, or materials may be identified in this chapter in order to describe an experimental procedure or concept adequately. Such identification is not intended to imply recommendation or endorsement by the National Institute of Standards and Technology, nor is it intended to imply that the entities, materials, or equipment are necessarily the best available for the purpose.

REFERENCES

1. Hastie, T., Tibshirani, R., and Friedman, J., *The Elements of Statistical Learning; Data Mining, Inference, and Prediction,* Springer–Verlag, New York, 2001.
2. Semmes, O.J. et al., Evaluation of serum protein profiling by surface-enhanced laser desorption/ionization time-of-flight mass spectrometry for detection of prostate cancer: I. Assessment of platform reproducibility, *Clinical Chemistry,* 51, 102, 2005.
3. Tibshirani, R. et al., Sample classification from protein mass spectrometry by "peak probability contrasts," *Bioinformatics,* 20, 3034, 2004.
4. Morris, J.S. et al., Feature extraction and quantification for mass spectrometry in biomedical applications using mean spectrum, *Bioinformatics,* 21, 1764, 2005.
5. Carlson, S.M. et al., Improving feature detection and analysis of surface-enhanced laser desorption/ionization-time of flight mass spectra, *Proteomics,* 5, 2778, 2005.
6. Vannucci, M., Sha, N., and Brown, P.J., NIR and mass spectra classification: Bayesian methods for wavelet-based feature selection, *Chemometrics and Intelligent Laboratory Systems,* 77, 139, 2005.
7. Ramsay, J.O. and Silverman, B.W., *Functional Data Analysis,* 2nd ed., Springer–Verlag, New York, 2005.
8. Bischoff, R. and Luider, T.M., Methodological advances in the discovery of protein and peptide disease markers, *Journal of Chromatography* B, 803, 27, 2004.
9. Cordingley, H.C. et al., Multifactorial screening design and analysis of SELDI-TOF ProteinChip array optimization experiments, *Biotechniques,* 34, 364, 2003.
10. Coombes, K.R. et al., Quality control and peak finding for proteomics data collected from nipple aspirate fluid by surface-enhanced laser desorption and ionization, *Clinical Chemistry,* 49, 1615, 2003.
11. Malyarenko, D.I. et al., Enhancement of sensitivity and resolution of surface-enhanced laser desorption/ionization time-of-flight mass spectrometric records for serum peptides using time-series analysis techniques, *Clinical Chemistry,* 51, 65, 2005.
12. Baggerly, K.A. et al., A comprehensive approach to the analysis of matrix-assisted laser desorption/ionization-time of flight proteomics spectra from serum samples, *Proteomics,* 3, 1667, 2003.
13. Baggerly, K.A., Morris, J.A., and Coombes, K.R., Reproducibility of SELDI-TOF protein patterns in serum: comparing datasets from different experiments, *Bioinformatics,* 5, 777, 2004.
14. Leurgans, S.E., Moyeed, R.A., and Silverman, B.W., Canonical correlation analysis when data are curves, *Journal of the Royal Statistical Society, Series B,* 55, 725, 1993.
15. Buckheit, J.B. et al., Modeling of progressive glomerular injury in humans with lupus nephritis, *American Journal of Physiology-Renal Physiology,* 273, F158, 1997.
16. Liggett. W., Cazares, L., and Semmes, O.J., A look at mass spectral measurement, *Chance,* 16(4), 24, 2003.
17. Liggett, W.S. et al., Measurement reproducibility in the early stages of biomarker development, *Disease Markers,* 20, 295, 2004.
18. Fung, E.T. and Enderwick, C., ProteinChip clinical proteomics: Computational challenges and solutions, *Computational Proteomics Supplement,* 32, S34, 2002.

19. Billheimer, D.A., A functional data approach to MALDI-TOF MS protein analysis. Poster presentation at the University of Florida Fifth Annual Winter Workshop, January 10–11, 2003. http://web.stat.ufl.edu/symposium/2003/fundat/Archive.
20. Li, J. et al., Proteomics and bioinformatics approaches for identification of serum biomarkers to detect breast cancer, *Clinical Chemistry*, 48, 1296, 2002.

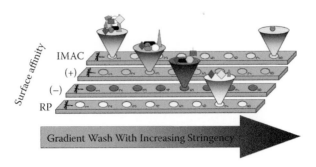

FIGURE 7.4

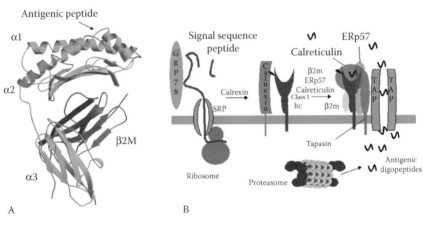

FIGURE 9.2

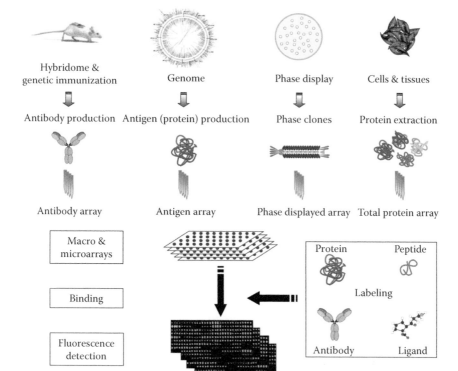

Hybridome &
genetic immunization

Genome

Phase display

Cells & tissues

Antibody production   Antigen (protein) production   Phase clones   Protein extraction

Antibody array

Antigen array

Phase displayed array   Total protein array

Macro &
microarrays

Binding

Fluorescence
detection

Protein   Peptide

Labeling

Antibody   Ligand

FIGURE 10.1

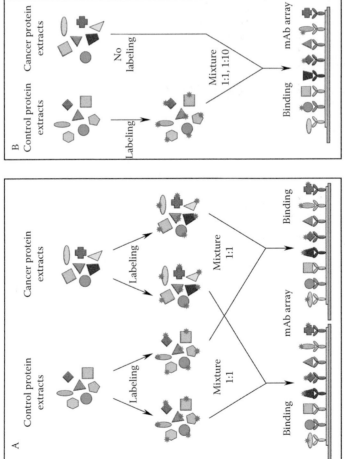

FIGURE 10.2

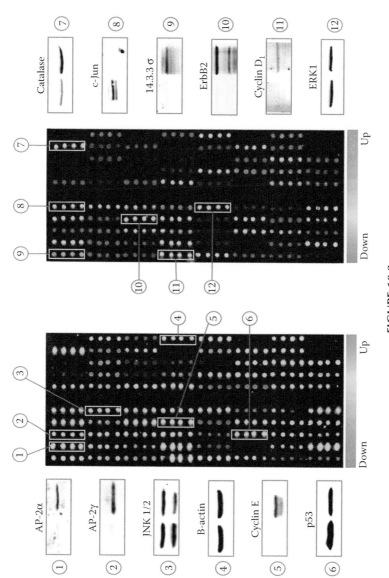

FIGURE 10.3

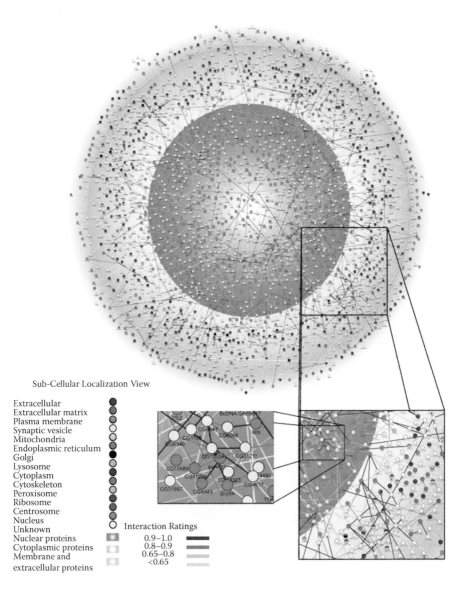

Sub-Cellular Localization View

Extracellular
Extracellular matrix
Plasma membrane
Synaptic vesicle
Mitochondria
Endoplasmic reticulum
Golgi
Lysosome
Cytoplasm
Cytoskeleton
Peroxisome
Ribosome
Centrosome
Nucleus
Unknown
Nuclear proteins
Cytoplasmic proteins
Membrane and
extracellular proteins

Interaction Ratings

0.9–1.0
0.8–0.9
0.65–0.8
<0.65

FIGURE 14.1

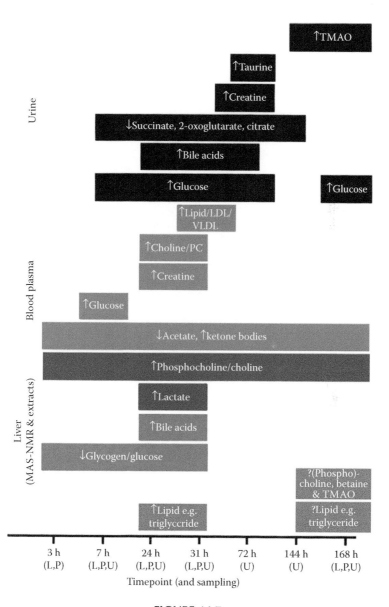

FIGURE 16.7

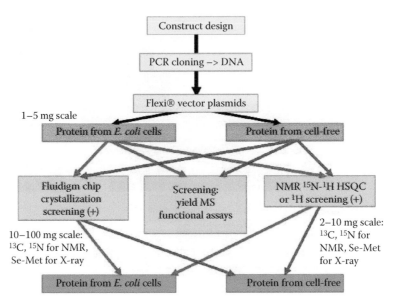

FIGURE 17.2

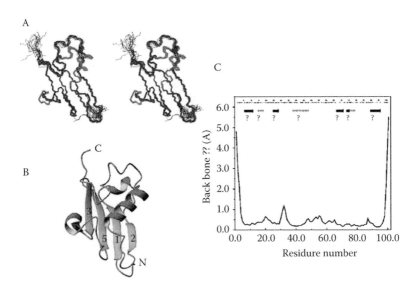

FIGURE 17.5

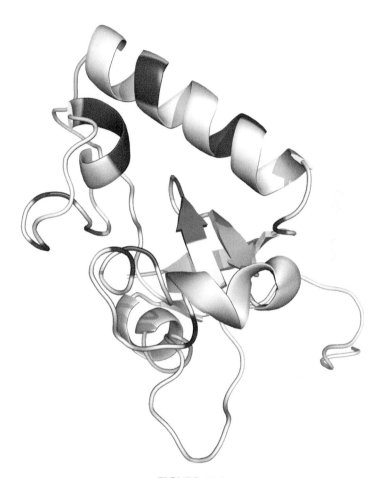

FIGURE 17.6

Part III

Protein–Protein (or Peptide) Interactions: Studies in Parallel and with Mixtures

9 Mass Spectrometric Applications in Immunoproteomics

Anthony W. Purcell, Nicholas A. Williamson, Andrew I. Webb, and Kim Lau

CONTENTS

9.1 INTRODUCTION

The mammalian immune system is a complex set of molecular and cellular responses that have evolved to protect the body from pathological changes. In most cases the recognition of these changes is fundamentally centered around the specific recognition of foreign molecules from pathogens or neoantigens found in cancerous cells. When a pathogenic change occurs, there is an initial response facilitated by the innate immune response, followed shortly after by the adaptive (or cellular) immune response, which combats infection or malignancy in an antigen specific manner.

Typically, studied antigens are protein based; however, lipids, carbohydrates, and other molecules can all function to induce immune responses. Protein antigens are proteolytically degraded in antigen presenting cells (APCs), which subsequently display the resulting peptides on the surface of the cell to the effector cells and molecules of the adaptive immune system. The recognition of peptide antigens by T lymphocytes is exquisitely specific and sensitive, and in most cases the sensitivity of this functional recognition *in vivo* greatly exceeds that of modern proteomics procedures. Nevertheless, mass spectrometric techniques have provided some of the best insights into the molecular events involved in antigen processing and recognition. This chapter will describe some of these discoveries and examine the potential for the latest proteomic technologies to further define this complex area of research and extend into investigations of other forms of antigen.

9.1.1 ANTIGEN IDENTIFICATION

Protein identification can be achieved in several ways using mass spectrometry coupled to a "front end" separation tool. Once the antigen is introduced into the mass spectrometer, the underlying premise for identification is the use of multistage (usually tandem) MS to generate fragment ions that can be used to derive structural information about the antigen. Peptide and protein antigens can be identified by virtue of the amino acid sequence information contained within MS/MS spectra of intact polypeptides or proteolytic fragments of the protein. When combined with the genome sequence, the map of tryptic peptides derived from a protein antigen can also be used to identify the protein reliably using a technique commonly referred to as peptide mass fingerprinting. In combination with more definitive MS/MS-based sequencing of individual proteolytic fragments, the identity of homologous proteins from organisms without a sequence genome can also be deduced. In the case of peptides, nested sets of ion fragments are formed, which allow amino acid sequence information to be derived (see fig. 9.1). For electrospray ionization (ESI)-based techniques, tandem mass spectrometry or MS/MS instrumentation can be used to generate and analyze fragments.

Frequently, a major difficulty surrounding immunoproteomics is the need to identify which antigens, out of the myriad derived from a pathogen or a tumor cell, are immunogenic. The most direct way to ascertain this is to use existing immune responses that may arise spontaneously in patients or in models of disease to probe the proteome. Techniques such as western blots, 2D immunoblots and T cell assays of the fractionated proteome have proven to be powerful tools to dissect out immunogenic proteins from those nonimmunogenic species.

The success of this approach is clearly illustrated in the identification of tumor antigens, in an approach called SEREX (serological expression of cDNA expression libraries). This technique involves the screening of complementary DNA (cDNA) expression libraries with serum from cancer patients that contains the antibody products of the immune system. The rationale behind this approach is that the best candidate tumor antigens should be immunogenic and elicit a spontaneous immune response.

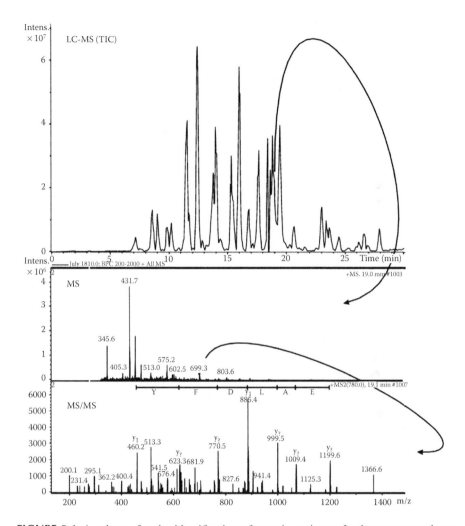

FIGURE 9.1 A schema for the identification of protein antigens. In the upper panel, an LC-MS total ion chromatogram (TIC) is shown from an in-gel tryptic digest of a spot from a two-dimensional gel. An MS spectrum of a peak of ion intensity is shown in the middle panel highlighting the presence of several species, each of which can be subjected to MS/MS using an automated algorithm. An example of an MS/MS spectrum and sequence assignment is shown in the bottom panel.

9.1.2 CYTOTOXIC T LYMPHOCYTES

All nucleated cells present fragments of their intracellular proteome on their surface (figs. 9.2A and 9.2B). Endogenous proteins (potentially including, for example, viral proteins) are degraded to oligopeptides predominantly in the cytoplasm through the action of a multicatalytic protease structure known as the proteasome. The proteasome can exist in several different forms, depending on the exposure of the APC to

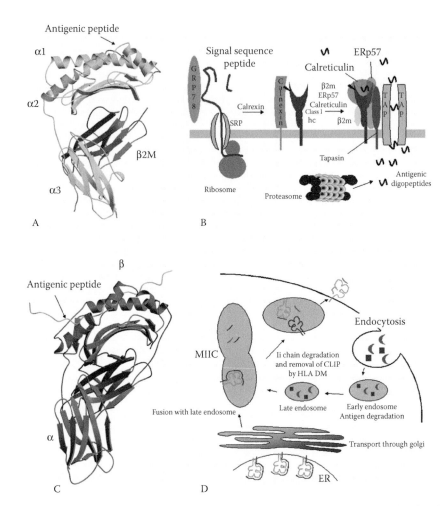

FIGURE 9.2A (See color insert.) The three-dimensional structure of a class I molecule. The class I heterodimer acts as a platform to which peptide antigen binds (shown in the green space filling model). The class I heavy chain (shown in blue ribbon form) has three domains; α1 and α2 form the peptide binding groove. This groove is lined with the highest density of polymorphic residues, which have an impact on ligand specificity of the MHC molecule. The α3 domain is the membrane proximal domain and lies adjacent to the monomorphic β2microglobulin molecule (shown in red ribbon form). This figure was generated from the PDB coordinates (accession number 1N2R) of HLA B*4403 bound to an endogenous peptide derived from HLA DPα chain. (Macdonald, W. A. et al. *J. Exp. Med.* 198, 679, 2003; Macdonald, W. et al. *FEBS Lett.* 27, 2002.)

FIGURE 9.2B (See color insert.) Antigen processing in the MHC class I pathway. 1. Protein antigen is degraded in the cytoplasm through the actions of the proteasome, a multisubunit protease complex with several defined proteolytic activities, some of which are induced through proinflammatory cytokines. 2. Peptides generated by the proteasome and by other cytosolic proteases are transported into the lumen of the ER in an ATP-dependent manner through the actions of the TAP (transporter associated with antigen processing) heterodimer. 3. Nascent class I heavy chain is targeted to the ER and stabilized by interactions with the

FIGURE 9.2B (continued)

chaperone calnexin. Once β2-microglobulin associates with the class I heavy chain, calnexin is exchanged for another ER resident chaperone, calreticulin. The association of the class I heterodimer with calreticulin is also associated with the recruitment of other members of the peptide loading complex (PLC), including tapasin and ERp57. ERp57 is a thiol oxidoreductase (Hughes, E. A. and Cresswell, P. *Curr. Biol.* 8, 709, 1998; Lindquist, J. A. et al. *EMBO J.* 17, 2186, 1998) involved in assuring correct disulfide bonding of the class I heavy chain. (Dick, T. P. and Cresswell, P. *Methods Enzymol.* 348, 49, 2002; Dick, T. P. et al. *Immunity* 16, 87, 2002; Farmery, M. R. et al. *J. Biol. Chem.* 275, 14933, 2000.) Tapasin is a 48-kDa glycoprotein that bridges peptide receptive class I heterodimers to the TAP heterodimer. (Grandea, A. G. et al. *Immunogenetics* 46, 477, 1997; Sadasivan, B. et al. *Immunity* 5, 103, 1996; Brocke, P. et al. *Curr. Opin. Immunol.* 14, 22, 2002.) 4. Once a peptide of sufficient affinity binds to the class I heterodimer, this complex dissociates from the PLC and is transported to the cell surface, where it may be recognized by CD8+ T cells (5).

FIGURE 9.2C (See color insert.) The three-dimensional structure of a class II molecule. The class II αβ heterodimer acts as a binding platform to which peptide antigen binds (shown in the green space filling model). Coordinates used to generate this figure represent the murine class II molecule I-A^k complexed to a hen egg lysozyme peptide (accession number 1IAK). (Fremont, D. H. et al. *Immunity* 8, 305, 1998.)

FIGURE 9.2D (See color insert.) Class II antigen processing pathway. Exogenous antigen is taken up by endocytosis and degraded in the early and late endosomes. The late endosome (containing antigenic peptides) fuses with class II rich transport vesicles (containing class II heterodimers associated with Ii) to form the MIIC compartment. In the MIIC, HLA DM catalyzes the removal of Ii-derived peptide (CLIP) from the antigen binding cleft of the αβ heterodimers, facilitating loading with antigenic peptides. This mature class II complex is then transported to the cell surface for scrutiny by CD4 + T helper cells.

pro-inflammatory stimuli [1–9]. These different forms of the proteasome engender alternate proteolytic activities and consequently produce a different array of peptide precursors for transport into the lumen of the endoplasmic reticulum (ER).

More recently, the interplay of the various forms of the proteasome with other peptidases has been elucidated, adding to the complexity of peptides destined for export to the cell surface, bound to major histocompatibility complex (MHC) class I molecules [10–12]. These peptide MHC class I complexes are then inspected by cytotoxic T cells expressing the CD8 co-receptor. The CD8 co-receptor facilitates recognition of the MHC class I bound peptide, and is expressed by cytotoxic T lymphocytes (CTL). It is through these CTL responses that virally infected cells, tumor cells, and sometimes even normal healthy cells are destroyed, clearing the virus or eradicating tumor cells from the host. In the case of normal tissue destruction the result is overt autoimmune disease such as that observed in the destruction of pancreatic β-cells in type 1 diabetes.

The structure of class I MHC molecules is well defined and consists of a polymorphic heavy chain, a monomorphic light chain (β-2 microglobulin), and an antigenic peptide (fig. 9.2A). The class I heavy chain has three extracellular domains (α1 and α2 domains that together form a binding cleft, and the membrane proximal α3 domain linked to a transmembrane domain and a short, nonsignaling, cytoplasmic tail). The antigen-binding groove accommodates an antigenic peptide typically 8–11 amino acid residues in length, although longer peptides have been reported [13–22].

Heavy chain residues that line the binding groove are the focus for the majority of class I MHC polymorphisms. These polymorphisms determine the specificity of allelic forms of MHC molecules via the formation of several conserved depressions or pockets (denoted A–F) that vary in composition and stereochemistry. The A and F pockets are located at either end of the cleft and contain conserved residues involved in hydrogen-bonding interactions with the N- and C-termini of the bound peptide respectively. These interactions effectively close off each end of the cleft encapsulating the termini of the bound peptide. The A pocket is frequently shallow, while the stereochemistry of the F pocket varies significantly and contributes to conserved interactions with the C-termini as well as to the specificity of the last amino acid residue of the bound peptide. The B, C, D, and E pockets contribute to the specificity of the central portion of the bound peptide. Our understanding of the specificity of different MHC class I molecules has increased enormously over the last decade; structural and biochemical studies of bound peptides have contributed to our knowledge of the binding characteristics of different MHC class I molecules and even closely related allotypes.

9.1.3 THE HUMORAL RESPONSE

To complement the cytotoxic T cell response, the adaptive immune system also elicits an antibody response. This humoral response is essential for the clearance of many pathogens. Antibodies recognize and bind to relatively short peptide sequences in the context of the intact antigen. In contrast to the T cell receptors (TcR) expressed on the surface of T lymphocytes, they frequently display conformational dependence for binding, since they recognize molecular surfaces—not extended antigen fragments bound to MHC molecules. The production of antibodies is facilitated by specialized APC such as B cells, dendritic cells (DCs), and macrophages. In B and Fc receptor positive cells, exogenous sources of antigen are endocytosed via receptor-mediated processes and degraded in the early and late endosomes. The late endosomes therefore provide a rich reservoir of antigenic peptides that are transported to the cell surface bound to MHC class II molecules for display and scrutiny by T cells.

However, it is critically important to recognize that this process is both mechanistically and physically distinct from the MHC class I processing pathway. MHC class II molecules are composed of two polymorphic polypeptide chains (α and β) forming an $\alpha\beta$ heterodimer that, like class I molecules, combines to form a binding cleft that accommodates peptide antigen (figs. 9.2C and 9.2D). Class II α and β chains are inserted cotranslationally into the lumen of the ER, where they associate to form nascent heterodimers [23]. These heterodimers are unstable in the absence of bound peptide and are stabilized through association with a chaperone known as invariant chain (Ii). This chaperone facilitates the formation of a multimeric structure consisting of three $\alpha\beta$-heterodimers, each of which is associated with an Ii molecule (i.e., $(Ii\alpha B)_3$) and occludes the peptide binding cleft, thereby preventing premature binding of endogenous ER-resident peptides.

The Ii is also important in trafficking nascent class II molecules to the endocytic route by virtue of an N-terminal sorting signal [24]. For antigenic peptides to be able to bind to class II molecules, Ii must be degraded to allow access to the antigen binding cleft. This process is mediated by cathepsin S, and the resulting proteolysis of the Ii–MHC complex [25] leaves a short peptide of the Ii protein (residues 81–104) bound to class II heterodimers. These class II-associated Ii peptides (known as CLIP) demonstrate promiscuous binding to MHC class II alleles and occlude the peptide binding cleft of these molecules [26,27]. In order to displace CLIP from the class II binding site, antigenic peptides must have a higher binding affinity than CLIP.

This peptide exchange is catalyzed by another MHC-encoded gene product, HLA DM [27,28]. This occurs in the MIIC (MHC class II compartments); the product of the fusion of MHC class II rich vesicles with the late endosomes. These compartments are rich in cathepsins and this intersection of the endosomal and class II processing pathways promotes removal of CLIP and loading of class II molecules with exogenous antigen fragments. The MHC class II/peptide complex is then transported to the cell surface for scrutiny by helper T cells expressing the CD4 co-receptor. Upon binding of a suitable helper T cell via the CD4 co-receptor/MHC class II/antigen complex, a number of co-stimulatory signals are passed between the two cells. B cells expressing specific MHC class II peptide complexes by virtue of the capture of antigen (via their clonally distributed surface immunoglobulin receptor) are triggered to differentiate into antibody-producing plasma cells by T helper cells.

The mode of binding and repertoire of peptide ligands bound by MHC class II molecules has also been analyzed by biochemical methods and x-ray crystallographic studies [29–34] and differs from the binding of peptides to class I molecules. The interactions that close the peptide binding cleft of class I molecules are not apparent in class II molecules, allowing the termini of the bound class II peptide to project out of the ends of the cleft. Hence, MHC class II peptides are typically longer than class I ligands, averaging around 13 amino acids in length, but can be considerably longer. Like class I molecules, polymorphic amino acid residues line the pockets of the binding cleft. Both structural and biochemical studies indicate that amino acid side chains at residues 1, 4, 6, and 9 of the class II-bound peptide typically interact with these pockets, conferring allelic specificity [29], and "anchor" the peptide into the cleft.

It has also been suggested that the binding of ligands to MHC class II molecules is more promiscuous than the binding of peptides to MHC class I molecules. This is due to their free termini and their ability to shift binding registers, making it more difficult to define anchor residues and to predict which peptides will be able to bind particular MHC class II molecules. This is further complicated by the observation that N- and C-terminal exopeptidase activities can trim the peptides bound to class II molecules during their transit from MIIC to the cell surface [35].

9.2 ROLE OF MASS SPECTROMETRY IN STUDYING TARGETS OF T CELL IMMUNITY

9.2.1 STUDYING THE DIVERSITY OF MHC-BOUND PEPTIDES: THE IMMUNOPROTEOME

It has been estimated that each MHC class I or class II allele may present as many as 10,000–100,000 different peptides on the surface of an APC, and only a very small proportion of these peptides (1–1000 specific MHC peptide complexes per cell) can be recognized by antigen-specific T cells [36–38]. Thus, functionally relevant T cell epitopes may be of extremely low abundance on the surface of the APC. Moreover, for humans, any individual may express up to six different class I (two different allotypes encoded by the HLA A, B, and C loci) and six different class II allotypes (two different allotypes encoded by the HLA DR, DQ, and DP loci). Thus, the resulting peptide landscape present on the surface of the APC may be extremely complex. This pool of peptides has been called the immunoproteome or immunopeptidome [39–49].

While a diverse array of peptide targets is the key to the immune defense, it makes the study of individual responses to a particular pathogen extraordinarily complex. Thus, a critical step in any comprehensive biochemical approach to assess the natural processing and presentation of candidate epitopes is to simplify or limit the peptide pool, preferably to the analysis of ligands derived from a single MHC allotype. For example, homozygous cell lines express a more limited number of MHC class I or class II alleles (three loci for each class), while mutant cell lines such as C1R express very low levels of endogenous class I molecules but support high-level expression of a single transfected class I molecule [50]. The simplified array of MHC molecules present on the surface of such cell lines makes them very attractive for examining endogenous peptides presented by individual class I alleles under normal physiological conditions [51–54] or during infection [55,56].

Another powerful method to simplify the immunoproteome is immunoaffinity chromatography. Thanks to the tremendous efforts that have gone into generating monoclonal antibodies that distinguish between the different MHC allotypes, the use of appropriate monoclonal antibodies allows isolation of a single MHC allele, and some antibodies can even select a subpopulation of MHC molecules with defined molecular or functional properties [16,57]. The use of immunoaffinity chromatography to isolate specific MHC molecules provides the most appropriate material for identifying individual peptide ligands restricted by a known MHC allele. It is also critical to the pool sequencing experiments and the peptide repertoire studies that focus on the analysis of ligands derived from a specific MHC molecule.

9.2.2 MHC-Binding Motifs

The peptide repertoire for any given MHC allotype will be constrained by the structural features of the binding cleft of that individual MHC molecule. Hence, while there may be many different peptides that bind, they will often share common structural features—a motif. These motifs describe the amino acids located at critical positions along the sequence of the antigenic peptide that are responsible for making highly conserved and energetically important contacts with pockets in the binding cleft of the class I and class II molecules. These conserved residues are therefore frequently described as anchor residues.

One particularly effective approach for defining the anchor residues, especially for class I ligands, is pool Edman sequencing [58]. Because the class I binding cleft constrains the overall length of the bound peptide, peptides isolated directly from immunoaffinity purified class I molecules can be loaded onto the Edman sequencer. While this approach does not identify individual peptides, it does reveal the position of conserved residues as inferred by the abundance or paucity of signal for particular amino acids in the various cycles of the Edman analysis [59]. When particular amino acids are favored in the sequences of the bound ligands, a significant increase in the signal observed for the given anchor amino acid is found in the corresponding cycle of Edman chemistry. Both dominant anchor residues (i.e., where the majority of MHC-bound peptides share a conserved amino acid at a distinct position in the ligand) and preferred or nonpreferred residues can be delineated using this technique.

This form of analysis is less amenable to the study of class II bound ligands because of the greater length heterogeneity of their ligands. Although pool sequencing is an excellent tool for assessing major changes in peptide specificity for different MHC alleles, it fails to distinguish between very closely related alleles that may have substantial overlap in their bound peptide repertoire. For example, we have recently demonstrated that two HLA B44 alleles share up to 95% of their ligands; only high-resolution peptide mapping studies using mass spectrometry were able to reveal these subtle but functionally important differences in ligand repertoire [18].

MHC-binding motifs have been used to predict T cell epitopes successfully, and listings of certain motifs are conveniently Web based (see information resources). However, the success rate for *de novo* prediction of T cell epitopes, even for well studied and abundant MHC alleles, is only about 60%. Moreover, there are numerous examples of atypical ligands possessing non-motif-based sequences or post-translationally modified ligands or of the failure of antigen processing to liberate the candidate peptides, which therefore restricts the predictive algorithms to a subset of T cell epitopes [60–69]. Furthermore, many T cell responses are focused on one or two immunodominant peptides selected from the numerous potential MHC ligands encoded within the pathogen's genome [70].

The participation of so few epitopes limits predictive studies, since markers of immunogenicity must take into account not only peptide binding characteristics but also the abundance and density of antigen present on the cell surface, the time of expression of the antigen during the infection or pathological process, correct processing and luminal transport of the epitope, and the available T cell repertoire in the host organism. Nonetheless, epitope prediction remains a popular first-screening method to identify candidate T cell determinants for subsequent biological validation [44,47,71–80], and predictive algorithms are frequently combined with *in vitro* MHC-binding assays to confirm experimentally that the predicted ligands bind to the targeted MHC molecule [73,81].

9.2.3 DISCOVERY OF T CELL EPITOPES

Several approaches have been used to isolate naturally processed and presented MHC-bound peptides directly from cells; these include analysis of peptides contained within cell lysates [82–84], isolation of peptides directly from the cell surface [85,86], and immunoaffinity purification of the MHC–peptide complexes from detergent-solubilized cell lysates [58,87]. Each approach has advantages, with the latter providing the best chance of epitope identification due to the additional specificity of the immunoaffinity chromatography step and subsequent simplification of the range of cellular peptides isolated. However, each method is based upon the

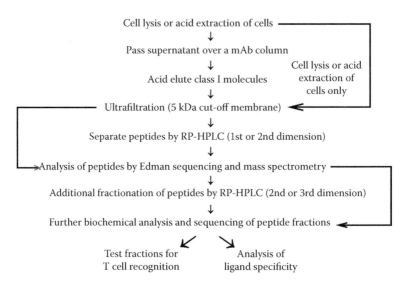

Cell lysis or acid extraction of cells
↓
Pass supernatant over a mAb column
↓   Cell lysis or acid
Acid elute class I molecules   extraction of
↓   cells only
Ultrafiltration (5 kDa cut-off membrane) ←
↓
Separate peptides by RP-HPLC (1st or 2nd dimension)
↓
→Analysis of peptides by Edman sequencing and mass spectrometry
↓
Additional fractionation of peptides by RP-HPLC (2nd or 3rd dimension)
↓
Further biochemical analysis and sequencing of peptide fractions ←

Test fractions for   Analysis of
T cell recognition   ligand specificity

FIGURE 9.3 Schematic representation of processes for isolating MHC-bound peptides. A variety of approaches can be taken to examine peptides bound to MHC molecules, with several optional steps highlighted by the various flow paths in the diagram, allowing additional specificity and purification of the starting material.

common features and assumptions that (1) upon cell lysis, peptides bound to MHC molecules are protected from intracellular and extracellular proteolysis because they are bound to the MHC receptor; and (2) treatment with acid dissociates bound peptides from the MHC complexes. The relationship between each approach is represented schematically in figure 9.3. This diagram highlights the variety of approaches that can be taken, and some of them are discussed next.

In the first approach, peptides are extracted from whole-cell lysates following treatment with an aqueous acid solution such as 1% trifluoroacetic acid (TFA). The presence of TFA also aids in the precipitation of larger proteins, leaving a complex mixture of intracellular and extracellular peptides, a proportion of which were bound to and protected by MHC molecules. Typically, these preparations are fractionated by reverse phase high-performance liquid chromatography (RP-HPLC) and screened with a functional assay to confirm the presence of a particular T cell epitope. These fractions can also be titrated into functional assays to allow relative quantitation of known T cell epitopes extracted from the surface of different cell types [82–84]. In some circumstances, the peptides are amenable to sequencing of individual components of the fractionated material by mass spectrometry.

An alternative to the acid lysis method utilizes a nonlytic approach for recovering cell surface-associated peptides. The cells are washed in an isotonic buffer containing citrate at pH 3.3; the acidic nature of this buffer facilitates dissociation of MHC-bound peptides from the cell surface without affecting cell viability [85]. The advantage of this technique is that the same cells may be harvested daily in an iterative approach for obtaining MHC-bound material. Although the specificity of this process is somewhat better for MHC-bound material than it is from whole-cell lysates, some form of biological assay is again usually necessary to locate the peptide of interest prior to attempting more definitive biochemical characterization.

9.2.4 VISUALIZING THE COMPLEX ARRAY OF PEPTIDES PRESENTED ON THE SURFACE OF APC

Our approach, discussed in detail next, has been to perform multidimensional chromatography followed by off-line matrix-assisted laser desorption/ionization (MALDI)-TOF mass spectrometry. The advantages of this approach are that the molecular complexity of the material can be appreciated as a function of a chromatographic index (in our case this is usually a combination of immunoaffinity chromatography to selectively isolate the MHC allotype of interest and microbore or capillary RP-HPLC) and that the majority of the fractionated material is available for functional assays or for further sequence interrogation using liquid chromatography (LC)-MS/MS and other mass spectrometry-based technologies. Notably, other groups have performed similar analyses using both MALDI and ESI mass spectrometry of fractionated material (e.g., see references 56 and 88 through 92) and displayed them in a variety of manners, including approximating the LC-MS data to a 2D gel-like format [91]. In our approach we chose to perform a systematic comparison of fractions. For close examination, individual spectra were displayed in a reflection mode [93] whereby the two compared spectra are superimposed on the m/z axis, but displayed in opposite polarity (fig. 9.4).

9.2.5 SEQUENCING MHC-BOUND PEPTIDES USING TANDEM MS AND RELATED APPROACHES

The challenges associated with resolving and sequencing individual peptides from a complex mixture of MHC-bound material is not unique to immunoproteomics. This type of experiment is analogous to the shotgun proteomics type approaches that generate complex mixtures of tryptic (or other proteolytic) fragments derived from a subset of the cellular proteome (e.g., Yates [94] and Chapter 11 in this book). Because MHC-bound peptides frequently have varied termini and the proteolytic specificities that generate them are quite diverse, confident assignment of the peptide sequence can be difficult. Similarly, it is rare to detect peptides derived from the same protein unless the study is related to infection, as when target antigen sequences are known or suspected [56,89,90,95].

These properties of MHC-bound peptides reduce confidence in their sequence assignments by MS/MS techniques, and dictate the requirements for additional screening algorithms in epitope identification strategies. For example, if the binding motif for the given allele is known, this frequently can act as an initial filter for assigning fragmentation spectra derived from immunoaffinity purified class I MHC molecules. Figure 9.5 shows the identification of prominent peptides derived from immunoaffinity purified HLA B*4402 molecules (binding motif XEXXXXXXF/Y) by postsource decay in MALDI-TOF MS (PSD-MALDI-TOF) and highlights the difficulties associated with the sequence assignment of class I bound peptide: Many do not strictly adhere to the canonical binding motifs, lack charged termini, and lack

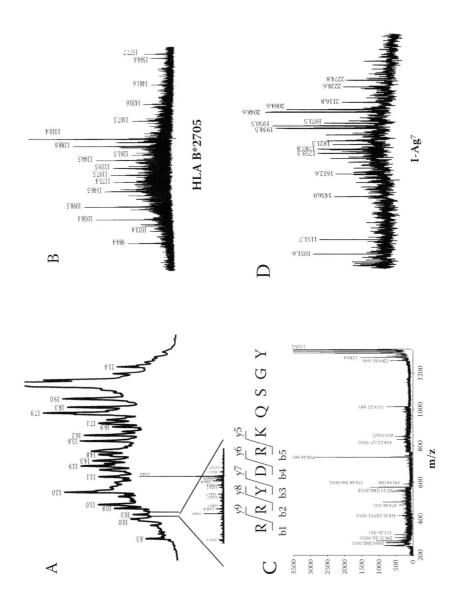

FIGURE 9.4 Examples of biochemical analyses of peptides eluted from MHC class I and class II molecules. Peptides eluted from (A) human class I (HLA B*2705) and (B) murine class II (I-Ag7) MHC molecules following one round of RP-HPLC separation and visualized by MALDI-TOF MS. (C) PSD-MALDI and (D) ESI-ion trap MS² based fragmentation and sequence assignment for B*2705-restricted self-peptides isolated from the surface of human B-lymphoblastoid cells, following two dimensions of optimized RP-HPLC separation. (Purcell, A. W. et al. *J. Immunol.* 166,1016, 2001; Purcell, A. W. and Gorman, J. *J. J. Immunol. Methods* 249, 17, 2001.) For MALDI-TOF MS 1-μL aliquots of the analyte were mixed with 1 μl of 2,5-dihydroxybenzoic acid and dried onto a sample stage for analysis by MALDI-QqTOF MS (QSTAR pulsar, Applied Biosystems). Selected ions were subjected to further analysis using MS/MS. Fragment ions generated in this way were manually assigned based on the known sequence of the parental peptide, and modified amino acid residues were identified within the sequence. PSD-MALDI was performed on a Bruker Reflex IV MALDI-TOF mass spectrometer. Postsource decay experiments were performed using 14 stepwise decrements in the reflectron potential and increasing the laser irradiance to optimize the production of fragment ions at each voltage. Assembly of the individual spectra on to a continuous mass scale was performed using FAST software routines within the Bruker XTOF software package. Identification of fragmented ion species was done by manually assigning C- and N-terminal ion series and comparing parent m/z and fragmentation data to database entries using MS-FIT routines available through the protein prospector program (http://prospector.ucsf.edu). Accurate parent ion mass and fragmentation data allowed assignment of peptide sequences in several instances. Peptide sequencing by ion trap ESI MS was performed on dried samples, which were resuspended in 3 μL of methanol/water (1/1) containing 0.1% formic acid. One microliter of each sample was subjected to peptide sequencing in an LCQ electrospray/ion trap mass spectrometer (Finnigan Thermoquest, San Jose, California). Collision energy and precursor ion resolution were individually optimized for each peptide to obtain the optimum fragmentation spectra. Putative peptide sequences were obtained by database comparison of the fragmentation spectra using the PEPSEARCH program (Bioworks package, Finnigan Thermoquest) followed by manual assignment of expected fragments from the highest scoring sequences. In several instances, the authenticity of these sequence assignments was confirmed by comparing the fragmentation spectrum with that of the corresponding retrospectively synthesized peptide.

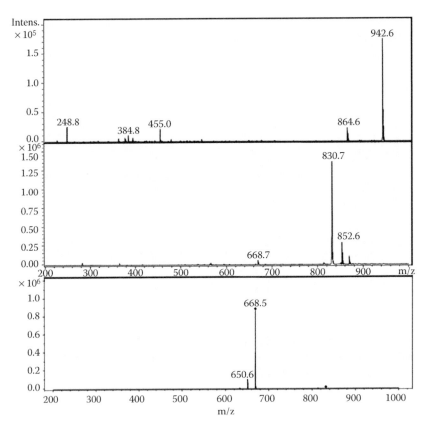

FIGURE 9.5 MS of a glycolipid related to α-galactosyl ceramide that binds to CD1d. Upper panel shows positive mode ionization, middle panel shows negative mode ionization, and bottom panel indicates symmetrical cleavage of the molecule in the gas phase. Additional fragmentation is required to glean more structural information using MS^n approaches. For MALDI-TOF MS, 1-µL aliquots of the analyte were mixed with 1 µL of 2,5-dihydroxybenzoic acid and dried onto a sample stage for analysis by MALDI-QqTOF MS (QSTAR pulsar, Applied Biosystems). Selected ions were subjected to further analysis using MS/MS analysis. Fragment ions generated in this way were manually assigned based on the known sequence of the parent peptide and modified amino acid residues identified within the sequence.

mobile protons to facilitate good fragmentation. In this example, the presence of a proline in the sequence favored internal fragment formation, further complicating sequence assignment using automated methods. In our experience, definitive identification of peptides often requires comparison to the fragmentation "fingerprint" of synthetic versions of the candidate sequence as well as confirmation of identical RP-HPLC retention behavior.

As proteomics instrumentation becomes more accessible to a more diverse array of researchers, so do the demands for robust techniques for protein and epitope identification. Our work has principally involved MALDI-TOF mass spectrometry for repertoire analysis, followed by a combination of PSD-MALDI-TOF, MALDI-MS/MS and nanoESI-Qq-TOF-MS/MS. Others have used different configurations of instrumentation to achieve the same end-point. The use of triple quadrupole, 3D ion trap and, more recently, Fourier transform MS by the Engelhardt/Hunt and other prominent groups has clearly demonstrated the power of these other techniques for epitope identification and repertoire analysis [7,38,54,61,68,91,96–120].

One limitation in analysis of MHC-bound peptides is operator bias in selecting peptides from extremely complex mixtures for MS/MS. Newer technologies, such as the MALDI-TOF-TOF-MS/MS instruments will allow much higher throughput analyses in an automated mode. This will enable many more peptide fractions to be analyzed without operator bias and make it possible to use higher resolution collection of HPLC fractions for analysis. The reduced peptide repertoires of narrower fractions will enhance the potential to characterize many more peptides due to the reduced likelihood of the coincidences of masses of peptides in wide fraction cuts and a lowered tendency for suppression of ionization of peptides. The flexibility of multiple MS/MS modes of other recently developed MS analyzers involving linear ion trap should also facilitate immunoproteomics.

9.3 T CELL RECOGNITION OF NONPEPTIDIC ANTIGENS

CD1 is the term used to describe a family of transmembrane glycoproteins found on various APCs that are able to present exogenous and endogenous lipid antigens to CD1 restricted T cells [121]. They are coded by a series of nonpolymorphic genes found on human chromosome 1q23.1 [122]. Despite not being linked to the MHC, the five isoforms of CD1 have a very similar structure to the MHC class I molecule. CD1 is expressed on the cell surface as a noncovalently linked heterodimer consisting of 43- to 49-kDa heavy chains and β_2m. The CD1 heavy chain consists of three extracellular domains (α1, α2, α3), where the α1 and α2 domains form the binding pocket of CD1 where lipid antigen binds [123].

There are five CD1 isoforms in humans: CD1a, CD1b, CD1c, CD1d, and CD1e. Based on their sequence homology, these five isoforms can be placed in two groups: group 1, which consists of CD1a, CD1b, and CD1c, and group 2, which consists of CD1d (CD1e has not yet been characterized) [124]. Group 1 CD1 molecules are able to present both exogenous and endogenous lipid antigen to CD1 restricted $\alpha\beta$ T cells, whereas group 2 CD1 molecules have been shown to present endogenous lipid antigen to NKT cells that have an invariant TCR (T cell receptor) consisting of a Vα24Jα18 chain [123]. The CD1 isoforms differ according to the signal motif found on their cytoplasmic tails and the hydrophobic binding pocket, which determines the lipid antigens they bind.

Although the structures of the hydrophobic binding pocket for all the CD1 isoforms have yet to be determined, the structure for CD1a [125] and CD1b [126] have been solved. While superficially similar to the MHC class I heavy chain, the antigen binding groove is deeper and narrower than the peptide binding groove of an MHC class I molecule, accommodating the hydrophobic regions of the lipid and allowing the hydrophilic regions of the lipid to remain exposed to contact a T cell receptor.

The different trafficking pathways of the various CD1 isoforms reflect the physiological roles they fulfill: CD1a to the sorting endosomes, CD1c to the early and late endosomes, and CD1b and CD1d to the late endosomes and lysosomes, allowing the sampling of lipid from different cellular locations. This has been seen with various lipids from mycobacteria; CD1b has been shown to present free mycolates, phosphatidylinositolmannosides, and glucose monomycolates and CD1d has been shown to present glycosylphosphatidylinositols [121]. CD1 is also able to bind self-diacylglycerols and self-sphingolipids, indicating that the CD1 molecules may also have a role in establishing tolerance against self-lipids [123] as well as presenting exogenous lipid antigens to the immune system. Recently, endogenous glycolipid ligands of CD1 molecules have been identified [127], thus paving the way forward for mass spectrometric interrogation of these molecules as ligands for this very interesting family of receptors. An example of an MS analysis of an α-galactosyl ceramide-related molecule bound to CD1d is shown in figure 9.5.

9.4 SUMMARY

Peptides and other small pathogen-derived molecules play an essential role in informing the immune system of infection or the transformation of cells in malignancy. As such, mass spectrometry is ideally suited to study these immune responses at a molecular level. Here we reviewed recent advances in studies of immune responses that have utilized mass spectrometry and associated technologies with a particular emphasis on the role of peptide antigens in T cell-mediated immunity. These antigenic peptides are liberated from the intact antigens through distinct proteolytic mechanisms and are subsequently transported to cell surfaces bound to chaperone-like receptors known as MHC molecules. These complexes are then scrutinized by T cells that express receptors with specificity for specific MHC–peptide complexes.

In normal uninfected cells, this process of antigen processing and presentation occurs continuously, with the resultant array of self-antigen derived peptides displayed on the surface of these cells. Changes in this cellular peptide array act to alert the immune system to changes in the intracellular environment that may be associated with infection, oncogenesis, or other abnormal cellular processes, resulting in a cascade of events that results in the elimination of the "abnormal" cell.

9.5 FUTURE PROSPECTS

The specificity and sensitivity of the immune response is unparalleled in biology. Proteomic analysis of the peptides and antigens involved in immune responses to pathogens and to abnormal or even normal tissues presents exciting and difficult challenges to the investigator. Underlying all these studies is the power of the immunological reagents and functional readouts that provide exquisite sensitivity, unmatched by the analytical techniques we have at hand. The use of these reagents as screening tools or as a fractionation technique will drive immunoproteomics research in the future. The use of antibodies and recombinant reagents such as MHC tetramers and other markers of immune effector cells provides opportunities to obtain large numbers of homogeneous cells from tissues and fluids using fractionation or isolation techniques such as flow cytometry and cell sorting and laser capture microscopy. When used in combination with more typical fractionation techniques, such as subcellular fractionation and affinity chromatography, the degree to which specific cells can be isolated and analyzed is improved dramatically.

The challenges that lie ahead in immunoproteomics are very similar to those of proteomics in general. Common problems that are frequently encountered include quantitation, data management and data mining or interpretation, and sample preparation and sensitivity, in addition to the problems associated with separating complex mixtures. All of these areas are being addressed. Several vendors are now making serious attempts to produce software packages that can interpret (or at least assist in interpretation of) and effectively manage the enormous amount of data that can be produced using modern proteomic technology. While these tools are far from perfect, the concerted efforts directed at these problems should see rapid advances in this area.

While the technology of mass spectrometry will continue to advance in sensitivity and resolution, our experience is that much of the potential of the current generation of MS technology remains untapped because of an inability to prepare samples adequately. Hence, over the next few years we expect to see a great diversification and increase in sophistication of the techniques used upstream of the mass spectrometer. Techniques such as laser capture microscopy, cell sorting, biosensor (SPR-MS), and continued developments in multidimensional liquid chromatography all hold great promise. Developments in these areas, possibly used in conjunction with isotope labeling techniques, will also have an impact on the issue of absolute quantitation of MHC-bound peptides.

Finally, proteomic analysis of peptide repertoires from cells and tissues of animal models of disease and from patients is one area that will become the focus of many active researchers in this area. The ability to examine peptides presented by tumor cells derived from biopsies of patients, for example, provides a portal to individualized immunotherapy based on antigen expression or epitope presentation as determined by analysis of MHC-bound peptides or other proteomic approaches to address tumor antigen expression.

INFORMATION RESOURCES

BioInformatics & Molecular Analysis Section (BIMAS) MHC-binding prediction	http://bimas.dcrt.nih.gov/molbio/hla_bind/
MHC class II ligand prediction (ProPred)	http://www.imtech.res.in/raghava/propred/index.html
MHC sequence polymorphism	http://www.anthonynolan.com/HIG/data.html; http://www.ebi.ac.uk/imgt/hla/
Analysis and prediction of antigen processing and presentation	http://www.ihwg.org/components/peptider.htm
SYFPEITHI[a] MHC ligand database and predictive algorithms	http://www.syfpeithi.de/

[a] Lists allele specific epitopes identified by direct mass spectrometric analyses as well as functional immunological studies.

REFERENCES

1. Kuckelkorn, U., Ruppert, T., Strehl, B., Jungblut, P. R., Zimny-Arndt, U., Lamer, S., Prinz, I., Drung, I., Kloetzel, P. M., Kaufmann, S. H., and Steinhoff, U., Link between organ-specific antigen processing by 20S proteasomes and CD8(+) T cell-mediated autoimmunity, *J. Exp. Med.* 195, 983, 2002.
2. Rivett, A. J., Bose, S., Brooks, P., and Broadfoot, K. I., Regulation of proteasome complexes by gamma-interferon and phosphorylation, *Biochimie* 83, 363, 2001.
3. Groettrup, M., Khan, S., Schwarz, K., and Schmidtke, G., Interferon-gamma inducible exchanges of 20S proteasome active site subunits: Why? *Biochimie* 83, 367, 2001.
4. Chen, W., Norbury, C. C., Cho, Y., Yewdell, J. W., and Bennink, J. R., Immunoproteasomes shape immunodominance hierarchies of antiviral CD8(+) T cells at the levels of T cell repertoire and presentation of viral antigens, *J. Exp. Med.* 193, 1319, 2001.
5. Niedermann, G., Geier, E., Lucchiari-Hartz, M., Hitziger, N., Ramsperger, A., and Eichmann, K., The specificity of proteasomes: Impact on MHC class I processing and presentation of antigens, *Immunol. Rev.* 172, 29, 1999.
6. Preckel, T., Fung-Leung, W. P., Cai, Z., Vitiello, A., Salter-Cid, L., Winqvist, O., Wolfe, T. G., Von Herrath, M., Angulo, A., Ghazal, P., Lee, J. D., Fourie, A. M., Wu, Y., Pang, J., Ngo, K., Peterson, P. A., Fruh, K., and Yang, Y., Impaired immunoproteasome assembly and immune responses in PA28-/- mice, *Science* 286, 2162, 1999.
7. Luckey, C. J., King, G. M., Marto, J. A., Venketeswaran, S., Maier, B. F., Crotzer, V. L., Colella, T. A., Shabanowitz, J., Hunt, D. F., and Engelhard, V. H., Proteasomes can either generate or destroy MHC class I epitopes: Evidence for nonproteasomal epitope generation in the cytosol, *J. Immunol.* 161, 112, 1998.
8. Nandi, D., Jiang, H., and Monaco, J. J., Identification of MECL-1 (LMP-10) as the third IFN-g-inducible proteasome subunit, *J. Immunol.* 156, 2361, 1996.
9. Momburg, F., Ortiz-Navarrete, V., Neefjes, J., Goulmy, E., van de Wal, Y., Spits, H., Powis, S. J., Butcher, G. W., Howard, J. C., Walden, P., et al., Proteasome subunits encoded by the major histocompatibility complex are not essential for antigen presentation, *Nature* 360, 174, 1992.
10. Kloetzel, P. M. and Ossendorp, F., Proteasome and peptidase function in MHC-class-I-mediated antigen presentation, *Curr. Opin. Immunol.* 16, 76, 2004.
11. Rock, K. L., York, I. A., and Goldberg, A. L., Post-proteasomal antigen processing for major histocompatibility complex class I presentation, *Nat. Immunol.* 5, 670, 2004.
12. Kloetzel, P. M., Generation of major histocompatibility complex class I antigens: Functional interplay between proteasomes and TPPII, *Nat. Immunol.* 5, 661, 2004.
13. Rojo, S., Garcia, F., Villadangos, J.A., and Lopez de Castro, J.A., Changes in the repertoire of peptides bound to HLA-B27 subtypes and to site-specific mutants inside and outside pocket B, *J. Exp. Med.* 177, 613, 1993.
14. Jardetzky, T. S., Lane, W. S., Robinson, R. A., Madden., D. R., and Wiley, D. C., Identification of self peptides bound to purified HLA-B27, *Nature* 353, 326, 1991.
15. Garcia, F., Marina, A., Albar, J. P., and Lopez de Castro, J. A., HLA-B27 presents a peptide from a polymorphic region of its own molecule with homology to proteins from arthritogenic bacteria, *Tissue Antigens* 49, 23, 1997.
16. Urban, R. G., Chicz, R. M., Lane, W. S., Strominger, J. L., Rehm, A., Kenter, M. J., UytdeHaag, F. G., Ploegh, H., Uchanska-Ziegler, B., and Ziegler, A., A subset of HLA-B27 molecules contains peptides much longer than nonamers, *Proc. Natl. Acad. Sci. U.S.A.* 91, 1534, 1994.

17. Frumento, G. H., Gawinowicz, M. A., Suciu-Foca, N., and Pernis, B., Sequence of a prominent 16-residue self-peptide bound to HLA-B27 in a lymphoblastoid cell line, *Cell. Immunol.* 152, 623, 1993.

18. Macdonald, W. A., Purcell, A. W., Mifsud, N., Ely, L. K., Williams, D. S., Chang, L., Gorman, J. J., Clements, C. S., Kjer-Nielsen, L., Koelle, D. M., Burrows, S. R., Tait, B. D., Holdsworth, R., Brooks, A. G., Lovrecz, G. O., Lu, L., Rossjohn, J., and McCluskey, J., A naturally selected dimorphism within the HLA-B44 supertype alters class I structure, peptide repertoire and T cell recognition, *J. Exp. Med.* 198, 679, 2003.

19. Miles, J. J., Elhassen, D., Borg, N. A., Silins, S. L., Tynan, F. E., Burrows, J. M., Purcell, A. W., Kjer-Nielsen, L., Rossjohn, J., Burrows, S. R., and McCluskey, J., CTL recognition of a bulged viral peptide involves biased TCR selection, *J. Immunol.* 175, 3826, 2005.

20. Tynan, F. E., Borg, N. A., Miles, J. J., Beddoe, T., El-Hassen, D., Silins, S. L., van Zuylen, W. J. M., Purcell, A. W., Kjer-Nielsen, L., McCluskey, J., Burrows, S. R., and Rossjohn, J., High resolution structures of highly bulged viral epitopes bound to major histocompatibility complex class I: Implications for T-cell receptor engagement and T-cell immunodominance, *J. Biol. Chem.* 280, 23900, 2005.

21. Probst-Kepper, M., Stroobant, V., Kridel, R., Gaugler, B., Landry, C., Brasseur, F., Cosyns, J.-P., Weynand, B., Boon, T., and Van den Eynde, B. J., An alternative open reading frame of the human macrophage colony-stimulating factor gene is independently translated and codes for an antigenic peptide of 14 amino acids recognized by tumor-infiltrating CD8 T lymphocytes, *J. Exp. Med.* 193, 1189, 2001.

22. Speir, J. A., Stevens, J., Joly, E., Butcher, G. W., and Wilson, I. A., Two different, highly exposed, bulged structures for an unusually long peptide bound to rat MHC class I RT1-Aa, *Immunity* 14, 81, 2001.

23. Arunachalam, B. and Cresswell, P., Molecular requirements for the interaction of class II major histocompatibility complex molecules and invariant chain with calnexin, *J. Biol. Chem.* 270, 2784, 1995.

24. Germain, R. N., Castellino, F., Han, R., Reis e Sousa, C., Romagnoli, P., Sadegh-Nasseri, S., and Zhong, G. M., Processing and presentation of endocytically acquired protein antigens by MHC class II and class I molecules, *Immunol. Rev.* 151, 5, 1996.

25. Villadangos, J. A., Presentation of antigens by MHC class II molecules: Getting the most out of them, *Mol. Immunol.* 38, 329, 2001.

26. Weenink, S. M. and Gautam, A. M., Antigen presentation by MHC class II molecules, Immunol. *Cell Biol.* 75, 69, 1997.

27. Jensen, P. E., Weber, D. A., Thayer, W. P., Chen, X., and Dao, C. T., HLA-DM and the MHC class II antigen presentation pathway, *Immunol. Res.* 20, 195, 1999.

28. Denzin, L. K. and Cresswell, P., HLA-DM induces CLIP dissociation from MHC class II alpha beta dimers and facilitates peptide loading, *Cell* 82, 155, 1995.

29. Rammensee, H. G., Chemistry of peptides associated with MHC class I and class II molecules, *Curr. Opin. Immunol.* 7, 85, 1995.

30. Falk, K., Rotzschke, O., Stevanovic, S., Jung, G., and Rammensee, H. G., Pool sequencing of natural HLA-DR, DQ, and DP ligands reveals detailed peptide motifs, constraints of processing, and general rules, *Immunogenetics* 39, 230, 1994.

31. Brown, J. H., Jardetzky, T. S., Gorga, J. C., Stern, L. J., Urban, R. G., Strominger, J. L., and Wiley, D. C., Three-dimensional structure of the human class II histocompatibility antigen HLA-DR1, *Nature* 364, 33, 1993.

32. Reinherz, E. L., Tan, K., Tang, L., Kern, P., Liu, J., Xiong, Y., Hussey, R. E., Smolyar, A., Hare, B., Zhang, R., Joachimiak, A., Chang, H. C., Wagner, G., and Wang, J., The crystal structure of a T cell receptor in complex with peptide and MHC class II, *Science* 286, 1913, 1999.

33. Scott, C. A., Peterson, P. A., Teyton, L., and Wilson, I. A., Crystal structures of two I-Ad-peptide complexes reveal that high affinity can be achieved without large anchor residues, *Immunity* 8, 319, 1998.

34. Fremont, D. H., Hendrickson, W. A., Marrack, P., and Kappler, J., Structures of an MHC class II molecule with covalently bound single peptides, *Science* 272, 1001, 1996.

35. Watts, C., Capture and processing of exogenous antigens for presentation on MHC molecules, *Annu. Rev. Immunol.* 15, 821, 1997.

36. Sette, A., Sidney, J., del Guercio, M. F., Southwood, S., Ruppert, J., Dahlberg, C., Grey, H. M., and Kubo, R. T., Peptide binding to the most frequent HLA-A class I alleles measured by quantitative molecular binding assays, *Mol. Immunol.* 31, 813, 1994.

37. Sette, A., Vitiello, A., Reherman, B., Fowler, P., Nayersina, R., Kast, W. M., Melief, C. J. M., Oseroff, C., Yuan, L., Ruppert, J., Sidney, J., Delguercio, M. F., Southwood, S., Kubo, R. T., Chesnut, R. W., Grey, H. M., and Chisari, F. V., The relationship between class I binding affinity and immunogenicity of potential cytotoxic T cell epitopes, *J. Immunol.* 153, 5586, 1994.

38. Crotzer, V. L., Christian, R. E., Brooks, J. M., Shabanowitz, J., Settlage, R. E., Marto, J. A., White, F. M., Rickinson, A. B., Hunt, D. F., and Engelhard, V. H., Immunodominance among EBV-derived epitopes restricted by HLA-B27 does not correlate with epitope abundance in EBV-transformed B-lymphoblastoid cell lines, *J. Immunol.* 164, 6120, 2000.

39. Zijlstra, A., Testa, J. E., and Quigley, J. P., Targeting the proteome/epitome, implementation of subtractive immunization, *Biochem. Biophys. Res. Commun.* 303, 733, 2003.

40. Ballot, E., Bruneel, A., Labas, V., and Johanet, C., Identification of rat targets of antisoluble liver antigen autoantibodies by serologic proteome analysis, *Clin. Chem.* 49, 634, 2003.

41. Bumann, D., Holland, P., Siejak, F., Koesling, J., Sabarth, N., Lamer, S., Zimny-Arndt, U., Jungblut, P. R., and Meyer, T. F., A comparison of murine and human immunoproteomes of *Helicobacter pylori* validates the preclinical murine infection model for antigen screening, *Infect. Immun.* 70, 6494, 2002.

42. Klade, C. S., Proteomics approaches towards antigen discovery and vaccine development, *Curr. Opin. Mol. Ther.* 4, 216, 2002.

43. Haas, G., Karaali, G., Ebermayer, K., Metzger, W. G., Lamer, S., Zimny-Arndt, U., Diescher, S., Goebel, U. B., Vogt, K., Roznowski, A. B., Wiedenmann, B. J., Meyer, T. F., Aebischer, T., and Jungblut, P. R., Immunoproteomics of *Helicobacter pylori* infection and relation to gastric disease, *Proteomics* 2, 313, 2002.

44. Maecker, B., von, B.-B., Anderson, K. S., Vonderheide, R. H., and Schultze, J. L., Linking genomics to immunotherapy by reverse immunology—"immunomics"—in the new millennium, *Curr. Mol. Med.* 1, 609, 2001.

45. Shalhoub, P., Kern, S., Girard, S., and Beretta, L., Proteomic-based approach for the identification of tumor markers associated with hepatocellular carcinoma, *Dis. Markers* 17, 217, 2001.

46. Thery, C., Boussac, M., Veron, P., Ricciardi-Castagnoli, P., Raposo, G., Garin, J., and Amigorena, S., Proteomic analysis of dendritic cell-derived exosomes: A secreted subcellular compartment distinct from apoptotic vesicles, *J. Immunol.* 166, 7309, 2001.

47. Ristori, G., Salvetti, M., Pesole, G., Attimonelli, M., Buttinelli, C., Martin, R., and Riccio, P., Compositional bias and mimicry toward the nonself proteome in immunodominant T cell epitopes of self and nonself antigens, *FASEB J.* 14, 431, 2000.

48. Romer, T. G. and Boyle, M. D., Application of immunoproteomics to analysis of post-translational processing of the antiphagocytic M protein of *Streptococcus*, *Proteomics* 3, 29, 2003.

49. Jungblut, P. R., Proteome analysis of bacterial pathogens, *Microbes Infect.* 3, 831, 2001.

50. Storkus, W. J., Howell, D. N., Salter, R. D., Dawson, J. R., and Cresswell, P., NK susceptibility varies inversely with target cell class I HLA antigen expression, *J. Immunol.* 138, 1657, 1987.

51. Boisgerault, F., Tieng, V., Stolzenberg, M. C., Dulphy, N., Khalil, I., Tamouza, R., Charron, D., and Toubert, A., Differences in endogenous peptides presented by HLA-B*2705 and B*2703 allelic variants. Implications for susceptibility to spondylarthropathies, *J. Clin. Invest.* 98, 2764, 1996.

52. Fruci, D., Butler, R. H., Greco, G., Rovero, P., Pazmany, L., Vigneti, E., Tosi, R., and Tanigaki, N., Differences in peptide-binding specificity of two ankylosing spondylitis-associated HLA-B27 subtypes, *Immunogenetics* 42, 123, 1995.

53. Paradela, A., Garcia-Peydro, M., Vazquez, J., Rognan, D., and Lopez de Castro, J. A., The same natural ligand is involved in allorecognition of multiple HLA- B27 subtypes by a single T cell clone: Role of peptide and the MHC molecule in alloreactivity, *J. Immunol.* 161, 5481, 1998.

54. Skipper, J. C., Kittlesen, D. J., Hendrickson, R. C., Deacon, D. D., Harthun, N. L., Wagner, S. N., Hunt, D. F., Engelhard, V. H., and Slingluff, C. L., Shared epitopes for HLA-A3-restricted melanoma-reactive human CTL include a naturally processed epitope from Pmel-17/gp100, *J. Immunol.* 157, 5027, 1996.

55. Ringrose, J. H., Yard, B. A., Muijsers, A., Boog, C. J., and Feltkamp, T. E., Comparison of peptides eluted from the groove of HLA-B27 from Salmonella infected and noninfected cells, *Clin. Rheumatol.* 15 Suppl 1, 74, 1996.

56. van Els, C. A., Herberts, C. A., van der Heeft, E., Poelen, M. C., van Gaans-van den Brink, J. A., van der Kooi, A., Hoogerhout, P., Jan ten Hove, G., Meiring, H. D., and de Jong, A. P., A single naturally processed measles virus peptide fully dominates the HLA-A*0201-associated peptide display and is mutated at its anchor position in persistent viral strains, *Eur. J. Immunol.* 30, 1172, 2000.

57. Purcell, A. W., The peptide-loading complex and ligand selection during the assembly of HLA class I molecules, *Mol. Immunol.* 37, 483, 2000.

58. Falk, K., Rötzschke, O., Stevanovic, S., Jung, G., and Rammensee, H.-G., Allele-specific motifs revealed by sequencing of self-peptides eluted from MHC molecules, *Nature* 351, 290, 1991.

59. Rammensee, H.-G., Friede, T., and Stevanovic, S., MHC ligands and peptide motifs: First listing, *Immunogenetics* 41, 178, 1995.

60. Andersen, M. H., Bonfill, J. E., Neisig, A., Arsequell, G., Sondergaard, I., Neefjes, J., Zeuthen, J., Elliott, T., and Haurum, J. S., Phosphorylated peptides can be transported by TAP molecules, presented by class I MHC molecules, and recognized by phospho-peptide-specific CTL, *J. Immunol.* 163, 3812, 1999.

61. Chen, Y., Sidney, J., Southwood, S., Cox, A. L., Sakaguchi, K., Henderson, R. A., Appella, E., Hunt, D. F., Sette, A., and Engelhard, V. H., Naturally processed peptides longer than nine amino acid residues bind to the class I MHC molecule HLA-A2.1 with high affinity and in different conformations, *J. Immunol.* 152, 2874, 1994.

62. Chen, W., Ede, N. J., Jackson, D. C., McCluskey, J., and Purcell, A. W., CTL recognition of an altered peptide associated with asparagine bond rearrangement. Implications for immunity and vaccine design, *J. Immunol.* 157, 1000, 1996.

63. Haurum, J. S., Arsequell, G., Lellouch, A. C., Wong, S. Y., Dwek, R. A., McMichael, A. J., and Elliott, T., Recognition of carbohydrate by major histocompatibility complex class I-restricted, glycopeptide-specific cytotoxic T lymphocytes, *J. Exp. Med.* 180, 739, 1994.

64. Kohler, J., Martin, S., Pflugfelder, U., Ruh, H., Vollmer, J., and Weltzien, H. U., Cross-reactive trinitrophenylated peptides as antigens for class II major histocompatibility complex-restricted T cells and inducers of contact sensitivity in mice—limited T cell receptor repertoire, *Eur. J. Immunol.* 25, 92, 1995.

65. Martin, S., Ruh, H., Hebbelmann, S., Pflugfelder, U., Rude, B., and Weltzien, H. U., Carrier-reactive hapten-specific cytotoxic T lymphocyte clones originate from a highly preselected T cell repertoire: Implications for chemical-induced self-reactivity, *Eur. J. Immunol.* 25, 2788, 1995.

66. Moulon, C., Vollmer, J., and Weltzien, H. U., Characterization of processing requirements and metal cross-reactivities in T cell clones from patients with allergic contact dermatitis to nickel, *Eur. J. Immunol.* 25, 3308, 1995.

67. Purcell, A. W., Chen, W., Ede, N. J., Gorman, J. J., Fecondo, J. V., Jackson, D. C., Zhao, Y., and McCluskey, J., Avoidance of self-reactivity results in skewed CTL responses to rare components of synthetic immunogens, *J. Immunol.* 160, 1085, 1998.

68. Skipper, J. C. A., Hendrickson, R. C., Gulden, P. H., Brichard, V., Vanpel, A., Chen, Y., Shabanowitz, J., Wolfel, T., Slingluff, C. L., Boon, T., Hunt, D. F., and Engelhard, V. H., An HLA-A2-restricted tyrosinase antigen on melanoma cells results from post-translational modification and suggests a novel pathway for processing of membrane proteins, *J. Exp. Med.* 183, 527, 1996.

69. Eisenlohr, L. C., Yewdell, J. W., and Bennink, J. R., Flanking sequences influence the presentation of an endogenously synthesized peptide to cytotoxic T lymphocytes, *J. Exp. Med.* 175, 481, 1992.

70. Yewdell, J. W. and Bennink, J. R., Immunodominance in major histocompatibility complex class I-restricted T lymphocyte responses, *Annu. Rev. Immunol.* 17, 51, 1999.

71. van-der-Burg, S. H., Klein, M. R., van-de-Velde, C. J., Kast, W. M., Miedema, F., and Melief, C. J., Induction of a primary human cytotoxic T-lymphocyte response against a novel conserved epitope in a functional sequence of HIV-1 reverse transcriptase, *AIDS* 9, 121, 1995.

72. Feller, D. C. and de la Cruz, V. F., Identifying antigenic T-cell sites, *Nature* 349, 720, 1991.

73. Andersen, M. H., Tan, L., Sondergaard, I., Zeuthen, J., Elliott, T., and Haurum, J. S., Poor correspondence between predicted and experimental binding of peptides to class I MHC molecules, *Tissue Antigens* 55, 519, 2000.

74. Schirle, M., Weinschenk, T., and Stevanovic, S., Combining computer algorithms with experimental approaches permits the rapid and accurate identification of T cell epitopes from defined antigens, *J. Immunol. Methods* 257, 1, 2001.

75. Verginis, P., Stanford, M. M., and Carayanniotis, G., Delineation of five thyroglobulin T cell epitopes with pathogenic potential in experimental autoimmune thyroiditis, *J. Immunol.* 169, 5332, 2002.

76. Lu, J. and Celis, E., Use of two predictive algorithms of the World Wide Web for the identification of tumor-reactive T-cell epitopes, *Cancer Res.* 60, 5223, 2000.

77. Schafer, J. R., Jesdale, B. M., George, J. A., Kouttab, N. M., and De Groot, A. S., Prediction of well-conserved HIV-1 ligands using a matrix-based algorithm, EpiMatrix, *Vaccine* 16, 1880, 1998.

78. Brusic, V., Rudy, G., Honeyman, G., Hammer, J., and Harrison, L., Prediction of MHC class II-binding peptides using an evolutionary algorithm and artificial neural network, *Bioinformatics* 14, 121, 1998.

79. Roberts, C. G., Meister, G. E., Jesdale, B. M., Lieberman, J., Berzofsky, J. A., and De Groot, A. S., Prediction of HIV peptide epitopes by a novel algorithm, *AIDS Res. Hum. Retroviruses* 12, 593, 1996.

80. Meister, G. E., Roberts, C. G., Berzofsky, J. A., and De Groot, A. S., Two novel T cell epitope prediction algorithms based on MHC-binding motifs; comparison of predicted and published epitopes from *Mycobacterium tuberculosis* and HIV protein sequences, *Vaccine* 13, 581, 1995.

81. Chang, L., Kjer-Nielsen, L., Flynn, S., Brooks, A. G., Mannering, S. I., Honeyman, M. C., Harrison, L. C., McCluskey, J., and Purcell, A. W., Novel strategy for identification of candidate cytotoxic T-cell epitopes from human preproinsulin, *Tissue Antigens* 62, 408, 2003.

82. Falk, K., Rötzschke, O., Deres, K., Metzger, J., Jung, G., and Rammensee, H.-G., Identification of naturally processed viral nonapeptides allows their quantification in infected cells and suggests an allele-specific T cell epitope forecast, *J. Exp. Med.* 174, 425, 1991.

83. Sijts, A. J., Neisig, A., Neefjes, J., and Pamer, E. G., Two *Listeria monocytogenes* CTL epitopes are processed from the same antigen with different efficiencies, *J. Immunol.* 156, 683, 1996.

84. Rotzschke, O., Falk, K., Deres, K., Schild, H., Norda, M., Metzger, J., Jung, G., and Rammensee, H. G., Isolation and analysis of naturally processed viral peptides as recognized by cytotoxic T cells, *Nature* 348, 252, 1990.

85. Storkus, W. J., Zeh, H. J., Salter, R. D., and Lotze, M. T., Identification of T-cell epitopes: Rapid isolation of class I-presented peptides from viable cells by mild acid elution, *J. Immunother.* 14, 94, 1993.

86. Storkus, W. J., Zeh, H. J., Maeurer, M. J., Salter, R. D., and Lotze, M. T., Identification of human melanoma peptides recognized by class I restricted tumor infiltrating T lymphocytes, *J. Immunol.* 151, 3719, 1993.

87. Rammensee, H. G., Falk, K., and Rotzschke, O., Peptides naturally presented by MHC class I molecules, *Annu. Rev. Immunol.* 11, 213, 1993.

88. Gregers, T. F., Fleckenstein, B., Vartdal, F., Roepstorff, P., Bakke, O., and Sandlie, I., MHC class II loading of high or low affinity peptides directed by Ii/peptide fusion constructs: Implications for T cell activation, *Int. Immunol.* 15, 1291, 2003.

89. Herberts, C. A., Stittelaar, K. J., van der Heeft, E., van Gaans-Van den Brink, J., Poelen, M. C., Roholl, P. J., van Alphen, L. J., Melief, C. J., de Jong, A. P., and van Els, C. A., A measles virus glycoprotein-derived human CTL epitope is abundantly presented via the proteasomal-dependent MHC class I processing pathway, *J. Gen. Virol.* 82, 2131, 2001.

90. Herberts, C. A., van Gaans-van den Brink, J., van der Heeft, E., van Wijk, M., Hoekman, J., Jaye, A., Poelen, M. C., Boog, C. J., Roholl, P. J., Whittle, H., de Jong, A. P., and van Els, C. C., Autoreactivity against induced or upregulated abundant self-peptides in HLA-A*0201 following measles virus infection, *Hum. Immunol.* 64, 44, 2003.

91. Luckey, C. J., Marto, J. A., Partridge, M., Hall, E., White, F. M., Lippolis, J. D., Shabanowitz, J., Hunt, D. F., and Engelhard, V. H., Differences in the expression of human class I MHC alleles and their associated peptides in the presence of proteasome inhibitors, *J. Immunol.* 167, 1212, 2001.

92. Hickman, H. D., Luis, A. D., Bardet, W., Buchli, R., Battson, C. L., Shearer, M. H., Jackson, K. W., Kennedy, R. C., and Hildebrand, W. H., Cutting edge: Class I presentation of host peptides following HIV infection, *J. Immunol.* 171, 22, 2003.

93. Spengler, B., Kirsch, D., Kaufmann, R., and Jaeger, E., Peptide sequencing by matrix-assisted laser-desorption mass spectrometry, *Rapid Commun. Mass Spectrom.* 6, 105, 1993.

94. Yates, J. R., Mass spectrometry. From genomics to proteomics, *Trends Genet.* 16, 5, 2000.

95. van der Heeft, E., ten Hove, G. J., Herberts, C. A., Meiring, H. D., van Els, C. A., and de Jong, A. P., A microcapillary column switching HPLC-electrospray ionization MS system for the direct identification of peptides presented by major histocompatibility complex class I molecules, *Anal. Chem.* 70, 3742, 1998.

96. Cox, A. L., Skipper, J., Chen, Y., Henderson, R. A., Darrow, T. L., Shabanowitz, J., Engelhard, V. H., Hunt, D. F., and Slingluff, C. L., Identification of a peptide recognized by five melanoma-specific human cytotoxic T cell lines, *Science* 264, 716, 1994.

97. Henderson, R. A., Michel, H., Sakaguchi, K., Shabanowitz, J., Appella, E., Hunt, D. F., and Engelhard, V. H., HLA-A2.1-associated peptides from a mutant cell line: A second pathway of antigen presentation, *Science* 255, 1264, 1992.

98. Hunt, D. F., Henderson, R. A., Shabanowitz, J., Sacaguchi, K., Michel, H., Sevilir, N., Cox, A. L., Appella, E., and Engelhard, V. H., Characterization of peptides bound to the class I MHC molecule HLA-A2.1 by mass spectrometry, *Science* 255, 1261, 1992.

99. Hunt, D. F., Michel, H., Dickinson, T. A., Shabanowitz, J., Cox, A. L., Sakaguchi, K., Appella, E., Grey, H. M., and Sette, A., Peptides presented to the immune system by the murine class II major histocompatibility complex molecule I-Ad, *Science* 256, 1817, 1992.

100. Sette, A., Ceman, S., Kubo, R. T., Sakaguchi, K., Appella, E., Hunt, D. F., Davis, T. A., Michel, H., Shabanowitz, J., Rudersdorf, R., et al., Invariant chain peptides in most HLA-DR molecules of an antigen-processing mutant, *Science* 258, 1801, 1992.

101. Henderson, R. A., Cox, A. L., Sakaguchi, K., Appella, E., Shabanowitz, J., Hunt, D. F., and Engelhard, V. H., Direct identification of an endogenous peptide recognized by multiple HLA-A2.1-specific cytotoxic T cells, *Proc. Natl. Acad. Sci. U. S. A.* 90, 10275, 1993.

102. Huczko, E. L., Bodnar, W. M., Benjamin, D., Sakaguchi, K., Zhu, N. Z., Shabanowitz, J., Henderson, R. A., Appella, E., Hunt, D. F., and Engelhard, V. H., Characteristics of endogenous peptides eluted from the class I MHC molecule HLA-B7 determined by mass spectrometry and computer modeling, *J. Immunol.* 151, 2572, 1993.

103. Sette, A., DeMars, R., Grey, H. M., Oseroff, C., Southwood, S., Appella, E., Kubo, R. T., and Hunt, D. F., Isolation and characterization of naturally processed peptides bound by class II molecules and peptides presented by normal and mutant antigen-presenting cells, *Chem. Immunol.* 57, 152, 1993.

104. Slingluff, C. L., Jr., Cox, A. L., Henderson, R. A., Hunt, D. F., and Engelhard, V. H., Recognition of human melanoma cells by HLA-A2.1-restricted cytotoxic T lymphocytes is mediated by at least six shared peptide epitopes, *J. Immunol.* 150, 2955, 1993.

105. Slingluff, C. L., Jr., Hunt, D. F., and Engelhard, V. H., Direct analysis of tumor-associated peptide antigens, *Curr. Opin. Immunol.* 6, 733, 1994.

106. Slingluff, C. L., Jr., Cox, A. L., Stover, J. M., Jr., Moore, M. M., Hunt, D. F., and Engelhard, V. H., Cytotoxic T-lymphocyte response to autologous human squamous cell cancer of the lung: Epitope reconstitution with peptides extracted from HLA-Aw68, *Cancer Res.* 54, 2731, 1994.

107. Appella, E., Padlan, E. A., and Hunt, D. F., Analysis of the structure of naturally processed peptides bound by class I and class II major histocompatibility complex molecules, *EXS* 73, 105, 1995.

108. den Haan, J. M., Sherman, N. E., Blokland, E., Huczko, E., Koning, F., Drijfhout, J. W., Skipper, J., Shabanowitz, J., Hunt, D. F., Engelhard, V. H., et al., Identification of a graft versus host disease-associated human minor histocompatibility antigen, *Science* 268, 1476, 1995.

109. den Haan, J. M., Bontrop, R. E., Pool, J., Sherman, N., Blokland, E., Engelhard, V. H., Hunt, D. F., and Goulmy, E., Conservation of minor histocompatibility antigens between human and nonhuman primates, *Eur. J. Immunol.* 26, 2680, 1996.

110. Fiorillo, M. T., Meadows, L., D'Amato, M., Shabanowitz, J., Hunt, D. F., Appella, E., and Sorrentino, R., Susceptibility to ankylosing spondylitis correlates with the C-terminal residue of peptides presented by various HLA-B27 subtypes, *Eur. J. Immunol.* 27, 368, 1997.

111. Hu, Q., Bazemore Walker, C. R., Girao, C., Opferman, J. T., Sun, J., Shabanowitz, J., Hunt, D. F., and Ashton-Rickardt, P. G., Specific recognition of thymic self-peptides induces the positive selection of cytotoxic T lymphocytes, *Immunity* 7, 221, 1997.

112. Meadows, L., Wang, W., den Haan, J. M., Blokland, E., Reinhardus, C., Drijfhout, J. W., Shabanowitz, J., Pierce, R., Agulnik, A. I., Bishop, C. E., Hunt, D. F., Goulmy, E., and Engelhard, V. H., The HLA-A*0201-restricted H-Y antigen contains a post-translationally modified cysteine that significantly affects T cell recognition, *Immunity* 6, 273, 1997.

113. Kittlesen, D. J., Thompson, L. W., Gulden, P. H., Skipper, J. C., Colella, T. A., Shabanowitz, J., Hunt, D. F., Engelhard, V. H., Slingluff, C. L., Jr., and Shabanowitz, J. A., Human melanoma patients recognize an HLA-A1-restricted CTL epitope from tyrosinase containing two cysteine residues: Implications for tumor vaccine development, *J. Immunol.* 160, 2099, 1998.

114. Skipper, J. C., Gulden, P. H., Hendrickson, R. C., Harthun, N., Caldwell, J. A., Shabanowitz, J., Engelhard, V. H., Hunt, D. F., and Slingluff, C. L., Jr., Mass-spectrometric evaluation of HLA-A*0201-associated peptides identifies dominant naturally processed forms of CTL epitopes from MART-1 and gp100, *Int. J. Cancer* 82, 669, 1999.

115. Zarling, A. L., Ficarro, S. B., White, F. M., Shabanowitz, J., Hunt, D. F., and Engelhard, V. H., Phosphorylated peptides are naturally processed and presented by major histocompatibility complex class I molecules *in vivo*, *J. Exp. Med.* 192, 1755, 2000.

116. Guimezanes, A., Barrett-Wilt, G. A., Gulden-Thompson, P., Shabanowitz, J., Engelhard, V. H., Hunt, D. F., Schmitt-Verhulst, A. M., and Engelhardt, V. H., Identification of endogenous peptides recognized by in vivo or in vitro generated alloreactive cytotoxic T lymphocytes: Distinct characteristics correlated with CD8 dependence, *Eur. J. Immunol.* 31, 421, 2001.

117. Nepom, G. T., Lippolis, J. D., White, F. M., Masewicz, S., Marto, J. A., Herman, A., Luckey, C. J., Falk, B., Shabanowitz, J., Hunt, D. F., Engelhard, V. H., and Nepom, B. S., Identification and modulation of a naturally processed T cell epitope from the diabetes-associated autoantigen human glutamic acid decarboxylase 65 (hGAD65), *Proc. Natl. Acad. Sci. U.S.A.* 98, 1763, 2001.

118. Pierce, R. A., Field, E. D., Mutis, T., Golovina, T. N., Von Kap-Herr, C., Wilke, M., Pool, J., Shabanowitz, J., Pettenati, M. J., Eisenlohr, L. C., Hunt, D. F., Goulmy, E., and Engelhard, V. H., The HA-2 minor histocompatibility antigen is derived from a diallelic gene encoding a novel human class I myosin protein, *J. Immunol.* 167, 3223, 2001.

119. Lippolis, J. D., White, F. M., Marto, J. A., Luckey, C. J., Bullock, T. N., Shabanowitz, J., Hunt, D. F., and Engelhard, V. H., Analysis of MHC class II antigen processing by quantitation of peptides that constitute nested sets, *J. Immunol.* 169, 5089, 2002.

120. Seamons, A., Sutton, J., Bai, D., Baird, E., Bonn, N., Kafsack, B. F., Shabanowitz, J., Hunt, D. F., Beeson, C., and Goverman, J., Competition between two MHC binding registers in a single peptide processed from myelin basic protein influences tolerance and susceptibility to autoimmunity, *J. Exp. Med.* 197, 1391, 2003.

121. Matsuda, J. L. and Kronenberg, M., Presentation of self and microbial lipids by CD1 molecules, *Curr. Opin. Immunol.* 13, 19, 2001.

122. Brigl, M. and Brenner, M. B., CD1: Antigen presentation and T cell function, *Annu. Rev. Immunol.* 22, 817, 2004.

123. Moody, D. and Porcelli, S. A., Intracellular pathways of CD1 antigen presentation, *Nat. Rev. Immunol.* 3, 11, 2003.

124. Lawton, A. P. and Kronenberg, M., The third way: Progress on pathways of antigen processing and presentation by CD1, Immunol. *Cell Biol.* 82, 295, 2004.

125. Zajonc, D. M., Elsliger, M. A., Teyton, L., and Wilson, I. A., Crystal structure of CD1a in complex with a sulfatide self antigen at a resolution of 2.15 angstroms, *Nat. Immunol.* 4, 808, 2003.

126. Gadola, S. D., Zaccai, N. R., Harlos, K., Shepherd, D., Castro-Palomino, J. C., Ritter, G., Schmidt, R. R., Jones, E. Y., and Cerundolo, V., Structure of human CD1b with bound ligands at 2.3 angstroms, a maze for alkyl chains, *Nat. Immunol.* 3, 721, 2002.

127. Zhou, D., Mattner, J., Cantu, C., III, Schrantz, N., Yin, N., Gao, Y., Sagiv, Y., Hudspeth, K., Wu, Y.-P., Yamashita, T., Teneberg, S., Wang, D., Proia, R. L., Levery, S. B., Savage, P. B., Teyton, L., and Bendelac, A., Lysosomal glycosphingo-lipid recognition by NKT cells, *Science* 306, 1786, 2004.

128. Macdonald, W., Williams, D. S., Clements, C. S., Gorman, J. J., Kjer-Nielsen, L., Brooks, A. G., McCluskey, J., Rossjohn, J., and Purcell, A. W., Identification of a dominant self-ligand bound to three HLA B44 alleles and the preliminary crystallo-graphic analysis of recombinant forms of each complex, *FEBS Lett.* 27, 2002.

129. Hughes, E. A. and Cresswell, P., The thiol oxidoreductase ERp57 is a component of the MHC class I peptide-loading complex, *Curr. Biol.* 8, 709, 1998.

130. Lindquist, J. A., Jensen, O. N., Mann, M., and Hammerling, G. J., ER-60, a chaperone with thiol-dependent reductase activity involved in MHC class I assembly, *EMBO J.* 17, 2186, 1998.

131. Dick, T. P. and Cresswell, P., Thiol oxidation and reduction in major histocompatibility complex class I-restricted antigen processing and presentation, *Methods Enzymol.* 348, 49, 2002.

132. Dick, T. P., Bangia, N., Peaper, D. R., and Cresswell, P., Disulfide bond isomerization and the assembly of MHC class I-peptide complexes, *Immunity* 16, 87, 2002.

133. Farmery, M. R., Allen, S., Allen, A. J., and Bulleid, N. J., The role of ERp57 in disulfide bond formation during the assembly of major histocompatibility complex class I in a synchronized semipermeabilized cell translation system, *J. Biol. Chem.* 275, 14933, 2000.

134. Grandea, A. G., Lehner, P. J., Cresswell, P., and Spies, T., Regulation of MHC class I heterodimer stability and interaction with TAP by tapasin, *Immunogenetics* 46, 477, 1997.

135. Sadasivan, B., Lehner, P. J., Ortmann, B., Spies, T., and Cresswell, P., Roles for calreticulin and a novel glycoprotein, tapasin, in the interaction of MHC class I molecules with TAP, *Immunity* 5, 103, 1996.

136. Brocke, P., Garbi, N., Momburg, F., and Hammerling, G. J., HLA-DM, HLA-DO and tapasin: Functional similarities and differences, *Curr. Opin. Immunol.* 14, 22, 2002.

137. Fremont, D. H., Monnaie, D., Nelson, C. A., Hendrickson, W. A., and Unanue, E. R., Crystal structure of I-Ak in complex with a dominant epitope of lysozyme, *Immunity* 8, 305, 1998.

138. Purcell, A. W., Gorman, J. J., Garcia-Peydro, M., Paradela, A., Burrows, S. R., Talbo, G. H., Laham, N., Peh, C. A., Reynolds, E. C., Lopez De Castro, J. A., and McCluskey, J., Quantitative and qualitative influences of tapasin on the class I peptide repertoire, *J. Immunol.* 166, 1016, 2001.

139. Purcell, A. W. and Gorman, J. J., The use of post-source decay in matrix-assisted laser desorption/ionization mass spectrometry to delineate T cell determinants, *J. Immunol. Methods* 249, 17, 2001.

10 Near-Infrared Fluorescence Detection of Antigen–Antibody Interactions on Microarrays

Vehary Sakanyan and Garabet Yeretssian

CONTENTS

10.1 INTRODUCTION

The remarkable progress in genome sequencing projects has shifted modern biology interests from virtual bioinformatic descriptions of genomes to a postgenomic focus on complex proteomes. To investigate protein behavior from the broad perspective of understanding the complexity of whole organisms, this fundamental transition requires new concepts to replace individual protein studies by high-throughput assays. Different methods, such as two-dimensional gel electrophoresis, yeast two-hybrid assays, mass spectrometry, and protein arrays, have been successfully exploited to analyze prokaryotic and eukaryotic proteomes. However, protein array technology has the advantages of being able to detect interactions directly with proteinaceous and nonproteinaceous molecules (including nucleic acids, carbohydrates, lipids, and small ligands) and to monitor various parameters, like relative

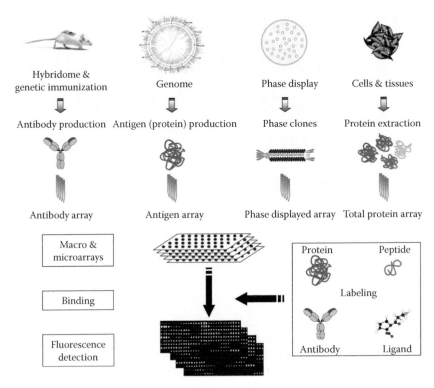

FIGURE 10.1 (See color insert.) Fluorescence detection of molecular interactions with high-throughput protein microarrays. A variety of antibody and antigen arrays and four main types of molecules employed for binding assays are shown. Both primary detection (labeled molecules) and secondary detection (labeled secondary binders) can be used to detect bound molecules on spots.

concentration, binding affinity, and post-translational modification of proteins. It thus provides the unique possibility of analyzing proteomes and developing various applications for biomedical purposes.

Protein arrays are ordered arrangements of protein samples on a solid support. Their utility lies in their scalability, flexibility, and suitability for performing high-throughput and multiplexed molecular interaction assays (fig. 10.1). A minute spot area with immobilized biomolecules on planar supports provides greater sensitivity for the detection of molecular interactions compared to other binding assays [1]. Another advantage of protein arrays is the possibility of diversifying their formats by using a variety of analytes in solution or immobilized as individually purified proteins or nonpurified cell extracts. An arbitrary definition of macroarrays and microarrays reflects the size rather than the number of spots (assay positions) on the array surface. The rapid appearance of protein array technology in the postgenomic field has been stimulated by the opportunity of using much of the equipment and many of the products developed for DNA arrays, such as robotic printers and immobilization surfaces, fluorescent dyes for probe labeling and fluorescence scanners, and data acquisition and data mining tools.

The performance of protein arrays depends greatly on detection systems. A variety of colorimetry, radioactivity, enzyme-catalyzed chemiluminescence- and fluorescence-based methods have been employed to visualize bound molecules in an array format. Colorimetric approaches are not adapted or sensitive enough to detect spots (assay positions) in high-throughput. Similarly, the low resolution of radioactive spots (less than 300 μm) as well as requirements in terms of staff protection, limits the use of radioactive detection. Chemiluminescence has the potential to amplify signal in an enzyme-catalyzed manner; however, the short duration of the signal and the laborious amplification steps reduce its utility. At present, state-of-the-art detection is based largely on fluorescence. Indeed, fluorescent signal is stable, and a large number of fluorescent dyes are commercially available. Detection can be carried out with a wide range of wavelengths by high-resolution scanners originally designed for DNA arrays. Furthermore, fluorescence scanners, which provide excitation and registration of emitted photons within a near-infrared (NIR) wavelength range, open up new perspectives to improve the performance of protein arrays.

Antibodies are relatively stable molecules that conserve an overall topology irrespective of the structure of captured antigenic determinants. They are therefore considered effective binders for protein array research [2]. In early studies, the feasibility of antibody microarray-based immunoassays to detect target proteins had already been demonstrated using competitive and noncompetitive approaches [3,4]. Antibody arrays enable extension of the multiplexed strategy of conventional immunoassays for molecular diagnostics and development of new high-throughput binding approaches to detect protein abundance in biological solutions. A logical consequence in this direction is the development of specific antigen arrays to search for disease-specific prognostic immune-response markers and to find efficient vaccines against different infections. The fabrication of antibody and antigen microarrays requires high-throughput production of the corresponding proteins. The current status of this field is discussed elsewhere [5,6].

The scope of this chapter is to present some of the major technical developments in protein array fabrication and NIR fluorescent detection and to emphasize recent achievements in antibody- and antigen-based array studies as a prerequisite to creating sensitive tools for basic and applied science and to increasing the information content of large-scale proteomic projects.

10.2 PROTEIN IMMOBILIZATION ON SOLID SUPPORTS

The fabrication of protein and antibody arrays is routinely based on the manual or robotic printing of samples onto a solid phase. The physicochemical parameters of the surface are important to ensure a high sensitivity in binding assays. Choice of surface is determined by the need for a low background, reproducibility of detected fluorescent signals, binding capacity, and protein structure preservation. A wide variety of materials have been developed and tested to immobilize proteins, and a comprehensive list of supports and chemical approaches to functionalize various surfaces is described elsewhere [7].

A simple way to anchor protein molecules is by physical adsorption by noncovalent hydrogen bonding or electrostatic, hydrophobic, and van der Waals interactions with flat or porous supports. The traditional glass slide coated with polylysine, widely used for the generation of DNA arrays, has been found to be suitable for protein and antibody arrays [8]. Noncovalent but high-affinity attachment has also been achieved by printing biotinylated proteins on streptavidin-coated surfaces [9,10].

Amino acids or post-translationally attached carbohydrate moieties in proteins possess the reactivity to bind, via covalent linkages of various residues, to chemically treated solid supports—in most cases, glass slides [11]. However, the random attachment of protein chains affects their conformation and orientation unpredictably, thereby decreasing the binding specificity and limiting the usefulness of proteins. In contrast, the oriented immobilization of proteins onto a flat surface increases the density of exposed molecules available to interact with partners, thereby improving the fluorescent intensity of the arrayed spots. A common way to generate arrays with a high density of immobilized recombinant proteins is to introduce affinity tag sequences at the N- or C-terminal ends and print purified tagged molecules on the affinity surface. This approach was used to orient His- or GST-tagged proteins on nickel-coated [12] or glutathione-functionalized [13] slides, respectively. It also allows the standardization of an experimental device by controlling the quantity of the printed proteins using indirect immunofluorescence with labeled antitag antibodies.

Good performance in the oriented immobilization of antibodies can be achieved with cross-linkers able to react with the amine- or thiol-groups of these proteins [14]. The highest sensitivity in antigen capture was observed with cross-linker-mediated binding of antibodies to a glass surface derivatized by epoxysilane. Alternatively, antibodies can be anchored to the solid support in a controlled way through reactive carbohydrate linked to particular amino acid residues in heavy and light protein chains [15,16]. A quantitative comparison of the density of differentially biotinylated immunoglobulin molecules immobilized in spots revealed that oriented immobilization of a full-length IgG or its Fab fragment onto a streptavidin-covered slide provides a 5- to 10-fold stronger signal due to the higher number of captured antigens compared with nonoriented immobilization [17].

Nevertheless, protein array technology seems to be moving from impermeable glass or silicon surfaces to other supports that assure better protein stability. Indeed, slides covered by a thin layer of gel or membrane have become attractive alternatives to functionalized flat surfaces. A film of gel composed of up to 95% water has a low threshold of autofluorescence and confers perfect three-dimensional orientation to be accessible to interaction partners, thereby increasing the sensitivity of detection [18]. However, the differences in diffusion rate of proteins into the contacting gel surface, related to molecular mass, solubility, and other properties, appears to limit the usefulness of the hydrogel-based arrays.

In contrast, nitrocellulose membranes lack this disadvantage and, due to ready availability, ease of use, and low background in chemiluminescence and fluorescence detection systems, show great utility for protein array-based analysis [19–22]. But, this support is relatively fragile, which limits numerous printings of weak protein solutions on the same spot. However, this disadvantage is largely compensated by the porous structure of nitrocellulose membranes because almost 200:1 ratio of the within-pore surface area to the contacting top surface area provides a much higher binding capacity for proteins [23]. Furthermore, pores form an appropriate microenvironment in which protein topology may be better maintained than on flat solid surfaces.

Nitrocellulose is also a convenient support to fabricate total protein arrays by printing cell membrane fractions or crude lysates [20,24–26] and has been largely employed to develop "reverse phase" arrays [27]. Recently, the availability of multi-pad formats (www.whatman.com/bioscience) has enlarged the experimental use of nitrocellulose membranes to include simultaneous binding assays with different probes. Furthermore, nitrocellulose and especially its new derivative, known as Fast-slide [28], provide a greater signal-to-noise ratio within the near-infrared wavelength range of fluorescence. This is essential in terms of the sensitivity and reproducibility of protein microarray assays.

10.3 NIR FLUORESCENCE DETECTION

Fluorescence labeling and detection has revolutionized many domains of biological sciences. Fluorescence detection enables the measurement of labeled molecules directly of compounds bound to labeled probes indirectly, with high sensitivity. It also allows real-time processes in cells and organisms to be followed. A wide variety of fluorescent dyes is available, among which amine-reactive fluorescent dyes are the most convenient for labeling proteins and for use as probes in array studies. Commonly used amine-reactive reagents such as succinimidyl esters (NHS ester) are preferred for bioconjugation to proteins because they form a stable amide bond between the dye and the protein. They react with nonprotonated aliphatic amine groups, including the N-terminal amine and the epsilon-amino group of lysine residues in proteins. Amine-reactive probes like cyanine derivatives are soluble in water and emit between 500 and 670 nm, a popular visible fluorescence range for most scanners designed for DNA arrays and more recently employed for protein arrays.

Fluorescence assays possess a critical sensitivity threshold for detecting a signal from arrayed spots that is largely determined by the number of captured labeled molecules in the miniaturized area. Hence, a general way to improve the sensitivity, as mentioned previously, is to use oriented immobilization of proteins on flat surfaces or three-dimensional supports to increase the density of molecules. The same goal can be achieved by signal amplification. Tyramide amplification has enhanced signal almost 50-fold via an enzyme-catalyzed activation of the bound molecules (www.probes.com). This enabled the detection of low-abundance IgE immunoglobulins raised against some allergens [29] and of variations in protein expression in immobilized tumor tissues [24]. The other method, referred to as rolling circle amplification, uses a small, circular, single-stranded DNA as a primer to hybridize a complementary oligonucleotide conjugated to an antibody of interest and to direct the formation of long concatamers by the polymerase chain reaction [30]. The fluorescent signal was amplified up to 1000-fold, sufficient to detect zeptomolar quantities of target proteins [31,32].

However, the intrinsic fluorescence from many biological molecules is high when visible light-absorbing fluorophores are used. In contrast, the near-infrared region of the spectrum provides a low noise response from various compounds, including polyaromatic hydrocarbons, plastic microplates, membrane supports, and many bio-

TABLE 10.1
NIR Fluorescent Dyes Employed for Protein Microarrays

Dye	Functionality	Relative Molecular Mass	Absorption (nm)	Emission (nm)	Molar extinction coefficient (M^{-1} cm^{-1})	Source
IRDye700	Amidite	981.15	681	712	170,000	LICOR
AlexaFluor680	NHS	1200	679	702	184,000	Molecular Probes
Cy5.5	NHS	1311.58	675	694	250,000	Amersham Biosciences
IRDye800	Amidite	999.3	787	812	275,000	LICOR
IRDye800CW	NHS	1166	774	789	240,000	LICOR
IRDye38	NCS	1067	778	806	179,000	LICOR

molecules [4]. Therefore, NIR fluorescence imaging has had an important impact on biomedical science by improving the visualization of *in vitro* and *in vivo* reactions with proteins and nucleic acids and even distinguishing normal and pathological processes in organisms.

LI-COR developed a two-diode, laser-based Odyssey infrared imaging system with exciting fluorescence emissions at 685 and 785 nm, allowing the detection of molecules labeled with high quantum infrared dyes (IRDyes) with high sensitivity on membrane supports (www.licor.com). The estimated sensitivity of IRDye-labeled antibodies was found to be in the very low picogram range on PVDF membrane western blots with a fluorophore detection limit of 10^{-18} mol [33]. The imaging system makes it easy to detect small changes in fluorescence signal and to follow, in particular, receptor internalization with antibodies directed against extracellular epitopes of G protein-coupled receptors [34].

We compared the sensitivity of chemiluminescence, visible, and NIR fluorescence for their ability to detect proteins in miniaturized spots on porous membranes [22,25]. A greater signal-to-noise ratio was observed with IRDye-labeled proteins, which provided the highest sensitivity and ability to detect attomole quantities of targets within almost a 4 log order linear concentration range of proteins arrayed as 125 μm diameter spots. Thus, it was possible to quantify weak and strong fluorescent signals on the same nitrocellulose membrane with an Odyssey imager. A list of IRDyes used in our protein array studies that provided a strong signal without amplification is shown in table 10.1. The high sensitivity, good linearity, and long stability of NIR signals provided repeatable and accurate binding measurements—crucial parameters for a microarray assay.

10.4 PROTEIN PROFILING WITH ANTIBODY ARRAYS

The measurement of gene expression in cells is of great interest to elucidate networks of protein/ligand/nucleic acid relationships in diseased and normal organisms. But DNA microarray-based assessment of mRNA does not necessarily reflect the abundance of the corresponding proteins in cells. Moreover, it does not give valuable information about post-translational modifications, which govern many cellular processes [35,36]. Therefore, alternative protein array-based approaches can provide not only data complementary to DNA microarrays but also, in some cases, precise and unique information about the functional state of proteins under normal and pathological conditions [37,38]. In this context, monoclonal antibodies (mAbs) are effective tools to capture and detect target proteins. Their usefulness for measuring protein levels on microarrays was mathematically formulated a decade ago [39]. Two major strategies based on the competition principle were proposed for high-throughput profiling of proteins on antibody microarrays, which use incubation with two analytes randomly labeled for further fluorescence detection of captured proteins.

In the two-color method, protein samples are labeled with two fluorophores, such as Cy3 and Cy5 (or Alexa Fluor680 and IRdye800CW in our studies), then mixed together in a 1:1 ratio and incubated with an antibody array (fig. 10.2). The balance of colors in spots indicates which of the two protein samples is present at the higher and at the lower concentrations [8,40]. A relative difference in protein abundance is measured by comparing the ratio of the fluorescence intensity of sample and reference protein spots. The differential expression of proteins is evaluated by taking into account normalized ratios within a statistical interval, above or below which proteins are considered up- or down-regulated, respectively. This method reports only on the relative protein concentration in samples without quantification of target proteins and appears to be less suitable for the detection of low-abundance proteins.

Several laboratories have successfully proven the power of antibody microarrays by employing the two-color method to study protein profiling and post-translational modifications [40–48]. Sreekumar and colleagues discovered novel radiation-regulated proteins by profiling proteins in LoVo colon carcinoma cells after treatment with ionizing radiation with arrays made up of 146 antibodies [40]. In order to decrease the effect of the bioconjugation bias towards particular proteins, oppositely labeled sample and reference protein extracts can be mixed and the normalized ratio values from two parallel experiments can provide more reliable information on profiling (see fig. 10.2A). In this way, differential expression was monitored by comparing deregulated pathways in hepatocellular carcinomas and normal liver protein samples with an array composed of 83 antibodies. Of the 33 proteins detected, 21 were up-regulated and 12 down-regulated in carcinoma biopsies from different patients [46].

Another one-color approach, this one using a "competitive displacement" strategy, has been described for the quantification of differential levels of proteins [49,50]. Two lysates, labeled "reference" and "unlabeled sample," are mixed in increasing ratios (1:1, 1:10, etc.) and put in competition (fig. 10.2B); the signal detected is inversely proportional to the concentration of the unlabeled target protein. Therefore, proteins of similar abundance mixed in the ratios 1:1 and 1:10 should give, respectively, 50 and 90% displacement that can be detected by the diminution of fluorescence intensity. For data analysis, the displacement level between labeled and unlabeled samples at a given ratio is subtracted from that obtained after competition between the same labeled reference and another unlabeled sample.

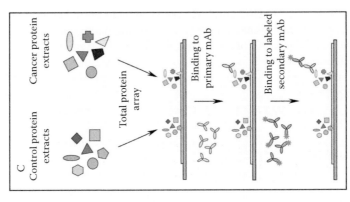

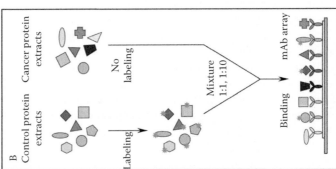

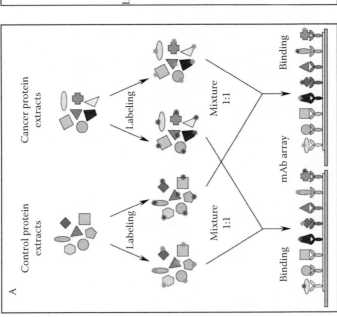

FIGURE 10.2 (See color insert.) Protein profiling with antibodies. (A) Two-color fluorescence detection with antibody microarrays. (B) Monocolor competitive displacement with antibody microarrays. (C) Total protein microarrays. For details, see text.

This simplified and cost-effective method appeared to be well suited for high-throughput profiling studies. However, no data were presented in the original study regarding the competitive nature of labeled and unlabeled proteins for immobilized antibodies. We compared the performance of the two-color and one-color "competitive displacement" methods in parallel experiments with breast cancer cell lines using NIR fluorescence for detection [48]. Differential expression was detected for four proteins in MDA MB-231 and SKBR3 cells by both methods. However, a more extensive study of other proteins indicated that the two-color method detects the difference in protein expression more precisely, as confirmed by western blot (fig. 10.3).

Indeed, bioconjugation of fluorescent dyes can dramatically change the properties of target proteins such as their folding, solubility, migration, and molecular recognition by antibodies. Hence, labeled molecules would not exhibit the same competitive behavior as their unlabeled analogs; this can explain the distortion in profiling data of some proteins by the displacement method. In the two-color method, the unpredicted effect of bioconjugation bias is less important since similar dyes are linked to proteins through the same reactive groups. Moreover, running two parallel assays with mutually exchanged dyes decreases artifacts significantly after normalization.

Antibody arrays can be used to understand complex networks involved in signal transduction in organisms. An array of antibodies, raised against phosphorylated and nonphosphorylated proteins, was used to study the phosphorylation state of the ErbB receptor tyrosine kinases in cancer cell lines A431, SKBR3, and MCF7 [51]. Using fluorescence detection, significant modulation in the levels of ErbB expression and tyrosine-phosphorylated EGFR and ErbB2 was observed in response to EGF activation in the different cell lines. The use of direct or indirect "sandwich" detection with two labeled probes by applying labeled lysates or labeled antibodies enabled the amounts of receptor proteins and their phosphorylated forms to be measured simultaneously. Recently, antibodies specific for post-translational modifications have been used to follow phosphorylation, acetylation, and ubiquitination of target proteins in cancer cells [52]. Different protein profiles have been detected for poly-ubiquitination and acetylation in response to trichostatin A treatment in HeLa cells and tyrosine-kinase phosphorylation initiated by EGFR in A431 cells using both visible and NIR fluorescence.

10.5 PROTEIN PROFILING WITH TOTAL PROTEIN ARRAYS

Total protein arrays, also referred to as "reverse phase protein arrays," consist of immobilized extracts of cells, tissues, or sera that are probed using a given antibody, followed by detection with a labeled secondary antibody (fig. 10.2C). This format allows numerous protein samples to be analyzed simultaneously with one or two antibodies under the same experimental conditions [24,27,53]. Such protein arrays were used to study cancer progression from normal prostate epithelium to intra-epithelial neoplasia and to invasive prostate cancer [24]. A remarkable reduction was observed in the ratio of phosphorylated ERK to total ERK during the longitudinal progression of the cancer. In a similar way, protein microarrays were employed to

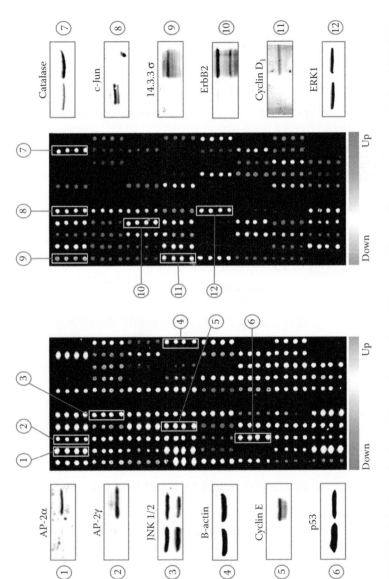

FIGURE 10.3 (See color insert.) Two-color protein profiling in cell lysates from breast tumor cell lines MDA MB-231 and SKBR3. Analytes were labeled with IRDye 800CW (red) and Alexa Fluor 680 (green). Circled numbers refer to position in the array of antibodies used in the corresponding western blot. (Adapted from Barry, R. et al. *J. Biomol. Screen.* 8(3), 257, 2003. With permission.)

monitor the kinetics of site-specific phosphorylation of different signaling proteins in Jurkat T lymphocytes [54].

The phosphorylation state of 62 signaling components was established in cells stimulated via CD3 and CD28 membrane receptors. In the same context, reverse phase protein arrays were widely used to study post-translational modifications of proteins at different stages of breast cancer progression in patients before and after therapy [55]. In earlier studies, detection from reverse phase protein arrays was performed by tyramide amplification [24,56,57], whereas in a more recent study a comparable sensitivity was ensured by NIR fluorescence with Alexa Fluor 680 goat antimouse and IRDye800CW goat antirabbit secondary antibodies [55]. NIR fluorescence also allowed the detection of a small carbohydrate molecule binding to GABA receptors on macrospots of immobilized membrane fractions of cell lysates using anticarbohydrate mAb [26].

10.6 ASSESSMENT OF IMMUNE RESPONSE WITH ANTIGEN ARRAYS

Many diagnostic approaches currently use conventional immunoassays and most popular among them is the enzyme-linked immunosorbent assay (ELISA), an outstanding multiplexed approach to assess enzymatic and binding parameters of proteins, including antigens and antibodies, in individual chambers of microtiter plates. Antigen microarrays provide an alternative to immunoassays by high-throughput monitoring of target molecules in a single assay. This is crucial for the precise comparison of binding affinity or parallel characterization of antigenic properties of numerous samples (for reviews, see Neuman de Vegvar and Robinson [58] and Sakanyan [59]). Miniaturized arrays on planar supports are also more economical in terms of the consumption of samples and reagents, the labor and time of analysis, and the cost per analyte. Moreover, the advantages of array and ELISA strategies can be combined by placing the arrays in wells [51,60–62] or by arraying antigens in separate chambers of commercial nitrocellulose-coated slides. These provide much versatility for performing binding assays simultaneously in a high-throughput and multiplexed fashion.

The feasibility of antigen microarrays was demonstrated by measuring the concentration of antibodies generated in patients against pathogen viruses and parasites as a result of a host immune defense [63–65]. Using internal calibration curves, a linear concentration-dependent response was observed with ~10% coefficient of variation between the arrayed slides, and the array data were in agreement with ELISA results [63]. The arrayed antigens of herpes simplex virus types 1 and 2, cytomegalovirus, the virus of rubella and *Toxoplasma gondii*, were able to detect specific IgG and IgM antibodies in the sera of patients. In another study, a panel of 430 chemically synthesized peptides and recombinant proteins was prepared to detect the immune response in rhesus monkeys immunized with genetically engineered vaccinia virus derivatives carrying antigenic determinants from a chimeric simian–human immunodeficiency virus [64]. This study indicated that an immune response was generated against immunodominant epitopes and some not yet characterized epitopes. More prolonged reactivity was monitored in immunized rather than nonimmunized animals, and surviving animals had a higher antibody titer against a wider spectrum of virus antigens. This first successful example of guiding vaccines with antigen microarrays emphasizes the utility of the high-throughput method for similar applications.

In our study, we applied microarrays to better understand the antigenic diversity of the human HIV-1 gp41 immunodominant epitope used for AIDS diagnosis [66] by evaluation of the immune response of mutant epitope sequences [65]. Mimetic peptides were selected in phage display libraries by consecutive panning with IgG from patients. Then, an arrayed panel of mimotopes was fabricated on a nitrocellulose membrane. The parallel assessment of peptides before and after "highly active anti-retroviral therapy" showed a good correlation between the binding affinity of mimotopes monitored by ELISA and array-based immunoassay with the sera of AIDS patients. However, microarray immunoassay supported by NIR fluorescence detection at 800 nm was found to reflect the binding prevalence better (i.e., the difference between reference and infected probes, which is of great value in the detection of suboptimal concentrations of antibodies in HIV-1 infected patients).

These data are encouraging in terms of the development of microarrays with a large panel of epitopes and mimotopes for the diagnosis of viral and perhaps bacterial infections that provoke levels of pathogenic agents in humans too low to be detected by convenient methods. In contrast, the generated antibodies are sufficient to provide a higher sensitivity for detection with arrayed antigens [67]. Antigen microarrays will also be useful to assess the antibody repertoire in patients infected by Epstein–Barr or hepatitis B and C viruses since acute, chronic, and convalescence forms of the corresponding infections are characterized by modulation of IgG and IgM immunoglobulin titer against particular viral antigens (reviewed in Neuman de Vegvar et al. [64] and Storch [68]).

Antigen microarrays have been employed to target other human diseases as well. Different sets of potential antigens and allergens (both peptides and proteins) were used to evaluate auto-antibodies in the sera of patients with systemic rheumatic diseases, autoimmune encephalomyelitis, or multiple sclerosis [69–71]. Recently, tumor-derived proteins have been arrayed and incubated with sera obtained from cancer patients and healthy controls [72–74]. Positive responses have been observed in prostate cancer and lung cancer patients, indicating the wider applicability of antigen microarrays to elicit specific antibodies in complex biological fluids.

The data accumulated during recent years show that antibody, antigen, and protein microarrays offer various strategies to elucidate the biological diversity and complexity of regulatory networks and biological responses to environmental actions with wide destinations for different domains of biomedical research (table 10.2).

10.7 IMPROVING BINDING SPECIFICITY ON PROTEIN MICROARRAYS

Although a minute spot area is best suited for sensitive and high-throughput analysis, the fluorescent signal intensity reflects the sum of all bound molecules, and nonspecific interactions can often dominate over specific binding. Therefore, nonspecific

---

TABLE 10.2
Possible Applications of Antibody and
Antigen Microarrays in Biomedicine

Destination of microarrays

Protein profiling and biomarker searching
Post-translational modifications and signal transduction
Immunoassays and diagnostic tools
Assessment of immune response
Prognostic of therapeutic treatment
Target discovery and guiding vaccines
Screening biomolecules and drug discovery
Cytotoxicity of leads

---

binding remains a critical issue in high-throughput dissection of antigen–antibody interactions. Improving the affinity and cross-reactivity of antibodies, as well as the quality and homogeneity of protein samples and labeled probes, is essential for the performance of protein microarrays. Different solutions have been proposed to overcome this problem.

The preparation of proteins from tissue samples requires optimal cell lysis, providing a constant yield and proportion of target proteins. Our experience shows that prolonged storage of extracts, even in the presence of protease inhibitors, causes changes in proteins and is misleading for the interpretation of results, as previously observed for analysis by two-dimensional electrophoresis [75]. Tissue heterogeneity in cancer biopsies can also be a source of differences in protein abundance from sample to sample. Laser capture microdissection enables cells to be depleted from a heterogeneous context and improves the isolation of the desired tumor cell population from cancer tissues [24]. This dramatically improves the isolation of defined cell populations from cancer tissues and therefore improves the statistical confidence in microarray data [24,76].

Similarly, protein enrichment by chromatographic fractionation of cell lysates before printing them on the surface improves experimental performance [20]. Consecutive centrifugation of bacterial cell lysates and filtration of the supernatant through hydrophobic filters eliminates the cell membrane protein fraction and insoluble inclusion bodies of overexpressed targets, thereby increasing the signal specificity of soluble proteins, as tested by competitive binding to arrayed antibodies [48]. Furthermore, nuclear protein subfractionation of cancer cell lines increases the yield of low-abundance targets and decreases possible cross-reactivity with nontarget molecules located in other cellular compartments, thus identifying up- and down-regulated transcription factors better, as compared with total cell lysate [48]. Similar enrichment of cytosolic, membrane-organelle and cytoskeleton proteomes can improve the profiling of proteins implicated in other diseases.

The quality of fluorescently labeled probes depends on the fluorescent dye used, as well as the number and location of target amino acid residues in protein chains. Therefore, random bioconjugation can lead to nonspecific aggregation of chemically modified probes with many proteins and affect precise measurements if additional standardized probes are not used. This inconvenience can be reduced by labeling the protein of interest in a cell-free system in which dye-conjugated puromycin at low concentrations is incorporated into the C-terminus [77,78] or dye-conjugated amber initiator tRNA marks the N-terminus [79,80]. Thus, wider application of these approaches might provide uniformly labeled antigen and antibody probes for microarray studies.

The application of a variety of sandwich approaches increases the specificity of the detection of target protein and hence ensures better quantitative information from complex analytes (for an overview, see Nielsen and Geierstanger [81]). However, a sandwich assay requires two antibodies, each of which must recognize a different epitope of the same protein. But, this is not always possible, especially for intra-cellular proteins. Besides, the introduction of one or two additional binding steps (some assays use consecutive primary mAb binding to a protein of interest captured by another arrayed mAb, then secondary mAb binding) followed by washing steps complicates the assay protocol and reduces the possibility of detecting protein interactions of varying affinity. Nevertheless, with or without signal amplification, the microsandwich approach has been successfully employed to monitor cytokines [21,32,61,82,83], cancer-specific antigens and growth factors [84], and, more recently, protein post-translational modifications [51,52].

Without contesting a negative effect of cross-reactivity on microarray data, it is appropriate to mention that a "nonspecific" signal can also be a positive sign, indi-cating discovery of a new epitope or binding partner of potential medical interest. In this respect the yeast proteome array, recently exploited for the evaluation of antibody cross-reactivity [85], suggests similar application of human protein microarrays for the identification of highly specific antibodies for studying the human proteome and for better understanding of the complexity of antigen–antibody interactions.

10.8 AUTOASSEMBLING PROTEIN MICROARRAYS

Snyder and coworkers used conventional gene cloning and recombinant protein expression with affinity purification methods to prepare the first whole-proteome array, composed of 5,800 of 6,200 yeast open reading frames [12]. Its functional exploitation allowed new calmodulin targets and phospholipid-binding proteins to be identified. However, there appears to be an alternative to the costly and time-consuming strategy of gene cloning in cells that might be widely used for other proteomes. Prokaryotic and eukaryotic cell-free systems for protein synthesis were developed a long time ago [86–88]. The best systems, with improved biosynthetic and energetic parameters, provide quite a high yield of many protein families and the production of "toxic" and more soluble proteins. Tagged proteins produced in cell-free systems and purified by affinity were used to prepare arrays and to study protein–protein and protein–DNA interactions, thereby significantly shortening an array-based binding protocol [6].

It is notable that the earliest reports on peptide arrays already demonstrated the possibility of *in situ* chemical peptide synthesis directly on solid supports instead of printing molecules [89,90]. This direct assembly of short peptide synthesis into

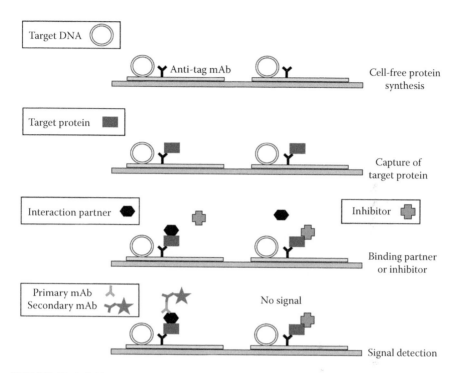

FIGURE 10.4 Self-assembled microarrays to detect protein interactions and the effect of inhibitors on protein interactions.

miniaturized spots ready for binding assays might have inspired the generation of similar arrays of larger polypeptides using biological systems. However, the first protein array prototype fabricated by an autoassembling strategy was obtained almost 10 years later by using cell-free protein synthesis. A human single-chain antiprogesterone antibody fragment fused to C-terminal his-tag was synthesized in a 96-well microplate and recovered by affinity on a Ni-NTA-coated surface of the same wells [91]. The antibody fragment exhibited antigen recognition ability as judged by several tests.

Recently, a new strategy of autoassembled protein microarrays has been developed that takes advantage of cell-free synthesized proteins and the high-affinity binding ability of antibodies [92]. The method is based on the immobilization within the same spot of the DNA template that encodes a C-terminal GST-tagged protein of interest as well as the anti-GST polyclonal antibody (fig. 10.4).

As the protein is released from ribosome during cell-free synthesis, it is captured by the polyclonal antibody. Its full-length expression is confirmed with an anti-GST monoclonal antibody that recognizes another epitope. The method is well suited to the detection of protein–protein interactions because the suspected partner protein is also cell-free synthesized in the same reaction. The bound partner, tagged with a hemaglutinin (HA) epitope, can be detected with the corresponding mAb using chemiluminescence. To illustrate the utility of the autoassembled array, it was tested for mapping binary interactions among 29 proteins of the human DNA replication complex. This elegant method confirmed many data obtained from immunoprecip-itation and yeast two-hybrid techniques and identified not yet established protein contacts. In our laboratory, we attempt to adapt the method to search for inhibitors of protein–protein interactions by using NIR fluorescence to increase the sensitivity of the screening (see fig. 10.4).

10.9 FUTURE PROSPECTS

Antibody- and antigen-based microarrays represent a new generation of promising molecular tools for the development of biomedical applications. The detection of up- and down-regulated proteins with antibody arrays is an urgent challenge for the characterization of disease-specific biomarkers that will enable the early diagnosis and prognosis of many diseases. Moreover, a linkage between profiling with antibody arrays and with total protein arrays might constitute a powerful approach for high-throughput and multiplexed monitoring of cellular regulatory and signaling networks as a prerequisite for an individualized therapy.

Engineered and recombinant antibodies with a higher affinity and lower cross-reactivity [93] will be made to replace the full-length antibodies currently showing a limited utility for high-throughput applications. Furthermore, an example of auto-assembled microarrays for studying protein interactions [92] is encouraging for the application of a similar array-based method to the systematic selection of mutant antibodies. Another challenge is the development of efficient approaches to simplify protein composition in serum, plasma, saliva, cerebrospinal fluids, and urine, in which the dynamic range of protein concentration might vary by up to 12 orders of magnitude [94], in order to study proteomes of complex biological fluids with microarrays. One of the main questions is still the organization and dissection of the huge amounts of data generated by antibody and antigen microarrays using statistical approaches and algorithms that should avoid a high sensitivity to noise.

Fluorescence remains the cornerstone of microarray technology, but it is not a universal panacea for the sensitive detection of molecular interactions. Alternative label-free methods, which rely on optical, electric, calorimetric, or acoustic detection principles, are under development to answer unresolved problems in proteomics. In the same context, only complementary high-throughput and multiplexed methods, combined with bioinformatic tools, will guarantee success in understanding the complexity of biological systems.

REFERENCES

1. Ekins, R. and Chu, F. In *Protein Arrays, Biochips, and Proteomics. The Next Phase of Genomic Discovery*, ed. Albala, J. S. and Humphery-Smith, I., Marcel Dekker, Inc., 2003, 81.
2. Kodadek, T. Protein microarrays: Prospects and problems. *Chem. Biol.*, 8(2), 105, 2001.
3. Ekins, R. P. Ligand assays: from electrophoresis to miniaturized microarrays. *Clin. Chem.*, 44(9), 2015, 1998.

4. Silzel, J. W. et al. Mass-sensing, multianalyte microarray immunoassay with imaging detection. *Clin. Chem.*, 44(9), 2036, 1998.

5. Bradbury, A. et al. Antibodies in proteomics II: screening, high-throughput characterization and downstream applications. *Trends Biotechnol.* 21(7), 312, 2003.

6. Sakanyan, V. High-throughput and multiplexed protein array technology: Protein–DNA and protein–protein interactions. *J. Chromatogr. B*, 815(1–2), 77, 2005.

7. Glökler, J. and Angenendt, P. Protein and antibody microarray technology. *J. Chromatogr. B Analyt. Technol. Biomed. Life Sci*, 797(1–2), 229, 2003.

8. Haab, B. B., Dunham, M. M., and Brown, P. O. Protein microarrays for highly parallel detection and quantitation of specific proteins and antibodies in complex solutions. *Genome Biol.*, 2(2), 1, 2001.

9. MacBeath, G. and Schreiber, S. Printing proteins as microarrays for high-throughput function determination. *Science*, 289(5485), 1760, 2000.

10. Ruiz-Taylor, L. A. et al. Monolayers of derivatized poly(L-lysine)-grafted poly(ethylene glycol) on metal oxides as a class of biomolecular interfaces. *Proc. Natl. Acad. Sci. USA*, 98(3), 852, 2001.

11. Kusnezow, W. and Hoheisel, J. D. Antibody microarrays: promises and problems. *BioTechniques*, 33 (suppl), S14, 2002.

12. Zhu, H. et al. Global analysis of protein activities using proteome chips. *Science*, 293(5537), 2101, 2001.

13. Kawahashi, Y. et al. *In vitro* protein microarrays for detecting protein–protein interactions: Application of a new method for fluorescence labeling of proteins. *Proteomics*, 3(7), 1236, 2003.

14. Kusnezow, W. et al. Antibody microarrays: an evaluation of production parameters. *Proteomics*, 3(3), 254, 2003.

15. Lu, B., Smyth, M. R., and O'Kennedy, R. Oriented immobilization of antibodies and its applications in immunoassays and immunosensors. *Analyst*, 121(3), 29R, 1996.

16. Vijayendran, R. A. and Leckband, D. E. A quantitative assessment of heterogeneity for surface-immobilized proteins. *Anal. Chem.*, 73(3), 471, 2001.

17. Peluso, P. et al. Optimizing antibody immobilization strategies for the construction of protein microarrays. *Anal. Biochem.*, 312(2), 113, 2003.

18. Arenkov, P. A. et al. Protein microchips: use for immunoassay and enzymatic reactions. *Anal. Biochem.*, 278(2), 123, 2000.

19. Ge, H. UPA, a universal protein array system for quantitative detection of protein–protein, protein–DNA, protein–RNA and protein–ligand interactions. *Nucleic Acids Res.*, 28(2), e3.i–e3.vii, 2000.

20. Madoz-Gurpide, J. et al. Protein based microarrays: A tool for probing the proteome of cancer and tissues. *Proteomics*, 1(10), 1279, 2001.

21. Huang, R. P. et al. Simultaneous detection of multiple cytokines from conditioned media and patient's sera by an antibody-based protein array system. *Anal. Biochem.*, 294(1), 55, 2001.

22. Ghochikyan, A. et al. Implication of the oligomerization domain of ArgR repressors from two evolutionary distant bacteria in the operator DNA-binding specificity. *J. Bacteriol.*, 184(23), 6602, 2002.

23. Tonkinson, J. L. and Stillman, B. A. Nitrocellulose: A tried and true polymer finds utility as a postgenomic substrate. *Front. Biosci.*, 7, c1, 2002.

24. Paweletz, C. P. et al. Reverse phase protein microarrays which capture disease progression show activation of prosurvival pathways at the cancer invasion front. *Oncogene*, 20(16), 1981, 2001.

25. Snapyan, M. et al. Dissecting DNA–protein and protein–protein interactions involved in bacterial transcriptional regulation by a sensitive protein array method combining a near-infrared fluorescence detection. *Proteomics*, 3(5), 647, 2003.

26. Saghatelyan, A. et al. Recognition molecule associated carbohydrate inhibits post-synaptic GABA(B) receptors: A mechanism for homeostatic regulation of GABA release in perisomatic synapses. *Mol. Cel. Neurosci.*, 24(2), 271, 2003.

27. Espina, V. et al. Protein microarrays: molecular profiling technologies for clinical specimens. *Proteomics*, 3(11), 2091, 2003.

28. Stillman, B. A. and Tonkinson, J. L. FAST slides: A novel surface for microarrays. *Biotechniques*, 29(3), 630, 2000.

29. Bacarese-Hamilton, T. et al. Detection of allergen-specific IgE on microarrays by use of signal amplification techniques. *Clin. Chem.*, 48(8), 1367, 2002.

30. Lizardi, P. M. et al. Mutation detection and single-molecule counting using isothermal rolling-circle amplification. *Nat. Genet.*, 19(3), 225, 1998.

31. Schweitzer, B. et al. Inaugural article: Immunoassays with rolling circle DNA amplification: A versatile platform for ultrasensitive antigen detection. *Proc. Natl. Acad. Sci. USA*, 97(19), 10113, 2000.

32. Schweitzer, B. et al. Multiplexed protein profiling on microarrays by rolling-circle amplification. *Nat. Biotech.*, 20(4), 359, 2002.

33. Schutz, A. R. et al. Highly sensitive detection of proteins on membranes with near-infrared fluorescence. Online www.licor.com.

34. Miller, J. W. Tracking G protein-coupled receptor trafficking using Odyssey Imaging. Online www.licor.com.

35. Andersen, L. and Seilhamer, J. A. Comparison of selected mRNA and protein abundances in human liver. *Electrophoresis*, 18(3–4), 533, 1997.

36. Gygi, S. P. et al. Correlation between protein and mRNA abundance in yeast. *Mol. Cell. Biol.*, 19(3), 1720, 1999.

37. MacBeath, G. Protein microarrays and proteomics. *Nat. Genet.*, 32 (suppl.), 526, 2002.

38. Hanash, S. Disease proteomics. *Nature*, 422 (6928), 226, 2003.

39. Ekins, R. and Chu, F. Multianalyte microspot immunoassay. The microanalytical "compact disk" of the future. *Ann. Biol. Clin.* (Paris), 50(5), 337, 1992.

40. Sreekumar, A. et al. Profiling of cancer cells using protein microarrays: Discovery of novel radiation-regulated proteins. *Cancer Res.*, 61(20), 7585, 2001.

41. Knezevic, V. et al. Proteomic profiling of the cancer microenvironment by antibody arrays. *Proteomics*, 1(10), 1271, 2001.

42. Belov, L. et al. Immunophenotyping of leukemias using a cluster of differentiation antibody microarray. *Cancer Res.*, 61(11), 4483, 2001.

43. Belov, L. et al. Identification of repertoires of surface antigens on leukemias using an antibody microarray. *Proteomics*, 3(11), 2147, 2003.

44. Miller, J. C. et al. The application of protein microarrays to serum diagnostics: Prostate cancer as a test case. *Dis. Markers*, 17(4), 225, 2001.

45. Miller, J. C. et al. Antibody microarray profiling of human prostate cancer sera: Antibody screening and identification of potential biomarkers. *Proteomics*, 3(1), 56, 2003.

46. Tannapfel, A. et al. Identification of novel proteins associated with hepatocellular carcinomas using protein microarrays. *J. Pathol.*, 201 (2), 238, 2003.

47. Hudelist, G. et al. Use of high-throughput protein array for profiling of differentially expressed proteins in normal and malignant breast tissue. *Breast Cancer Res. Treatment*, 86(3), 281, 2004.

48. Yeretssian, G. et al. Competition on nitrocellulose-immobilized antibody arrays: From bacterial protein binding assay to protein profiling in breast cancer cells. *Mol. Cell Proteomics*, 4(5), 605, 2005.

49. Barry, R. et al. Competitive assay formats for high-throughput affinity arrays. *J. Biomol. Screen.*, 8(3), 257, 2003.

50. Barry, R. and Soloviev, M. Quantitative protein profiling using antibody arrays. *Proteomics*, 4(12), 3717, 2004.

51. Nielsen, U. B. et al. Profiling receptor tyrosine kinase activation by using Ab microarrays. *Proc. Natl. Acad. Sci. USA*, 100(16), 9330, 2003.

52. Ivanov, S. S. et al. Antibodies immobilized as arrays to profile protein post-translational modifications in mammalian cells. *Mol. Cell Proteomics*, 3(8), 788, 2004.

53. Simone, N. L. et al. Laser capture microdissection: Beyond functional genomics to proteomics. *Mol. Diagn.*, 5(4), 301, 2000.

54. Chan, S. M. et al. Protein microarrays for multiplex analysis of signal transduction pathways. *Nat. Med.*, 10(12), 1390, 2004.

55. Calvert, V. S. et al. Development of multiplexed protein profiling and detection using near infrared detection of reverse-phase protein microarrays. *Clin. Proteomics*, 1(1), 81, 2004.

56. Nishizuka, S. et al. Proteomic profiling of the NCI-60 cancer cell lines using new high-density reverse-phase lysate microarrays. *Proc. Natl. Acad. Sci. USA*, 100(24), 14229, 2003.

57. Wulfkuhle, J. D. et al. Signal pathway profiling of ovarian cancer from human tissue specimens using reverse-phase protein microarrays. *Proteomics*, 3(11), 2085, 2003.

58. Neuman de Vegvar, H. E. and Robinson, W. H. Microarray profiling of antiviral antibodies for the development of diagnostics, vaccines, and therapies. *Clin. Immunology*, 111(2), 196, 2004.

59. Sakanyan, V. Puces à protéines: Nouvelle approche du diagnostic des maladies infectieuses. *Antibiotiques* (France), 6, 185, 2004.

60. Mendoza, L. et al. High-throughput microarray-based enzyme-linked immunosorbent assay (ELISA). *Biotechniques*, 27(4), 778, 1999.

61. Moody, M. D. et al. Array-based ELISA for high-throughput analysis of human cytokines. *Biotechniques*, 31(1), 186, 2001.

62. Perrin, A. et al. A combined oligonucleotide and protein microarray for the codetection of nucleic acids and antibodies associated with human immunodeficiency virus, hepatitis B virus, and hepatitis C virus infections. *Anal. Biochem.*, 322(2), 148, 2003.

63. Mezzasoma, L. et al. Antigen microarrays for serodiagnosis of infectious diseases. *Clinic. Chem.*, 48(1), 121, 2002.

64. Neuman de Vegvar, H. E. et al. Microarray profiling of antibody responses against simian–human immunodeficiency virus: Postchallenge convergence of reactivities independent of host histocompatibility type and vaccine regimen. *J. Virol.*, 77(20), 11125, 2003.

65. Arnaud, M-C. et al. Array assessment of phage displayed peptide mimics of human immunodeficiency virus type 1 gp41 immunodominant epitope: binding to antibodies of infected individuals. *Proteomics*, 4(7), 1959, 2004.

66. Gnann, J. W. Jr., et al. Diagnosis of AIDS by using a 12-aminoacid peptide representing an immunodominant epitope of the human immunodeficiency virus. *J. Infect. Dis.*, 156(2), 261, 1987.

67. Roehrig, J. T. et al. Persistence of virus-reactive serum immunoglobulin M antibody in confirmed West Nile virus encephalitis cases. *Emer. Infect. Dis.*, 9(3), 376, 2003.

68. Storch, G. A. Diagnosis virology. *Clin. Infect. Dis.*, 31(3), 739, 2000.

69. Joos, T. O. et al. A microarray enzyme-linked immunosorbent assay for autoimmune diagnostics. *Electrophoresis*, 21(13), 2641, 2000.
70. Robinson, W. H. et al. Autoantigen microarrays for multiplex characterization of autoantibody responses. *Nat. Med.*, 8(3), 295, 2002.
71. Robinson, W. H. et al. Protein microarrays guide tolerizing DNA vaccine treatment of autoimmune encephalomyelitis. *Nat. Biotechnol.*, 2003, 21(9), 1033, 2002.
72. Yan, F. et al. Protein microarrays using liquid phase fractionation of cell lysates. *Proteomics*, 3(7), 1228, 2003.
73. Bouwman, K. et al. Microarrays of tumor cell derived proteins uncover a distinct pattern of prostate cancer serum immunoreactivity. *Proteomics*, 3(11), 2200, 2003.
74. Qiu, J. et al. Development of natural protein microarrays for diagnosing cancer based on an antibody response to tumor antigens. *J. Proteome Res.*, 3(2), 261, 2004.
75. Ahmed, N. and Rice, G. E. Strategies for revealing abundance proteins in two-dimensional protein maps. *J. Chromatogr. B*, 815(1–2), 39, 2005.
76. Liotta, L. and Petricoin, E. Molecular profiling of human cancer. *Nat. Rev. Genet.*, 1(1), 48, 2000.
77. Miyamoto-Sato, E. et al. Specific bonding of puromycin to full-length protein at the C-terminus. *Nucleic Acids Res.*, 28(5), 1176, 2000.
78. Doi, N. et al. Novel fluorescence labeling and high-throughput assay technologies for *in vitro* analysis of protein interactions. *Genome Res.*, 12(3), 487, 2002.
79. Gite, S. et al. Ultrasensitive fluorescence-based detection of nascent proteins in gels. *Anal. Biochem.*, 279(2), 218, 2000.
80. Mamaev, S. et al. Cell-free N-terminal protein labeling using initiator suppressor tRNA. *Anal. Biochem.*, 326(1), 25, 2004.
81. Nielsen, U. B. and Geierstanger, B. H. Multiplexed sandwich assays in microarray format. *J. Immun. Methods*, 290(1–2), 107, 2004.
82. Wang, C. C. et al. Array-based multiplexed screening and quantitation of human cytokines and chemokines. *J. Proteome Res.*, 1(4), 337, 2002.
83. Tam, S. W. et al. Simultaneous analysis of eight human Th1/Th2 cytokines using microarrays. *J. Immunol. Methods*, 261(1–2), 157, 2002.
84. Woodbury, R. L., Varnum, S. M., and Zangar, R. C. Elevated HGF levels in sera from breast cancer patients detected using a protein microarray ELISA. *J. Proteome Res.*, 1(3), 233, 2002.
85. Michaud, G. A. et al. Analyzing antibody specificity with whole proteome micro-arrays. *Nat. Biotechnol.*, 21(12), 1509, 2003.
86. De Vries, J. K. and Zubay, G. DNA-directed peptide synthesis. II. The synthesis of the alpha-fragment of the enzyme beta-galactosidase. *Proc. Natl. Acad. Sci. USA*, 57(4), 1010, 1967.
87. Roberts, B. E. and Paterson, B. M. Efficient translation of tobacco mosaic virus RNA and rabbit globin 9S RNA in a cell-free system from commercial wheat germ. *Proc. Natl. Acad. Sci. USA*, 70(8), 2330, 1973.
88. Pelham, H. R. and Jackson, R. J. An efficient mRNA-dependent translation system from reticulocyte lysates. *Eur. J. Biochem.*, 67(1), 247, 1976.
89. Fodor, S. P. et al. Light-directed, spatially addressable parallel chemical synthesis. *Science*, 251(4995), 767, 1991.
90. Frank, R. Spot-synthesis: An easy technique for the positionally addressable, parallel chemical synthesis on a membrane support. *Tetrahedron*, 48(42), 9217, 1992.
91. He, M.Y. and Taussig, M. J. Single step generation of protein arrays from DNA by cell-free expression and *in situ* immobilization (PISA method). *Nucleic Acids Res.*, 29(15), e73, 2001.

92. Ramachandran, N. et al. Self-assembling protein microarrays. *Science*, 305(5680), 86, 2004.

93. Hudson, P.J. and Souriau, C. Engineered antibodies. *Nat. Med.*, 9(1), 129, 2003.

94. Corthals, G. L. et al. The dynamic range of protein expression: a challenge for proteomic research. *Electrophoresis*, 21(6), 1104, 2000.

11 Application of Shotgun Proteomics to Transcriptional Regulatory Pathways

Amber L. Mosley and Michael P. Washburn

CONTENTS

11.1 INTRODUCTION

11.1.1 TRANSCRIPTION AND PROTEOMICS

The application of mass spectrometry to studies of biological pathways has become an indispensable tool in elucidating the function and regulation of proteins in the context of a cell. These studies have allowed for the identification of protein interaction partners as well as the rapid identification of multiple post-translational modifications such as phosphorylation and acetylation. This type of analysis can be extremely useful for the study of complex biological processes such as the regulation of transcription in the nucleus. In recent years, the application of microarray technology has led to the ability to identify genome-wide changes in mRNA levels in response to changes in environmental stimuli and the mutation or deletion of individual target proteins. These studies have been useful in the identification of novel members of signaling pathways and target genes, but do not give a complete picture of the state of the cell under various conditions. With the application of mass spectrometry to complex biological mixtures in combination with microarray technologies, we can now gain a more complete picture of the effects of gene expression on the overall abundance and modification of proteins under different cellular conditions.

11.1.2 CHALLENGES

The use of mass spectrometry for the identification of transcriptional regulatory proteins within complex mixtures provides a unique challenge, since many of the proteins of interest are present at very low levels in the cell. This problem has been illustrated in yeast with the analysis of the global levels of protein expression, which was determined using genome-wide protein fusions with a tandem affinity purification (TAP) epitope tag [1]. After the number of protein molecules present in each cell was determined, it was discovered that transcription factors were expressed at much lower levels than the average protein [1]. When these protein abundance measurements were compared with a number of proteomic datasets [2–5], it was found that these low-abundance proteins were rarely detected in complex mixtures using proteomic methods.

A recent study by Graumann et al. showed that the lower the abundance is, the lower the percentage of proteins that are detected, even with multiple proteomic analyses on a duplicated protein extract [6]. Their study found that of the proteins determined to be expressed at less than 1,000 molecules per cell by Ghaemmaghami et al. [1] (approximately 1,068 proteins), only 15.3% of those proteins were identified in 9 *multi*dimensional *p*rotein *i*dentification *t*echnology (MudPIT) runs. Therefore, in order to analyze transcriptional regulatory proteins using mass spectrometry techniques, the ability to detect these low-abundance proteins must be drastically improved. This could be accomplished using a variety of different techniques, including the analysis of protein samples that are highly purified in order to remove many highly abundant proteins, which can easily mask low-abundance proteins, or through the development of more sensitive techniques and/or instruments able to identify proteins present at low levels in highly complex mixtures.

11.1.3 USE OF SHOTGUN PROTEOMICS IN ANALYSIS OF TRANSCRIPTIONAL REGULATORY PATHWAYS

A number of methods can be used to analyze protein mixtures using mass spectrometry. After isolation of proteins from a biological system through simple extraction or subsequent affinity purification, the samples are separated in order to decrease the number of different proteins/peptides introduced to the mass spectrometer at once (fig. 11.1). A common way that this is achieved is through the separation of intact proteins by either 1-D or 2-D gel electrophoresis (fig. 11.1). Following the separation, the corresponding bands for the proteins of interest are excised and then digested into a peptide mixture. This mixture is then loaded onto a column where the peptides are separated using a chromatographic resin such as reverse phase (RP) and then analyzed by mass spectrometry (fig. 11.1).

As mentioned earlier, the identification of low-abundance proteins in complex mixtures requires a highly sensitive technique that displays a low bias towards the abundance of any given protein. For example, when complex mixtures are analyzed, the use of systems such as 2-D gel electrophoresis can hinder the identification of low-abundance proteins prior to analysis by mass spectrometry due to a number of factors, including the dynamic range of proteins that are visualized when using protein-binding dyes and the additional loss of proteins during recovery from sodium dodecyl sulfate polyacrylamide gel electrophoresis (SDS-PAGE) gels following electrophoresis. Low-abundance proteins can also be masked when analyzing mixtures by gel electrophoresis if they comigrate with high-abundance proteins or protein contaminants such as keratin. Therefore, methods such as MudPIT analysis that do not require large amounts of post-isolation sample handling are a more optimal choice for the identification of transcriptional regulatory proteins and other low-abundance proteins in complex mixtures.

Shotgun proteomics predominantly involves the use of strong cation exchange (SCX) coupled to RP chromatography for separation of complex peptide mixtures and tandem mass spectrometry (MS/MS) for the identification of peptides (reviewed in Swanson and Washburn [7]). Three primary configurations have been implemented using SCX/RP/MS/MS approaches (reviewed in Swanson and Washburn). In one configuration, SCX is carried out and fractions of peptides are collected. Each individual fraction is then subjected to RP/MS/MS analysis, typically via an autosampler. An advantage of this approach includes the ability to use nonvolatile salts like KCl. But, this approach requires more manual sample handling due to the disconnected SCX and RP steps.

In a second configuration, which is an online configuration, SCX fractions are trapped on an RP column, typically washed with salt-free buffer, and then eluted onto an RP analytical column where RP/MS/MS is then carried out. An advantage of this approach is also the ability to use nonvolatile salts like KCl. In a third configuration known as the MudPIT approach, which is also an online configuration, a biphasic column containing SCX and RP materials is placed in line with a tandem mass spectrometer for SCX/RP/MS/MS analysis. An advantage of this approach is the lack of dead volume between the SCX and RP steps. Each of the three approaches is beginning to be increasingly adopted by the proteomic community, and each approach has advantages and disadvantages (reviewed in Swanson and Washburn [7]).

The shotgun proteomics approach, commonly referred to as MudPIT (reviewed in Paoletti et al. [8] and Delahunty and Yates [9]), allows for the direct loading of any given protein sample, whether it is a whole-cell lysate or a purified complex with many subunits (fig. 11.1). After isolation of the proteins, the samples are typically prepared by digestion with a standard protease such as trypsin and then loaded onto a microcapillary column that has been packed with reverse phase and strong cation exchange resins. This allows for the separation of peptides based on their charge and their hydrophobicity. The columns are then connected in-line to a quaternary high-performance liquid chromatography (HPLC) pump and coupled to

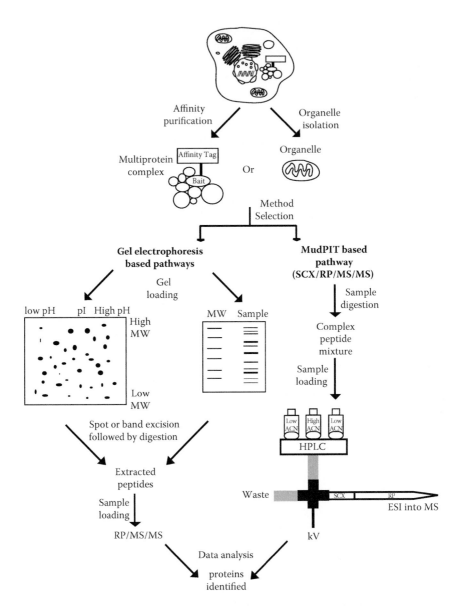

FIGURE 11.1 Schematic representation of isolation and sample processing steps for proteomic analysis of protein mixtures/complexes. The general approaches used for proteomic analyses of protein complexes and organelles are shown in this figure. A protein complex or organelle from a population of cells is prepared and subjected to a gel electrophoresis-based proteomic analysis pathway or a shotgun proteomic approach such as a MudPIT (*multi*dimensional *p*rotein *i*dentification *t*echnology)-based proteomic pathway. In the gel electrophoresis-based pathway, organelle analysis is typically carried out using two-dimensional gel electrophoresis, where proteins are separated by isoelectric point and molecular weight. Gel spots are then excised and proteins digested into peptides. On the other hand, multiprotein complexes are

FIGURE 11.1 (continued)

typically analyzed using one-dimensional gel electrophoresis. Gel bands are then excised and proteins digested into peptides. The extracted peptides are then analyzed using reverse phase chromatography and tandem mass spectrometry (RP/MS/MS) followed by data analysis to determine the proteins present in the mixture. In a MudPIT-based pathway, protein complexes or organelles are immediately digested into peptides and the subsequent complex peptide mixture is loaded onto a biphasic microcapillary column containing strong cation exchange (SCX) and reverse phase (RP) packing materials. The packed, loaded, and washed column is then placed in-line with an HPLC and tandem mass spectrometer. Chromatographic elution of peptides into the tandem mass spectrometer then leads to data analysis, and the proteins from the original sample are identified.

an ion trap mass spectrometer. The samples are ionized within the column and sprayed directly into the tandem mass spectrometer. This process allows for a high level of separation of the peptides within the mixture, and severely limits the amount of postisolation sample handling. This prevents sample contamination and loss of proteins during postisolation separation, which can occur during protein isolation from a gel. It is important to note, however, that when highly complex mixtures are analyzed, the chance of co-elution of highly abundant proteins with lower abundance proteins such as transcription factors is increased. This again stresses the point that in order to analyze the subset of proteins involved in gene regulation in the nucleus, it is important to enrich the sample for the proteins of interest in order to obtain the most information possible.

In this chapter, we provide an overview of the current progress that has been made towards the identification of transcriptional regulatory and other nuclear proteins using proteomic methods. This review will highlight the need for more sensitive separation and detection methods in order to obtain high-quality information when trying to study transcriptional regulatory proteins and their complexes, especially in the context of a complicated mixture.

11.2 CURRENT ANALYSES

11.2.1 Identification of Transcriptional Regulators from Complex Mixtures

After comparison of the protein abundance measurements obtained in the Ghaem-maghami et al. study with other proteomic analyses of whole-cell lysates in yeast [1], it was determined that low-abundance proteins such as transcription factors and cell-cycle proteins were rarely detected using current proteomic methods. In fact, with further characterization of the *Saccharyomyces cerevisiae* proteins quantified in the screen, it appears that transcription factors also display a lower protein abundance distribution than a variety of other nuclear proteins, which show the same abundance trend as the total protein population (fig. 11.2). Proteins involved in chromatin remodeling and DNA damage sensing and repair processes, as annotated in the *S. cerevisiae* genome database, all display the same abundance distribution as that seen with the total protein distribution (fig. 11.2).

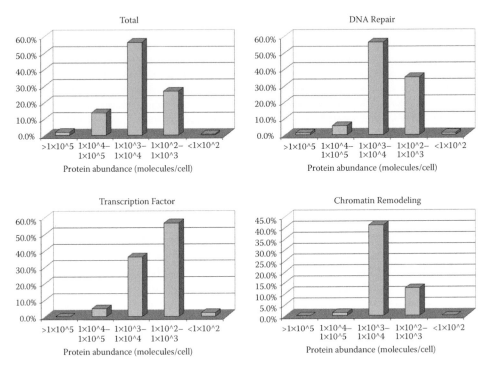

FIGURE 11.2 Abundance distribution of different functional groups of proteins in *S. cerevisiae*. Abundance measurements were obtained from http://yeastgfp.ucsf.edu/. Function groups of proteins were based on classifications from the *Saccharyomyces* genome database (http://www.yeastgenome.org/). The total distribution of yeast protein abundance is compared to the distribution of protein abundance of DNA repair proteins, transcription factors, and chromatin remodeling proteins. Note that the graphs are from most abundant on the left-hand side to least abundant on the right-hand side.

When the abundance data obtained by Ghaemmaghami et al. were compared to previous studies in which complex mixtures of proteins were analyzed by MudPIT or two-dimensional gel electrophoresis, it was discovered that these processes were strongly skewed towards the identification of higher abundance proteins [1]. In fact, their analysis determined that proteins present at less than 5,000 molecules per cell were only detected 19% of the time in some of these analyses [4,5]. When taking these analyses into account with regard to transcriptional regulatory proteins, proteins such as Spt4, Spt5, and Spt6, which are abundant transcriptional elongation regulators, were reproducibly identified in the analysis of whole-cell lysates performed by Washburn et al. [10]. In addition, a number of other relatively abundant transcriptional regulators, such as the co-repressor Tup1 and the general RNA polymerase II transcription factor Anc1, were also identified [10]. However, these proteins are all expressed at levels of over 2,000 molecules per cell, which increases their chances of being identified in these analyses.

A major challenge when analyzing complex mixtures by MudPIT and other shotgun proteomic methods that couple liquid chromatography and mass spectrometry is that a large number of peptides coelute during each chromatographic step. When this occurs, the number of available peptides is much greater than the number of ions that can be analyzed to generate tandem mass spectra. Since the ions are sorted based on their intensity, it would be expected that peptides present at a higher concentration in the cell—and therefore a higher concentration in the peptide mixture—would be more likely to be selected for further analysis. Liu et al. actually investigated this phenomenon during MudPIT analysis and came to the conclusion that the selection of more abundant peptides over low-abundance peptides leads to the ability to identify the relative abundance of a protein by looking at the spectral count [11]. This has the potential to provide a fairly straightforward way to obtain the relative abundance of proteins under different conditions. Improving the chromatographic methods, in order to limit the number of peptides that coelute, can approach the problem of being unable to identify low-abundance proteins in these complex mixtures. In addition, it may be possible to use instruments that are more sensitive to the identification of low-abundance proteins by having the ability to generate a larger number of tandem mass spectra.

Although a number of other proteomes have been analyzed by MudPIT and other shotgun proteomic methods, the correlation between the detection of the proteins and their relative abundance within those species is not readily available. It is very likely though that the analysis of those proteomes resulted in an over-representation of highly abundant proteins and an underrepresentation of low-abundance proteins like transcription factors. The results from the analyses of the *S. cerevisiae* proteome indicate that the current methods used to identify proteins in complex mixtures are strongly skewed towards the identification of high-abundance proteins and therefore do not reveal a truly complete picture of the status of a cell under any given condition. The majority of the studies discussed previously come from experiments in lower organisms. These problems will most likely be compounded in higher organisms, which have more complex proteomes, so that the only way to analyze changes in transcriptional regulatory proteins would be to study them in a more enriched population, which will be discussed next.

11.2.2 ENRICHMENT OF NUCLEAR PROTEINS USING SUBCELLULAR FRACTIONATION

In order to identify a large number of proteins in a complex mixture, it may be necessary to enrich the protein samples in order to identify proteins involved in a certain process like transcription. In fact, the isolation of cellular organelles has long been a first step in the biochemical characterization of proteins with specific functions in those organelles. This type of biochemical fractionation, in order to enrich proteins of interest, has already been used by a number of groups to narrow down the proteins identified by 2-D electrophoresis in various organelles (fig. 11.1), including the endoplasmic reticulum, the Golgi apparatus, mitochondria, lysosomes, peroxisomes, and the nucleus; a number of the proteins resolved by 2-D gels have been identified by mass spectrometry [12–18]. The potential for these studies is attractive, since it decreases the complexity of the cell and the fractionation of organelles has already been optimized in a number of different cell types.

To date, only a few examples of proteomic analysis of nuclear proteins using two-dimensional electrophoresis, with the subsequent identification of protein spots followed by mass spectrometry, have been reported in the literature. This is likely due to the large number of high-abundance proteins that are present in the nucleus, including various cellular chaperones and metabolic enzymes as well as abundant structural proteins such as the lamins and the histones, which provide the building blocks for chromatin structure in the nucleus. In fact, the histone proteins, which are some of the most highly conserved proteins in eukaryotes, are expressed at a concentration between 6×10^5 and 3×10^4 molecules per cell in *S. cerevisiae*, as determined by genome-wide analysis [1].

Because so many high-abundance proteins are present in the nucleus, it is not surprising that an analysis of the *Arabidopsis* nuclear proteome under control and cold-stress conditions only led to the identification of 184 proteins by 2D electrophoresis followed by matrix-assisted laser desorption/ionization time of flight (MALDI-TOF) MS [16]. Of the proteins identified, approximately 39.7% were proteins involved in signaling and gene regulation and 3.8% were proteins involved in DNA repair, which was higher than that observed after purification from other organelles in *Arabidopsis* [16]. Our analysis of the proteins annotated in SGD or in one of the genome-wide GFP localization screens in *S. cerevisiae* found that approximately 2,000 proteins are localized to, and therefore likely play a role in, the nucleus. Also, it is expected that this number would only increase in more complex organisms such as *Arabidopsis*. Bae et al. also noted that many of the proteins in the basic region of the 2D gel, which should be enriched in DNA binding proteins, were present at low levels [16].

In another analysis of nuclear proteins from proliferating and differentiated human colonic intestinal epithelial cells by 2D electrophoresis, 60 different proteins were identified by MALDI-TOF MS [18]. This analysis was focused on the identification of changes in the expression of proteins from the two different cell types. Of the 60 proteins identified, 13% of the proteins were classified as DNA/RNA binding proteins and 1.7% were classified as transcription factors (which corresponded to only one protein). In another study by Shakib et al., the effects of hypoxia on the relative quantity of nuclear proteins from rat kidney fibroblasts were studied [17]. In this analysis, nuclei were pelleted by centrifugation and the nuclear proteins were extracted with high concentrations of KCl (~700–800 m*M*) [17]. The extracted proteins were then separated by 2D electrophoresis, and a total of 791 spots were resolved by 2D electrophoresis and monitored for changes in their relative abundance. Following their separation, the 51 most abundant proteins that showed changes in abundance on the 2D gels were isolated and identified by mass spectrometry. In this study, 11 transcription factors were identified by mass spectrometry, as well as a number of transcriptional cofactors. As seen in the other two studies that we have discussed, the analysis also identified a number of metabolic enzymes.

Although the three preceding analyses were performed with nuclei from very diverse species under very different cellular conditions, a number of similar proteins from related protein classes were identified. This includes a number of protein chaperones, multiple ribosomal proteins, and various metabolic enzymes. Because a number of proteins in these families were found to be highly abundant in *S. cerevisiae*, it is possible that their identification in these different organisms is also a result of their high expression levels in those cells. It also demonstrates that highly abundant proteins such as the cytosolic eukaryotic initiation factors, of which a homolog was identified in all three of these studies, can also easily contaminate organelle preparations due to their extremely high levels in the cell. Although it is possible that these proteins also play a role in the nucleus, it is important to verify their localization by other methods. These studies illustrate the challenges involved in the identification of low-abundance transcriptional regulators in complex mixtures by 2D electrophoresis, and stresses that more sensitive methods for isolation and detection of these proteins are needed in order to use proteomic methods to study transcription factor complexes and dynamics.

11.2.3 PROTEOMICS AND SUBNUCLEAR COMPONENTS

In order to study highly specialized processes in the cell, it is imperative that the sample used for proteomic analysis is as enriched as possible for the proteins of interest. Along these lines, a number of studies have been performed that take the idea of subcellular fractionation one step further. These studies employ methods that isolate various multiprotein clusters in the nucleus, which we will refer to broadly as subnuclear components but include the nuclear pore complex, the nucleolus, Cajal bodies, promyelocytic leukemia (PML) bodies, interchromatin granules, and the clastosome (for review, see Lamond and Sleeman [19] and Spector [20]). These specialized compartments likely form to facilitate some of the many functions that occur in the nucleus, including nucleo-cytoplasmic transport, DNA transcription, DNA replication, DNA repair, ribosome biogenesis, and pre-mRNA processing. Many of these are intimately associated. Although it has been well documented that these concentrated areas of proteins exist within the nucleus, for many of these compartments neither their composition nor their exact role in nuclear function is well understood. For this reason and because many of these compartments can be enriched using biochemical methods, they are prime targets for proteomic analysis.

Nuclear pore complexes (NPCs) are located throughout the nuclear membrane and play an essential role in regulating nucleo-cytoplasmic transport. NPCs consist of a number of transmembrane proteins as well as a number of stably and transiently associated proteins. In 2000, Rout et al. sought to determine the components of the yeast nuclear pore complex using proteomic methods [21]. After purification of the yeast NPC, the proteins were separated using various HPLC methods in order to decrease the complexity of the sample prior to separation by 1-D electrophoresis on SDS-PAGE gels. This biochemical fractionation helped to increase the number of unique protein bands that could be isolated from the SDS-PAGE gels. After trypsin digestion of the isolated proteins, the samples were analyzed by MALDI-TOF mass spectrometry. Through their exhaustive analysis of the biochemical fractions, they were able to identify 174 proteins, of which 40 were confirmed to be associated with the nuclear pore.

In 2002, Cronshaw et al. continued the characterization of the NPC by purifying the complex from rat liver nuclei in order to gain insight into the components of the mammalian NPC [22]. Following the enrichment of the nucleoporins from the rat liver nuclei, the complex mixture of proteins was subjected to reverse phase chromatography followed by analysis of the fractions by 1-D SDS-PAGE and MALDI-quadrapole-quadrapole time of flight or MALDI-ion trap mass spectrometry. Although the mammalian NPC complex had been shown to be significantly larger than the NPC in yeast, the authors were able to identify a similar number of proteins from their analysis (~94 total). Through their verification of the proteomic results, it was determined that approximately 30 distinct proteins make up the mammalian NPC, including a few with no discernible yeast homolog [22]. The proteomic analysis was able to detect six novel nucleoporins not previously shown to associate with the nuclear pore.

Studies on the purified nuclear pore provided much insight into the differences between the proteins that make up the yeast and mammalian NPC complex. These studies, however, were very focused on the identification of nuclear pore components and therefore did not investigate the presence of other proteins in the nuclear envelope that may also play important roles in regulating functions in the nucleus. In 2003, Schirmer et al. undertook a slightly different approach in order to gain a compre-hensive analysis of the nuclear envelope isolated from mouse liver nuclei [23]. In this study, the 30 NPC components isolated in the Cronshaw et al. [22] study were detected as well as a number of other nuclear transmembrane proteins. The authors employed MudPIT analysis of the isolated proteins, which provided the afore-mentioned advantages such as limited postisolation sample handling and loss of yield due to chromatography and isolation of proteins from SDS-PAGE gels—of particular concern when studying membrane proteins.

The authors also used an approach referred to as "subtractive proteomics," which took into consideration the high probability of contamination of the nuclear envelope fraction by microsomal membranes. This approach was possible because purification of nuclear-free microsomal membranes is possible, whereas isolation of nuclear envelopes without contamination with microsomal membranes is much less likely. Proteins were therefore only considered to be positively associated with the nuclear envelope if the identified proteins were not detected in samples from isolated microsomal membranes, which lacked any associated nuclear envelope. Analyzing these different protein populations using MudPIT resulted in the identification of 337 proteins that were unique to the nuclear envelope fractions [23]. This study helped to identify a large number of membrane proteins not previously known to associate with the nuclear envelope in previous studies, indicating that 13 integral membrane proteins were in the nuclear envelope, with this study identifying 67 additional proteins.

We will now turn our focus from the nuclear membrane and its complexes to the nucleoplasm and one of its most highly conserved subcompartments, the nucle-olus. The nucleolus is one of the most commonly studied subnuclear components and has recently been the focus of a number of proteomic studies. Nucleoli form around ribosomal RNA gene clusters in the nucleus and serve as centers for the synthesis of rRNA and the subsequent assembly of ribosomal subunits. These regions are transcribed specifically by RNA polymerase I and contain a variety of specific proteins involved in this highly specialized process. These structures have been isolated from nuclei from a number of different species, including *Arabidopsis* and human cells (for review, see Lam et al. [24]).

In 2002, Andersen et al. isolated nucleoli from HeLa cells that were verified using electron microscopy [25]. The proteins were then resolved by 1D or 2D gel electrophoresis and analyzed by MALDI-TOF or nanoelectrospray tandem mass spectrometry. Using the combination of these methods, the authors were able to identify 271 proteins from their nucleoli isolation, 30% of which encoded novel or uncharacterized proteins with unknown localization patterns. Recently, Andersen et al. took their analysis of the nucleolar proteome one step further by using proteomic methods to determine the dynamics of nucleolar proteins when cells were exposed to three different metabolic inhibitors that had previously been shown to affect nucleolar structure [26]. For this analysis, proteins from the nucleoli isolation were separated by one-dimensional gel electrophoresis, followed by multiple runs of nanoscale liquid chromatography mass spectrometry on an ion trap-Fourier transform mass spectrometer. In this study, the authors were able to detect changes in the levels of 489 proteins that were isolated with the nucleoli by making use of stable isotope labeling (using ^{15}N-Arginine).

A number of other subcomponents of the nucleus have also been studied using proteomic methods. This includes interchromatin granules (ICGs), which are found in proximity to a large amount of the DNA in the nucleus. A number of proteins have been localized to these structures using immunofluorescent and immunoelectron microscopy methods, including a number of pre-mRNA splicing factors. It is thought that these regions of the nucleus may serve as storage or assembly sites for splicing factors such that they would be readily accessible upon the activation of transcription (for review, see Lamond and Spector [27]). Proteomic analysis of ICGs from mice was performed using liquid chromatography and tandem mass spectrometry [28]. The analysis resulted in the identification of 146 ICG proteins, including a number of known splicing factors and 32 novel protein candidates of unknown function.

These studies clearly show that the biochemical purification and characterization of these different nuclear compartments can result in the identification of novel protein components, which, it is hoped, will lead to a new understanding of their function in the nucleus. At this time, it is still very difficult to isolate and characterize the transcriptionally active portion of the nucleus that, for the most part, resides in the nucleoplasm. For this reason, characterization of transcriptional regulatory proteins remains a technically challenging area for proteomic study. The nucleus contains a very dynamic subset of proteins that range in abundance from the very abundant histone proteins to some transcription factors present at less than 100 copies per cell in yeast, for which the identification and study of their dynamic behavior would likely prove to be of great interest.

11.2.4 ANALYSIS OF TRANSCRIPTIONAL REGULATORY COMPLEXES FOLLOWING AFFINITY PURIFICATION

As discussed previously, highly abundant proteins can act to shield lower abundance proteins when analyzing complex mixtures by mass spectrometry. For this reason, the analysis of low-abundance proteins, such as transcriptional regulatory proteins, requires that the samples be enriched for the proteins of interest. Up to this point, we have discussed a variety of ways in which researchers have achieved the enrichment of proteins of interest, including organelle (in this case nuclear) purification and subnuclear purification. When looking at the wide variety of proteins that play important roles in the nucleus, however, we can see that there are still quite a number of highly abundant proteins found in the nucleus, including histones and some metabolic enzymes. For this reason, a highly successful approach to the proteomic analysis of transcription factors and transcriptional regulatory proteins includes purification of these proteins, along with their associated proteins, using systems such as immunopurification or tandem affinity purification (TAP) [29,30]. When purifying multiple subunit complexes, proteomic analysis can reveal valuable information, such as identification of all of the complex subunits, the modification state of the proteins in the complex, and even the stoichiometry of the complex. In fact, using approaches such as stable isotope labeling, one could even use proteomic approaches to study the dynamics of a protein complex under different conditions.

The combination of TAP [29,30] and mass spectrometry analysis has become a common method for use in the determination of the subunits that make up a protein complex. It is interesting to note, however, that the method by which the protein complex is analyzed can have a great impact on the results. For instance, gel-based analysis of protein complexes using 1D and 2D electrophoresis in order to separate proteins prior to mass spectrometry can cause some complex components to be overlooked. This includes, in many cases, very small proteins that may not be resolved on SDS-PAGE gels and low-abundance proteins that may not be present at a 1:1 ratio with other protein components. For instance, Lee et al. used MudPIT analysis to study the composition of the well-defined SAGA histone acetyltransferase complex and the Swi/Snf complex from *S. cerevisiae* [31]. Although a number of components of these complexes had been previously identified by mass spectrometry analysis, MudPIT analysis identified Sgf11p as a novel component of SAGA, and Rtt102 as a novel component of Swi/Snf and RSC. This information was also independently discovered previously by two other laboratories using the MudPIT approach [6,32].

It is interesting to note here that although previous data had indicated that the SAGA complex is recruited to active promoters *in vivo* by sequence-specific transcription factors [33], no significant amount of any transcription factor copurified with the SAGA complex. There are many possible explanations for this observation, including that the sequence-specific factors are likely only associated with a small subpopulation of SAGA and that, since their expression level is also very low, they are likely not identified in the mass spectrometry analysis. It is possible, however, that increased enrichment or more sensitive detection methods would be able to facilitate the identification of low-level associated proteins. In fact, the isolation of transcription cofactors associated with transcription factors has been demonstrated when using the transcription factor as the bait for the purification.

In studies by Brand et al., the transcription factor MafK was isolated with a single-step immunoprecipitation using MafK-specific polyclonal antibodies from differentiated and undifferentiated murine erythroleukemia cells where MafK is active and inactive, respectively [34]. Using cysteine-reactive ICAT reagents to label the differentiated samples, the researchers combined the two immunoprecipitations and analyzed them by microcapillary reverse phase liquid chromatography electrospray ionization tandem mass spectrometry. A total of 103 proteins were identified as potential MafK interacting proteins; a number of them were confirmed as bona fide interacting proteins by other methods. This analysis allowed for the identification of two different multiprotein complexes that associate with MafK before and after differentiation and contribute to its ability to regulate β-major globin gene [34]. These studies illustrate that the association of a transcription factor with cofactors such as Swi/Snf or SAGA, which play very diverse roles in the cell, can be accomplished by purifying the complexes by using the lowest abundance protein as bait.

One challenge that this type of sensitive proteomic analysis presents to researchers is to determine whether the additional subunits are stable components of the complex, transiently associated with the complex, or only associated with a subpopulation of the complex. These issues can be addressed, however, using various biochemical characterizations of the complexes. This type of analysis has also been commonly performed and can be very valuable in determining different functions of a complex. For instance, Sato et al. analyzed purified mediator complexes using MudPIT [35]. In their study, they were able to purify different populations of mediator, using biochemical fractionation, which contained or did not contain a kinase module containing CDK11. These different forms of mediator have been hypothesized to play different functional roles in the cell, although their exact function has not been described.

Another exciting role for proteomic analysis of transcriptional regulatory proteins is the identification of dynamic modifications. The transcription factor Met4 is required to regulate the levels of S-adenosylmethionine in *S. cerevisiae*. It has been shown in previous work that Met4 was modified by ubiquitination, although the exact nature of the modification was unknown. Studies by Flick et al. focused on determining the sites of ubiquitin modification on this transcription factor, which is expressed at a level of 1,300 molecules per cell in yeast [36]. Through purification of Met4 and MudPIT analysis of the digested peptides, it was determined that Met4 was modified at K48 by a single ubiquitin molecule.

Studies on purified protein complexes using proteomic methods can also be combined with other technologies in order to increase the amount of information that one can obtain about their function. Recent studies by Tackett et al. illustrate this point, with their analysis of protein complexes that form boundaries around silenced regions of chromatin [37]. Following purification of these complexes and characterization of their associated subunits, the DNA associated with these complexes was also purified and then analyzed by microarray analysis. These studies allowed the authors to gain insight into not only the composition of the complex, but also its localization within the *S. cerevisiae* genome, which supported the hypothesized function of the complex.

11.3 FUTURE PROSPECTS

A major challenge for the development of new proteomic tools and methods is to determine ways to identify low-abundance proteins in complex mixtures. Low-abundance proteins, such as transcription factors and other signaling molecules, play a very dynamic role in the function of the cell. Being able to better monitor their association with different protein complexes and their post-translational modification state would greatly increase the amount of information that we can gain from proteomic analysis of complex mixtures. In the meantime, however, the analysis of low-abundance proteins works very well using current proteomics techniques when the proteins are highly enriched within a protein population.

Proteomic analysis of protein complexes and other complex biological mixtures has increased our ability to understand how proteins cooperate with each other within the context of the cell. Further advances that improve the identification of proteins within a wide dynamic range of protein expression levels will greatly improve our ability to understand dynamic biological pathways. This can be accomplished with the development of more sensitive separation methods for protein mixtures and with improved sensitivity of instrumentation. For example, improvements in mass spectrometry are continually occurring with novel instrumentation systems occasionally introduced. If these systems prove more sensitive than pre-existing systems, proteomic analysis of low-abundance proteins from complex mixtures should improve. With new areas of proteomic research, such as the ability to perform quantitative proteomic analysis of protein groups under different cellular conditions (for review, see Julka and Regnier [38]), proteomics has the potential to provide a snapshot of the cell, including information on protein quantities and post-translational modification state.

REFERENCES

1. Ghaemmaghami, S., Huh, W.K., Bower, K., Howson, R.W., Belle, A., Dephoure, N., O'Shea, E.K., and Weissman, J.S. Global analysis of protein expression in yeast. *Nature* 425, 737–741, 2003.
2. Gygi, S.P., Rochon, Y., Franza, B.R., and Aebersold, R. Correlation between protein and mRNA abundance in yeast. *Mol Cell Biol* 19, 1720–1730, 1999.
3. Futcher, B., Latter, G.I., Monardo, P., McLaughlin, C.S., and Garrels, J.I. A sampling of the yeast proteome. *Mol Cell Biol* 19, 7357–7368, 1999.
4. Washburn, M.P., Koller, A., Oshiro, G., Ulaszek, R.R., Plouffe, D., Deciu, C., Winzeler, E., and Yates, J.R., III. Protein pathway and complex clustering of correlated mRNA and protein expression analyses in *Saccharomyces cerevisiae*. *Proc Natl Acad Sci U S A* 100, 3107–3112, 2003.
5. Washburn, M.P., Wolters, D., and Yates, J.R., III. Large-scale analysis of the yeast proteome by multidimensional protein identification technology. *Nat Biotechnol* 19, 242–247, 2001.
6. Graumann, J., Dunipace, L.A., Seol, J.H., McDonald, W.H., Yates, J.R., III, Wold, B.J., and Deshaies, R.J. Applicability of tandem affinity purification MudPIT to pathway proteomics in yeast. *Mol Cell Proteomics* 3, 226–237, 2004.
7. Swanson, S.K., and Washburn, M.P. The continuing evolution of shotgun proteomics. *Drug Discovery Today* 10, 719–725, 2005.
8. Paoletti, A.C., Zybailov, B., and Washburn, M.P. Principles and applications of multi-dimensional protein identification technology. *Expert Rev Proteomics* 1, 275–282, 2004.
9. Delahunty, C., and Yates, J.R., III. Protein identification using 2D-LC-MS/MS. *Methods* 35, 248–255, 2005.

10. Washburn, M.P., Ulaszek, R., Deciu, C., Schieltz, D.M., and Yates, J.R., III. Analysis of quantitative proteomic data generated via multidimensional protein identification technology. *Anal Chem* 74, 1650–1657, 2002.

11. Liu, H., Sadygov, R.G., and Yates, J.R., III. A model for random sampling and estimation of relative protein abundance in shotgun proteomics. *Anal Chem* 76, 4193–4201, 2004.

12. Wu, C.C., MacCoss, M.J., Mardones, G., Finnigan, C., Mogelsvang, S., Yates, J.R., III, and Howell, K.E. Organellar proteomics reveals Golgi arginine dimethylation. *Mol Biol Cell* 15, 2907–2919, 2004.

13. Sickmann, A., Reinders, J., Wagner, Y., Joppich, C., Zahedi, R., Meyer, H.E., Schonfisch, B., Perschil, I., Chacinska, A., Guiard, B., Rehling, P., Pfanner, N., and Meisinger, C. The proteome of *Saccharomyces cerevisiae* mitochondria. *Proc Natl Acad Sci USA* 100, 13207–13212, 2003.

14. Bagshaw, R.D., Mahuran, D.J., and Callahan, J.W. A proteomic analysis of lysosomal integral membrane proteins reveals the diverse composition of the organelle. *Mol Cell Proteomics* 4, 133–143, 2005.

15. Kikuchi, M., Hatano, N., Yokota, S., Shimozawa, N., Imanaka, T., and Taniguchi, H. Proteomic analysis of rat liver peroxisome: presence of peroxisome-specific isozyme of Lon protease. *J Biol Chem* 279, 421–428, 2004.

16. Bae, M.S., Cho, E.J., Choi, E.Y., and Park, O.K. Analysis of the *Arabidopsis* nuclear proteome and its response to cold stress. *Plant J* 36, 652–663, 2003.

17. Shakib, K., Norman, J.T., Fine, L.G., Brown, L.R., and Godovac-Zimmermann, J. Proteomics profiling of nuclear proteins for kidney fibroblasts suggests hypoxia, meiosis, and cancer may meet in the nucleus. *Proteomics*, 5, 2819–2838, 2005.

18. Turck, N., Richert, S., Gendry, P., Stutzmann, J., Kedinger, M., Leize, E., Simon-Assmann, P., Van Dorsselaer, A., and Launay, J.F. Proteomic analysis of nuclear proteins from proliferative and differentiated human colonic intestinal epithelial cells. *Proteomics* 4, 93–105, 2004.

19. Lamond, A.I., and Sleeman, J.E. Nuclear substructure and dynamics. *Curr Biol* 13, R825–828, 2003.

20. Spector, D.L. Macromolecular domains within the cell nucleus. *Annu Rev Cell Biol* 9, 265–315, 1993.

21. Rout, M.P., Aitchison, J.D., Suprapto, A., Hjertaas, K., Zhao, Y., and Chait, B.T. The yeast nuclear pore complex: Composition, architecture, and transport mechanism. *J Cell Biol* 148, 635–651, 2000.

22. Cronshaw, J.M., Krutchinsky, A.N., Zhang, W., Chait, B.T., and Matunis, M.J. Proteomic analysis of the mammalian nuclear pore complex. *J Cell Biol* 158, 915–927, 2002.

23. Schirmer, E.C., Florens, L., Guan, T., Yates, J.R., 3rd, and Gerace, L. Nuclear membrane proteins with potential disease links found by subtractive proteomics. *Science* 301, 1380–1382, 2003.

24. Lam, Y.W., Trinkle-Mulcahy, L., and Lamond, A.I. The nucleolus. *J Cell Sci* 118, 1335–1337, 2005.

25. Andersen, J.S., Lyon, C.E., Fox, A.H., Leung, A.K., Lam, Y.W., Steen, H., Mann, M., and Lamond, A.I. Directed proteomic analysis of the human nucleolus. *Curr Biol* 12, 1–11, 2002.

26. Andersen, J.S., Lam, Y.W., Leung, A.K., Ong, S.E., Lyon, C.E., Lamond, A.I., and Mann, M. Nucleolar proteome dynamics. *Nature* 433, 77–83, 2005.

27. Lamond, A.I., and Spector, D.L. Nuclear speckles: A model for nuclear organelles. *Nat Rev Mol Cell Biol* 4, 605–612, 2003.

28. Saitoh, N., Spahr, C.S., Patterson, S.D., Bubulya, P., Neuwald, A.F., and Spector, D.L. Proteomic analysis of interchromatin granule clusters. *Mol Biol Cell* 15, 3876–3890, 2004.

29. Puig, O., Caspary, F., Rigaut, G., Rutz, B., Bouveret, E., Bragado-Nilsson, E., Wilm, M., and Seraphin, B. The tandem affinity purification (TAP) method: A general procedure of protein complex purification. *Methods* 24, 218–229, 2001.

30. Rigaut, G., Shevchenko, A., Rutz, B., Wilm, M., Mann, M., and Seraphin, B. A generic protein purification method for protein complex characterization and proteome exploration. *Nat Biotechnol* 17, 1030–1032, 1999.

31. Lee, K.K., Prochasson, P., Florens, L., Swanson, S.K., Washburn, M.P., and Workman, J.L. Proteomic analysis of chromatin-modifying complexes in *Saccharomyces cerevisiae* identifies novel subunits. *Biochem Soc Trans* 32, 899–903, 2004.

32. Powell, D.W., Weaver, C.M., Jennings, J.L., McAfee, K.J., He, Y., Weil, P.A., and Link, A.J. Cluster analysis of mass spectrometry data reveals a novel component of SAGA. *Mol Cell Biol* 24, 7249–7259, 2004.

33. Brown, C.E., Howe, L., Sousa, K., Alley, S.C., Carrozza, M.J., Tan, S., and Workman, J.L. Recruitment of HAT complexes by direct activator interactions with the ATM-related Tra1 subunit. *Science* 292, 2333–2337, 2001.

34. Brand, M., Ranish, J.A., Kummer, N.T., Hamilton, J., Igarashi, K., Francastel, C., Chi, T.H., Crabtree, G.R., Aebersold, R., and Groudine, M. Dynamic changes in transcription factor complexes during erythroid differentiation revealed by quantitative proteomics. *Nat Struct Mol Biol* 11, 73–80, 2004.

35. Sato, S., Tomomori-Sato, C., Parmely, T.J., Florens, L., Zybailov, B., Swanson, S.K., Banks, C.A., Jin, J., Cai, Y., Washburn, M.P., Conaway, J.W., and Conaway, R.C. A set of consensus mammalian mediator subunits identified by multidimensional protein identification technology. *Mol Cell* 14, 685–691, 2004.

36. Flick, K., Ouni, I., Wohlschlegel, J.A., Capati, C., McDonald, W.H., Yates, J.R., and Kaiser, P. Proteolysis-independent regulation of the transcription factor Met4 by a single Lys 48-linked ubiquitin chain. *Nat Cell Biol* 6, 634–641, 2004.

37. Tackett, A.J., Dilworth, D.J., Davey, M.J., O'Donnell, M., Aitchison, J.D., Rout, M.P., and Chait, B.T. Proteomic and genomic characterization of chromatin complexes at a boundary. *J Cell Biol* 169, 35–47, 2005.

38. Julka, S., and Regnier, F. Quantification in proteomics through stable isotope coding: a review. *J Proteome Res* 3, 350–363, 2004.

12 Electrophoretic NMR of Protein Mixtures and Its Proteomic Applications

Qiuhong He, Sunitha B. Thakur, and Jeremy Spater

CONTENTS

12.1 ABSTRACT

Electrophoretic nuclear magnetic resonance (ENMR) has the potential of becoming an important proteomic technique. The method offers possible structural characterization of protein interactions in biological signaling processes. The proteomic analyses currently employed offer separation-based methods using two-dimensional gel electrophoresis or mass spectrometry to identify proteins. In these proteomic analyses, information on protein structures is lost and therefore the methods cannot be used to characterize protein conformational changes in biochemical reactions.

Although mass spectrometry is a powerfully sensitive technique to identify proteins from their peptide fragments, a reliable proteomic analysis based on this method alone is subject to the limited accuracy for characterizing protein structure with bioinformatic analysis. Therefore, novel proteomic techniques are needed to identify proteins accurately as well as study their three-dimensional structural changes during protein signaling events.

X-ray crystallography and multidimensional NMR spectroscopy are two methods that are widely used for determining the three-dimensional structures of pure proteins or strongly bound protein complexes. For multicomponent protein systems with weak protein interactions, NMR spectroscopy holds an advantage over x-ray crystallography for studying functional protein interactions *in situ* in aqueous solutions. However, severe signal overlap has often prevented the full structural determination of coexisting proteins in solution. Conventional NMR investigations of dynamic protein interactions are limited to those multicomponent systems with resolvable chemical shifts between proteins and small molecules. In this chapter, we will illustrate the potential of ENMR in proteomic analysis with efficient characterization of proteomic proteins and protein interactions in solution. Several important ENMR techniques will be reviewed. These include structural characterization of protein conformational changes in biochemical reactions; capillary-array ENMR techniques to reduce heat-induced convection; application of the maximum entropy method to resolve protein signals without truncation artifacts; and development of microcoil ENMR to improve signal sensitivity.

12.2 INTRODUCTION

A new era of molecular medicine is arriving, as is demonstrated by the increasing amount of genomic and proteomic research being conducted today. Effective analysis of genomic and proteomic information from serum or cell extracts of tissue specimens can now be performed on individual patients. As the disease protein profiles become available for individual patients to tailor drug therapies, the idea of "personalized medicine" may become a reality in the future [1]. Currently, proteomic techniques are mostly based on separation methods using two-dimensional gel electrophoresis [2] and mass spectrometry (e.g., MALDI-TOF) [3,4]. In the latter, bioinformatic analysis is typically required in order to piece together mass spectral information from protein fragments and to identify the proteomic proteins.

The major technical challenges reside not only in the extraction of reliable disease markers from the overwhelming proteomic data matrix for the purposes of clinical diagnosis, but also in the detection of protein conformational changes in proteomic samples for studying biological signaling events. Novel high-throughput proteomic technologies are needed in order to identify proteins and protein interactions reliably in normal and pathological conditions. To address some of these issues, we have proposed a proteomic electrophoretic NMR (ENMR) approach [5]. By sorting the NMR signals of individual proteins in the dimension of electrophoretic flow, the ENMR technique may identify proteins based on their signature NMR spectral patterns. The technique is potentially applicable to characterizing protein interactions in biological signaling processes.

X-ray crystallography and multidimensional NMR spectroscopy are the two primary methods used to determine the three-dimensional structures of pure proteins or strongly bound protein complexes. For multicomponent protein systems, NMR spectroscopy has an advantage in studying functional protein interactions *in situ* in aqueous solutions, without the requirement of protein crystallization. However, the characterization of proteomic samples is difficult using conventional NMR methods, since severe NMR signal overlap from different protein components has prevented full structural determination of coexisting proteins. Most NMR investigations of weakly interacting protein systems are limited to those with resolvable chemical shifts between proteins and small molecules.

In ENMR experiments, the NMR signals from different proteins or protein conformations in solution can be displayed at distinct resonance frequencies in a new dimension of electrophoretic flow [6–14], making it possible to study multiple protein structures and identify proteins in proteomic samples from body fluid, cell extracts, and tissue specimens. These ENMR proteomic profiles may be useful for disease diagnosis and prognosis. The critical technical developments in this direction include multidimensional ENMR to characterize protein mixtures; capillary array ENMR (CA-ENMR) [15] and convection-compensated ENMR (CC-ENMR) [16], to study proteins in biological buffer conditions; high-resolution ENMR signal processing using maximum entropy method (MEM) [40]; and the development of capillary-microcoil ENMR probes to meet the demands of signal sensitivity and flow resolution in clinical proteomics. Structural visualization of protein reaction interfaces (or "active pockets") by multidimensional ENMR may be used to monitor protein conformational changes during biological signaling processes [5]. ENMR applications in proteomics may give new insights into the molecular mechanisms of human diseases, leading to discoveries of novel diagnostic markers and drug targets and design of effective therapeutic interventions.

12.3   BASIC PRINCIPLES OF ELECTROPHORETIC NMR

Measurement of electrophoretic flow of ionic species was first attempted by Packer et al. in 1972 [17], followed by the first successful demonstration by Holz et al. in the 1980s using a horizontal-bore electromagnetic NMR spectrometer [18–21]. Johnson et al. later pioneered the development of the high-resolution 1D and 2D ENMR techniques on a vertical-bore superconducting NMR spectrometer [7–9,13,14,22–24]. More recently, He et al. extended 2D ENMR into multidimensional ENMR in order to simultaneously examine structures of coexisting proteins in aqueous solution and their structural changes during protein interactions [5,10,11]. To study protein samples in high-salt biological buffer solutions, we developed CA-ENMR and CC-ENMR techniques to remove the severe spectral artifacts due to heat-induced convection [15,16]. The reader should consult several reviews on this novel technique and its historical landmarks [5,12,25–29]. Diffusion and flow measurements using NMR were established in Stejskal and Tanner's analysis of spin dynamics [30,31]. In the classic Bloch equations [32], modified with flow and diffusion terms, the flow modulation of the nuclear spin magnetization is described as:

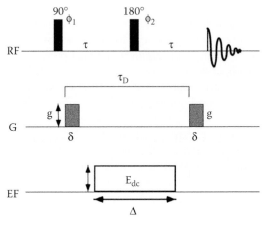

Phase cycling procedures:

ϕ_1:   +x −x −x +x +y −y −y +y
ϕ_2:   +y −y −y +y +x −x −x +x
ACQ: +x −x −x +x +y −y −y +y

FIGURE 12.1 The spin–echo ENMR pulse sequence.

$$\frac{\partial \vec{M}}{\partial t} = \gamma \vec{M} \times \vec{B} - \frac{\left(M_x \hat{x} + M_y \hat{y}\right)}{T_2} + \frac{\left(M_0 - M_z\right)\hat{z}}{T_1} - \nabla \cdot \vec{v}\vec{M} + \nabla \cdot \tilde{D} \cdot \nabla \vec{M} \quad (12.1)$$

where $\vec{v}$ is the velocity vector and $\tilde{D}$ is the diffusion tensor. The magnetic field gradient $\vec{G}(t)$ is uniform throughout the sample, nuclear magnetization $\vec{M} = M_x \hat{x} + M_y \hat{y} + M_z \hat{z}$ is in an external magnetic field, and $\vec{B} = \left[B_0 + \left(\vec{r} \cdot \vec{G}\right)\right]\hat{z}$, assuming the spin–lattice (T_1) and spin–spin relaxation (T_2) processes obey first-order kinetics. In a Hahn spin–echo-based ENMR experiment [33] using pulsed field gradients (fig. 12.1),

$$\vec{G}(t) = \begin{cases} \vec{g}_0 & 0 < t < t_1 \\ \vec{g}_0 + \vec{g} & t_1 < t < t_1 + \delta < \tau \\ \vec{g}_0 & \text{for} \quad t_1 + \delta < t < t_1 + \tau_D > \tau \\ \vec{g}_0 + \vec{g} & t_1 + \tau_D < t < t_1 + \tau_D + \delta < 2\tau \\ \vec{g}_0 & t_1 + \tau_D + \delta < t \end{cases} \quad (12.2)$$

with B_0 field inhomogeneity $\vec{g}_0 \to 0, \delta \to 0$ and the area of $\delta\vec{g}$ remains finite, where $\vec{g}$ and δ are the amplitude and duration of the pulsed magnetic field gradient. In ENMR experiments, an applied DC electric field between the pulsed magnetic field gradients drives ionic particles to move in the direction of the electric field (fig. 12.1). Thus, the solution of equation 12.1 at the echo maximum $(\tau' = 2\tau)$ is [9,12,23]:

$$M_i(2\tau) = M_i(0) \exp[-K^2 D_i (\tau_D - \tfrac{1}{3}\delta) - \frac{2\tau}{T_{2i}}] \exp(-iK\mu_i E_{dc}\Delta_E) . \quad (12.3)$$

where E_{dc} and Δ_E are the amplitude and duration of the pulsed DC electric field and μ_i is the electrophoretic mobility of the ith molecular component in the solution. The electric field is

$$E_{dc} = \frac{I_e}{\sigma A}$$

where

σ is the conductivity of the electrolyte solution
A is the cross-sectional area of the sample tube
I_e is the DC electric current
$K = \gamma\delta g$ is the magnetization helix pitch

The electrophoretic migration of ionic charges can be measured by a modified spin-echo procedure in which the DC electric field pulse is inserted between the gradient pulses and applied along the gradient direction $\hat{z}$. Using a cylindrical tube, ions flowing from one position to another along $\hat{z}$ during the time delay Δ_E experience a coherent phase change of $\phi_E = Kv\Delta_E$, which is recorded as a phase modulation of the spin-echo signal as a function of electric field strength. Molecular diffusion introduces an exponential decay of the ENMR signal, where D_i is the diffusion coefficient of the ith species. Since electrophoretic mobility is a molecular parameter, NMR signals of a migrating molecule can be displayed at the same frequency in a new dimension of electrophoretic flow by Fourier transformation with respect to the electric field amplitude E_{dc} or electric field duration Δ in a 2D ENMR experiment.

12.4 ONE-DIMENSIONAL ELECTROPHORETIC NMR OF AMINO ACIDS, PEPTIDES, AND PROTEINS

ENMR techniques were developed using small charged molecules including amino acids, metabolites, and small peptides [6,10,12,16]. To demonstrate the feasibility of ENMR for studying biological samples, we acquired a series of 1D ENMR data from a sample containing a mixture of 1 mM Ala-Gly-Gly and 1 mM ethylenediamine ($NH_2CH_2CH_2NH_2$) in D_2O (fig. 12.2) [6,12]. The distinctive chemical shifts of the three molecules allowed us to monitor their electrophoretic migrations in an electric field simultaneously. As the electric current was increased incrementally from 0 to 0.45 mA, the peptide and the protonated diamine generated different electrophoretic flow oscillating patterns. From the oscillating frequencies, the electrophoretic mobilities of the two charged molecules were determined as μ(peptide) $= 2.8 \times 10^{-4}$ cm^{2}V^{-1}s^{-1} and μ (amine) $= 4.5 \times 10^{-4}$ cm^{2}V^{-1}s^{-1}. The water signal amplitude remained constant, suggesting negligible heating and electroosmotic effects in this experiment.

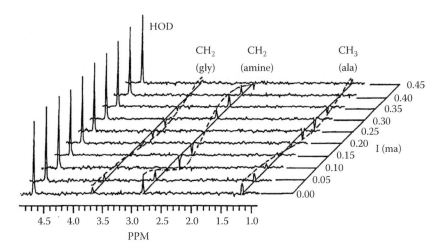

FIGURE 12.2 ENMR spectra acquired on a Bruker 250-MHz spectrometer, for a mixture of 1 m*M* Ala-Gly-Gly and 1 m*M* ethylenediamine in D$_2$O. (Johnson, C.S., Jr., and He, Q. In *Advanced Magnetic Resonance*, vol. 13, ed. W.S. Warren. San Diego: Academic Press, 1989, 131–159. With permission.)

Another ENMR experiment was successfully carried out using the transmembrane segment S4 peptide of the *shaker* K$^+$ channel protein (2.8 kDa). The S4 peptide, which contains 23 amino acid residues [ILR$^+$VIR$^+$LVR$^+$VFR$^+$IFK$^+$LSR$^+$HSK$^+$GL], with six basic amino acids, is believed to control the gating of excitable ion channels including sodium, potassium, and calcium channels. Our experimental design was based on a hypothesis that the gating current of the *shaker* potassium channel originates from conformational changes and charge movement of S4 in cell membranes. Depending on the number of charges in S4 and its backbone flexibility, the interaction of the electric field with charges in S4 may induce conformational changes. The stable conformations of S4 in the presence and absence of an electric field may be determined by high-resolution ENMR.

Experimentally, we tested the conformational difference of the positively charged S4 peptide segments dissolved in D$_2$O as well as in a mixture of solvent CF$_3$CF$_2$OH and D$_2$O at a 4:1 ratio in order to mimic the cell membrane environment. The conductivity of the S4 solution was $\kappa = 0.16$ mΩ^{-1} in mixed solvent. The different NMR spectral patterns of S4 were observed in the two solvent conditions (fig. 12.3). The S4 in a 4:1 mixture of CF$_3$CF$_2$OH and D$_2$O gave an NMR spectrum of the folded S4 conformation (fig. 12.3a), which may be similar to that in the cell membrane environment. The S4 in D$_2$O appeared to be denatured (fig. 12.3b). Both conformations migrated in the electric field, as observed by the electrophoretic cosinusoidal interferograms (fig. 12.3).

The ENMR experiments were carried out at 25°C using an electric field generator of 1 kV (maximum). Two Pt-electrodes were inserted into the S4 solution on two ends of a large U-shaped glass tube connected to an electric wire descending from the top of the magnet of a Bruker AM 500 MHz spectrometer. The S4 migration in the electric field, for both solvent conditions, indicated that S4 segment in *shaker*

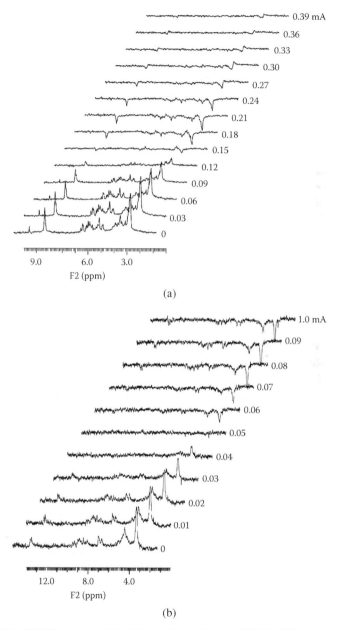

FIGURE 12.3 ENMR spectra of the S4 peptide acquired at 25°C in different solvent conditions with the stimulated-echo ENMR sequence. (a) ENMR of S4 in the mixed solvent of CF_3CF_2OD and D_2O at a volume ratio of 4:1. $\tau = 1.535$ ms, $\tau_D = 0.6102$ s. The electric current changes from 0 to 0.4 mA. Other parameters are the same. (b) S4 peptide in D_2O (0.5 mM). The electric current was incremented from 0 to 1 mA with a duration $\Delta = 210.05$ ms. Gradients $g_1 = g_2 = 48.72$ G cm^{-1}, $\delta_1 = 1$ ms, $\delta_2 = 8$ ms, T_2 delay $\tau = 2.525$ ms, diffusion delay $\tau_D = 214.6218$ ms, and T_1 delay $T_R = 5$ s.

K^+ channel could move in lipid membranes, responding to the voltage gating of membrane potential. The S4 movement in lipid membranes, which could couple to any S4 conformational changes, may be responsible for the voltage gated opening of the *shaker* K^+ ion channels. The migration rates of S4 and its mutations may be studied by ENMR in lipid membrane mimics using the 4:1 mixture of CF_3CF_2OH and D_2O.

A series of proteins were investigated in order to determine the size and concentration limits in ENMR studies. The electrophoretic motion of the positively charged lysozyme (14.3 kDa) was detected in a D_2O solution at 3.0 mM concentration (pH = 7.89, pI = 11). Similar flow oscillation patterns were also observed for the negatively charged pepsin (34.4 kD, 1.0 mM in D_2O, pH = 4.19, pI = 2.86 at 4°C). The 66-kDa bovine serum albumin (BSA, pI = 5.85) had a very slow electrophoretic flow rate in D_2O; however, the negatively charged BSA migrated when the solution pH changed to 9.72 following ethylenediamine treatment. The largest protein we investigated was the 480-kDa urease treated with 1 mM ethylenediamine. Because of the signal superposition in the narrow proton chemical shift range, only 20 μM (not 20 mM) urease in D_2O was necessary for observation of the electrophoretic motion of this protein. As in conventional NMR experiments of proteins, the ENMR for protein structure determination requires a protein concentration in the millimolar range.

The net charge of a protein can be tuned in order to obtain the desired electrophoretic interferograms. For example, at pH = 6.52, no observable electrophoretic motion was detected for 1.09 mM ubiquitin (8.6 kDa) in D_2O near its isoelectric point (fig. 12.4a). However, after adding ethylenediamine (pK_{a1} = 10.7 and pK_{a2} = 7.5) to the solution, the ubiquitin became negatively charged (pH = 10.4) and migrated in the electric field (fig. 12.4b).

12.5   ENMR OF PROTEINS IN BIOLOGICAL BUFFER SOLUTIONS

In the earlier days of ENMR, when large glass tubes were used as sample chambers, the study of proteins in biological buffer solutions of high salt concentration was impossible due to the heat-induced convection. For such protein solutions, the electric conductivity (σ) is high and the electric field ($E_{dc} = I_e/\sigma A$) is too low to drive the electrophoretic motion of proteins in the large U-tube glass chambers of fixed cross-sectional area (A). The high electric field required for this purpose would be accompanied by a relatively large electric current (I_e) and, subsequently, by large heat-induced convection that would prevent accurate ENMR measurements of protein migration. To control sample temperature, heat is removed in ENMR experiments by cooling air or liquid outside the sample tube [11,23,34]. Since heat is more effectively removed at the edge of the tube rather than the tube center, a temperature gradient is established, which in turn produces a density gradient to cause bulk convective motion of the solution. Obviously, the electrophoretic flow superimposed on an irregular convective flow is very difficult, if not impossible, to measure. Two ENMR techniques—capillary array electrophoretic NMR (CA-ENMR) [15] and convection compensated ENMR (CC-ENMR) [16]—were designed to address this issue.

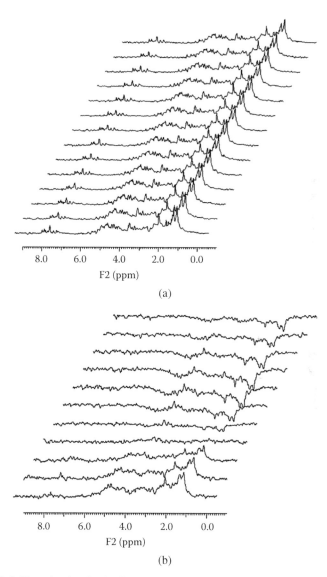

FIGURE 12.4 The stimulated echo ENMR spectra of ubiquitin. (a) Without treatment with ethylenediamine, the neutral ubiquitin does not move in the electric field (signal does not oscillate); (b) after treatment with 10 mM ethylenediamine, the charged ubiquitin molecules (0.5 mM) move in the electric field (signal oscillates). (He, Q. et al. *J Am Chem Soc*, 120, 1341–1342, 1998. With permission.)

In CA-ENMR, a capillary array sample chamber (fig. 12.5) is constructed to break the convective current loops and the electrical eddy current because of the limited physical space available for molecular motion. In the resulting smaller cross-sectional area of the sample chamber, the strength of the electric field increases at the same output voltage (~1 kV maximum). The electrophoretic oscillations of

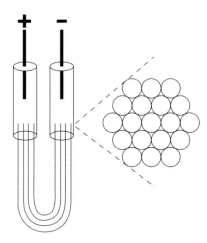

FIGURE 12.5 Schematics of a CA-ENMR sample cell. Physical barriers are imposed to block convective flow. (He, Q. et al. *J Magn Reson*, 141, 355–359, 1999. With permission.)

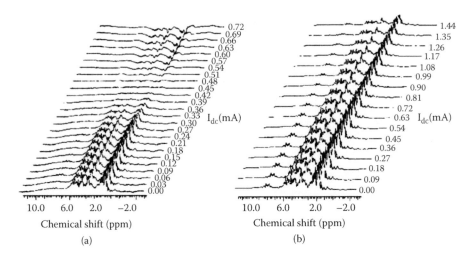

FIGURE 12.6 Electrophoretic interferograms of 1 m*M* lysozyme in a D_2O solution of 50 m*M* NaH_2PO_4. Data were obtained at 25°C with (a) a 12-bundle capillary array U-tube (I.D.= 250 μm) that gave electrophoretic signal oscillation; and (b) a conventional glass U-tube that could not produce the same electrophoretic oscillation curve due to a weak electric field. (He, Q. et al. *J Magn Reson*, 141, 355–359, 1999. With permission.)

lysozyme were observed for the ENMR measurements. Dramatically reduced convection in high-salt conditions was demonstrated using a solution of 1 m*M* lysozyme in 50 m*M* NaH_2PO_4/D_2O (fig. 12.6). The sample conductivity is $\sigma = 4.21$ mS•cm⁻¹ in the pH = 5.89 solution. Alternatively, the CC-ENMR method was developed to sensitize electrophoretic motion in the presence of bulk convective flow [16]. Gradient moment nulling and switching polarity of the applied DC electric field to generate electrophoretic signal oscillations accomplished this task. The method was

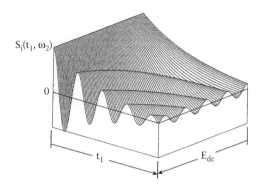

$S_i(t_1, \omega_2)$

0

t_1

E_{dc}

FIGURE 12.7 The ENMR signal intensity versus the duration (t_1) and the amplitude (E_{dc}) of the electric field. (He, Q. and Johnson, C.S. Jr. *J Magn Reson*, 81, 435, 1989. With permission.)

demonstrated using a high-salt solution containing 100 mM L-aspartic acid and 100 mM 4,9-dioxa-1,12-dodecanediamine in D_2O.

12.6 TWO-DIMENSIONAL ENMR SIGNAL SEPARATION OF COEXISTING PROTEINS IN SOLUTION

A decade ago, the principle of 2D ENMR was demonstrated using a spin-echo sequence on a high-resolution superconducting NMR spectrometer (fig. 12.1) [9]. A new dimension of electrophoretic flow was introduced as the second dimension; the chemical shift was displayed in the first dimension [6–9,12]. Signals from different ionic species can be separated in the new flow dimension according to different electrophoretic mobilities. Two different approaches were developed to generate 2D ENMR spectra (fig. 12.7) by Fourier transformation of the oscillating electrophoretic interferogram with respect to the duration of electric field t_1 (at a fixed electric field amplitude E_{dc}) or the electric field amplitude E_{dc} (at its fixed duration t_1).

12.6.1 MEASUREMENT OF ELECTROPHORETIC MOBILITY DISTRIBUTIONS BY 2D ENMR

The first approach was successfully tested experimentally with an oil/water micro-emulsion using a spin-echo 2D ENMR sequence (fig. 12.1). In the electrophoretic flow dimension, the cyclohexane oil core presented the same resonance frequency as the hydrophilic head group of the ionic surfactant because they belong to the same microemulsion droplet [9]. Resonances S2 and S3 from the surfactant hydro-phobic tail diminished due to fast T_2-relaxation, and the water signal disappeared due to diffusion broadening. When we replaced cyclohexane with 1,3,5-triisopropyl-benzene (TIPB) in the oil core, we observed the split TIPB doublet signal in the dimension of electrophoretic flow due to spin J-coupling interaction (fig. 12.8a) [8]. This is the first experimental evidence showing that 2D ENMR may be used to obtain information for determining the chemical structures of migrating molecules, including proteins (e.g., chemical shifts and J-coupling constants, etc.).

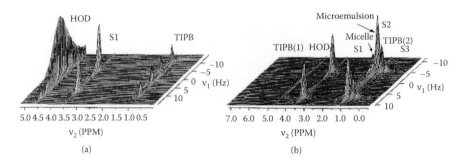

FIGURE 12.8 (a) A 2D ENMR spectrum of an oil/water microemulsion made by mixing 1,3,5-triisopropylbenzene (TIPB), Brij 30, sodium dodecyl sulfate (SDS), and D_2O. (b) A 2D stimulated-echo ENMR spectrum from the same microemulsion. The conductivity $\kappa = 0.121$ mS•cm^{-1}. S2 and S3 show two resonance components that have the same chemical shift; one comes from the micelle and the other from the microemulsion droplets. (He, Q. and Johnson, C.S., Jr. *J Magn Reson*, 85, 181–185, 1989. With permission.)

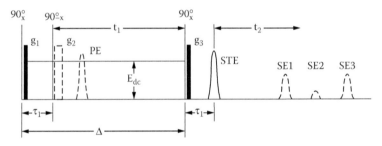

FIGURE 12.9 The stimulated-echo ENMR sequence designed to suppress resonance splitting due to spin scalar interactions and to recover the signal that has a fast T_2 relaxation. Primary (PE) and secondary (SE1, SE2, and SE3) echoes that may interfere with the stimulated echo (STE) are suppressed by gradient g_2 or by phase cycling procedures. The second 90° pulse stores half of the magnetization helix in the z-direction so that spin relaxes by T_1, rather than by T_2, during the ENMR experiment. The last 90° pulse converts the nondetectable z-magnetization into an observable transverse magnetization. Gradient g_1 labels the spin positions and g_2 refocuses the recovered transverse magnetization into a stimulated echo. (He, Q. et al. *J Magn Reson*, 91, 654–658, 1991. With permission.)

To measure electrophoretic mobility, this J-splitting effect in the flow dimension was removed (fig. 12.8b) using the stimulated-echo ENMR sequence (fig. 12.9) [8]. As compared with the spin-echo ENMR sequence, the stimulated-echo ENMR sequence employed an additional 90° pulse that flipped half of the magnetization helix into the z-direction, in which spins relaxed with the longer relaxation time T_1. The last 90° pulse was applied to convert the nonobservable z-magnetization into a detectable stimulated echo in the transverse plane. The dephasing gradient g_1 was balanced by the refocusing gradient g_3, whereas the spoiler gradient g_2 dephased the undesirable signals. Using the stimulated echo ENMR sequence, we recovered the missing S2 and S3 surfactant resonances from the hydrophobic tails (fig. 12.8b). Interestingly, the recovered S2 and S3 resonances displayed a slower flowing component in the second dimension (fig. 12.8b). The fast migrating component had a diffusion coefficient $D = 1.1 \times 10^{-6}$ cm^{2}s^{-1}, corresponding to an effective hydrodynamic radius of 23 Å; the slower flowing microemulsion had a diffusion coefficient $D = 2.6 \times 10^{-6}$ cm^{2}s^{-1}, corresponding to a hydrodynamic radius of 9 Å. These data suggest that the slow moving component is a micelle formed from the excess of the nonionic surfactant Brij 30 in the microemulsion solution [8].

The second type of 2D ENMR was demonstrated by Fourier transformation of the electrophoretic interferogram with respect to the electric field amplitude at constant duration of electric field [7,22]. Since the evolution time was fixed for diffusion and spin relaxation (fig. 12.7), the line broadening in the flow dimension due to these processes was removed. Therefore, line width in the flow dimension can be used to measure a continuous distribution of electrophoretic mobility. This was demonstrated by using the polydisperse unilamellar phospholipid vesicles (about 0.22 μm) obtained by mixing 30 mL mixed egg phosphatidylcholine (EPC) and dioleoyl phosphotidylglycerol (DOPG) (EPC:DOPG = 4:1), with or without 0.3 M glucose in D$_2$O [7]. These data were acquired with the two-dimensional stimulated ENMR sequence (fig. 12.9). The vesicles with narrow size distribution (the ones without glucose inside) gave an electrophoretic interferogram of lipid protons with a slight damping as the amplitude of the electric field was increased (fig. 12.10a). Conversely, the vesicles with a high degree of polydispersity (the ones with glucose inside) gave a dramatic damping in the interferogram, due to the interfering glucose signals carried by variously sized migrating vesicles (fig. 12.10b). Fourier transformation of the electrophoretic interferogram displayed the distribution of the electrophoretic mobilities (fig. 12.10c).

12.6.2 Two-Dimensional ENMR of Proteins and Protein Mixtures

Electrophoretic mobility measurements of lysozyme (14 kDa) and BSA (67 kDa) were successful in the early years of 1D and 2D ENMR development, using the stimulated echo ENMR sequence (fig. 12.9) [6]. For example, the 2D stimulated echo ENMR spectrum was acquired for a 0.2-mM lysozyme solution in D$_2$O (fig. 12.11), giving an electrophoretic mobility of lysozyme of 2.7×10^{-4} cm^{2}V^{-1}s^{-1} (25°C). In this experiment, the 2D data matrix was mapped by incrementing the duration of the electric field, which was responsible for line broadening in the flow dimension by the fast T_2-relaxation of lysozyme. This line broadening effect can be removed by incrementing the amplitude of the electric field or by using the CT-ENMR technique that will be described later. Although the truncated ENMR data for small molecules were successfully analyzed using the linear prediction algorithm [13,14], the maximum entropy method (MEM) gives superior results for protein ENMR data, as is shown in the next section.

A 2D ENMR study of noninteracting protein mixtures of BSA and ubiquitin in D$_2$O demonstrated that the overlapping protein signals can be separated by their electrophoretic mobilities in the flow dimension [11]. The experiment proved the feasibility of simultaneous structural determination of proteins using three-dimensional electrophoretic NMR [10]. Maximum entropy method (MEM) or Fourier transformation (FFT) was applied in the electrophoretic flow dimension for a comparison, whereas

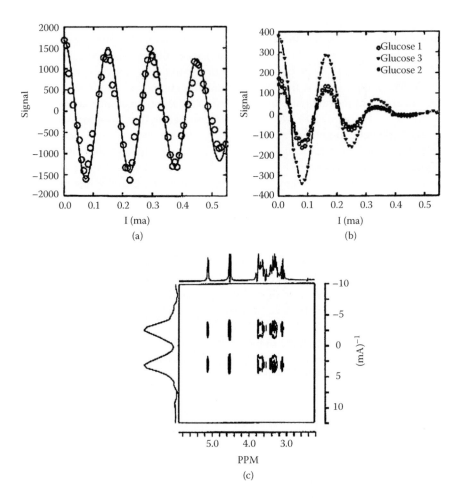

FIGURE 12.10 Mapping the electrophoretic mobility distributions of phospholipid vesicles. (a) The interferogram of the lipid protons (3.11 ppm) from a vesicle shows a narrow size distribution. (b) The interferogram of glucose trapped inside a different vesicle shows a heterogeneous size distribution. (c) The 2D ENMR data display mobility distribution in the electrophoretic flow dimension from the more polydisperse vesicle. (He, Q. et al. *J Magn Reson*, 91, 654–658, 1991. With permission.)

Fourier transformation was applied in the chemical shift dimension (fig. 12.12). No physical separation of the proteins was necessary to obtain NMR spectra of BSA and ubiquitin, different from the proteomic separation methods by gel electrophoresis. The method can be employed to study protein interactions and biological signaling events in proteomic ensembles of biomacromolecules in aqueous solution.

12.6.3 ENHANCED 2D ENMR RESOLUTION BY MAXIMUM ENTROPY METHOD

In our current ENMR experiments, the available electric field is limited to about 100 V•cm^{-1} before the onset of heat-induced conduction. The truncation in electrophoretic

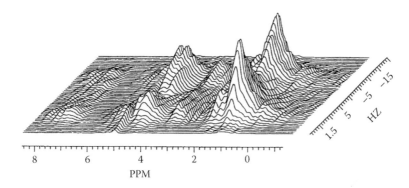

FIGURE 12.11 Two-dimensional stimulated-echo ENMR spectrum of 0.2 mM lysozyme in D$_2$O. (He, Q. Ph.D. thesis, University of North Carolina at Chapel Hill, 1990.)

interferograms often generates spectroscopic artifacts in the 2D ENMR spectra, with reduced protein signal resolution in the flow dimension. Therefore, we have tested a few ENMR signal processing methods, including linear prediction, wavelet denoising algorithms, and the maximum entropy method (MEM), to obtain high-resolution ENMR spectra of proteins. Among these, the MEM was the most effective and easiest to use for removing electrophoretic truncation artifacts. This was demonstrated by using the data matrix (1024 × 21) from the ubiquitin and BSA mixture solution previously analyzed by the FFT method (fig. 12.12). MEM analysis in the electrophoretic flow dimension dramatically improved the spectral line shapes and signal resolution (fig. 12.12a) [40]. When the MEM spectrum was compared to the two-dimensional NMR spectrum obtained by FFT (fig. 12.12b), it was observed that the spectral line broadening in the flow dimension (from severe truncation artifacts in the FFT spectrum) could be removed with MEM analysis. The electrophoretic mobilities of BSA and ubiquitin were measured as 1.9×10^{-4} and 8.7×10^{-5} cm^{2}V^{-1}S^{-1}, respectively, by the MEM analysis. The improved signal resolution in the flow dimension is critical for high-resolution ENMR analysis of proteomic samples.

12.7 THREE-DIMENSIONAL ENMR

Two-dimensional NMR spectra of individual proteins in solution mixtures can be separated in the dimension of electrophoretic flow without physical separation of the proteins (fig. 12.13). Three-dimensional ENMR was developed to simultaneously obtain the NMR structural parameters of multiprotein components in solution. To prove the principle, we have developed the 3D EP-correlation spectroscopy (EP-COSY) method for obtaining conventional 2D COSY spectra of several molecules in a single ENMR experiment (fig. 12.14). An electric field pulse was applied between the labeling magnetic field gradients to drive the electrophoretic flow of ionic species. The COSY-type chemical shift correlation was generated in the first two dimensions to observe chemical shift evolutions and spin J-coupling connectivities in the evolution period (t_1) and the detection period (t_2). The electrophoretic motion of molecules modulated their COSY resonances differently as the electric field was

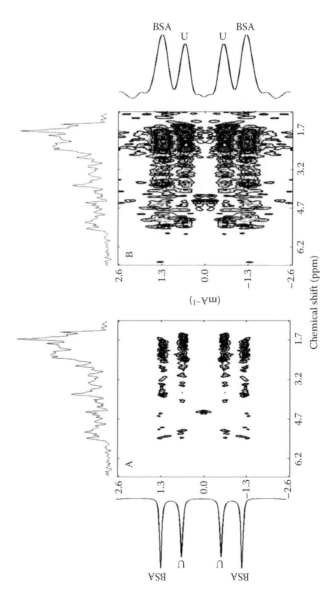

FIGURE 12.12 Two-dimensional electrophoretic NMR spectrum of a BSA and ubiquitin (U) protein mixture employing one-dimensional frequency domain transformation of electrophoretic interferograms using (a) maximum entropy method with optimized AR model order 10 (Thakur, S.B. and He, Q.J., *Magn Reson*, 183, 32–40, 2006); and (b) FFT method with 256 time point zero filling in the electrophoretic flow dimension. Horizontal and vertical projections are also presented in the chemical shift and flow dimensions. Other experimental parameters were published previously. (He, Q. et al. *J Am Chem Soc*, 120, 1341–1342, 1998.) (With modification)

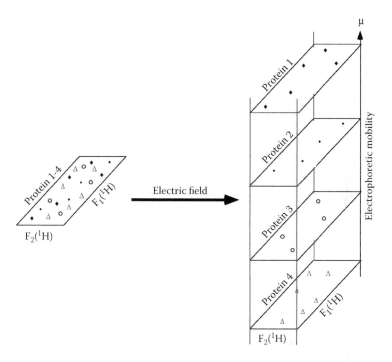

FIGURE 12.13 Two-dimensional NMR spectra of four different proteins (represented by ♦, •, ○, and Δ, respectively) are sorted by their electrophoretic mobilities (μ) in a single 3D ENMR experiment. (He, Q. et al. *J Am Chem Soc*, 120, 1341–1342, 1998. With permission.)

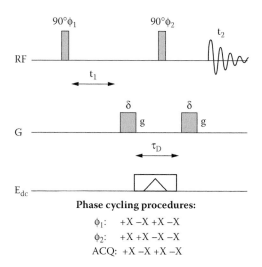

FIGURE 12.14 Pulse sequence for 3D electrophoretic COSY. E_{dc} and Δ are the amplitude and duration of the applied electric field pulse. (He, Q. et al. *J Magn Reson*, 147, 361–365, 2000. With permission.)

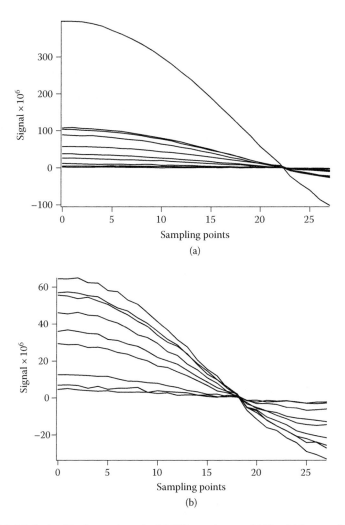

FIGURE 12.15 In the 3D electrophoretic COSY experiment of 100 m*M* L-aspartic acid and 100 m*M* 4,9-dioxa-1,12-dodecanediamine in D₂O, the electrophoretic interferograms from resonances of the two molecules were sorted into two groups by their electrophoretic migration rates. (He, Q. et al. *J Magn Reson*, 147, 361–365, 2000. With permission.)

progressively increased in the third flow dimension (fig. 12.15). If the 3D EP-COSY sequence was applied to a proteomic sample solution, the 2D NMR spectra of different protein components would then be displayed at different frequencies in the flow dimension.

A 3D EP-COSY data matrix (256 × 156 × 28) was acquired from a solution mixture containing 100 m*M* L-aspartic acid and 148 m*M* 4, 9-dioxa-1, 12-dodecanediamine in D₂O. The experiment was performed on a Bruker 500 MHz NMR spectrometer equipped with an actively shielded magnetic field gradient in the z-axis.

Uncoated U-shaped CA-ENMR sample cells were used after treatment with 1 M HCl, deionized water, and 1 M NaOH. The migration rates of L-aspartic acid and 4, 9-dioxa-1, 12-dodecanediamine were the net result of electrophoretic motion and bulk electroosmotic flow of the solution. Twenty-eight COSY spectra, each containing resonances from both molecules, were obtained at different E_{dc} in increments from 0 to 24.5 V•cm^{-1}. Two electrophoretic cosinusoidal oscillation frequencies were obtained that differentiated the COSY resonances of L-aspartic acid and 4, 9-dioxa-1, 12-dodecanediamine (figs. 12.16a and 12.16b). In the 3D EP-COSY matrix, the chemical shifts and J-coupling constants can be measured for the two molecules in separate COSY planes (fig. 12.16).

12.8 EX-ENMR CHARACTERIZATION OF PROTEIN REACTION INTERFACES

ENMR experiments can be designed to visualize protein conformational changes during protein interactions in the presence of other proteomic molecules. By applying a DC electric field, multicomponent protein interactions can be studied by using intermolecular nuclear Overhauser effects (NOEs) to identify interface residues in polypeptide chains or, by mapping the altered chemical shifts, molecular dynamic parameters and residual dipolar coupling patterns [35]. These exchange ENMR (Ex-ENMR) methods have potential for use in high-throughput structural mapping of protein reaction interfaces [5].

For a two-site protein exchange system, $A(\text{in protein A}) \underset{k_{BA}}{\overset{k_{AB}}{\rightleftharpoons}} B(\text{in protein B})$, where k_{AB} and k_{BA} are the chemical exchange rates. The ENMR signals from interacting residues appear at the average migration rates at $v_{exchange} = {}^1/_2(v_a + v_b)$ in the flow dimension when the chemical exchange rate is much faster than the difference of the two protein migration rates ($k_{AB} = k_{BA} \gg {}^1/_2|v_a - v_b|$), assuming that the two states are equally populated and the reactive protein A and protein B have different electrophoretic migration rates, v_a or v_b. In the slow exchange limit when $k_{AB} = k_{BA} \ll {}^1/_2|v_a - v_b|$, the signals of the interacting proteins can be distinguished at v_a or v_b in the electrophoretic flow dimension. In practice, many protein reactions have a fast exchange limit on the ENMR time scale (ca. 50–1000 ms). Therefore, the exchanging spin resonances would be extracted into separate 2D ENMR planes in a 3D ENMR spectrum for interacting proteins in a proteomic sample solution.

In most situations, protein interactions involve at least two proteins: $A + B \rightleftharpoons AB$. To predict the spectral outcome in a 3D ENMR experiment, the two-site exchange model needs to be applied twice to the two exchange reactions $A \rightleftharpoons AB$ and $B \rightleftharpoons AB$. When A and AB or B and AB are equally populated ($A:AB = 1:1$ or $B:AB = 1:1$), ENMR planes will appear at the average resonance frequencies of ${}^1/_2(v(A)+ v(AB))$ and ${}^1/_2(v(B)+ v(AB))$ in the flow dimension. If the two states are not equally populated, the more populated molecule will contribute more to its exchange spectral plane, with a location near its intrinsic migration rate in the flow dimension. We can obtain NMR resonances of reaction interfaces in both proteins by comparing NMR spectra from solutions containing different populations

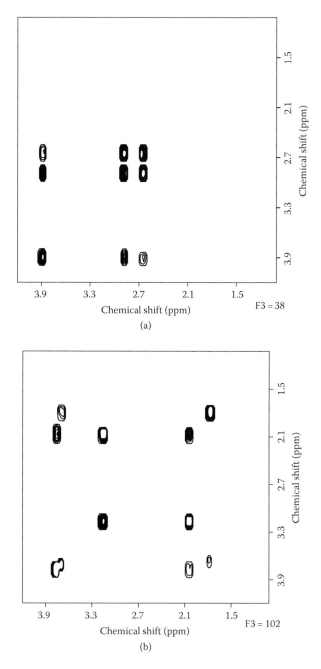

FIGURE 12.16 The two 2D COSY planes from the 3D EP-COSY matrix of (a) L-aspartic acid; and (b) 4, 9-dioxa-1, 12-dodecanediamine in D$_2$O solution. Each 2D COSY plane displays the COSY spectrum of a component molecule. (c) A control 2D COSY spectrum from the mixture using the conventional method. (He, Q. et al. *J Magn Reson*, 147, 361–365, 2000. With permission.)

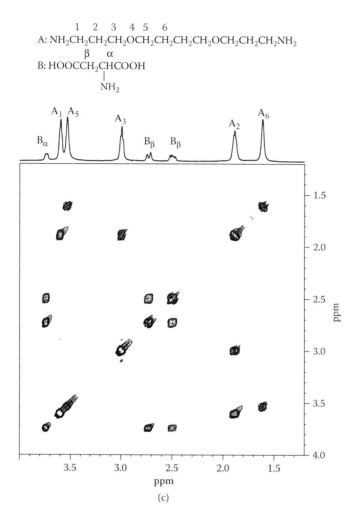

FIGURE 12.16 (continued)

of interacting molecules. Thus, the 3D conformation of the active sites can be determined for the interacting proteins in proteomic samples. In practice, this can be achieved by titrating a protein (or a drug) into the solution of its interaction partners (fig. 12.17). This ENMR approach can map multiple reactive proteins in a signaling network without selective labeling.

12.9 MICROCOIL ENMR USING A SINGLE HORIZONTAL CAPILLARY

The low intrinsic sensitivity for NMR signal detection imposes a challenge for proteomic ENMR. To address this issue, we are constructing a microsolenoidal-coil ENMR probe using a single capillary, which will achieve unprecedented mass

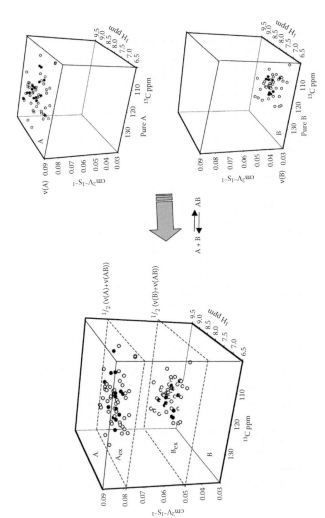

FIGURE 12.17 The effective electrophoretic migration rates of protein A and protein B change when they are involved in a chemical reaction using $A + B \rightleftharpoons AB$ as a model system. The schematic 3D exchange electrophoretic ^{1}H-^{15}N HSQC spectrum gives 2D NMR spectral planes of A_{ex} and B_{ex}, which differ from that of the pure protein spectra A or B. The simulation plots a case when proteins A, B and their complex AB are equally populated. The protein resonances involved in the chemical exchange are represented with filled circles and those not involved with open circles. (He, Q. and Song, X. In *Separation Methods in Proteomics*, ed. Smejkal, G.B. and Lazarev, A. New York: Taylor & Francis Group, 2005, 489–504. With permission.)

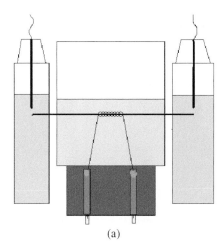

(a)

FIGURE 12.18 (a) The schematics of a microcoil ENMR probe and capillary sample chamber with three reservoirs: two flanking sample reservoirs containing two Pt electrodes and one central reservoir containing susceptibility-matching fluid to cover the microcoil and the capillary. (b) Sample reservoirs were 1-mL syringes; matching reservoir was a 3-mL syringe. Reservoirs were permanently connected and sealed with epoxy glue. (c) High-resolution NMR spectrum of a solution of 50% 2-propanol and 50% saline. Obtained using horizontal capillary microcoil ENMR on a Bruker Avance 500 MHz NMR spectrometer. A 600-MHz Varian probe (a gift from Varian, Inc.) was modified to tune and match the microcoil at 500 MHz. The broad peak on the left was from sample contamination.

sensitivity [36] (fig. 12.18a and 12.18b). Excellent NMR signal resolution can be obtained by immersing the microcoil-capillary assembly into a perfluorocarbon susceptibility matching fluid FC-43 (fig. 12.18c). The capillary is positioned horizontally so that bubbles generated from water hydrolysis can escape in the flanking reservoirs without disturbing the solution in the receiver region. Similar to the CA-ENMR method, the single capillary produces a stronger electric field compared to large glass tubes, thus allowing us to study proteomic samples with high salt concentrations. A preliminary test of this microcoil-ENMR apparatus is in progress on a Bruker Avance 11.7 T NMR spectrometer equipped with three-axial gradients (fig. 12.18).

12.10 CONSTANT-TIME MULTIDIMENSIONAL ENMR FOR HORIZONTAL SAMPLE TUBES

When horizontal ENMR sample cells are used, the electric current (I_e) applied perpendicularly to B_0 may produce an additional magnetic field gradient. The related diffusion attenuation and undesirable phase changes of the xy-magnetization could be superimposed on the electrophoretic phase modulations in the flow dimension. Since increasing the amplitude of the electric field (EF) induces the magnetic field gradient, special hardware compensations and careful signal processing algorithms are required for reliable ENMR measurements. To remove signal distortions from

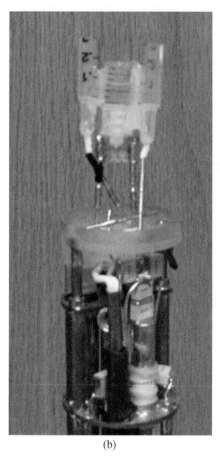

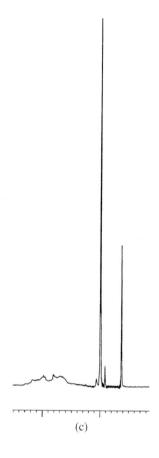

(b)   (c)

FIGURE 12.18 (continued)

the gradient induced by the DC electric field, we developed a novel constant-time multidimensional (CT-ENMR) method (fig. 12.19a) [37]. A constant electric field (E_{dc}) was applied to the xy-magnetization. The electrophoretic signal modulations were obtained by incrementing the duration of the electric field (t_E) while simultaneously decrementing the flanking time delays (t_v) between the gradient pulses and the electric field pulse. The diffusion time delay (T) between the two gradient pulses is constant (fig. 12.19a); therefore, the relaxation and diffusion decays of the ENMR signal in a conventional ENMR method (fig. 12.19b) were removed (fig. 12.19c). The electric current induced constant spectral phase modulations can be corrected by routine NMR signal processing procedures.

In the special case of the 2D STE-ENMR experiment (fig. 12.9) [8], the electric field pulse is applied between the second and third 90° pulses when the magnetization is stored in the z-direction. The electric current-induced magnetic field gradient does not alter the pattern of electrophoretic phase modulation, but simply dephases the remaining magnetization in the xy-plane (as a spoil gradient). Therefore, the electric field can be applied in any orientation relative to the B_0 field. The CT-ENMR strategy

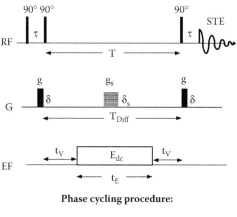

Phase cycling procedure:

ϕ_1:   +x +x +x +x +y +y +y +y +x +x +x +x−y −y −y −y

ϕ_2:   −x −x −x −x −y −y −y −y +x +x +x +x +y +y +y +y

ϕ_3:   +x +y −x −y +y −x −y +x −x −y +x +y −y +x +y −x

ACQ: +x +y −x −y +y −x −y +x −x −y +x +y −y +x +y −x

(a)

FIGURE 12.19 (a) The CT-ENMR from a modified STE-ENMR pulse sequence. (b) A control experiment with conventional time-incrementing STE-ENMR was carried out under similar experimental conditions as CT-ENMR. Data were acquired at 25°C from a sample solution containing 100 mM L-asp and 100 mM 4,9-dioxa-1,12-dodecanediamine in D_2O. A severe signal decay due to diffusion and relaxation was observed in the flow dimension as the duration of the electric field pulse was increased from 8.510 to 1008.510 ms. The diffusion time, t_{diff}, and relaxation time delay, T, were increased from 12.68 to 1012.68 ms and from 8.61 to 1008.61 ms, respectively. $K = 570.2$ cm^{-1}, $E_{dc} = 57.7$ V/cm. (c) A CT-ENMR spectral interferogram obtained as a function of the duration of electric field pulse. The duration of the DC electric field pulse (t_E) was increased stepwise from 8.51 to 1008.51 ms while the time delay, t_v, was decreased stepwise from 500 ms to 5 μs in 26 steps, synchronized with the t_E increase. Constant diffusion and T_1 relaxation time delays were obtained as $T_{diff} = 1012.58$ ms and $T = 1008.51$ ms, respectively. The electrophoretic interferogram was obtained at constant amplitudes, without diffusion and relaxation decays. (Li, E. and He, Q. *J Magn Reson*, 156, 1–6, 2002. With permission.)

will be the method of choice to acquire ENMR data using the horizontal microcoil ENMR probe and the capillary array sample chambers [37].

12.11   FUTURE PROSPECTS

Multidimensional electrophoretic NMR introduces another dimension, of electrophoretic flow, to separate individual 1D and 2D protein spectra, which contain the conventional chemical shift and spin coupling parameters for the structural characterization of proteins. The electrophoretic mobilities of mixed amino acids, peptides, proteins, microemulsions, and phospholipid vesicles have been simultaneously measured in the flow dimension. The ENMR technique can be applied to studying protein interactions in proteomic samples. Work is in progress to improve signal sensitivity using microcoil ENMR probes with a single capillary to resolve signals of

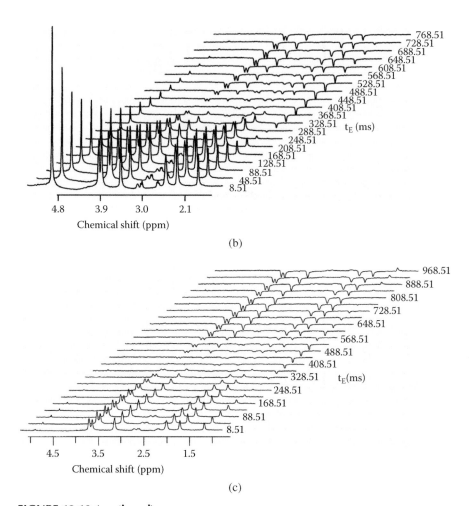

FIGURE 12.19 (continued)

migrating anions and cations [13,14]. Removal of the heat-induced convection artifacts is the most important recent advance towards ENMR analysis of proteomic samples. We have previously demonstrated that the effective electrophoretic mobility measured in an n-quantum ENMR experiment is n times the real electrophoretic mobility measured in a single-quantum ENMR experiment ($\mu_{eff} = n\,\mu$) [6,12,38]. Therefore, selective excitation of molecules into n-quantum states would differentiate these molecules from those staying in SQ modes with similar electrophoretic mobilities.

The ENMR approach has the intrinsic advantage of being able to identify the chemical nature of the proteins in proteomic samples. The ENMR technique is applicable in the structural characterization of protein interactions and biological signaling processes. With successful NMR measurements of protein abundance *in vitro* and *in vivo* [39], the clinical significance of ENMR applications to biomarker discovery is immense. In addition, proteins that undergo structural changes during protein interactions can be detected in their native conformations. Therefore, proteomic ENMR signatures associated with abnormal signaling networks and patient responses to drug therapies may be discovered for early diagnosis, staging and prognosis of human diseases. Cutting-edge nanotechnology may be employed to improve the detection sensitivity of ENMR in proteomic analysis.

We are currently working on studying cancer proteomics by ENMR. It is well known that cancer cells develop altered signaling networks of proteins, DNA/RNA, and other biomolecules that allow uncontrolled growth and metastasis. The ensemble shifts in a neoplastic proteome are heavily studied with mass spectrometry and gel electrophoresis techniques. Again, these methods show high promise, but require fractionation and do not report on the abundance of proteins in their native conformations. We will perform "proof of principle" studies on the application of novel ENMR technologies to the discovery of composite proteomic biomarker signatures in cancer. For example, identification of protein conformations that can serve as chemotherapeutic targets may be achieved to prevent chemotaxis of the migrating tumor cells homing to distant organs or to reduce cancer drug resistance.

ACKNOWLEDGMENTS

This work was supported in part by grants from the National Institutes of Health (RR12774-01, GM/OD55209, R21CA80906, R21EB001756, and R01CA109471-01A1), Susan G. Komen Breast Cancer Foundation (IMG0100117), the National Science Foundation (NSF MCB-9707550), the NSF Research Experience for Undergraduates (REU) program at the University of Pittsburgh (PHY-0244105), and the American Chemical Society Petroleum Research Funds (PRF# 32308-G4). We are grateful for the generous gift of the 600 MHz NMR probe body from Varian, Inc.

REFERENCES

1. Jones, M.B., Krutzsch, H., Shu, H., Zhao, Y., Liotta, L.A., Kohn, E.C., Petricoin, E.F., III, Proteomic analysis and identificaiton of new biomarkers and therapeutic targets for invasive ovarian cancer. *Proteomics*, 2, 76–84, 2002.
2. Taylor, C.F., Paton, N.W., Garwood, K.L., Kirby, P.D., Stead, D.A., Yin, Z., Deutsch, E.W., Selway, L., Walker, J., Riba-Garcia, I. et al., A systematic approach to modeling, capturing, and disseminating proteomics experimental data. *Nat Biotechnol*, 21, 247–254, 2003.
3. Koopmann, J., Zhang, Z., White, N., Rosenzweig, J., Fedarko, N., Jagannath, S., Canto, M.I., Yeo, C.J., Chan, D.W., Goggins, M., Serum diagnosis of pancreatic adenocarcinoma using surface-enhanced laser desorption and ionization mass spectrometry. *Clin Cancer Res*, 10, 860–868, 2005.
4. Li, J., Zhang, Z., Rosenzweig, J., Wang, Y.Y., Chan, D.W., Proteomics and bioinformatics approaches for identification of serum biomarkers to detect breast cancer. *Clin Chem*, 48(8), 1296–1304, 2002.
5. He, Q., Song, X., Electrophoretic nuclear magnetic resonance in proteomics: Toward high-throughput structural characterization of biological signaling processes. In *Separation Methods in Proteomics*, ed. Smejkal, G.B., Lazarev, A. New York: Taylor & Francis Group, 489–504, 2005.

6. He, Q., Electrophoretic nuclear magnetic resonance. Ph.D. thesis. University of North Carolina at Chapel Hill, 1990.

7. He, Q., Hinton, D.P., Johnson, C.S., Jr., Measurement of mobility distributions for vesicles by electrophoretic NMR. *J Magn Reson*, 91, 654–658, 1991.

8. He, Q., Johnson, C.S., Jr., Stimulated echo electrophoretic NMR. *J Magn Reson*, 85, 181–185, 1989.

9. He, Q., Johnson, C.S., Jr., Two-dimensional electrophoretic NMR for the measurement of mobilities and diffusion in mixtures. *J Magn Reson*, 81, 435, 1989.

10. He, Q., Lin, W., Liu, Y., Li, E., Three-dimensional electrophoretic NMR correlation spectroscopy. *J Magn Reson*, 147, 361–365, 2000.

11. He, Q., Liu, Y., Nixon, T., High-field electrophoretic NMR of mixed proteins in solution. *J Am Chem Soc*, 120, 1341–1342, 1998.

12. Johnson, C.S., Jr., He, Q., Electrophoretic NMR. In *Advances in Magnetic Resonance*, ed. Warren, W.S., vol. 13. San Diego: Academic Press, 131–159, 1989.

13. Morris, K.F., Johnson, C.S., Jr., Mobility-ordered 2D NMR spectroscopy for the analysis of ionic mixtures. *J Magn Reson*, Series A, 100, 67–73, 1993.

14. Morris, K.F., Johnson, C.S., Jr., Mobility-ordered two-dimensional nuclear magnetic resonance spectroscopy. *J Am Chem Soc*, 114, 776–777, 1992.

15. He, Q., Liu, Y., Sun, H., Li, E., Capillary array electrophoretic NMR of proteins in biological buffer solutions. *J Magn Reson*, 141, 355–359, 1999.

16. He, Q., Wei, Z., Convection compensated electrophoretic NMR. *J Magn Reson*, 150(2), 126–131, 2001.

17. Packer, K.J., Rees, C., Tomlinson, D.J., Studies of diffusion and flow by pulsed NMR techniques. *Adv Mol Relaxation Processes*, 3, 119–131, 1972.

18. Holz, M., Lucas, O., Muller, C., NMR in the presence of an electric current. Simultaneous measurements of ionic mobilities, transfer numbers, and self-diffusion coefficients using an NMR pulsed-gradient experiment. *J Magn Reson*, 58, 294–305, 1984.

19. Holz, M., Muller, C., Direct measurement of single ionic drift velocities in electrolyte solutions. An NMR method. *Ber Bunsenges Phys Chem*, 86, 141–147, 1982.

20. Holz, M., Muller, C., NMR measurement of internal magnetic field gradients caused by the presence of an electric current in electrolyte solutions. *J Magn Reson*, 40, 595–599, 1980.

21. Holz, M., Muller, C., Wachter, A.M., Modification of the pulsed magnetic field gradient method for the determination of low velocities by NMR. *J Magn Reson*, 69, 108–115, 1986.

22. Hinton, D.P., Johnson, C.S., Jr., Diffusion coefficients, electrophoretic mobilities, and morphologies of charged phospholipid vesicles by pulsed field gradient NMR and electron microscopy. *J Colloid Interface Sci*, 173, 364–371, 1995.

23. Saarinen, T.R., Johnson, C.S., Jr., High-resolution electrophoretic NMR. *J Am Chem Soc*, 110, 3332, 1988.

24. Wu, D., Chen, A., Johnson, C.S., Jr., Flow imaging by means of 1D pulsed-field-gradient NMR with applications to electroosmotic flow. *J Magn Reson Series A*, 115, 123–126, 1995.

25. Griffiths, P.C., Paul, A., Hirst, N., Electrophoretic NMR studies of polymer and surfactant systems. *Chem Soc Rev*, 35(2), 134–145, 2006.

26. Holz, M., Electrophoretic NMR. *Chem Soc Rev*, 23, 165–174, 1994.

27. Holz, M., Field-assisted diffusion studies by electrophoretic NMR. In *Diffusion in Condensed Matter—Methods, Materials, Models*, ed. Heitjans, P., Karger, J. Berlin: Springer, 721–746, 2005.

28. Johnson, C.S., Jr., Electrophoretic NMR. In *Encyclopedia of NMR*, ed. Grant, D. New York: Wiley, 1886–1895, 1995.
29. Johnson, C.S., Jr., Transport odered 2D ENMR spectroscopy. In *Nuclear Magnetic Resonance Probes of Molecular Dynamics*, ed. Tycko, R. Dordrecht: Kluwer Academic Publishers, 455–488, 1994.
30. Stejskal, E.O., Use of spin echoes in a pulsed magnetic-field gradient to study anisotropic, restricted diffusion and flow. *J Chem Phys*, 43, 3597–3603, 1965.
31. Stejskal, E.O., Tanner, J.E., Spin diffusion measurements: Spin echoes in the presence of a time-dependent field gradient. *J Chem Phys*, 42(1),288–292, 1965.
32. Bloch, F., Nuclear induction. *Phys Rev*, 70, 460–474, 1946.
33. Hahn, E.L., Spin echoes. *Phys Rev*, 80, 580–594, 1950.
34. Holz, M., Seiferling, D., Mao, X., Design of a new electrophoretic NMR probe and its application to $^7Li^+$ and $^{133}Cs^+$ mobility studies. *J Magn Reson Series A*, 105, 90–94, 1993.
35. Zuiderweg, E.R.P., Mapping protein–protein interactions in solution by NMR spectroscopy. *Biochemistry*, 41(1), 1–7, 2002.
36. Olson, D.L., Peck, T.L., Webb, A.G., Magin, R.L., Sweedler, J.V., High-resolution microcoil ^1H NMR for mass-limited, nanoliter-volume samples. *Science*, 270, 1967–1970, 1995.
37. Li, E., He, Q., Constant-time multidimensional electrophoretic NMR. *J Magn Reson*, 156, 1–6, 2002.
38. He, Q., Shungu, D.C., van Zijl, P.C.M., Bhujwalla, Z.M., Glickson, J.D., Single scan *in vivo* lactate editing with complete lipid and water suppression by selective multiple quantum coherance transfer (Sel-MQC) with application to tumors. *J Magn Reson B*, 106, 203–211, 1995.
39. He, Q., Xu, R.Z., Shkarin, P., Pizzorno, G., Lee-French, C.H., Rothman, D.L., Shungu, D.C., Shim, H., Magnetic resonance spectroscopic imaging of tumor metabolic markers for cancer diagnosis, metabolic phenotyping, and characterization of tumor microenvironment. *Dis Markers*, 19, 69–94, 2004.
40. Thakur, S.B., He, Q., High flow-resolution mobility estimations in 2D-ENMR of proteins using maximum entropy method (MEM-ENMR). *J Magn Reson*, 183, 32–40, 2006.

Part IV

Chemical Proteomics: Studies of Protein–Ligand Interactions in Pools and Pathways

13 Characterizing Proteins and Proteomes Using Isotope-Coded Mass Spectrometry

Uma Kota and Michael B. Goshe

CONTENTS

13.1 INTRODUCTION

Proteome analysis has increasingly relied on mass spectrometry (MS) to identify and elucidate protein function because it is has proven to be an effective and versatile analytical tool for protein characterization. In addition to detecting and enumerating the proteins expressed in an organism, the quantification of differences in the protein profiles of cells, tissues, or body fluids of different origins or states is increasingly being recognized as a key objective of mass spectrometry-based proteomics research. Such differential analysis of protein expression provides a more accurate and comprehensive view of the dynamic changes that occur within a cell under different conditions, as compared to mRNA expression analysis using cDNA microarrays. This is because regulation of protein expression is not solely dependent upon regulation of the expression of genes coding for that protein, but also involves a number of post-translational modifications (PTMs) that play a more decisive role in cellular regulation.

The conventional method used for initial protein detection, characterization and quantification has been two-dimensional (2D) gel analysis [1–3]. During two-dimensional polyacrylamide gel electrophoresis (2D-PAGE) proteins are separated by isoelectrophoresis in the first dimension and sodium dodecyl sulfate (SDS) PAGE in the second dimension as illustrated in figure 13.1. The separated proteins are visualized by implementing a myriad of staining techniques to produce a "spotted" pattern that is reflective of the protein abundance profile for a given sample. When two samples are compared, the differences in staining intensities for defined spots are quantified by densitometry. For stained spots of interest, gel slices are excised and in-gel proteolytically digested (normally using trypsin) to produce peptides extracted from the gel matrix and analyzed by mass spectrometry.

Although there have been some significant advancements in 2D-PAGE analysis to improve reproducibility and sensitivity of quantitative measurements, which make it extremely useful as an initial screening technique, there still remain analytical limitations for its application in comprehensive proteomic analysis [4]. Highly acidic and basic proteins as well as very small and very large proteins cannot be effectively separated. Hydrophobic membrane-bound proteins or proteins of low abundance are usually not detected. The occurrence of multiple protein forms per spot or multiple spots per protein—typically a result of PTMs—makes quantitative analysis difficult and tends to produce low sequence coverage of in-gel digested proteins for MS analysis.

During the past five years, alternatives to 2D-PAGE-based proteomic quantitative analysis have been developed that rely on liquid chromatography-tandem mass spectrometry (LC/MS/MS) to detect and quantify constituent peptides of enzymatically digested proteins (fig. 13.2). These LC/MS/MS measurements can circumvent the problems encountered with 2D-PAGE-MS analysis by providing enhanced proteome coverage while enabling simultaneous identification and quantification to be performed [5]. Although the sensitivity and accuracy of MS measurements are improving, the immense complexity of biological samples, as well as of the proteome itself, requires that the proteins and/or peptides be fractionated prior to MS analysis to facilitate more accurate measurements. This fractionation can include isolation of

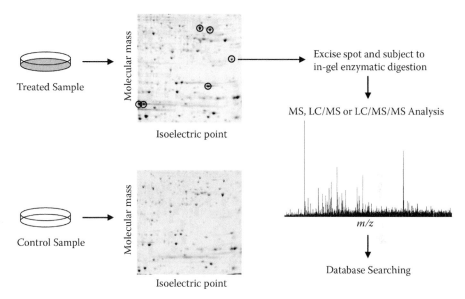

FIGURE 13.1 Two-dimensional gel-based approach to identify and quantify protein abundance changes for proteomic analysis. Proteins obtained from control and treated samples are separated based on their isoelectric point in the first dimension (isoelectric focusing) and their molecular mass in the second dimension (SDS-PAGE). Spot intensity detected by differential protein staining is used to quantify protein abundance. Spots of varying intensity, reflecting a measurable change in protein abundance (a few are indicated with circles), are excised, and the proteins contained in each gel slice are subjected to in-gel proteolysis. The resulting peptides are extracted and analyzed by mass spectrometry. Based on the mass spectral data acquired, a variety of database-searching algorithms are used to identify the peptides, which, in turn, identify the proteins present in the sample.

cellular organelles or protein complexes coupled with a combination of labeling and liquid chromatography techniques [6].

One of the more important and continually evolving techniques to facilitate the identification and quantification of proteins in proteomic mixtures is the use of isotope-coded mass spectrometry [7]. This strategy takes advantage of specifically labeling proteins and peptides using a variety of chemical, metabolic, and enzymatic stable isotope coding strategies. By covalently labeling with different mass tags, MS-based quantitation of protein abundances and their PTMs can be performed.

13.2 ISOTOPE CODING AND MS DETECTION FOR RELATIVE PROTEIN QUANTIFICATION

Isotope coding of proteins is the process by which proteins (or their corresponding peptides) are labeled with a combination of stable isotopes used to differentiate control from treated samples while permitting relative quantification between proteins of two distinct proteomes to be determined. These "chemical tagging" or "chemical labeling" strategies involve the modification of functional groups of amino acid side chains and

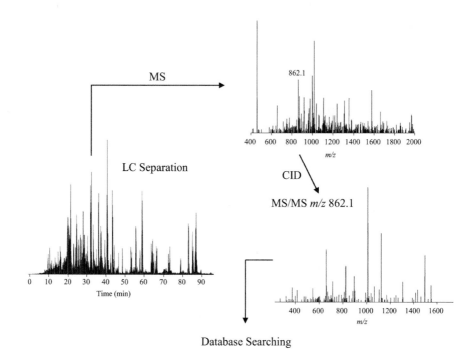

FIGURE 13.2 Mass spectrometry-based proteomics using liquid chromatography-tandem mass spectrometry (LC/MS/MS). Proteins obtained from a given sample are proteolytically digested, and the resulting peptides are separated by reversed-phase LC. Using electrospray ionization, the eluting peptides are detected and produce a mass spectrum (MS). Detected peptide ions at measured mass-to-charge (m/z) ratios of sufficient intensity (an example is shown as m/z 862.1) are selected for collision-induced dissociation (CID), which fragments the peptide to produce a product ion spectrum (i.e., MS/MS spectrum). The CID fragmentation of the intact peptide preferentially occurs at amide bonds to generate N-terminal fragments (b ions) and C-terminal fragments (y ions) at specific m/z ratios that provide structural information regarding amino acid sequence and sites of modification. Matching the b- and y-ion patterns to a peptide sequence present in a translated genomic database is used to identify the protein present in the sample and is performed using a variety of database-searching algorithms.

various PTMs in proteins and peptides [7,8]. In this gel-free approach, chemical modifications are used to attach an affinity tag (e.g., a biotin moiety) to the functional group of interest and permit the sample to be purified by affinity chromatography. Affinity tagging combined with stable isotope labeling of proteins/peptides allows reduction of sample complexity using chromatography while providing a means for relative quantitation using MS analysis.

The principal steps to quantify protein abundance using stable-isotope labeling are illustrated in figure 13.3. Two proteome samples that are to be compared are digested separately and individually labeled with chemically identical but mass-differentiated stable isotope tags. One sample is labeled with an isotopically "light" tag (containing ^{1}H, ^{12}C, ^{14}N, or ^{16}O atoms) and the other sample with the "heavy" tag (containing, ^{2}H, ^{13}C, ^{15}N, or ^{18}O atoms). After isotope labeling, the samples are combined and analyzed by mass spectrometry. The mass difference between the two samples (due to the incorporation of different isotopes) produces readily measurable changes in mass-to-charge (m/z) ratios. Because light and heavy forms serve as mutual internal standards, the relative intensities of the mass-differentiated forms should accurately reflect the ratios of the peptides (and therefore the proteins) in the original samples.

Currently, most proteomic studies using stable isotope coding strategies rely on LC/MS/MS employing electrospray ionization (ESI) to identify and quantify constituent peptides of enzymatically digested proteins. This is in contrast to the two-dimensional gel approach where quantification is performed by measuring stained spots using densitometry, and identification is performed by in-gel digestion and peptide mass fingerprinting using matrix-assisted laser desorption/ionization (MALDI) time-of-flight (TOF) mass spectrometry. With ESI, the peptides are ionized in a liquid-to-gas phase transition using an applied voltage, whereas in MALDI the transition occurs from a solid-to-gas phase facilitated by matrix molecules. Consequently, peptides do not necessarily ionize with the same efficiency with each method; this affects the ability to measure the abundance of isotope-coded peptides present within a given sample. The advantage of ESI is the improved ionization efficiency of labeled peptides, which increases the dynamic range of quantitative peptide measurements provided by a single analysis; however, some peptides will only be detected by MALDI. Although both methods provide distinct advantages, ESI is more attractive due to efficient coupling of LC separations to MS instrumentation; the precision of quantitative proteomic measurements is on the order of 20% error using a variety of mass analyzers [7].

13.2.1 ISOTOPE LABELING OF SPECIFIC AMINO ACID RESIDUES

13.2.1.1 Cysteinyl Residues

For years, cysteinyl residues (Cys-residues) have been attractive sites of chemical modification because they are relatively rare residues for most proteins and can be targeted for specific modification due to the distinctive pH of the side chain sulfhydryl group (pK_a 8.3). Consequently, Cys-residues have been targeted for amino acid-specific labeling with stable isotopes for proteomic analysis. One of the first uses of cysteinyl specific tags for proteomic analysis was the isotope-coded affinity tag (ICAT) approach developed by Aebersold and coworkers [9]. The original ICAT reagents consisted of (1) a biotin moiety (the affinity tag); (2) a polyether linker that serves as the isotope-coded region in the tag (containing eight ^1H (d_0) atoms or ^2H (d_8) atoms); and (3) a thiolate-reactive iodoacetyl group that allows the specific attachment of the label to Cys-residues of proteins (fig. 13.4A). The biotin moiety allows specific isolation of Cys-peptides by affinity chromatography using immobilized avidin, thereby significantly reducing the complexity of the peptide mixture prior to LC/MS/MS analysis. Since its introduction, the ICAT approach has been used to perform quantitative proteomic measurements [10–14].

As the first generation of d_0/d_8-ICAT reagents became commercially available and used in a number of quantitative proteomics applications, several analytical

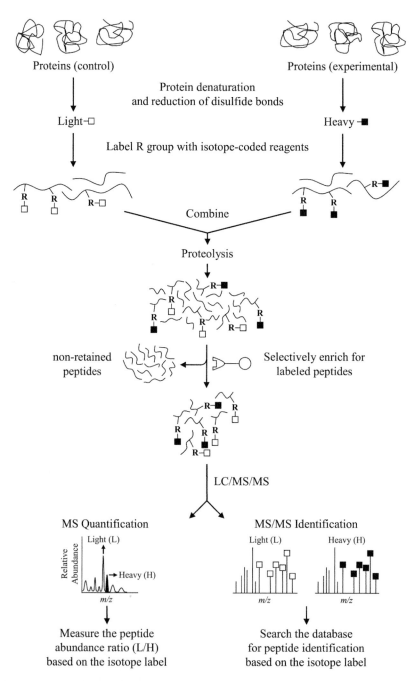

FIGURE 13.3 Stable isotope-coded mass spectrometry for relative protein quantification. Two protein samples, control and experimental, are individually labeled with a "light" (□) or "heavy" (■) isotope-coded reagent, respectively, by covalent modification to a preferred functional group (R). The two samples are then combined in a 1:1 ratio and proteolytically digested.

FIGURE 13.3 (continued)
The labeled peptides are selectively enriched using affinity chromatography or an alternative selection technique based on the incorporated functionality of the labeling reagent. The isotope-coded peptides are subjected to liquid chromatography-tandem mass spectrometry analysis (LC/MS/MS). Labeled peptides are identified by matching the MS/MS spectra (lower right) against a translated genomic database. By integrating the signal intensity for each isotope-coded peptide (lower left), the ratio of protein abundances (L/H) between the two samples can be determined. In this manner, the control acts as an internal reference to normalize the level of protein abundance for comparative analysis. The increase or decrease in the abundance ratio provides information regarding protein expression or degradation.

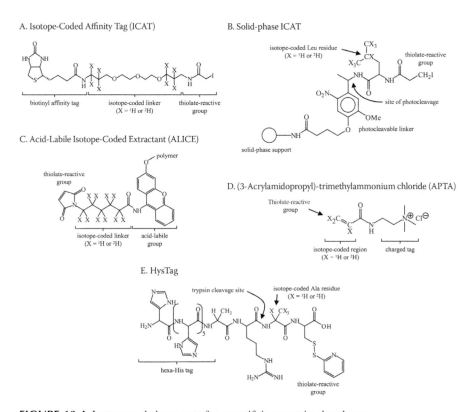

FIGURE 13.4 Isotope-coded reagents for quantifying protein abundance.

issues pertinent to the d_0/d_8-ICAT labeling strategy for proteomic quantitation became apparent [7]. Some of these include:

- Complications in quantitation due to a ^{2}H isotope effect result in differential elution between the d_0-ICAT and d_8-ICAT peptides during liquid chromatographic separation [15].
- Enrichment of non-Cys-peptides via nonspecific binding occurs during affinity chromatography. There could also be irreversible adsorption of some Cys-peptides during affinity chromatography [16].

- The hydrophobic biotin moiety of the ICAT label influences the retention behavior of the peptides during reverse-phase chromatographic separation; this can cause a relatively narrow elution zone of all tagged peptides [17].
- The ICAT label could undergo fragmentation due to collision-induced dissociation (CID) and interfere with the accurate identification of modified peptides [18].

Because of these studies with d_0/d_8-ICAT reagents, more effective ICAT labeling strategies were developed. The problem of differential elution using the $^1H/^2H$ coding system was overcome by employing $^{12}C/^{13}C$ isotope coding as demonstrated for cysteinyl reactive tags similar to ICAT in which the $^{12}C/^{13}C$-labeled peptides did not exhibit chromatographic fractionation (i.e., a shift in retention) [19,20]. This coding change significantly improves the determination of relative abundances of tagged peptides from any point in the elution profile. It was also observed that the d_0/d_8-ICAT reagents produced label-specific fragments upon CID [18], so a smaller tag can reduce the potential for label dissociation and enhance the identification of larger peptides or those containing multiple Cys-labeled residues.

As a result, improved ICAT reagents called "cleavable ICAT" have been made commercially available (Applied Biosystems, Inc., www.appliedbiosystems.com). These tags use $^{12}C/^{13}C$ coding while still providing a biotin moiety and a thiolate-reactive functionality. After removal of nonretained peptides, the labeled peptides bound to the immobilized avidin are selectively released by acid cleavage. This results in a smaller tag being attached to the peptide while providing a 9-Da mass difference between the ^{12}C and ^{13}C labels. The application of non-isotope-coded biotin affinity tags have also been used for the isolation of Cys-peptides in a number of proteomic studies [21–23]; this can aid in identification of low-abundance proteins and serve as a good model for method development prior to using more expensive isotope-coded reagents.

Although the biotin–avidin interaction has been exploited for affinity chromatography of Cys-labeled peptides, a more effective isolation is immobilization to solid-phase supports. The "solid-phase ICAT" reported by the Aebersold laboratory [24] used an isotope-coded tag containing a photocleavable linker immobilized on glass beads via an amide bond (fig. 13.4B). With this reagent, Cys-peptides are captured and tagged with the d_0/d_7 label and subsequently released by UV irradiation. When compared to the d_0/d_8-ICAT labeling method, the solid-phase approach for stable isotope tagging is comparatively simpler, more efficient, and more sensitive, as demonstrated on measuring galactose-induced changes in protein abundance in *Saccharomyces cerevisiae*.

Similar approaches to the solid-phase ICAT strategy have also been introduced by other laboratories. ALICE (acid-labile isotope-coded extractants), developed by Qiu et al. [25], consists of (1) a thiolate-reactive maleimido group; (2) a perproteo (d_0) or perdeutero (d_{10}) 6-aminocaproic acid linker; and (3) a nonbiological polymer with an acid-labile attachment (fig. 13.4C). The isotopomers of this linker are connected to the polymeric resin via an acid-labile amide. Once the Cys-peptides have been captured, they are cleaved off the resin with 5% trifluoroacetic acid. As reported, this isotope coding method was able to capture a Cys-peptide at 90% efficiency at neural pH and was successfully implemented to demonstrate the isolation and relative quantitation of standard protein mixtures.

In a method developed by Shi et al. [26], Cys-peptides are captured by a solid-phase tag that includes a thiolate-reactive iodoacetyl group encoded with d_0/d_3-alanine attached to the resin by an acid-cleavable linker. A major drawback of this reagent is that it requires highly acidic conditions (50% trifluoroacetic acid) for labeled Cys-peptide cleavage off the resin, thus limiting its use for proteins or peptides containing labile PTMs. Another cysteinyl-specific isotope-coding reagent called the element-coded affinity tag contains a chelate binding moiety that can be loaded with different rare earth elements [27]. The rare earth elements are heavy elements whose mass defects produce tagged peptide mass values that are not normally shared by molecules that contain only light elements. The differentially labeled Cys-peptides are enriched by a special column containing antibodies that recognize the metal chelate moiety.

Using a variant of the solid-phase tagging method, Cys-peptides can also be enriched by covalent chromatography that promotes covalent disulfide bond formation between the thiolate side chains of Cys-peptides to thiopropyl-Sepharose [28,29]. Following disruption of disulfide bridges with 2,2'-dipyridyl disulfide, the experimental and control samples were digested with trypsin and acylated with succinic anhydride. Cys-peptides were then selected from the acylated digest by disulfide exchange with sulfhydryl groups on the thiopropyl Sepharose gel. Captured Cys-peptides were eluted under reducing conditions and alkylated with iodoacetic acid prior to their chromatographic separation by reversed-phase LC. The fractions collected were analyzed by MALDI-MS. Differential labeling of peptides from experimental and control samples with d_0- and d_4-succinic anhydride, respectively, allowed the quantification of the relative concentration of each peptide species between the two samples. This method significantly reduces the complexity of the protein digest and greatly simplifies database searching. It was successfully applied to proteins obtained from *Escherichia coli* cell lysates to determine which proteins were upregulated when the expression of plasmid proteins was induced.

Alternative strategies to reduce sample complexity utilizing Cys-peptide capture are continually being developed. Ren et al. established a tagging method that utilizes a quaternary amine moiety [30]. Following reduction of disulfide bonds, Cys-residues of proteins are tagged with (3-acrylamidopropyl)-trimethylammonium chloride (APTA) to generate labeled peptides (fig. 13.4D). After trypsin digestion, APTA-labeled peptides are enriched by strong cation exchange chromatography (SCXC). Using model peptides and the protein transferrin for method development, it was determined that the main advantage of this charged tag was to increase the ionization efficiency in positive ion mode for ESI and for MALDI.

In another method, a "HysTag" has been synthesized and used to identify and quantify proteins from enriched plasma membrane preparations from mouse fore- and hindbrain [31]. The derivatized peptide tag (fig. 13.4E) consists of a hexa-histidinyl (His_6) sequence, a tryptic cleavage site, a d_0/d_4-coded alanyl residue, and a pyridyl disulfide moiety. The pyridyl moiety is used to tag the Cys-residues in the protein via disulfide exchange. The differentially tagged proteins are initially digested with endoprotease Lys-C and enriched by immobilized metal affinity chromatography (IMAC) or SCXC by virtue of the His_6 sequence. A second digestion with trypsin reduces the mass of the label to produce a dipeptide tag containing the d_4/d_0-coded alanyl residue.

As described in this section, a wide variety of stable isotope labeling techniques can be used to analyze cysteinyl proteins and perhaps additional innovative labeling and enrichment methods will be developed to examine this protein class.

13.2.1.2   Lysyl Residues

While the thiolate group of Cys-residues will continue to remain one of the preferred targets for isotope coding, the ε-amino group of lysyl residues (Lys residues) can also be targeted for specific modification using reagents such as O-methylisourea or 2-methoxy-4,5-dihydro-1H-imidazole that do not react with the amino-termini (N-termini) of peptides [32]. In a method termed "quantitation using enhanced signal tags" (QUEST) [33], differential amidination of Lys-residues using S-methylthio-acetimidate or S-methyl thiopropionimidate results in a 14 Da mass difference due to an additional methylene group in the larger tag that can be measured by MALDI-MS. Although QUEST was only applied to model protein samples, alternative reagents containing stable isotope coding could be used for LC/MS/MS.

Mass-coded abundance tagging (MCAT), described by Cagney and Emili [34], uses differential guanidination of C-terminal Lys-residues on tryptic peptides without requiring two distinct isotope coding reagents. Differential guanidination (i.e., the conversion of lysyl residues to homoarginyl residues) is carried out using O-methyl isourea for one sample while the other sample remains untreated, thus leading to a mass tag differential that can be detected with LC/MS/MS. This approach was validated by monitoring protein abundance changes in extracts obtained from a control strain of S. cerevisiae and a strain expressing an exogenous recombinant protein.

13.2.1.3   Tryptophanyl Residues

Tryptophan, like cysteine, is a relatively rare amino acid that has been exploited for affinity chromatographic prefractionation of tryptophanyl peptides from complex protein digests. Kuyama et al. [35] describe differential isotope labeling using 2-nitrobenzenesulfonyl chloride coded with either $^{12}C_6$ or $^{13}C_6$ stable isotopes. The mass differential of 6 Da between samples serves as a mass signature for all tryptophan-containing peptides in a pool of proteolytic digests to facilitate protein identification through peptide mass mapping. This approach was used to compare the protein expression profiles in rat sera between a normal (control) and a hyperglycemic rat.

13.2.2   N-Terminal and C-Terminal Isotope Labeling

As shown, modification of specific amino acid residues present in a protein or peptide has several advantages. It allows simplification of the affinity matrix used for enrichment and permits the presence of certain amino acids to be applied as a constraint in database searches; both lead to more accurate identification of low-abundance proteins. However, these labeling approaches could lead to reduced sequence and proteome coverage since only those proteins possessing the designated modified amino acid residues will be detected. Tagging of the N- and C-termini of peptides has an advantage in isotope-coded mass spectrometry because it does not rely on the presence of any specific residue for labeling. Since all peptides released by proteolytic digestion are susceptible to N- and C-terminal modification procedures, more sophisticated separation steps like multidimensional chromatography or high-resolution MS are required due to the higher sample complexity [8].

Several methods for N-terminal tagging are reported in the literature, many of which have been successfully applied to standard protein mixtures and also used in relative quantitation of proteins by incorporating stable isotope labeling techniques. The "global internal standard technology" (GIST) describes several methods for stable isotope labeling of amino groups. Most methods involve using N-acetoxysuccinimide [36–38] or succinic anhydride [28,29] and their respective deuterated isotopomers.

Prefractionation techniques are used to reduce sample complexity and include IMAC for selecting histidinyl peptides, disulfide exchange chromatography for selecting cysteinyl peptides, or lectin affinity chromatography to enrich glycosylated peptides. N-terminal isotope labeling has also been performed with d_0/d_3-acetic anhydride to quantify relative abundance changes in neuropeptides in mice [39]. Münchbach et al. [40] implemented a two-step tagging strategy to label the N-termini of peptides isotopically. First, the proteins are treated with succinic anhydride to block all Lys residues prior to digestion with Asp(Glu)-C protease. Second, the peptides are specifically labeled at the N-termini with d_0- or d_4-nicotinoyloxysuccinimide.

Stable isotope coding of the C-termini of peptides was demonstrated by converting carboxylate groups (along with the carboxylate side chains of aspartyl and glutamyl residues) to their corresponding methyl esters [41,42]. The labeling was accomplished using d_0- or d_3-methanol to aid in *de novo* sequencing and to verify results from MS/MS database searching. A possible limitation of this reaction is the partial hydrolysis of the esters during chromatography when using acidic mobile phase.

13.2.3 METABOLIC STABLE ISOTOPE LABELING

In addition to modifying proteins and peptides with isotope-coded reagents chemically, isotope coding can also be performed metabolically during *in vivo* translation. This can be accomplished by either growing cultured cells on stable isotope-labeled media [43–48] or by supplementing the growth media with stable isotope-coded amino acids [49–53]. Although *in vivo* labeling has several advantages, this technique is not amenable for many types of samples, particularly those obtained from body fluids and tissues.

13.2.3.1 Stable Isotope-Coded Amino Acids for *in Vivo* Labeling

In the "stable isotope labeling by amino acids in cell culture" (SILAC) method designed by Mann and coworkers, cells are metabolically labeled with perproteo- and perdeutero-amino acids that permit relative protein abundance measurements. A combination of d_0/d_3-leucine has been used for the relative quantitation of changes in protein abundance during the process of muscle cell differentiation [54]. A more extensive study regarding the quantitative aspects of measuring protein abundance was reported [55] where proteins from *S. cerevisiae* grown in the presence of d_0- and d_{10}-leucine were separated by 2D-PAGE. Protein spots were excised, in-gel digestion performed, and the extracted peptides analyzed by MALDI-TOF-MS. Despite the complexity of the resulting MS spectrum (which would be alleviated if LC were used), the measurements displayed standard errors of approximately 25%, values consistent with those previously reported for other stable isotope labeling techniques [7].

Chen and colleagues have used stable isotope-coded amino acids in *in-vivo* labeling to enhance identification of recombinant proteins from *E. coli* using d_2-glycine, d_3-methionine [49], or d_2-tyrosine mass tags [51]. This group also implemented d_0/d_3-serine coding and immunoprecipitation to identify protein phosphorylation in a histone protein (H2A.X) from human skin fibroblast cells in response to low-dose radiation [56]. Other amino acids, such as [$^{13}C_6$]arginine and [$^{13}C_6$]lysine, have been used to label the proteomes of *E. coli* and *S. cerevisiae* [57]. The application of $^{12}C_6/^{13}C_6$-lysine coding has been utilized in conjunction with Fourier transform ion cyclotron resonance (FTICR) mass spectrometry to assess global proteomic tryptic digestion efficiency [53] and to discriminate between peptides containing a lysine to glutamine amino acid substitution [52]. The implementation of stable isotope labeling using specific amino acids seems to have its greatest application in characterizing abundance changes of specific proteins of interest as opposed to comprehensive proteome analysis.

13.2.3.2 $^{14}N/^{15}N$ Stable Isotope Coding for *in Vivo* Labeling

Metabolic stable isotope labeling for comparative quantification of proteins from cell cultures can also be performed by growing cells in ^{14}N-minimal media or ^{15}N-enriched media, which provides for complete proteome coverage. Yates and coworkers utilized this labeling technique to label the proteome of *S. cerevisiae* and assess the dynamic range of quantitation using multidimensional protein identification technology (MudPIT) [45]. MudPIT is a method for separation of peptides obtained from proteolytic digestion of proteins using microscale two-dimensional liquid chromatography (SCXC and reversed-phase chromatography) coupled to tandem mass spectrometry; this allows for the relatively rapid analysis of a large proteome. Based on their results, any protein abundance changes greater than 30% covering one order of magnitude can be confidently measured with an ion trap mass spectrometer.

Smith and coworkers have used metabolic labeling for measuring changes in protein abundance at the intact protein level using FTICR-MS [58] and $^{14}N/^{15}N$ labeling in conjunction with a thiolate-specific affinity labeling reagent to isolate and quantify Cys-peptides from bacterial (*Deinococcus radiodurans*) and mammalian (mouse) proteomes [43]. The Smith group is also developing methods for global quantitative proteomics using peptides as unique biomarkers and the $^{14}N/^{15}N$ coding system for internal standardization in the quantification of protein abundance changes [59,60]. The continued development of new FTICR-MS technology [61] coupled with high mass measurement accuracy and advanced LC separation platforms will continue to increase the coverage and precision of quantitative global proteome measurements.

13.2.3.3 $^{16}O/^{18}O$ Stable Isotope Coding Using Proteolysis

Enzymatic digestion of proteins can be used to code proteome samples isotopically by performing the proteolysis in the presence of "normal" water (predominately as the $H_2^{16}O$ isotope) or "heavy" water (enriched in $H_2^{18}O$). After one protein sample is digested in normal water and the other in heavy water, the two peptide samples are mixed, processed, and analyzed by MS, where the ratios of the relative intensities of the ^{16}O-labeled peptide peak to the ^{18}O-labeled peptide peak are used to quantify the protein abundance between samples. This isotope coding strategy differs from chemical labeling (sections 13.3.1 and 13.4.1) and metabolic labeling (section 13.5.1) since it can accommodate many types of proteomic investigations, such as serum and tissue samples, and provides for complete proteome coverage. Because the peptides are labeled at the C-terminus during protease cleavage, artifacts inherent during chemically stable isotope labeling are avoided, an asset for MS measurements.

Enzymatically stable isotope labeling with $^{16}O/^{18}O$ water has been used to characterize proteins and proteomes on a variety of levels. This labeling strategy has been implemented to examine the C-terminal regions of rat liver proteins [62], identify disulfide bond networks of specific proteins [63], characterize cross-linked peptides in protein cross-linking experiments [64], and probe the Rad6-Rad18 complex involved in DNA repair [65]. It has also been used to evaluate proteomic sample preparation methods like tryptic digestion efficiency, solid-phase extraction, and absorptive losses and recovery after lyophilization [66], thus demonstrating the versatility of this labeling technique.

Because it can be readily applied to a wide variety of samples, enzymatic digestion in the presence of $H_2^{16}O$ or $H_2^{18}O$ has been frequently implemented in isotope-coded proteomic analyses. Fenselau and coworkers have used enzymatic labeling to study serotypes Ad2 and Ad5 of adenovirus [67]. Using high-performance liquid chromatography (HPLC) separation of peptides in which collected fractions were desalted and then analyzed by MALDI-FTICR-MS, ratios between proteins (Ad2 labeled with ^{18}O and Ad5 with ^{16}O during proteolysis) could be determined with a precision greater than 25% when multiple peptides were used to quantify the corresponding precursor protein (^{18}O incorporation being approximately 90% complete). Smith and coworkers have used the technique to study changes in the human plasma proteome upon administration of lipopolysaccharides by reversed-phase LC-FTICR-MS [68].

To facilitate more confident identification of completely degraded or newly synthesized proteins while improving relative protein abundance measurements, inverse stable isotope coding of two distinct samples can be performed in a manner analogous to the dye labeling techniques used in transcriptome analysis. Wang and coworkers have performed "inverse labeling" by conducting two converse isotope coding experiments in parallel ([light]label-control × [heavy]label-experimental and [heavy]label-control × [light]label-experimental) with $^{16}O/^{18}O$ enzymatic [69] and $^{14}N/^{15}N$ metabolic labeling [46]. When samples from each experiment are compared, an inverse labeling pattern reflecting a characteristic mass shift is observed between the two parallel analyses for proteins that are differentially expressed.

This strategy allows more effective screening of peptides that dramatically change (threefold or greater) since ambiguities associated with extreme changes in protein abundance are eliminated. It can be employed with other isotope labeling strategies described in this chapter and may assist in identifying potential protein markers and targets of disease states as well as provide new insights regarding drug action and toxicity. Although ^{18}O-labeling is a simple procedure, incomplete incorporation of ^{18}O would limit the mass shift between the two samples and lead to significant overlap of the isotopic distribution for the light and heavy forms [8]. Recently, quantitative ^{18}O labeling has been achieved by using immobilized trypsin after in-solution digestion in $H_2^{18}O$ to promote exchange of both oxygen atoms in the C-termini of peptides [70]. Combining $^{16}O/^{18}O$ labeling with other separations like SCXC and/or reverse-phase LC prior to TOF-MS or FTICR-MS analysis will increase the coverage of proteins quantified with this technique.

13.3 ANALYSIS AND QUANTIFICATION OF POST-TRANSLATIONAL MODIFICATIONS USING ISOTOPE CODING

Protein function is highly modulated by post-translational modifications (PTMs). The complexity of these modifications, which vary between organisms and during modulation of cellular function, present the most challenging aspect of proteomic analysis. Because of their diversity, there are many methods to enrich and label particular PTMs for quantitative analysis using stable isotope-coded mass spectrometry. In this chapter, we will focus on just two of the more important PTMs: phosphorylation and glycosylation.

13.3.1 PHOSPHORYLATION

The reversible phosphorylation of proteins plays a major role in many vital cellular processes by modulating protein activity that propagates signals within cellular pathways and signal transduction networks [71,72]. Because phosphorylation is dynamic, the sites of phosphorylation cannot be predicted by an organism's genome and thus require proteomic measurements (termed "phosphoproteomics") using mass spectrometry to identify sites of and quantification of changes in protein phosphorylation. Typical phosphoproteins tend to be of low abundance and challenge the dynamic range of present analytical technologies. Characterizing phosphorylation events, even for a modest number of proteins, is an analytical challenge. If more efficient peptide capture and fractionation techniques could be effectively employed, the MS detection of phosphopeptides would be enhanced. To address these issues, various isotope coding methods have been developed to determine the phosphorylation states of proteins qualitatively and quantitatively [73–75].

One approach to increase MS detection of phosphoproteins is to enrich sample preparations from proteomic mixtures by utilizing noncovalent binding and recognition of phosphate groups. Enrichment of phosphoproteins from complex mixtures can be performed by affinity chromatography using immobilized antibodies specific for phosphotyrosyl (pTyr), phosphoseryl (pSer), and phosphothreonyl (pThr) residues. Another technique more amenable to LC/MS/MS analysis is the use of immobilized metal affinity chromatography (IMAC) for phosphopeptide enrichment.

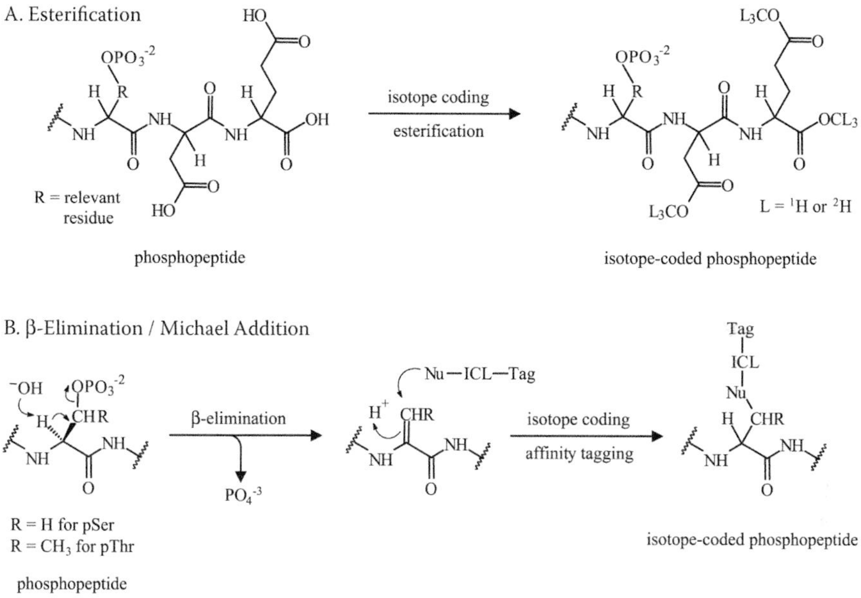

A. Esterification

isotope coding

esterification

R = relevant residue

phosphopeptide

L = ^{1}H or ^{2}H

isotope-coded phosphopeptide

B. β-Elimination / Michael Addition

β-elimination

isotope coding

affinity tagging

R = H for pSer
R = CH$_3$ for pThr

phosphopeptide

isotope-coded phosphopeptide

FIGURE 13.5 Isotope coding for phosphoproteomics. Phosphopeptides obtained from proteolytic digests can be isotopically labeled using two strategies. (A) Esterification using d_0/d_3-methanol. After global proteome digestion, the carboxylic acid groups of peptides (Asp, Glu, and C-terminus) are esterified in acidic methanol. Comparative proteomics is performed by differential esterification using d_0-methanol (L = ^{1}H) and d_3-methanol (L = ^{2}H) to label control and experimental samples, respectively. This neutralization of the carboxylate functionalities facilitates immobilized metal affinity chromatography (IMAC) enrichment of phosphopeptides for subsequent MS analysis. (B) Beta-elimination/Michael addition using isotope-coded reagents. Phosphoseryl (pSer, R = H) and phosphothreonyl (pThr, R = CH$_3$) peptides are subjected to β-elimination in order to remove the phosphate moieties and create electrophilic α, β-unsaturated double bonds that can be selectively labeled by reagents that contain a reactive nucleophile (Nu), an isotope-coded linker (ICL) for quantifying relative changes in protein phosphorylation, and an affinity tag or reactive functionality (Tag) to enable selective enrichment of labeled peptides for MS analysis.

IMAC relies on the affinity of phosphate groups for certain metal ions (e.g., Fe^{3+} or Ga^{3+}) bound to tethered chelating reagents present on solid-phase supports. After proteolytic digestion, phosphopeptides are isolated by IMAC and subsequently analyzed using LC/MS/MS.

However, nonspecific binding of aspartyl- and glutamyl-containing peptides occurs and can complicate downstream analysis. Hunt and coworkers have addressed this issue by converting the carboxylic acid groups of peptides to their corresponding methyl esters in order to attain higher specific binding of phosphopeptides [76], a modification amenable to stable isotope labeling with d_0/d_3-methanol [77] (fig. 13.5A). A possible limitation for effective quantification of phosphorylation using this isotope coding method is the partial hydrolysis of the esters during various chromatographic separations that would produce a mixture of differentially esterified forms for each phosphopeptide.

The IMAC procedure was successfully applied in a number of phosphoproteomic studies. Ficarro et al. [77] mapped more than 60 phosphorylated sequences in a whole protein digest from capacitated sperm by MS/MS. He et al. [78] used d_0/d_3-methanol coding and IMAC to compare enriched phosphopeptides from *in vitro* cultured human lung cells under control and starvation conditions. Salomon et al. [79] applied this method to identify pTyr sites during the activation of human T cells and measure changes in phosphorylation upon treatment of chronic myelogenous leukemia cells with the inhibitor of the oncogenic BCR-ABL kinase activity. Brill et al. [80] used IMAC to assign 70 tyrosine-phosphorylated peptides from human T cells' lysate. Ren et al. [81] utilized IMAC columns loaded with Cu^{2+} and N-acetylation of peptides to enrich for histidinyl peptides from protein digests, and Bieber et al. [82] used immobilized metal-affinity pipette tips to enrich for phosphoproteins obtained from human saliva.

Isotope coding by $^{14}N/^{15}N$ metabolic labeling was also used to study phosphorylation as reported by Chait and coworkers [83]. This method determines the relative abundance of the phosphorylation state of proteins by measuring the intensity ratio (^{14}N:^{15}N) of the detected unphosphorylated and phosphorylated peptides. Since the presence of other peptides makes analysis of only the phosphopeptides more difficult, the combination of IMAC and $^{14}N/^{15}N$ metabolic labeling could prove rather useful in phosphoproteomics.

In addition to their low abundance, phosphoproteins present unique challenges to mass spectrometry analysis. Peptide fragmentation for MS/MS analysis is almost exclusively performed using CID. However, CID of phosphopeptides typically results in the loss of the phosphate moiety and can prevent unequivocal identification of the phosphorylated residue [84]. The use of electron capture dissociation [85] instead of CID for phosphopeptide analysis [86], in which the phosphate moiety remains intact during peptide fragmentation, holds much promise for IMAC-based LC/MS/MS phosphoproteomics.

The strategy of using chemical labeling of phosphate groups to enable enrichment of phosphopeptides through covalent modification has also been explored. One approach utilizes high-affinity avidin–biotin coupling to immobilized supports, allowing removal of nonphosphorylated peptides during washing. The method reported by Chait and coworkers [87] involves a base-catalyzed β-elimination of the phosphate group from pSer and pThr residues and subsequent Michael addition of ethanedithiol (EDT). The new thiolate moiety serves as a linker for the attachment of a biotinylated affinity tag containing a maleimide group, thereby allowing avidin affinity chromatography to be used to isolate the modified phosphopeptides. The main disadvantages of this phosphoproteomic method are that it is not applicable to tyrosine phosphorylation and the maleimide group undergoes partial hydrolysis, resulting in two products for each modified peptide.

Another approach using covalent chromatography based on disulfide exchange (similar to that described in section 13.2.1.1) captured phosphopeptides on a solid-phase support [88]. Following base-catalyzed β-elimination of the phosphate groups and subsequent Michael addition of dithiothreitol (DTT) for pSer and pThr peptides, the DTT-labeled peptides were covalently attached via disulfide exchange to an immobilized thiolate resin. The covalently linked peptides were released using an excess of DTT and analyzed by MALDI-MS. The application of this strategy was demonstrated by the analysis of the *in vitro* phosphorylation of bovine synapsin I by Ca^{2+}/calmodulin-dependent kinase II.

In the solid-phase method developed by Zhou et al. [89], phosphopeptide capture is independent of the nature of the phosphorylated side chain, making it applicable to studying tyrosine phosphorylation. This method involves blocking amino groups of proteolytically digested peptides using tBoc chemistry. This is followed by carbodiimide catalyzed condensation of ethanolamine with the phosphate and carboxylate groups of the peptide to form phosphoramidate and amide bonds, respectively. Phosphate groups are regenerated by treatment with diluted trifluoroacetic acid that selectively cleaves ethanolamine from the phosphate groups. Following the addition of the cystamine to the regenerated phosphate group and reduction of the disulfide bond to release a free thiolate group, the phosphopeptides are captured with glass beads containing a thiolate-reactive immobilized iodoacetamide functionality. Cleavage of phosphoramidate bonds with concentrated trifluoroacetic acid releases the captured phosphopeptides and simultaneously removes the tBoc-protecting groups from the amines while the modified carboxylate groups remain intact. This covalent coupling of the phosphopeptides allows stringent washing conditions, resulting in highly enriched mixtures of phosphopeptides. Although we have not used this method, using isotopic variants of ethanolamine to block the carboxylate groups of the proteins (similar to the isotope coding strategy used in IMAC) could help quantify the relative phosphopeptide abundance.

Stable isotope coding using affinity tags to quantify protein levels, as demonstrated with ICAT, has been extended to phosphoproteins (fig. 13.5B) as described by Goshe et al. with the development of a phosphoprotein isotope-coded affinity tag (PhIAT) [90,91]. In this method, Cys-residues are blocked by performic acid oxidation, and the phosphate groups of pSer and pThr residues are removed by hydroxide ion-mediated β-elimination to produce thiolate-reactive sites (i.e., α, β-unsaturated double bonds). When the relative phosphorylation states of phosphopeptides from two distinct samples are compared, these thiolate-reactive sites are modified with isotopic versions of 1,2-ethanedithiol (EDT) that contain four alkyl hydrogens (d_0-EDT) or four alkyl deuteriums (d_4-EDT).

Once isotopically labeled, the d_0/d_4-EDT-labeled residues are biotinylated using (+)-biotinyl-iodoacetamidyl-3,6-dioxoctanediamine. The phosphorylated peptides are purified and concentrated using affinity chromatography and quantified by MS analysis. Identification of the PhIAT-labeled peptides is performed by LC/MS/MS. The relative stability of the PhIAT label during CID enables the localization of the site of phosphorylation to be identified from the product ion spectrum and permits the state of phosphorylation to be quantified [91]. The method was initially developed on casein phosphoproteins and was used to label the soluble phosphoproteins present in a complex protein mixture from *S. cerevisiae*. A PhIAT reagent, which contains a nucleophilic sulfhydryl and an isotopic label covalently linked to a biotin moiety, was synthesized and could be implemented to reduce the pSer and pThr derivatization into a one-step process [90].

The PhIAT approach for phosphoprotein analysis has several limitations as shared with the ICAT approach (section 13.2.1.1). Due to the use of immobilized avidin during affinity chromatography, there is difficulty in removing all nonspecifically bound peptides, and sample recovery is reduced because of irreversible binding of a subpopulation of the biotinylated peptides. In addition, the use of deuterium atoms in isotope coding causes differential elution of isotopomers during LC, and peptides containing multiple pSer and pThr residues are difficult to fragment and analyze using CID due to the mass of the additional PhIAT labels.

To address these issues, the Goshe and Smith labs have reported an improved solid-phase-based version of PhIAT, termed phosphoprotein isotope-coded solid-phase tag (PhIST) [92]. The tagging strategy is similar to the PhIAT approach: β-elimination of the phosphate moiety followed by the Michael addition of 1,2-ethanedithiol. However, the biotin affinity tag is replaced by an isotope-coded solid-phase reagent containing light ($^{12}C_6$, ^{14}N) or heavy ($^{13}C_6$, ^{15}N) stable isotopes and a photocleavable linker that is used to capture and label the phosphopeptides in a single step. The captured peptides are released from the solid-phase support by UV photocleavage and analyzed by LC/MS/MS. The efficiency and sensitivity of the PhIST labeling approach for identification of phosphopeptides from mixtures were determined using a mixture of casein proteins and applied to the quantification of soluble phosphoproteins from a human breast cancer cell line.

Unfortunately, the reaction conditions employed for the β-elimination/Michael addition of some protein samples can be substoichiometric and difficulties associated with the low solubility of the EDT compound in aqueous solutions can cause protein precipitation [73]. Recently, Goshe and coworkers have addressed these issues by improving their PhIST approach to yield quantitative labeling of pSer and pThr residues using β-elimination and Michael addition of (R,R)-dithiothreitol as the thiolate linker [93]. In addition, the improved PhIST labeling method using [$^{12}C_6$, ^{14}N]leucine and [$^{13}C_6$, ^{15}N]leucine isotope-coded solid-phase reagents was applied to an *in vitro* model of Parkinson's disease [94].

13.3.2 Glycosylation

Protein glycosylation is a common post-translational modification where carbohydrates are covalently attached to seryl or threonyl residues (O-linked glycosylation) or asparagyl residues (N-linked glycosylation). It has significant effects on protein folding, stability, and structure and consequently affects protein function [95]. Glycoproteins are prevalent in the plasma membrane, secreted proteins, and proteins present in body fluids (e.g., blood, serum, cerebrospinal fluid, saliva, and breast milk) and play a vital role in biological processes such as molecular recognition and inter- and intra-cellular signaling. Several approaches to characterize glycoproteins have been used successfully, including fast atom bombardment, MALDI, and ESI using a wide variety of mass analyzers. However, the identification of glycopeptides in complex mixtures as encountered in proteomic studies still remains a challenge due to the poor ionization efficiency and rapid degradation of glycopeptides. Many strategies have been developed to analyze the structure of the glycans and their modification sites, including derivatization techniques of the reducing ends and protection of the functional groups [96].

For proteomic studies, glycoproteins can be enriched from complex mixtures by lectin affinity chromatography. Lectins are a diverse group of plant proteins that bind to glycan moieties of glycosylated proteins and peptides. Many studies have shown that the combination of "tagging" strategies using stable isotope coding with lectin affinity chromatography allows the simultaneous enrichment and relative quantitation of glycosylated proteins. Geng et al. [37] used lectin affinity chromatography to select for glycosylated peptides from a tryptic digest of a complex mixture of proteins that were further fractionated by reversed-phase LC. For quantification, peptides from tryptic digests were acylated using N-acetoxysuccinimide and compared to internal standard peptides modified with the trideuterated analogue.

A similar approach combining lectin affinity selection and MALDI was used to analyze different types of glycoproteins in complex mixtures derived from either human blood serum or a cancer cell line [97]. In this study, deglycosylation with peptide-N-glycosidase F (PNGase F) enabled N-type glycoproteins of unknown structure to be identified. PNGase F is a glucosidase, specific for N-linked glycans, that hydrolyzes the β-aspartylglycosylamine bond of asparagine-linked glycopeptides that generates an aspartyl residue. Sample digestion with PNGase F in the presence of $H_2^{16}O$ or $H_2^{18}O$ has also been used to specifically label the carboxylate side chains of newly formed aspartyl residues with ^{16}O or ^{18}O, as demonstrated on the identification of glycopeptides from lactoferrin, mammaglobin, and the ion pairs of fetuin glycopeptides [98].

The isotope-coded glycosylation-site-specific tagging (IGOT) strategy described by Kaji et al. [99] was used to characterize N-linked high mannose and hybrid type N-glycans from a protein extract of *Caenorhabditis elegans*. Glycoproteins were enriched on a ConA lectin-affinity column and then subjected to trypsin digestion. The resulting mixture of glycosylated peptides was purified by a second affinity step using the same column. N-glycosylation sites were specifically labeled during digestion with PNGase F in the presence of either $H_2^{16}O$ or $H_2^{18}O$ and identified by 2D-LC/MS/MS analysis. A similar approach using lectin affinity chromatography and $^{16}O/^{18}O$ enzymatic labeling of asparagyl residues carrying glycan moieties identified 87 proteins and 33 glycosylation sites in the human bile proteome [100]. Fenselau and coworkers have also applied inverse isotope labeling (described in section 13.2.3.3) to characterize N-linked glycoproteins [101] in which sequential $^{16}O/^{18}O$ enzymatic labeling using Glu-C and PNGase F enabled the site of N-glycosylation to be determined based on the differences in peptide isotopic patterns.

A slightly different approach for the identification of glycosylation sites by Aebersold and his group [102] involves a solid-phase enrichment of glycoproteins. The specific capture of glycoproteins is based on the periodate oxidation of the vicinal diol groups to aldehydes present in carbohydrate-containing residues. The newly formed aldehyde functionalities are then covalently coupled to immobilized hydrazine groups present on an agarose support. On-resin tryptic digestion produces nonglycosylated peptides that are present in solution, which are washed away, to leave the glycopeptides retained on the solid support. Isotope coding is achieved by differentially labeling each immobilized glycopeptide sample with d_0/d_4-succinic anhydride. Following isotope coding, the peptides are released from the solid-phase support by PNGase F treatment and analyzed by LC/MS/MS. This method was applied to analyze human serum proteins and enabled identification of 145 glycosylation sites. It allows a broad range of N-glycosylated isoforms of glycoproteins to be selected but cannot be used to differentiate between them.

O-linked glycosylation sites have been analyzed in a manner analogous to the analysis of N-glycosylated peptides for various proteins. As mentioned by Goshe et al. [90], the PhIAT method could be applied to identify and quantify O-linked glycoproteins by utilizing a combination of phosphatases and glycanases. The method of Wells et al. [103] utilizes base-catalyzed β-elimination to cleave the O-linked N-acetylglucosamine (O-GlcNAc) modifications on seryl and threonyl residues, followed by Michael addition of DTT or biotin pentylamine to modify these specific sites for subsequent enrichment using disulfide exchange covalent chromatography or avidin affinity chromatography, respectively. The methodology was validated by mapping several previously known residues containing the O-GlcNAc modification on synapsin I and was used to identify O-GlcNAc modification sites on proteins obtained from a purified nuclear pore complex preparation. By incorporating isotope-coded DTT reagents or utilizing the improved PhIST approach [93,94], differential states of O-linked glycosylation sites could be quantified.

13.4   ABSOLUTE QUANTIFICATION USING INTERNAL STABLE ISOTOPE-CODED STANDARDS

The use of multidimensional chromatography coupled with tandem mass spectrometry allows the identification of a large number of proteins from complex biological mixtures. However, due to a number of variables, the intensity of a peptide ion signal is not necessarily an accurate representation of the amount of peptide in a sample and thus requires a method of normalization. According to stable isotope dilution theory, two peptides with identical chemical structure that differ in isotopic composition are thought to generate the same response detected by the mass spectrometer such that their relative ion signals reflect the relative concentrations in the sample. The stable isotope coding techniques described thus far can be used to determine the relative concentration of many proteins in one sample versus another by measuring the abundance of isotopically labeled proteolytic peptides. Although these stable isotope coding techniques are useful in characterizing changes occurring from one sample to another, they do not provide information regarding the absolute abundance levels contained in each sample due to the lack of true internal standards.

The absolute concentration of a protein in the cellular pool is an important quantitative measurement that has several important biological applications. For example, in functional proteomics the determination of the absolute amount of a specific enzyme can be used to determine its specific activity. The measure of the relative abundance of a given enzyme in different tissues can be used to determine its differential activity and provide insight regarding its biological role. Another application of absolute quantification is in the area of mRNA/cDNA expression profiles using microarrays. Following the identification of candidate genes as disease markers during the study of expression profiles at the mRNA level, the next logical step involves measuring the abundance levels of the corresponding proteins in the available samples. Since techniques such as ELISA are expensive, elaborate, and time consuming to develop, the application of isotope-coded internal standards in conjunction with MS analysis would allow rapid, economical, and sensitive detection and quantification of these candidate disease markers.

Several methods have been developed that enable the absolute quantification of proteins to be measured. One method involves the use of the "visible" ICAT (VICAT) reagent [104] to determine the absolute abundance of the human group V phospholipase A_2 [105]. This reagent covalently attaches to cysteinyl-thiolates or thioacetylated amino groups. Protein mixtures are proteolytically digested and tagged with the VICAT reagent. A known quantity of the internal standard tagged with the isotopically heavy version of the VICAT reagent (^{14}C-VICAT$_{SH}$) is doped into the peptide sample labeled with the light version of VICAT (^{12}C-VICAT$_{SH}$). An isoelectric focusing (IEF) marker, prepared by treating the same synthetic peptide as that used for the internal standard but labeled with a ^{14}C-VICAT$_{SH}$ reagent that contains a shorter diamino linker, is added to the biological sample. The IEF marker enables tracking of the peptide of interest during isoelectric focusing in order to identify regions of the gel for excision and subsequent peptide elution. The biotin tag allows enrichment of VICAT-labeled peptides by affinity chromatography using immobilized streptavidin, and the nitrobenzyl photocleavable linker allows the VICAT-labeled peptides to be photochemically cleaved from the immobilized support. The released peptides contain the isotope-coded portion of the tag to enable quantification by MS analysis.

A method of isotope dilution using synthetic isotope-coded peptides as internal standards for absolute protein abundance measurements by MS has also been developed and is referred to as AQUA (absolute quantification) [106]. In the AQUA approach, peptides identical to their native counterparts formed by proteolysis are synthesized with stable isotope-coded amino acids that provide a detectable mass shift. Prior to use, the synthetic peptides are evaluated by LC/MS/MS analysis to provide qualitative information regarding peptide retention during reversed-phase LC, ionization efficiency, and fragmentation propensity during CID in order to eliminate any errors resulting from peptide mass degeneracy. The basic strategy involves prefractionation of whole-cell lysates by SDS/PAGE, excision of the bands of interest, and in-gel proteolytic digestion in the presence of the AQUA internal peptide standard. However, other nongel-based fractionation techniques could be employed prior to adding the AQUA peptide. Peptides are separated by reversed-phase LC and analyzed by MS using selected reaction monitoring to selectively detect the peptides of interest and increase the sensitivity for low-abundance measurements. Since a known quantity of the synthetic peptide is added, the measured ratio of the synthetic to endogenous peptide intensity can be used to determine the absolute amount of the peptide present in the sample, which is reflective of absolute protein abundance.

The AQUA internal standards can also be synthesized with covalent modifications to probe PTMs, such as adding phosphate groups to peptides in order to measure *in vivo* phosphorylation events. The AQUA strategy has been used to quantify two low-abundance yeast proteins involved in gene silencing, quantify cell-cycle dependent phosphorylation of Ser-1126 of human separase protein, and identify the kinase that phosphorylates at Ser-1501 of separase [106]. In a similar manner, membrane proteins have been quantified by comparing their concentration to that of a chemically synthesized peptide present in the solvent-accessible region of the protein [107].

13.5   ISOBARIC TAGS FOR ABSOLUTE AND RELATIVE QUANTIFICATION

The various isotope-coded labeling methods described thus far used differential mass labeling for determining the absolute or relative abundance of proteins. However, inherent limitations are imposed by mass-difference labeling techniques. For these measurements, a binary set of reagents is used and precludes simultaneous differential comparison of more than two samples (e.g., time course studies, different conditions, several controls in the same study) while increasing MS complexity; this is further exacerbated when comparing data from multiple binary labeling experiments.

These factors have led to the development of a new quantification technique—isobaric tags for relative and absolute quantitation (iTRAQ)—that is an enhancement of the ICAT approach [108]. The iTRAQ labeling strategy uses a set of four amine-reactive isobaric reagents, enabling four different conditions to be multiplexed together in one experimental analysis. Each isobaric tag consists of a reporter group (based on N-methylpiperazine), a mass balance group (carbonyl group), and an amine-reactive group (N-hydroxysuccinimidyl ester). The overall mass of each of the four tags is kept constant by using different isotopic enrichments of ^{13}C, ^{15}N, and ^{18}O within the reporter and mass balance groups. The amine reactive group derivatizes lysyl side chains and the N-terminus of all peptides obtained from a protein digest.

When four samples are individually labeled with one of the four isobaric tags, combined, and analyzed by LC/MS/MS, the labeled peptides of the same amino acid sequence have identical mass and appear as a single unresolved precursor ion at the same m/z in the MS spectrum. Upon CID, the amide linkage within the tag fragments in a manner similar to backbone peptide amide bonds to produce a neutral loss of the mass balance group and low molecular mass reporter ions at m/z 114.1, 115.1, 116.1, and 117.1 that are unique to the tag used to label the peptides of each sample. Measurement of the intensity of these reporter ions enables relative quantification of the peptides in each digest and the generated b- and y-ions are used for identification of the labeled peptides and hence their corresponding proteins.

In addition to allowing simultaneous analysis of up to four distinct biological samples, the iTRAQ labeling approach has several other advantages. Unlike the ICAT approach, labeling is global and enhances overall proteome coverage while retaining important information regarding PTMs. The use of isobaric tags improves the overall sensitivity compared to mass-difference labeling and reduces ambiguity due to differential labeling of peptides. The multiplexing ability of this labeling strategy allows replicates of a given sample to be performed in a single experiment and hence increases the statistical relevance needed for quantitative measurements.

Absolute quantification can be achieved by using known amounts of internal standards tagged with one of the four-plex reagents.

The benefits of this multiplexed protein quantification approach were first demonstrated by comparing the global protein expression of proteins in wild-type and two mutant yeast strains that are defective in the nonsense-mediated mRNA decay [108]. The iTRAQ reagents are commercially available (Applied Biosystems, Inc., www.appliedbiosystems.com) and are being used in a variety of applications, such as to quantify the protein expression in *E. coli* expressing rhsA elements [109] and to assess the effects of diurnal variation on the composition of human parotid saliva [110]. In an application of PTM analysis, Zhang et al. have used the iTRAQ fragment ion ratios to quantify tyrosyl phosphorylation of specific residues simultaneously on dozens of key proteins in a time-resolved manner, downstream of epidermal growth factor receptor activation [111]. Undoubtedly, the use of iTRAQ combined with additional fractionation techniques such as SCXC and enrichment techniques for phosphopeptides using IMAC or titanium oxide will further promote the use of this labeling strategy.

13.6 CONCLUSION

The application of chemical, metabolic, or enzymatic incorporation of stable isotopes into peptides or proteins provides a platform to quantify protein abundances and their post-translational modifications using mass spectrometry. With the onset of more advanced separation and mass spectrometry innovations, more comprehensive biological investigations using isotope coding in global and in targeted proteomic analyses may be achieved. The difficulties in stable isotope labeling, particularly for phosphorylation, have prompted measurements of protein abundance to be performed without resorting to isotope coding by comparing MS signal intensities between the two samples [112–115] or using database identification scores [116] that would complement stable isotope-coded mass spectrometry. The ability to quantify changes in protein expression and modification, especially for those events occurring at low stoichiometries, will continue to be an analytical challenge for proteomics and systems biology for many years to come.

13.7 FUTURE PROSPECTS

The variety of methods described in this chapter using chemical, metabolic, or enzymatic incorporation of stable isotopes into peptides and proteins provides a platform to tailor a global or a targeted approach to proteomic analysis and quantification, including the characterization of post-translational modifications. With the continued, rapid advancement of mass spectrometry instrumentation and the innovations for more advanced chromatographic separations, the ability to quantify the changes of low-abundance proteins and their modifications may soon be realized. Although the use of isotope-coded mass spectrometry as a discovery tool for elucidating novel expression pathways and signaling networks is promising, there is an overwhelming demand to enhance data analysis and confidence in order for the results of any isotope-coding measurement to be considered valid.

Currently, most MS data are analyzed by the specialized software provided by the vendor who provided the instrument. However, a number of open source tools are available for further validation and analysis, such as ASAPRatio [117] to quantify isotopically labeled peptides and proteins and ZoomQuant [118] to analyze the mass spectra of ^{18}O labeled peptides generated by ion trap mass spectrometers. Efforts such as "the transproteomic pipeline"—a platform that enables a uniform analysis of product ion spectra generated from a variety of different instruments and database searching programs [119]—illustrate the trend to consolidate MS data and provide a means for assessing accuracy and validation in quantitative proteomic measurements. In this manner, more comprehensive global proteome analysis in complex systems will be achieved and "mining" of large datasets for biologically relevant information can be performed.

ACKNOWLEDGMENTS

The authors would like to thank the research agencies of North Carolina State University and the North Carolina Agricultural Research Service for continued support of biological mass spectrometry research. Portions of this work were supported by grants from the National Science Foundation (MCB-0419819) and the United States Department of Agriculture (NRI 2004-35304-14930 and NRI 2005-35604-15420).

REFERENCES

1. Patton, W. F., Proteome analysis. II. Protein subcellular redistribution: Linking physiology to genomics via the proteome and separation technologies involved, *J Chromatogr B Biomed Sci Appl*, 722, 203, 1999.
2. Yan, J. X., Sanchez, J. C., Binz, P. A., Williams, K. L., and Hochstrasser, D. F., Method for identification and quantitative analysis of protein lysine methylation using matrix-assisted laser desorption/ionization—time-of-flight mass spectrometry and amino acid analysis, *Electrophoresis*, 20, 749, 1999.
3. Newsholme, S. J., Maleeff, B. F., Steiner, S., Anderson, N. L., and Schwartz, L. W., Two-dimensional electrophoresis of liver proteins: Characterization of a drug-induced hepatomegaly in rats, *Electrophoresis*, 21, 2122, 2000.
4. Lill, J., Proteomic tools for quantitation by mass spectrometry, *Mass Spectrom Rev*, 22, 182, 2003.
5. Aebersold, R., and Mann, M., Mass spectrometry-based proteomics, *Nature*, 422, 198, 2003.
6. Panisko, E. A., Conrads, T. P., Goshe, M. B., and Veenstra, T. D., The postgenomic age: Characterization of proteomes, *Exp Hematol*, 30, 97, 2002.
7. Goshe, M. B., and Smith, R. D., Stable isotope-coded proteomic mass spectrometry, *Curr Opin Biotechnol*, 14, 101, 2003.
8. Leitner, A., and Lindner, W., Current chemical tagging strategies for proteome analysis by mass spectrometry, *J Chromatogr B Anal Technol Biomed Life Sci*, 813, 1, 2004.

9. Gygi, S. P., Rist, B., Gerber, S. A., Turecek, F., Gelb, M. H., and Aebersold, R., Quantitative analysis of complex protein mixtures using isotope-coded affinity tags, *Nat Biotechnol*, 17, 994, 1999.

10. Griffin, T. J., Han, D. K., Gygi, S. P., Rist, B., Lee, H., Aebersold, R., and Parker, K. C., Toward a high-throughput approach to quantitative proteomic analysis: Expression-dependent protein identification by mass spectrometry, *J Am Soc Mass Spectrom*, 12, 1238, 2001.

11. Han, D. K., Eng, J., Zhou, H., and Aebersold, R., Quantitative profiling of differentiation-induced microsomal proteins using isotope-coded affinity tags and mass spectrometry, *Nat Biotechnol*, 19, 946, 2001.

12. Gygi, S. P., Rist, B., Griffin, T. J., Eng, J., and Aebersold, R., Proteome analysis of low-abundance proteins using multidimensional chromatography and isotope-coded affinity tags, *J Proteome Res*, 1, 47, 2002.

13. Griffin, T. J., Gygi, S. P., Ideker, T., Rist, B., Eng, J., Hood, L., and Aebersold, R., Complementary profiling of gene expression at the transcriptome and proteome levels in *Saccharomyces cerevisiae*, *Mol Cell Proteomics*, 1, 323, 2002.

14. Yu, L. R., Johnson, M. D., Conrads, T. P., Smith, R. D., Morrison, R. S., and Veenstra, T. D., Proteome analysis of camptothecin-treated cortical neurons using isotope-coded affinity tags, *Electrophoresis*, 23, 1591, 2002.

15. Zhang, R., Sioma, C. S., Wang, S., and Regnier, F. E., Fractionation of isotopically labeled peptides in quantitative proteomics, *Anal Chem*, 73, 5142, 2001.

16. Moseley, M. A., Current trends in differential expression proteomics: Isotopically coded tags, *Trends Biotechnol*, 19, S10, 2001.

17. Li, J., Steen, H., and Gygi, S. P., Protein profiling with cleavable isotope-coded affinity tag (cICAT) reagents: The yeast salinity stress response, *Mol Cell Proteomics*, 2, 1198, 2003.

18. Borisov, O. V., Goshe, M. B., Conrads, T. P., Rakov, V. S., Veenstra, T. D., and Smith, R. D., Low-energy collision-induced dissociation fragmentation analysis of cysteinyl-modified peptides, *Anal Chem*, 74, 2284, 2002.

19. Zhang, R., and Regnier, F. E., Minimizing resolution of isotopically coded peptides in comparative proteomics, *J Proteome Res*, 1, 139, 2002.

20. Zhang, R., Sioma, C. S., Thompson, R. A., Xiong, L., and Regnier, F. E., Controlling deuterium isotope effects in comparative proteomics, *Anal Chem*, 74, 3662, 2002.

21. Goshe, M. B., Blonder, J., and Smith, R. D., Affinity labeling of highly hydrophobic integral membrane proteins for proteome-wide analysis, *J Proteome Res*, 2, 153, 2003.

22. Hoang, V. M., Conrads, T. P., Veenstra, T. D., Blonder, J., Terunuma, A., Vogel, J. C., and Fisher, R. J., Quantitative proteomics employing primary amine affinity tags, *J Biomol Tech*, 14, 216, 2003.

23. Zhao, Y., Zhang, W., and Kho, Y., Proteomic analysis of integral plasma membrane proteins, *Anal Chem*, 76, 1817, 2004.

24. Zhou, H., Ranish, J. A., Watts, J. D., and Aebersold, R., Quantitative proteome analysis by solid-phase isotope tagging and mass spectrometry, *Nat Biotechnol*, 20, 512, 2002.

25. Qiu, Y., Sousa, E. A., Hewick, R. M., and Wang, J. H., Acid-labile isotope-coded extractants: A class of reagents for quantitative mass spectrometric analysis of complex protein mixtures, *Anal Chem*, 74, 4969, 2002.

26. Shi, Y., Xiang, R., Crawford, J. K., Colangelo, C. M., Horvath, C., and Wilkins, J. A., A simple solid phase mass tagging approach for quantitative proteomics, *J Proteome Res*, 3, 104, 2004.

27. Whetstone, P. A., Butlin, N. G., Corneillie, T. M., and Meares, C. F., Element-coded affinity tags for peptides and proteins, *Bioconjug Chem*, 15, 3, 2004.

28. Wang, S., and Regnier, F. E., Proteomics based on selecting and quantifying cysteine containing peptides by covalent chromatography, *J Chromatogr A*, 924, 345, 2001.

29. Wang, S., Zhang, X., and Regnier, F. E., Quantitative proteomics strategy involving the selection of peptides containing both cysteine and histidine from tryptic digests of cell lysates, *J Chromatogr A*, 949, 153, 2002.

30. Ren, D., Julka, S., Inerowicz, H. D., and Regnier, F. E., Enrichment of cysteine-containing peptides from tryptic digests using a quaternary amine tag, *Anal Chem*, 76, 4522, 2004.

31. Olsen, J. V., Andersen, J. R., Nielsen, P. A., Nielsen, M. L., Figeys, D., Mann, M., and Wisniewski, J. R., HysTag—A novel proteomic quantification tool applied to differential display analysis of membrane proteins from distinct areas of mouse brain, *Mol Cell Proteomics*, 3, 82, 2004.

32. Peters, E. C., Horn, D. M., Tully, D. C., and Brock, A., A novel multifunctional labeling reagent for enhanced protein characterization with mass spectrometry, *Rapid Commun Mass Spectrom*, 15, 2387, 2001.

33. Beardsley, R. L., and Reilly, J. P., Quantitation using enhanced signal tags: A technique for comparative proteomics, *J Proteome Res*, 2, 15, 2003.

34. Cagney, G., and Emili, A., *De novo* peptide sequencing and quantitative profiling of complex protein mixtures using mass-coded abundance tagging, *Nat Biotechnol*, 20, 163, 2002.

35. Kuyama, H., Watanabe, M., Toda, C., Ando, E., Tanaka, K., and Nishimura, O., An approach to quantitative proteome analysis by labeling tryptophan residues, *Rapid Commun Mass Spectrom*, 17, 1642, 2003.

36. Chakraborty, A., and Regnier, F. E., Global internal standard technology for comparative proteomics, *J Chromatogr A*, 949, 173, 2002.

37. Geng, M., Ji, J., and Regnier, F. E., Signature-peptide approach to detecting proteins in complex mixtures, *J Chromatogr A*, 870, 295, 2000.

38. Ji, J., Chakraborty, A., Geng, M., Zhang, X., Amini, A., Bina, M., and Regnier, F., Strategy for qualitative and quantitative analysis in proteomics based on signature peptides, *J Chromatogr B Biomed Sci Appl*, 745, 197, 2000.

39. Che, F. Y., and Fricker, L. D., Quantitation of neuropeptides in Cpe(fat)/Cpe(fat) mice using differential isotopic tags and mass spectrometry, *Anal Chem*, 74, 3190, 2002.

40. Munchbach, M., Quadroni, M., Miotto, G., and James, P., Quantitation and facilitated *de novo* sequencing of proteins by isotopic N-terminal labeling of peptides with a fragmentation-directing moiety, *Anal Chem*, 72, 4047, 2000.

41. Goodlett, D. R., Keller, A., Watts, J. D., Newitt, R., Yi, E. C., Purvine, S., Eng, J. K., von Haller, P., Aebersold, R., and Kolker, E., Differential stable isotope labeling of peptides for quantitation and *de novo* sequence derivation, *Rapid Commun Mass Spectrom*, 15, 1214, 2001.

42. Li, X. J., Zhang, H., Ranish, J. A., and Aebersold, R., Automated statistical analysis of protein abundance ratios from data generated by stable-isotope dilution and tandem mass spectrometry, *Anal Chem*, 75, 6648, 2003.

43. Conrads, T. P., Alving, K., Veenstra, T. D., Belov, M. E., Anderson, G. A., Anderson, D. J., Lipton, M. S., Pasa-Tolic, L., Udseth, H. R., Chrisler, W. B., Thrall, B. D., and Smith, R. D., Quantitative analysis of bacterial and mammalian proteomes using a combination of cysteine affinity tags and 15N-metabolic labeling, *Anal Chem*, 73, 2132, 2001.

44. Smith, R. D., Pasa-Tolic, L., Lipton, M. S., Jensen, P. K., Anderson, G. A., Shen, Y., Conrads, T. P., Udseth, H. R., Harkewicz, R., Belov, M. E., Masselon, C., and Veenstra, T. D., Rapid quantitative measurements of proteomes by Fourier transform ion cyclotron resonance mass spectrometry, *Electrophoresis*, 22, 1652, 2001.
45. Washburn, M. P., Ulaszek, R., Deciu, C., Schieltz, D. M., and Yates, J. R., III, Analysis of quantitative proteomic data generated via multidimensional protein identification technology, *Anal Chem*, 74, 1650, 2002.
46. Wang, Y. K., Ma, Z., Quinn, D. F., and Fu, E. W., Inverse ^{15}N-metabolic labeling/mass spectrometry for comparative proteomics and rapid identification of protein markers/targets, *Rapid Commun Mass Spectrom*, 16, 1389, 2002.
47. Krijgsveld, J., Ketting, R. F., Mahmoudi, T., Johansen, J., Artal-Sanz, M., Verrijzer, C. P., Plasterk, R. H., and Heck, A. J., Metabolic labeling of *C. elegans* and *D. melanogaster* for quantitative proteomics, *Nat Biotechnol*, 21, 927, 2003.
48. Vogt, J. A., Schroer, K., Holzer, K., Hunzinger, C., Klemm, M., Biefang-Arndt, K., Schillo, S., Cahill, M. A., Schrattenholz, A., Matthies, H., and Stegmann, W., Protein abundance quantification in embryonic stem cells using incomplete metabolic labelling with ^{15}N amino acids, matrix-assisted laser desorption/ionisation time-of-flight mass spectrometry, and analysis of relative isotopologue abundances of peptides, *Rapid Commun Mass Spectrom*, 17, 1273, 2003.
49. Chen, X., Smith, L. M., and Bradbury, E. M., Site-specific mass tagging with stable isotopes in proteins for accurate and efficient protein identification, *Anal Chem*, 72, 1134, 2000.
50. Veenstra, T. D., Martinovic, S., Anderson, G. A., Pasa-Tolic, L., and Smith, R. D., Proteome analysis using selective incorporation of isotopically labeled amino acids, *J Am Soc Mass Spectrom*, 11, 78, 2000.
51. Hunter, T. C., Yang, L., Zhu, H., Majidi, V., Bradbury, E. M., and Chen, X., Peptide mass mapping constrained with stable isotope-tagged peptides for identification of protein mixtures, *Anal Chem*, 73, 4891, 2001.
52. Berger, S. J., Lee, S.-W., Anderson, G. A., Pasa-Tolic, L., Tolic, N., Shen, Y., Zhao, R., and Smith, R. D., High-throughput global peptide proteomic analysis by combining stable isotope amino acid labeling and data-dependent multiplexed-MS/MS, *Anal Chem*, 74, 4994, 2002.
53. Wierenga, S. K., Zocher, M. J., Mirus, M. M., Conrads, T. P., Goshe, M. B., and Veenstra, T. D., A method to evaluate tryptic digestion efficiency for high-throughput proteome analyses, *Rapid Commun Mass Spectrom*, 16, 1404, 2002.
54. Ong, S. E., Blagoev, B., Kratchmarova, I., Kristensen, D. B., Steen, H., Pandey, A., and Mann, M., Stable isotope labeling by amino acids in cell culture, SILAC, as a simple and accurate approach to expression proteomics, *Mol Cell Proteomics*, 1, 376, 2002.
55. Jiang, H., and English, A. M., Quantitative analysis of the yeast proteome by incorporation of isotopically labeled leucine, *J Proteome Res*, 1, 345, 2002.
56. Zhu, H., Hunter, T. C., Pan, S., Yau, P. M., Bradbury, E. M., and Chen, X., Residue-specific mass signatures for the efficient detection of protein modifications by mass spectrometry, *Anal Chem*, 74, 1687, 2002.
57. Martinovic, S., Veenstra, T. D., Anderson, G. A., Pasa-Tolic, L., and Smith, R. D., Selective incorporation of isotopically labeled amino acids for identification of intact proteins on a proteome-wide level, *J Mass Spectrom*, 37, 99, 2002.

58. Pasa-Tolic, L., Jensen, P. K., Anderson, G. A., Lipton, M. S., Peden, K. K., Martinovic, S., Tolic, N., Bruce, J. E., and Smith, R. D., High throughput proteome-wide precision measurements of protein expression using mass spectrometry, *J Am Chem Soc*, 121, 7949, 1999.

59. Lipton, M. S., Pasa-Tolic, L., Anderson, G. A., Anderson, D. J., Auberry, D. L., Battista, J. R., Daly, M. J., Fredrickson, J., Hixson, K. K., Kostandarithes, H., Masselon, C., Markillie, L. M., Moore, R. J., Romine, M. F., Shen, Y., Stritmatter, E., Tolic, N., Udseth, H. R., Venkateswaran, A., Wong, K. K., Zhao, R., and Smith, R. D., Global analysis of the *Deinococcus radiodurans* proteome by using accurate mass tags, *Proc Natl Acad Sci USA*, 99, 11049, 2002.

60. Smith, R. D., Anderson, G. A., Lipton, M. S., Pasa-Tolic, L., Shen, Y., Conrads, T. P., Veenstra, T. D., and Udseth, H. R., An accurate mass tag strategy for quantitative and high-throughput proteome measurements, *Proteomics*, 2, 513, 2002.

61. Pasa-Tolic, L., Harkewicz, R., Anderson, G. A., Tolic, N., Shen, Y., Zhao, R., Thrall, B., Masselon, C., and Smith, R. D., Increased proteome coverage for quantitative peptide abundance measurements based upon high performance separations and DREAMS FTICR mass spectrometry, *J Am Soc Mass Spectrom*, 13, 954, 2002.

62. Kosaka, T., Takazawa, T., and Nakamura, T., Identification and C-terminal characterization of proteins from two-dimensional polyacrylamide gels by a combination of isotopic labeling and nanoelectrospray Fourier transform ion cyclotron resonance mass spectrometry, *Anal Chem*, 72, 1179, 2000.

63. Wallis, T. P., Pitt, J. J., and Gorman, J. J., Identification of disulfide-linked peptides by isotope profiles produced by peptic digestion of proteins in 50% (18)O water, *Protein Sci*, 10, 2251, 2001.

64. Huang, B. X., Dass, C., and Kim, H. Y., Probing conformational changes of human serum albumin due to unsaturated fatty acid binding by chemical cross-linking and mass spectrometry, *Biochem J*, 10, 10, 2004.

65. Back, J. W., Notenboom, V., de Koning, L. J., Muijsers, A. O., Sixma, T. K., de Koster, C. G., and de Jong, L., Identification of cross-linked peptides for protein interaction studies using mass spectrometry and [18]O labeling, *Anal Chem*, 74, 4417, 2002.

66. Stewart, II, Thomson, T., and Figeys, D., [18]O labeling: a tool for proteomics, *Rapid Commun Mass Spectrom*, 15, 2456, 2001.

67. Yao, X., Freas, A., Ramirez, J., Demirev, P. A., and Fenselau, C., Proteolytic [18]O labeling for comparative proteomics: Model studies with two serotypes of adenovirus, *Anal Chem*, 73, 2836, 2001.

68. Qian, W. J., Monroe, M. E., Liu, T., Jacobs, J. M., Anderson, G. A., Shen, Y., Moore, R. J., Anderson, D. J., Zhang, R., Calvano, S. E., Lowry, S. F., Xiao, W., Moldawer, L. L., Davis, R. W., Tompkins, R. G., Camp, D. G., II, and Smith, R. D., Quantitative proteome analysis of human plasma following in vivo lipopolysaccharide administration using ^{16}O/^{18}O labeling and the accurate mass and time tag approach, *Mol Cell Proteomics*, 7, 7, 2005.

69. Wang, Y. K., Ma, Z., Quinn, D. F., and Fu, E. W., Inverse [18]O labeling mass spectrometry for the rapid identification of marker/target proteins, *Anal Chem*, 73, 3742, 2001.

70. Brown, K. J., and Fenselau, C., Investigation of doxorubicin resistance in MCF-7 breast cancer cells using shot-gun comparative proteomics with proteolytic [18]O labeling, *J Proteome Res*, 3, 455, 2004.

71. Cohen, P., Signal integration at the level of protein kinases, protein phosphatases and their substrates, *Trends Biochem Sci*, 17, 408, 1992.

72. Cohen, P., The regulation of protein function by multisite phosphorylation—A 25 year update, *Trends Biochem Sci*, 25, 596, 2000.

73. Mann, M., Ong, S. E., Gronborg, M., Steen, H., Jensen, O. N., and Pandey, A., Analysis of protein phosphorylation using mass spectrometry: Deciphering the phosphoproteome, *Trends Biotechnol*, 20, 261, 2002.

74. Conrads, T. P., Issaq, H. J., and Veenstra, T. D., New tools for quantitative phospho-proteome analysis, *Biochem Biophys Res Commun*, 290, 885, 2002.

75. Loughrey Chen, S., Huddleston, M. J., Shou, W., Deshaies, R. J., Annan, R. S., and Carr, S. A., Mass spectrometry-based methods for phosphorylation site mapping of hyperphosphorylated proteins applied to Net1, a regulator of exit from mitosis in yeast, *Mol Cell Proteomics*, 1, 186, 2002.

76. Ficarro, S. B., McCleland, M. L., Stukenberg, P. T., Burke, D. J., Ross, M. M., Shabanowitz, J., Hunt, D. F., and White, F. M., Phosphoproteome analysis by mass spectrometry and its application to *Saccharomyces cerevisiae*, *Nat Biotechnol*, 20, 301, 2002.

77. Ficarro, S., Chertihin, O., Westbrook, V. A., White, F., Jayes, F., Kalab, P., Marto, J. A., Shabanowitz, J., Herr, J. C., Hunt, D. F., and Visconti, P. E., Phosphop-roteome analysis of capacitated human sperm. Evidence of tyrosine phosphorylation of a kinase-anchoring protein 3 and valosin-containing protein/p97 during capacita-tion, *J Biol Chem*, 278, 11579, 2003.

78. He, T., Alving, K., Feild, B., Norton, J., Joseloff, E. G., Patterson, S. D., and Domon, B., Quantitation of phosphopeptides using affinity chromatography and sta-ble isotope labeling, *J Am Soc Mass Spectrom*, 15, 363, 2004.

79. Salomon, A. R., Ficarro, S. B., Brill, L. M., Brinker, A., Phung, Q. T., Ericson, C., Sauer, K., Brock, A., Horn, D. M., Schultz, P. G., and Peters, E. C., Profiling of tyrosine phosphorylation pathways in human cells using mass spectrometry, *Proc Natl Acad Sci USA*, 100, 443, 2003.

80. Brill, L. M., Salomon, A. R., Ficarro, S. B., Mukherji, M., Stettler-Gill, M., and Peters, E. C., Robust phosphoproteomic profiling of tyrosine phosphorylation sites from human T cells using immobilized metal affinity chromatography and tandem mass spectrometry, *Anal Chem*, 76, 2763, 2004.

81. Ren, D., Penner, N. A., Slentz, B. E., Mirzaei, H., and Regnier, F., Evaluating immobilized metal affinity chromatography for the selection of histidine-containing peptides in comparative proteomics, *J Proteome Res*, 2, 321, 2003.

82. Bieber, A. L., Tubbs, K. A., and Nelson, R. W., Metal ligand affinity pipettes and bioreactive alkaline phosphatase probes: tools for characterization of phosphorylated proteins and peptides, *Mol Cell Proteomics*, 3, 266, 2004.

83. Oda, Y., Huang, K., Cross, F. R., Cowburn, D., and Chait, B. T., Accurate quantitation of protein expression and site-specific phosphorylation, *Proc Natl Acad Sci USA*, 96, 6591, 1999.

84. DeGnore, J. P., and Qin, J., Fragmentation of phosphopeptides in an ion trap mass spectrometer, *J Am Soc Mass Spectrom*, 9, 1175, 1998.

85. McLafferty, F. W., Horn, D. M., Breuker, K., Ge, Y., Lewis, M. A., Cerda, B., Zubarev, R. A., and Carpenter, B. K., Electron capture dissociation of gaseous mul-tiply charged ions by Fourier-transform ion cyclotron resonance, *J Am Soc Mass Spectrom*, 12, 245, 2001.

86. Syka, J.E., Coon, J.J., Schroeder, M.J., Shabanowitz, J., and Hunt, D.F., Peptide and protein sequence analysis by electron transfer dissociation mass spectrometry. *Proc Natl Acad Sci USA*, 101, 9528, 2004.

87. Oda, Y., Nagasu, T., and Chait, B. T., Enrichment analysis of phosphorylated proteins as a tool for probing the phosphoproteome, *Nat Biotechnol*, 19, 379, 2001.

88. McLachlin, D. T., and Chait, B. T., Improved beta-elimination-based affinity purification strategy for enrichment of phosphopeptides, *Anal Chem*, 75, 6826, 2003.

89. Zhou, H., Watts, J. D., and Aebersold, R., A systematic approach to the analysis of protein phosphorylation, *Nat Biotechnol*, 19, 375, 2001.

90. Goshe, M. B., Conrads, T. P., Panisko, E. A., Angell, N. H., Veenstra, T. D., and Smith, R. D., Phosphoprotein isotope-coded affinity tag approach for isolating and quantitating phosphopeptides in proteome-wide analyses, *Anal Chem*, 73, 2578, 2001.

91. Goshe, M. B., Veenstra, T. D., Panisko, E. A., Conrads, T. P., Angell, N. H., and Smith, R. D., Phosphoprotein isotope-coded affinity tags: Application to the enrichment and identification of low-abundance phosphoproteins, *Anal Chem*, 74, 607, 2002.

92. Qian, W. J., Goshe, M. B., Camp, D. G., 2nd, Yu, L. R., Tang, K., and Smith, R. D., Phosphoprotein isotope-coded solid-phase tag approach for enrichment and quantitative analysis of phosphopeptides from complex mixtures, *Anal Chem*, 75, 5441, 2003.

93. Soderblom, E. J., Cawthon, D., Duhart, H., Xu, Z. A., Slikker Jr., W., Ali, S. F., and Goshe, M. B., An improved labeling method using phosphoprotein isotope-coded solid-phase tags for neuronal cell applications, *Int J Neuroprotec Neuroregen*, 1(2), 91–97, 2005.

94. Cawthon, D., Soderblom, E. J., Xu, Z. A., Duhart, H., Slikker Jr., W., Ali, S. F., and Goshe, M. B., Quantitative analysis of phosphoproteins in a Parkinson's disease model using phosphoprotein isotope-coded solid-phase tags, *Int J Neuroprotec Neuroregen*, 1(2), 98–106, 2005.

95. Dwek, R. A., Glycobiology: "Towards understanding the function of sugars," *Biochem Soc Trans*, 23, 1, 1995.

96. Dell, A., and Morris, H. R., Glycoprotein structure determination by mass spectrometry, *Science*, 291, 2351, 2001.

97. Geng, M., Zhang, X., Bina, M., and Regnier, F., Proteomics of glycoproteins based on affinity selection of glycopeptides from tryptic digests, *J Chromatogr B Biomed Sci Appl*, 752, 293, 2001.

98. Xiong, L., and Regnier, F. E., Use of a lectin affinity selector in the search for unusual glycosylation in proteomics, *J Chromatogr B Anal Technol Biomed Life Sci*, 782, 405, 2002.

99. Kaji, H., Saito, H., Yamauchi, Y., Shinkawa, T., Taoka, M., Hirabayashi, J., Kasai, K., Takahashi, N., and Isobe, T., Lectin affinity capture, isotope-coded tagging and mass spectrometry to identify N-linked glycoproteins, *Nat Biotechnol*, 21, 667, 2003.

100. Kristiansen, T. Z., Bunkenborg, J., Gronborg, M., Molina, H., Thuluvath, P. J., Argani, P., Goggins, M. G., Maitra, A., and Pandey, A., A proteomic analysis of human bile, *Mol Cell Proteomics*, 3, 715, 2004.

101. Reynolds, K. J., Yao, X., and Fenselau, C., Proteolytic [18]O labeling for comparative proteomics: Evaluation of endoprotease Glu-C as the catalytic agent, *J Proteome Res*, 1, 27, 2002.

102. Zhang, H., Li, X. J., Martin, D. B., and Aebersold, R., Identification and quantification of N-linked glycoproteins using hydrazide chemistry, stable isotope labeling and mass spectrometry, *Nat Biotechnol*, 21, 660, 2003.

103. Wells, L., Vosseller, K., Cole, R. N., Cronshaw, J. M., Matunis, M. J., and Hart, G. W., Mapping sites of O-GlcNAc modification using affinity tags for serine and threonine post-translational modifications, *Mol Cell Proteomics*, 1, 791, 2002.

104. Bottari, P., Aebersold, R., Turecek, F., and Gelb, M. H., Design and synthesis of visible isotope-coded affinity tags for the absolute quantification of specific proteins in complex mixtures, *Bioconjug Chem*, 15, 380, 2004.

105. Lu, Y., Bottari, P., Turecek, F., Aebersold, R., and Gelb, M. H., Absolute quantification of specific proteins in complex mixtures using visible isotope-coded affinity tags, *Anal Chem*, 76, 4104, 2004.

106. Gerber, S. A., Rush, J., Stemman, O., Kirschner, M. W., and Gygi, S. P., Absolute quantification of proteins and phosphoproteins from cell lysates by tandem MS, *Proc Natl Acad Sci USA*, 100, 6940, 2003.

107. Barnidge, D. R., Dratz, E. A., Martin, T., Bonilla, L. E., Moran, L. B., and Lindall, A., Absolute quantification of the G protein-coupled receptor rhodopsin by LC/MS/MS using proteolysis product peptides and synthetic peptide standards, *Anal Chem*, 75, 445, 2003.

108. Ross, P. L., Huang, Y. N., Marchese, J. N., Williamson, B., Parker, K., Hattan, S., Khainovski, N., Pillai, S., Dey, S., Daniels, S., Purkayastha, S., Juhasz, P., Martin, S., Bartlet-Jones, M., He, F., Jacobson, A., and Pappin, D. J., Multiplexed protein quantitation in *Saccharomyces cerevisiae* using amine-reactive isobaric tagging reagents, *Mol Cell Proteomics*, 3, 1154, 2004.

109. Kunal Aggarwal, L. H. C., and Lee, K. H., Quantitative analysis of protein expression using amine-specific isobaric tags in *Escherichia coli* cells expressing rhsA elements, *Proteomics*, 5, 2297, 2005.

110. Hardt, M., Witkowska, H. E., Webb, S., Thomas, L. R., Dixon, S. E., Hall, S. C., and Fisher, S. J., Assessing the effects of diurnal variation on the composition of human parotid saliva: Quantitative analysis of native peptides using iTRAQ reagents, *Anal Chem*, 77, 4947, 2005.

111. Zhang, Y., Wolf-Yadlin, A., Ross, P. L., Papin, D. J., Rush, J., Lauffenburger, D. A., and White, F. M., *Mol Cell Proteomics*, 4, 1240, 2005.

112. Chelius, D., and Bondarenko, P. V., Quantitative profiling of proteins in complex mixtures using liquid chromatography and mass spectrometry, *J Proteome Res*, 1, 317, 2002.

113. Bondarenko, P. V., Chelius, D., and Shaler, T. A., Identification and relative quantitation of protein mixtures by enzymatic digestion followed by capillary reversed-phase liquid chromatography-tandem mass spectrometry, *Anal Chem*, 74, 4741, 2002.

114. Wang, W., Zhou, H., Lin, H., Roy, S., Shaler, T. A., Hill, L. R., Norton, S., Kumar, P., Anderle, M., and Becker, C. H., Quantification of proteins and metabolites by mass spectrometry without isotopic labeling or spiked standards, *Anal Chem*, 75, 4818, 2003.

115. Chelius, D., Zhang, T., Wang, G., and Shen, R. F., Global protein identification and quantification technology using two-dimensional liquid chromatography nanospray mass spectrometry, *Anal Chem*, 75, 6658, 2003.

116. Liu, H., Sadygov, R. G., and Yates, J. R., III, A model for random sampling and estimation of relative protein abundance in shotgun proteomics, *Anal Chem*, 76, 4193, 2004.

117. Li, X. J., Zhang, H., Ranish, J. A., and Aebersold, R., Automated statistical analysis of protein abundance ratios from data generated by stable-isotope dilution and tandem mass spectrometry, *Anal Chem*, 75, 6648, 2003.

118. Halligan, B. D., Slyper, R. Y., Twigger, S. N., Hicks, W., Olivier, M., and Greene, A. S., ZoomQuant: An application for the quantitation of stable isotope labeled peptides, *J Am Soc Mass Spectrom*, 16, 302, 2005.

119. Keller, A., Eng, J., Zhang, N., Li, X. J., and Aebersold, R., A uniform proteomics MS/MS analysis platform utilizing open XML file formats, *Mol Syst Biol.*, 1, E1, Aug 2, 2005.

14 Surface Plasmon Resonance Biosensors' Contributions to Proteome Mapping

Rebecca L. Rich and David G. Myszka

CONTENTS

14.1 INTRODUCTION

With expanding interest in proteomics, we are faced with the daunting task of literally connecting all the "dots." Even the protein maps of relatively simple organisms such as *Drosophila melanogaster* reveal tens of thousands of potential interactions, which

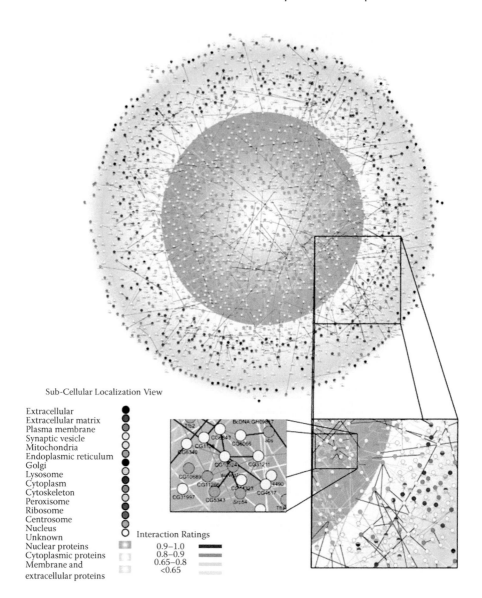

Sub-Cellular Localization View

Extracellular
Extracellular matrix
Plasma membrane
Synaptic vesicle
Mitochondria
Endoplasmic reticulum
Golgi
Lysosome
Cytoplasm
Cytoskeleton
Peroxisome
Ribosome
Centrosome
Nucleus
Unknown
Nuclear proteins
Cytoplasmic proteins
Membrane and
extracellular proteins

Interaction Ratings
0.9–1.0
0.8–0.9
0.65–0.8
<0.65

FIGURE 14.1 (See color insert.) Protein interaction map that depicts ~30% of the proteins and ~10% of the interactions identified in *Drosophila melanogaster*. Proteins are grouped by predicted subcellular localization. In addition, proteins are coded according to protein family and interactions. The most probable interactions are joined by the darkest lines. (Reprinted from Giot, L. et al., *Science*, 302, 1727, 2003. With permission from the American Association for the Advancement of Science © 2003.)

are represented in most proteomics maps by tiny dots interconnected by lines (fig. 14.1) [1]. But, of course, proteins are not dots and, more importantly, the connections between them are not simply lines. Proteins are dynamic, and their relationships are complex. They interact with themselves, other proteins, nucleic acids, substrates, cofactors, membranes, and other biomolecules. It is the dynamics of protein interactions that define their activity but, unfortunately, dynamics is the one thing that is often missing from protein interaction maps. If we intend to put proteome maps to use, we will need to supplement these static representations of pathways with kinetic information that reveals the time scale of binding events.

Fortunately, surface plasmon resonance (SPR) biosensors have the potential to provide access to the dynamic information about protein interactions. SPR biosensors monitor in real time how binding partners interact, and they do it without labeling, which saves time and money. SPR can be used qualitatively to screen protein panels to identify binding pairs or quantitatively to provide detailed information on reaction kinetics. Since SPR biosensors were commercially released 17 years ago, they have become well established as a robust, valid, and versatile technology for characterizing protein interactions [2,3].

Biosensors are routinely used on a pairwise scale to interpret binding mechanisms and to help correlate structure with function [4,5]. In addition, biosensors can be used as biophysical tools to better understand how environmental factors (e.g., buffer components, pH, temperature, and cofactors) influence an interaction [5]. Within multimolecular complexes, biosensors help to define which components are required for complex formation, as well as the order of assembly and disassembly [6–9]. Also, in interaction networks and pathways, biosensors can establish the stability of each complex [10] and indicate the *in situ* protein concentrations necessary for binding to occur.

14.2 TECHNICAL ASPECTS OF SPR

14.2.1 HOW SPR WORKS

In a typical protein/protein interaction study, one binding partner (ligand) is immobilized on a surface and the other partner (analyte) in solution interacts with this ligand (fig. 14.2A). In the most frequently used SPR instruments, the analyte is flowed across the ligand surface and, as protein complexes form and break down, the detector monitors the intensity of the light reflected from the surface (fig. 14.2B). Changes in the intensity spectra (fig. 14.2C) are monitored over time to produce a binding response referred to as a "sensorgram" (fig. 14.2D). The binding response is flat prior to injection of analyte and increases as analyte binds to the ligand; if the association phase is long enough, the interaction will reach equilibrium, which is indicated by a plateau in response. Finally, it decreases as buffer is flowed over the surface and analyte dissociates from the ligand surface.

14.2.2 OBTAINING KINETIC PARAMETERS

In general, biomolecular binding events can be described by straightforward interaction models, either a simple 1:1 model or a 1:1 model that includes a transport parameter to account for analyte diffusion to the surface (fig. 14.3A). To obtain precise kinetic rate constants, we collect binding data for a range of analyte concentrations flowed over a ligand surface and globally fit the chosen interaction model to the entire set of sensorgrams (fig. 14.3B). The overlay of the model fit with the

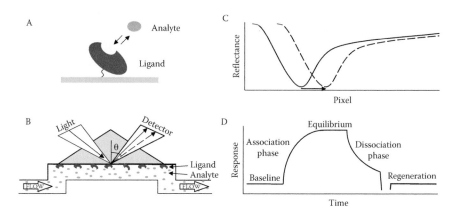

FIGURE 14.2 SPR biosensor analysis of proteins. (A) Assay design. One binding partner (analyte) in solution is flowed across the other partner (ligand), which is immobilized on the biosensor chip surface. (B) Cartoon of the flow cell and detector used in the Biacore 2000 and 3000 instruments. Ligand is immobilized on the top surface of the flow cell. Analyte is injected through the flow cell and across this surface. As the analyte binds to the ligand, the refractive index of the buffer near the surface changes. This induces a shift in the angle of refracted light recorded by the SPR detector. (C) Raw spectral data. The dark line depicts the baseline spectrum before the analyte binds; the dashed line depicts the spectral shift induced by analyte binding to the ligand surface. The amount of bound analyte corresponds to the shift of the spectrum's minimum (arrow). (D) Features of a typical binding cycle. Prior to injection of the analyte, the baseline is flat. The response intensity increases during the association phase as the analyte binds to the ligand. A plateau in response during the association phase indicates the ligand/analyte interaction has achieved equilibrium. During the dissociation phase, buffer is flowed over the ligand surface and the response decreases as analyte dissociates from the surface. If necessary, a short injection of a regeneration solution (usually weak acid or base) can be flowed over the ligand surface to strip away remaining analyte.

response data indicates how well the chosen model describes the binding interaction. Alternatively, we can test analytes in a higher-throughput format to identify binders and compare binding rates. For example, the kinetic screen of six analytes (shown in fig. 14.3C) revealed four binders, each of which interacts with the ligand with distinctive kinetic profiles (the greatest differences are detectable in the dissociation phase). Together, the two complementary analysis formats highlighted in figure 14.3B and 14.3C illustrate the biosensor's unique role in interpreting the dynamics of protein/protein interactions.

14.2.3 USING A SURFACE

Critics of SPR technology have long argued that immobilizing one binding partner on a surface affects the analyte/ligand interaction. In our experience, however, we find that when the experiments are set up properly, the kinetic rate constants and affinity constants determined from the surface-based biosensor assay match solution-based binding constants (e.g., obtained from isothermal calorimetry, stopped-flow fluorescence, analytical ultracentrifugation, KinExA, or electromobility shift assays).

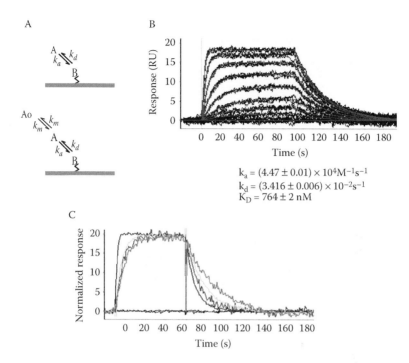

$$k_a = (4.47 \pm 0.01) \times 10^4 M^{-1} s^{-1}$$
$$k_d = (3.416 \pm 0.006) \times 10^{-2} s^{-1}$$
$$K_D = 764 \pm 2 \text{ nM}$$

FIGURE 14.3 Applying interaction models to biosensor data. (A) Simple 1:1 interaction model (top) and 1:1 model that includes a mass transport parameter (bottom). A represents the analyte, B is the immobilized ligand, and Ao is the analyte in bulk solution. (B) Rigorous kinetic analysis of an enzyme/inhibitor interaction. Triplicate injections of the inhibitor (at 0, 0.078, 0.156, 0.313, 0.625, 1.25, 2.50, 5.00, and 10.0 μM) were flowed over the immobilized enzyme. Binding responses (black lines) were globally fit to a 1:1 interaction model that included a mass transport parameter (gray lines) to yield rate constants. (C) Kinetic screen of six inhibitors (all tested at 4 μM) flowed over the enzyme surface. Two inhibitors did not bind to the surface. The four binders exhibit a range of kinetic profiles.

This correlation has been demonstrated for several different systems ranging from protein/protein [11,12] and antibody/antigen [8,13,14] interactions to DNA [15] and even small-molecule (<500 Da) inhibitor/enzyme interactions [16,17]. Surface- and solution-based binding constants agree because, in the biosensor assay, molecules are not actually affixed directly to the flat chip surface. Instead, the chip is coated with a carboxy-dextran hydrogel and the ligand is suspended in solution, tethered to this hydrophilic matrix. The matrix helps maintain the entropic properties of the system so that the immobilized protein, for example, can rotate and experience translational motion to some degree. More importantly, potential binding partners can approach this protein from all directions because it is suspended in three-dimensional space.

The biosensor surface, rather than being a drawback, is in fact a significant asset for studying interactions that occur at surface/solution interfaces, as well as for interpreting the relationships within multicomponent complexes. The biosensor chip can be used to imitate membrane- and other surface-based biology. For example, we can use SPR to characterize analytes binding to membrane proteins reconstituted within a lipid layer [6,9] or to a framework of collagen or other rigid extracellular matrix components [18], or to examine how fibrils grow to form plaques [19,20]. In addition, using the biosensor surface permits multicomponent complexes to be constructed via sequential injections of each component, thereby providing the opportunity to acquire kinetic data for each step of complex formation.

14.2.4 ADDITIONAL ADVANTAGES OF SPR

While the SPR biosensor's ability to monitor binding events in real time without reagent labels is its hallmark, this technology has several other important advantages over other methods to examine protein/protein interactions. Unlike other interaction technologies that examine interactions within only a relatively narrow kinetic/affinity range, this single technology measures a wide span of rate constants (k_a = 100 to 10^8 $M^{-1}s^{-1}$, k_d = 1 to 10^{-6} s^{-1}) that correspond to dissociation constants from 10^{-2} to 10^{-14} M. The ability of biosensors to monitor complex formation in the presence of free material makes it possible to measure very weak or transient interactions that would otherwise be missed by techniques that require washing steps. Also, for some systems, protein ligands can be extracted from crude samples, thereby eliminating purification/concentration steps. Today's biosensors can track multiple binding pairs simultaneously, and most instruments are highly automated. Automation improves data quality and increases sample throughput. Currently available platforms (e.g., Biacore S51 and FLEXchip) can analyze several hundred interactions per day.

14.2.5 LIMITATIONS OF SPR

SPR biosensors have limitations like any other biophysical technique. Unlike fluorescence- or radioligand-based detection methods, biosensors are not particularly sensitive instruments. SPR-based studies therefore require protein preparations that are highly active. Obtaining active material is the most common challenge we face when performing biosensor analyses. The biosensor's sensitivity also limits the size of molecules we can study. Historically, the technology was utilized to study only macromolecular interactions, but with improvements in hardware, experimental technique, and data processing software, the analysis of small molecule analytes (500–200 Da) is now fairly routine [23]. The two greatest challenges experimentally have to do with how the ligand is immobilized and how any bound complexes are regenerated. There are in fact a number of methods for addressing both issues by using a variety of surface attachment chemistries or testing regeneration conditions empirically. In the end, if the molecules of interest are active to begin with, the success rate of biosensor analysis is high.

14.3 SPR'S ESTABLISHED AND DEVELOPING ROLES IN PROTEOMICS

14.3.1 IDENTIFYING BINDING PARTNERS

Often, the first step in characterizing a protein is finding its binding partners. Traditionally, this has required stepwise fractionation of complex solutions (e.g., sera or cell lysates) to isolate putative binders. By coupling SPR with mass spectrometry (MS) (using the MS-compatible sample preparation/recovery unit available in Biacore 3000 platform) we can efficiently extract and identify binding partners from these mixtures. First, the sensor chip, which has a larger contact surface than standard chips (to maximize analyte capture), serves as a miniature affinity chromatography column: a mixture is flowed across the ligand surface, a binding partner is captured by the ligand, and the surface is washed to remove nonspecifically bound material. Then, the isolated binding partner is eluted directly onto a MALDI target and identified using MS. Experimental details of these methods are outlined in Buijs and Natsume [31]. Using this integrated approach, Zhukov et al. isolated and identified a kinase binding partner from bovine brain extract [30].

14.3.2 INTERPRETING BINDING MECHANISMS

The ability to determine binding kinetics is the bread and butter of biosensor technology. Kinetic data provide mechanistic information that is not available from structural analyses or equilibrium assays. For example, it was unclear until recently how agonist and antagonist ligands, which bind with similar affinities at the same site in estrogen receptor (ER), induce opposite biological effects. The difference is in the kinetics: as a class, antagonists bind ER 500–1,000 times more slowly than agonists do (fig. 14.4A) [21]. This difference in k_a suggested that agonists and antagonists induce or trap different conformational states of the receptor. Additional studies revealed the antagonist-bound conformation cannot recognize coactivator proteins that fully activate ER [22]. These functional studies complemented structural data (fig. 14.4B) to provide the basis for interpreting the ligand-binding mechanisms of ER and other nuclear receptors.

14.3.3 INTERFACE MAPPING

Kinetic parameters also contribute to defining binding interfaces. For example, in their structure/activity characterization of transferrin (Tf) binding to its receptor (TfR), Gianetti et al. examined how pH and iron affect the mutant and wild-type receptor's activity [5]. From a visual inspection of the sensorgram sets shown in figure 14.4C, it is apparent how different mutations and buffer conditions affected the receptor's Tf recognition. For example, the slow decay in response at >120 sec in the sensorgrams obtained for the wild-type TfR/Tf interaction indicate this complex dissociates slowly. The square-shaped responses for holo-Tf binding to R651A TfR, on the other hand, indicate this complex dissociates within seconds (this same mutation eliminates apo-Tf binding completely). Fitting the entire panel of kinetic data to interaction models yielded thermodynamic information about the binding interface residues. Coupled with structural characterization, this revealed that the residues required for apo- and holo-transferrin binding are overlapping but not identical (fig. 14.4D).

14.3.4 MULTICOMPONENT COMPLEX ASSEMBLY

Figure 14.5A illustrates the two-step assembly of the multisubunit interleukin-2 receptor (IL-2R). First, IL-2 binds to IL-2R α and β subunits; second, the IL-2$_\gamma$

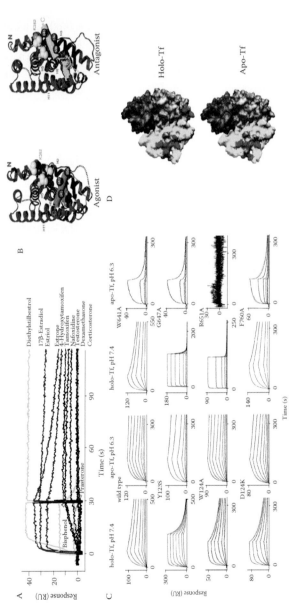

FIGURE 14.4 Functional and structural analyses of binding pairs. (A) Kinetic screen of ligands binding to ER-α ligand-binding domain. Responses were generated from the injection of nonestrogen agonists (gray), estrogen agonists, antagonists, and control ligands (all shown in black). (B) Crystal structures of agonist- and antagonist-bound ER-α ligand-binding domain. (C) Apo- and holo-transferrin (Tf) binding to immobilized wild-type and mutant transferrin receptor (TfR). Binding responses (black lines) are superimposed with the fit of the chosen interaction model (gray lines). (D) Summary of effects of TfR substitutions for binding holo- and apo-Tf. (Panels A and B reproduced from Rich, R.L. et al. *Proc. Natl. Acad. Sci. USA*, 99, 8562, 2002, and Brzozowski, A.M. et al., *Nature*, 389, 753, 1997, respectively. With permission from the National Academy of Sciences, USA, and the Nature Publishing Group © 2002 and 1997. Panels C and D adapted from Giannetti, A. et al., *PLoS Biol.*, 1, 341, 2003. With permission from PLoS journals and Giannetti et al. © 2003.)

subunit binds to the IL2R$_{\alpha\beta}$/IL-2 complex. Using the biosensor chip as a mimic of the cell surface, Ciardelli and coworkers determined the order of component addition and the kinetics of each step in assembly of the IL-2R/IL-2 complex [7,8]. In one set of experiments, these researchers examined how IL-2 binds to surfaces of α alone, β alone, and the αβ heterodimer (cartoons in fig. 14.5, panels B through D). Binding data shown in figure 14.5B through 14.5D demonstrate that IL-2 bound to the individual α and β monomer surfaces with rapid dissociation rates and lower affinities compared to the αβ dimer surface. These data established that the αβ subunit exists as a preformed heterodimer on the cell surface and α and β together bind IL-2 [8].

In a second set of experiments, these researchers tested γ binding to different permutations of the IL-2/IL-2R$_{\alpha\beta}$ complex preassembled on the biosensor chip surface. The binding data shown in figure 14.5E demonstrate that γ binds to IL-2/IL-2R$_{\alpha\beta}$ (γ also bound to IL-2/IL-2R$_{\beta}$, but not to surfaces lacking either IL-2 or IL-2R$_{\beta}$ or both). Combined, these kinetic studies determined the roles of the α, β, and γ subunits within the IL-2/IL-2R complex: α contributes to IL-2 binding; β participates in binding of IL-2 and the γ subunit; and γ binds after the initial IL-2/IL-2R$_{\alpha\beta}$ complex has formed (fig. 14.5F).

14.3.5 CONFIRMATORY PROTEOMICS MAPPING

From yeast two-hybrid and GST pull-down assays, Sundquist and coworkers constructed a protein network of human class E proteins involved in human multivesicular body (MVB) biogenesis (fig. 14.6A) [10]. They are now using SPR to (1) establish which of the postulated, but previously uncharacterized, interactions are valid; (2) obtain kinetic and affinity parameters for each confirmed interaction; and (3) identify critical residues within binding pairs. For example, a kinetic screen of potential AIP1/CHMP interactions indicated AIP1 binds to CHMP4 proteins, but not to CHMP proteins from other families. Figure 14.6B depicts a more in-depth analysis of one interaction (AIP1 + TSG101): wild-type TSG101 binds AIP1 with affinity of ~140 μM, but mutation of one residue (M95A) completely disrupts this interaction. Refining this map not only defines the roles of the individual proteins within the native MVB biogenesis pathway, but is also proving relevant to HIV research. This pathway is hijacked by the virus as it buds from the host cell, but release of HIV is arrested by deletion or mutation of eight different proteins within this network. Continued analysis of this pathway should aid in understanding and inhibiting HIV and related viral pathogens.

14.4 NEW SPR TECHNOLOGIES AFFECTING PROTEOMICS

The most commonly used commercial SPR biosensor technology is manufactured by Biacore AB in Uppsala, Sweden. Their most widely used instruments, Biacore 2000 and Biacore 3000, have four flow cells that are arranged in series. This allows three different reactants to be immobilized on three of the surfaces, while one surface serves as an internal reference to correct for bulk refractive index changes, nonspecific binding, and instrument drift (fig. 14.7A). While these instruments are widely

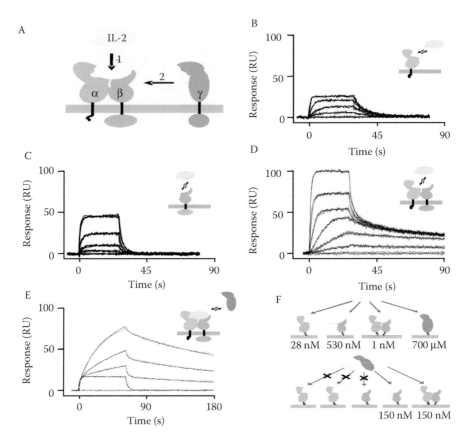

FIGURE 14.5 SPR-based characterization of a multicomponent complex: interleukin-2 (IL-2) interaction with three interleukin-2 receptor (IL-2R) subunits α, β, and γ. (A) Cartoon of the two steps involved in complex assembly. (B) Kinetic analysis of IL-2 binding to immobilized IL-2R$_\alpha$ to yield $k_a = 1.1 \times 10^7$ $M^{-1}s^{-1}$, $k_d = 0.30$ s^{-1}, and $K_D = 28$ nM. (C) Kinetic analysis of IL-2 binding to immobilized IL-2R$_\beta$ to yield $k_a = 5.8 \times 10^5$ $M^{-1}s^{-1}$, $k_d = 0.31$ s^{-1}, and $K_D = 530$ nM. (D) Kinetic analysis of IL-2 binding to immobilized heterodimeric IL-2R$_{\alpha\beta}$ to yield $k_a = 1.7 \times 10^7$ $M^{-1}s^{-1}$, $k_d = 0.018$ s^{-1}, and $K_D = 1.1$ nM. (E) Kinetic analysis of IL-2R$_\gamma$ binding to immobilized IL-2/IL-2R$_{\alpha\beta}$ to yield $k_a = 3.4 \times 10^4$ $M^{-1}s^{-1}$, $k_d = 5.1 \times 10^{-3}$ s^{-1}, and $K_D = 150$ nM. (F) Summary of IL-2/IL-2R interactions examined using SPR. In panels B and C, the binding responses (black lines) are superimposed with the fit (gray lines) of a partially mass transport-limited 1:1 interaction model. In panels D and E, the binding responses are superimposed with the fit of interaction models described in references 8 and 7, respectively. (The data shown in panels B through D are reproduced from Myszka, D.G. et al., *Prot. Sci.*, 5, 2468, 1996, and the data in panel E are reproduced from Liparoto, S.F. et al., *Biochemistry*, 41, 2543, 2002. With permission from the Protein Society and the American Chemical Society, respectively, © 1996 and 2002.)

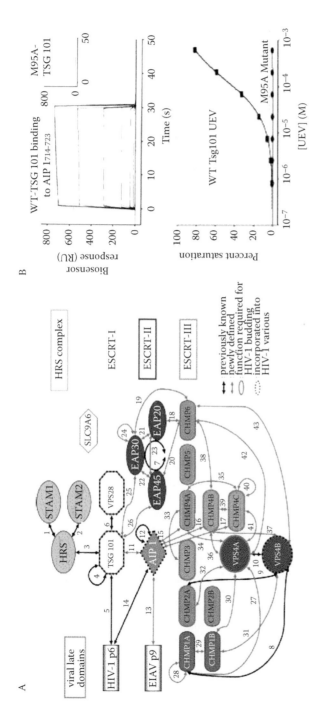

FIGURE 14.6 SPR characterization of a protein network. (A) Network of 22 putative human class E proteins and their interactions implicated in multivesicular body biogenesis. (B) Top panel: binding responses for concentration series (0–600 μM) of wild-type (main panel) or M95A mutant (inset) TSG101 UEV binding to an AIP1$_{714-723}$ surface. Bottom panel: binding isotherms for the AIP1$_{714-723}$/wt TSG101 UEV (squares, $K_D = 142 \pm 0.5$ μM) and AIP1$_{714-723}$/M95A TSG101 UEV (ovals) interactions. (Reproduced from von Schwedler, U.K. et al., *Cell*, 114, 701, 2003. With permission from Elsevier © 2003.)

used (data shown in the examples described so far in this chapter were obtained using them), the interest in high-throughput screening systems and the proteomics revolution is driving the development of higher-capacity, more-specialized SPR biosensor technology [28].

14.4.1 BIACORE T100

The T100 incorporates Biacore AB's latest developments in instrumentation intended for general biophysical applications. This platform, released in 2005, has a flow cell format similar to the 2000 and 3000 platforms. Unlike the complex three-dimensional flow paths of these earlier versions, however, the transitions between flow cells in the T100 occur within the same plane. This simpler path better maintains the integrity of the sample plug across the four flow cells and allows for efficient system washing between analyte injections. Other important features of the T100 include improved instrument control and data processing software, an in-line buffer degasser, and a buffer selector that provides for automated analysis of binding events in up to four different buffer conditions. Papalia et al.'s recent Biacore T100-based characterization of a Fab panel demonstrated how this platform can contribute unique information to a protein therapeutic discovery program [27].

14.4.2 BIACORE S51

Biacore S51, released in 2001, incorporated several design changes that dramatically improved the biosensor's sensitivity. The most significant change was the development of the Y-shaped flow cell, which contains two reaction spots and a reference spot (fig. 14.7B). The internal reference spot is sandwiched between the two reaction spots and each reaction spot can be addressed independently by adjusting flow through the two inlets. Sampling the reaction and reference spots in parallel ensures that the analyte concentration remains constant and reduces systematic microfluidics-dependent noise (e.g., changes in response that arise from pump and/or valve switching), which in turn improves the data quality and therefore the reliability of low-intensity responses.

As illustrated by the data shown in figure 14.7C, the increased sensitivity permits Biacore S51 to generate interpretable signals of less than 1 RU [23]. This platform can therefore be used to analyze protein preparations of lower activity. For example, a sample preparation that is 100% pure may produce the response levels shown in the top panel of figure 14.7C, whereas a preparation of only 3% active protein would produce response levels like those shown in the bottom panel. The similar kinetic parameters obtained from these two surfaces confirm that ultralow binding responses are valid.

Alternatively, if the protein preparation is mostly active and of high purity, this instrument's sensitivity means that we can immobilize the ligand at very low densities, which requires less material for immobilization and minimizes mass transport effects (mass transport influences binding responses when diffusion of the analyte to the ligand surface is comparable to or slower than analyte binding to the ligand). Also, another benefit of the S51's parallel sampling is that it allows measurement of faster on and off rates (up to 10^7 $M^{-1}s^{-1}$ and 2 s^{-1}, respectively). Finally, the S51's advanced robotics and microfluidics reduce the sampling time per analyte. For example, we

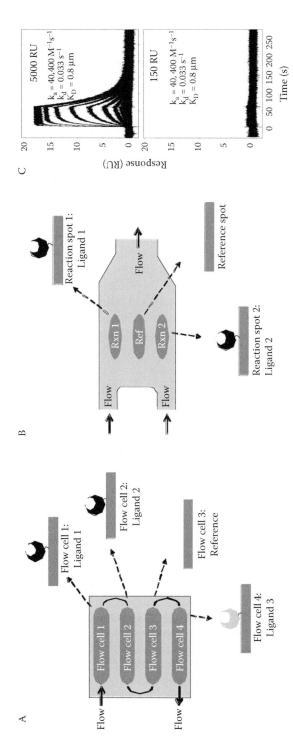

FIGURE 14.7 Analyte flow path configurations in Biacore instruments. (A) Cartoon of the biosensor chip in Biacore 2000 and 3000 instruments. Ligands can be immobilized on the surfaces of three flow cells (the unmodified flow cell serves as a reference). Analyte is flowed serially through the four flow cells. (B) Flow cell schematics from Biacore S51. Ligands can be immobilized within two reaction spots of the flow cell and the central spot serves as a reference. Analyte is flowed in parallel across the three spots. (C) Data from Biacore S51: kinetic analysis of an inhibitor binding to an enzyme immobilized at densities of 150 (spot 1) and 5000 RU (spot 2). Applying a 1:1 interaction model (gray lines) to the binding responses (black lines) from the two surfaces yielded the same kinetic rate constants.

can now test a 384-well plate of potential binders against two immobilized proteins in 24 hours.

14.4.3 BIACORE FLEXCHIP

Traditionally, biosensor analyses have been limited by the number of flow cells or spots available for ligand immobilization (typically, up to four). Characterizing protein binding interfaces, as well as identifying binding partners, would be more efficient if SPR was applied in a multiplexed format. Biacore's FLEXchip system is the first commercially available high-density array SPR platform. This instrument's detector is capable of monitoring analyte ($\geq \sim$3000 Da) binding up to 400 ligands at one time. Ligands are immobilized on the surface of a single, large (1 cm^2) flow cell and unmodified areas between the ligand spots are used for reference. Figure 14.8A shows a 12 × 13 array of protein and reference spots within the FLEXchip flow cell. Since in this configuration every ligand spot has a neighboring reference spot, high-quality kinetic information can be obtained across the entire surface (fig. 14.8B).

Also, the FLEXchip's novel fluidics allows analyte solution to be recirculated over the chip surface. This allows collection of association phase data for hours (fig. 14.8C) rather than minutes, which is a limitation in other biosensor platforms. By recirculating the analyte, we can acquire interpretable data from very low-response surfaces and/or analyze much lower analyte concentrations. This feature is also proving particularly critical for rigorous analysis of high-affinity (e.g., $K_D \sim$ picomolar) interactions [24].

Demonstrating FLEXchip's utility in epitope mapping, Stern and coworkers examined antibody binding to an array of peptides derived from a major histocompatibility complex class II protein to identify critical contacts within the binding interface [25,26]. While FLEXchip technology significantly decreases the analysis time needed for mapping studies, the drawback is that ligand immobilization uses solid-pin spotting. These spotting methods require ligands that are unaffected by drying or adsorption onto a surface and that are available in pure and concentrated preparations. This limits current array applications to analysis of peptides, oligonucleotides, antibodies, and other stable proteins. Flow-based sample-delivery methods under development will facilitate immobilization of a wider range of ligands (i.e., ligands that must remain in solution, require more complex linking chemistries, or need to be captured from crude mixtures) [29].

14.4.4 BIACORE A100

Biacore AB's most recent technology development is a system that allows for parallel processing of multiple analytes at the same time. This parallel array platform has four flow cells addressed with independent flow systems. Within each flow cell are five spots: four reaction spots and one reference spot (similar to Biacore S51). This configuration provides for internal referencing and thereby produces high-quality data even at ultralow response levels. This unit may be used in a 4 × 4 format that is optimal for biophysical characterization. For example, four analyte solutions would be injected through the flow cells to evaluate how different buffer conditions affect

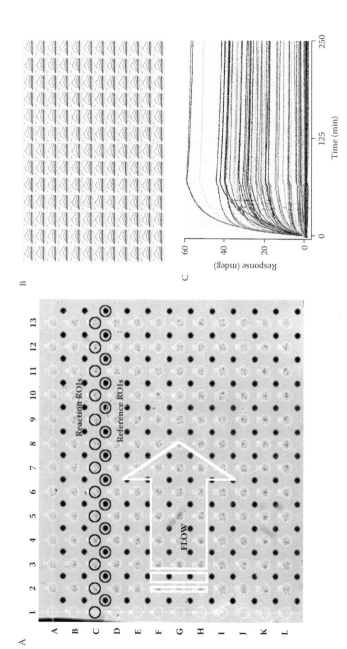

FIGURE 14.8 Flow cell configuration and data from Biacore FLEXchip. (A) Ligand spotting within the FLEXchip flow cell. In this 12 × 13 array, immobilized protein spots (reaction ROIs [regions of interest]) are circled in white and the reference ROIs appear as dark circles. One row each of reaction and reference ROIs is highlighted. Binding responses are recorded for each ROI as analyte is flowed across the surface. (B) Kinetic analysis of analyte binding to 156 protein spots demonstrates the reproducibility of data collected across the flow cell surface. (C) One-hour recirculation of analyte flowed across 100 different ligand spots. (Panels B and C reproduced from Cannon, M.J. and Myszka, D.G. In *Methods for Structural Analysis of Protein Pharmaceuticals*, ed. Jiskoot, W. and Crommelin, D.J.A. Arlington, VA: AAPS Press, 2005, and Rich, R.L. and Myszka, D.G., *Drug Discovery Today: Technol.*, 1, 301, 2004, respectively. With permission from AAPS Press and Elsevier © 2005 and 2004.)

the ligand/analyte interactions against four different targets. Alternatively, in a 4×16 screening format, four different analytes would be flowed over 16 different ligands immobilized in the four flow cells to identify binders and obtain kinetic information. Typical applications for this platform will include mAb screening [32] and higher-throughput small-molecule analyses.

14.5 SUMMARY

To understand the relationships within a proteome fully requires not only identifying all combinations of binding partners, but also determining the kinetics of each interaction (and, of course, establishing whether the partners are simultaneously present *in vivo*). SPR is now routinely applied to examine protein binding pairs and its impact in the analysis of multisubunit complexes is ever increasing. Where this technology is poised to make the greatest breakthrough contributions, however, may be on the larger, proteome-wide scale. By revealing the time scale of binding events in networks and pathways, SPR biosensors will provide critical dynamic information to enrich ongoing protein mapping projects.

14.6 FUTURE PROSPECTS

Developments in SPR technology will most likely focus on increasing the biosensor's sampling rate, lowering its detection limits, and expanding its range of applications. For example, Biacore A100, with its versatile assay design features and parallel analysis capabilities, represents the next wave of developments: platforms that effectively couple high sensitivity and high throughput. In addition, FLEXchip's next generation of chip surfaces, which may be coated with carboxymethyldextran, should improve immobilization efficiency and ligand stability (the charged dextran layer promotes ligand immobilization by concentrating the ligand at the surface and provides a solution-like environment for the tethered ligand). Also, combining the FLEXchip's flow cell format and recirculating fluidics with Biacore AB's proprietary dextran surface chemistries and flow-based immobilization methods will greatly extend the utility of this instrument. These ongoing advances in biosensor technology assure that SPR will become increasingly applied in a variety of formats to interpret the proteome.

REFERENCES

1. Giot, L. et al., A protein interaction map of *Drosophila melanogaster*, *Science*, 302, 1727, 2003.
2. Rich, R.L. and Myszka, D.G., Survey of the year 2004 commercial optical biosensor literature, *J. Molecular Recognition*, 18, 431, 2005.
3. Rich, R.L. and Myszka, D.G., Survey of the year 2005 commercial optical biosensor literature, *J. Molecular Recognition*, 19, 478, 2006.
4. Clackson, T. et al., Structural and functional analysis of the 1:1 growth hormone:receptor complex reveals the molecular basis for receptor affinity, *J. Molecular Biol.*, 277, 1111, 1998.

5. Giannetti, A. et al., Mechanism for multiple ligand recognition by the human transferrin receptor, *PLoS Biology*, 1, 341, 2003.
6. Karlsson, O.P. and Löfas, S., Flow-mediated on-surface reconstitution of G-protein coupled receptors for applications in surface plasmon resonance biosensors, *Anal. Biochem.*, 300, 132, 2002.
7. Liparoto, S.F. et al., Analysis of the role of the interleukin-2 receptor γ chain in ligand binding, *Biochemistry*, 41, 2543, 2002.
8. Myszka, D.G. et al., Kinetic analysis of ligand binding to interleukin-2 receptor complexes created on an optical biosensor surface, *Prot. Sci.*, 5, 2468, 1996.
9. Stenlund, P. et al., Capture and reconstitution of G protein-coupled receptors on a biosensor surface, *Anal. Biochem.*, 316, 243, 2003.
10. von Schwedler, U.K. et al., The protein network of HIV budding, *Cell*, 114, 701, 2003.
11. Leder, L. et al., A mutational analysis of the binding of staphylococcal enterotoxins B and C3 to the T cell receptor beta chain and major histocompatibility complex class II, *J. Exp. Med.*, 187, 823, 1998.
12. Yoo, S. et al., Molecular recognition in the HIV-1 capsid/cyclophilin A complex, *J. Mol. Biol.*, 269, 780, 1997.
13. Drake, A.W., Myszka, D.G., and Klakamp, S.L., Characterizing high-affinity antigen/antibody complexes by kinetic- and equilibrium-based methods, *Anal. Biochem.*, 328, 35, 2004.
14. Joss, L. et al., Interpreting kinetic rate constants from optical biosensor data recorded on a decaying surface, *Anal. Biochem.*, 261, 203, 1998.
15. Myszka, D.G., Jonsen, M.D., and Graves, B.J., Equilibrium analysis of high affinity interactions using BIACORE, *Anal. Biochem.*, 265, 326, 1998.
16. Day, Y.S. et al., Direct comparison of binding equilibrium, thermodynamic, and rate constants determined by surface- and solution-based biophysical methods, *Prot. Sci.*, 11, 1017, 2002.
17. Myszka, D.G. et al., The ABRF-MIRG'02 study: Assembly state, thermodynamic, and kinetic analysis of an enzyme/inhibitor interaction, *J. Biomol. Tech.*, 14, 247, 2003.
18. Humtsoe, J.O. et al., A streptococcal collagen-like protein interacts with the $\alpha_2\beta_1$ integrin and induces intracellular signaling, *J. Biol. Chem.*, 280, 13848, 2005.
19. Cannon, M.J. et al., Kinetic analysis of b-amyloid fibril elongation, *Anal. Biochem.*, 328, 67, 2004.
20. Myszka, D.G., Wood, S.J., and Biere, A.L., Analysis of fibril elongation using surface plasmon resonance biosensors, *Methods Enzymol.*, 309, 386, 1999.
21. Rich, R.L. et al., Kinetic analysis of estrogen receptor/ligand interactions, *Proc. Natl. Acad. Sci. USA*, 99, 8562, 2002.
22. Katsamba, P.S. et al., Label-free characterization of nuclear receptor/co-regulator peptide interactions in an array format, *Proc. Jap. Pept. Soc.* In press.
23. Myszka, D.G. Analysis of small-molecule interactions using Biacore S51 technology, *Anal. Biochem.*, 329, 316, 2004.
24. Navratilova, I., Eisenstein, E., and Myszka, D.G., Measuring long association phases using Biacore, *Anal. Biochem.*, 344, 295, 2005.
25. Baggio, R. et al., Induced fit of an epitope peptide to a monoclonal antibody probed with a novel parallel surface plasmon resonance assay, *J. Biol. Chem.*, 280, 4188, 2005.
26. Carven, G.J. et al., Monoclonal antibodies specific for the empty conformation of HLA-DR1 reveal aspects of the conformational change associated with peptide binding, *J. Biol. Chem.*, 279, 16561, 2004.

27. Papalia, G.A. et al., High-resolution characterization of antibody fragment/antigen interactions using Biacore® T100, *Anal. Biochem.* 359, 112, 2006.
28. Rich, R.L. and Myszka, D.G. Higher-throughput, label-free, real-time molecular interaction analysis, *Anal. Biochem.* In press.
29. Chang-Yen, D.A. et al., A novel PDMS microfluidie spotter for fabrication of protein chips and microarrays, *J. Microelectromech. Syst.* 15, 1145, 2006.
30. Zhukov, A. et al., Integration of surface plasmon resonance with mass spectrometry: automated ligand fishing and sample preparation for MALDI MS using a Biacore 3000 biosensor, *J. Biomol. Technol.*, 15, 112, 2004.
31. Buijs, J. and Natsume, T., Combining surface plasmon resonance with mass spectrometry: Identifying binding partners and characterizing interactions. In *Purifying Proteins for Proteomics*, ed. Simpson, R.J., Cold Spring Harbor Laboratory Press, Cold Spring Harbor, NY, 2004, chap 24.
32. Säfsten, P. et al., Screening antibody-antigen interactions in parallel using Biacore A100. *Anal. Biochem.*, 353, 181, 2006.

15 Application of In-Cell NMR Spectroscopy to Investigation of Protein Behavior and Ligand–Protein Interaction inside Living Cells

Volker Dötsch

CONTENTS

15.1 INTRODUCTION

Of all the biophysical methods currently available for obtaining high-resolution structural information of biological macromolecules, nuclear magnetic resonance (NMR) is the only one that can provide this information in solution under near-physiological conditions. Recently, we and others have extended the scope of NMR applications by investigating conformation and dynamics of biological macro-molecules not only under near physiological conditions but also directly inside living cells [1–8]. These investigations are made possible by the noninvasive character of the NMR technique. The aim of these in-cell NMR experiments is not to determine three-dimensional structures. For a full structure determination, which takes many days of measurement time, an *in vitro* sample of a purified protein provides far better conditions. The strength of in-cell NMR experiments lies in their ability to monitor changes in the conformation, binding status, and dynamic properties relative to the *in vitro* state of the macromolecule.

The most frequently used biophysical technique employed to study proteins inside cells is fluorescence. Labeling of macromolecules with fluorophores can reveal their intracellular localization, their rotational as well as translational diffusion, and, in the form of FRET measurements, also binding events. The enormous importance of fluorescence techniques in cell biology is based on their sensitivity, which even enables the observation of single macromolecules under certain conditions. Compared with fluorescence, NMR spectroscopy is very insensitive, requiring minimal concentrations of the investigated macromolecule in the micro- to millimolar range. However, the unique advantage of NMR spectroscopy over all other techniques is its ability to detect structural changes, measure dynamic parameters, and investigate binding events, with an unparalleled structural resolution. The only other technique that offers similarly detailed information is x-ray crystallography; however, it requires three-dimensional crystals and cannot be applied to investigations of macromolecules in cells.

NMR experiments can be used to detect binding between two molecules in general and can also precisely define where on the surface of a macromolecule the binding site is located [9–11]. In particular, for the investigation of proteins with potential drugs, obtaining information about the exact binding site is absolutely crucial. This ability to monitor binding events with high structural resolution is based on the fact that NMR spectroscopy allows us to observe distinct signals from individual protons. The exact resonance position of a given proton or NMR active nucleus in general—called chemical shift—is a very sensitive function of its exact magnetic environment. Changes in the local environment due to structural rearrangements or binding events result in changes in the magnetic environment, thus leading to changes in the resonance position. Through comparison of the chemical shifts of a particular protein in a spectrum measured with a purified *in vitro* sample with the corresponding chemical shifts obtained with an in-cell sample of the same protein inside living cells, it is possible to determine whether the conformation of the protein under both conditions is the same or if differences exist. In this way, NMR spectroscopy can be used for very detailed investigations of conformational changes. Since binding events also change the local magnetic environment of the binding site, they result in detectable changes of the chemical shifts and can thus be monitored by NMR spectroscopy (fig. 15.1).

One example that has revealed that conformational differences between the *in vitro* state and the *in vivo* state of a protein exist and can be detected by in-cell NMR experiments is the investigation of the intracellular behavior of the protein FlgM by the research group of Gary Pielak [6]. FlgM is a 97-residue protein that regulates flagellar synthesis by binding the transcription factor σ in bacteria. *In vitro* NMR and circular dichroism investigations have demonstrated that this protein is intrinsically disordered. Comparison of an amide proton detected NMR spectra of a purified *in vitro* sample with a corresponding spectrum of FlgM inside living *Escherichia coli* bacteria has revealed significant differences, suggesting that FlgM gains some secondary structure inside living bacteria.

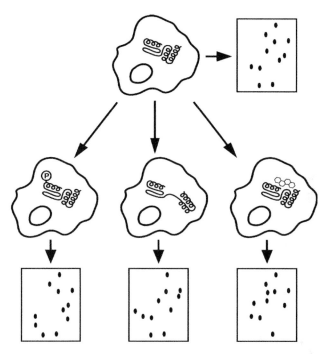

FIGURE 15.1 Examples of potential applications of in-cell NMR experiments. Post-translational modifications, conformational rearrangements, and binding of small molecules change the magnetic environment of the observed nuclei and lead to changes in their chemical shifts. Schematic NMR spectra are shown next to each cell. (Reprinted with permission from Serber, Z. and Dötsch, V., *Biochemistry*, 40, 14317, 2001.)

Pielak and coworkers simulated the *in vivo* conditions in the *in vitro* sample by adding 400g/L of BSA or 450g/L of ovalbumin. Adding high concentrations of these proteins resulted in the same spectral features that were observed in the in-cell spectra. In order to test whether the formation of secondary structure in FlgM is dependent on the presence of high concentrations of other proteins or the same effect could also be produced by small molecules, they added 450 g/L glucose to the *in vitro* sample. This experiment resulted again in the formation of secondary structure in FlgM. Based on these results, they concluded that the structural differences between the *in vitro* and *in vivo* forms of the protein are induced by molecular crowding [6,12–14]. These results have a significant impact on the interpretation of experiments with intrinsically disordered proteins.

Another example of the application of in-cell NMR experiments to the investigation of the behavior of proteins inside cells is FKBP [15,16]. Originally, we were unable to observe the backbone amide proton NMR signals of FKBP in in-cell NMR experiments despite the fact that this small, 10-kDa protein is highly expressed in *E. coli* and behaves well *in vitro*. Disruption of the bacterial membrane by adding lysozyme to the bacterial slurry in the sample, however, allowed us to observe the normal two-dimensional HSQC spectrum of FKBP [4]. In general, NMR resonances become undetectable if the molecule tumbles very slowly in solution. The slower the tumbling is, the broader the resonance lines will become until they are no longer distinguishable from the surrounding noise.

FKBP is a peptidyl-prolyl isomerase and the disappearance of its resonance lines in in-cell NMR experiments suggests that FKBP is involved in larger complexes inside the bacterial cell with tumbling rates that are sufficiently slow to broaden its resonance lines beyond the detection limit. Disrupting the cellular membrane releases FKBP from these complexes and makes it observable. Methyl groups show a very high intrinsic rotation rate [9,17,18], thus allowing their observation in NMR experiments even with slowly tumbling proteins or complexes undetectable in amide proton-based experiments [19]. Methyl group-based NMR experiments enabled us to detect the NMR signals of two of the three methionine methyl groups of the protein in in-cell NMR experiments [4]. These results are in agreement with the interpretation that FKBP is never free inside the bacterial cytoplasm, but instead is constantly interacting with other proteins, trying to act as a peptidyl-prolyl isomerase.

NMR spectroscopy is also a very well established technique to investigate the dynamics of proteins, and many different techniques have been developed to study dynamics ranging from picosecond motions to millisecond motions [20]. We have used longitudinal and transverse relaxation experiments to investigate the dynamics of the bacterial protein NmerA. NmerA is a key factor in the most common bacterial pathways for detoxification of mercurials [21] and is expressed at levels up to 6% of the total soluble protein of the cell upon induction of the mer operon by mercury compounds [22,23]. The bacterial cytoplasm is therefore the natural environment for this protein. Interestingly, our experiments have indicated that the dynamics of amino acids in the metal binding loop of the protein differs between an *in vitro* sample and an in-cell sample. While the cause of this result is still subject to an ongoing investigation, the example shows that dynamic differences between the *in vitro* and *in vivo* forms of a protein can exist and can be investigated by in-cell NMR spectroscopy.

15.2 BINDING STUDIES

The examples described in the previous section demonstrate how NMR spectroscopy can be used as a tool to study the behavior of proteins inside living cells and that significant differences between the *in vitro* and *in vivo* behavior of a protein can exist. Furthermore, the sensitivity of the chemical shift to changes in the chemical environment of a nucleus can also be used to study binding events such as the interaction of proteins with small molecules or with metal ions [10]. This feature has made high-resolution liquid-state NMR spectroscopy a widely used tool for screening protein–drug interactions in the pharmaceutical industry. *In vitro* NMR screens have, however, the disadvantage that an interaction that is observed might not occur in the same way *in vivo*. Binding of a drug to its intended target protein *in vivo* could, for example, be prevented due to the inability of the drug to cross the cellular membrane, its rapid metabolism inside the cell, its binding to other cellular components with higher affinity than to its intended target, or differences in the target protein's conformation between its *in vitro* and *in vivo* states.

Using NMR spectroscopy to monitor the interaction of a protein inside a cell with a drug that is added to the extracellular environment can in principle overcome these problems. Chemical shift changes observed in in-cell NMR experiments indicate that the drug molecules can cross the cellular membrane and interact with the target protein. A good example of the investigation of a particular drug-binding event was reported by Hubbard et al. [7]. By observing virtually identical chemical shift changes in *in vitro* and in-cell spectra of CheY upon adding the drug BRL-16492PA, they could show that the drug that binds to the bacterial two-component signal transduction protein CheY *in vitro* also binds to the same protein inside the bacterial cytoplasm.

In addition to amide protons, which are normally used to monitor drug–protein interactions due to their high chemical shift dispersion, methyl groups are excellent indicators for interaction [9]. Interestingly, an investigation has found that within a set of 191 crystal structures of protein–ligand complexes, 92% of the ligands had a heavy atom within 6 Å of a methyl group while only 82% had a heavy atom within the same distance of an amide proton [24]; this makes methyl groups very attractive alternatives to amide protons for monitoring protein–drug interactions. As an example, we had investigated the interaction of calmodulin with the drug phenoxy-benzamine hydrochloride in living bacteria. Calmodulin binds to several drugs that mimic its interaction with peptides [25]. The main site of interaction is a hydrophobic pocket lined with methionines, which makes the methionine-methyl group an excellent indicator for drug–calmodulin interactions.

While no differences in chemical shift could be detected between an *in vitro* sample and the in-cell sample, upon addition of the drug to a bacterial culture expressing calmodulin, some of the peaks in the in-cell spectrum showed increased line broadening, which is indicative of a weak interaction with the drug [4] (fig. 15.2). Phenoxybenzamine binds to calmodulin only in the calcium-bound form of the protein. Analysis of the backbone amide-based spectra of calmodulin in bacterial cells, however, revealed chemical shifts characteristic of the apo-form of the protein, showing that the intracellular Ca^{2+} concentration is not high enough to make the calcium-bound form the major conformation in the bacterial cytoplasm [3]. This result explains the observed weak interaction between calmodulin and phenoxyben-zamine. In order to investigate if and how much of the drug had been taken up by the bacteria, we removed the cells from the sample by centrifugation and measured a proton 1D spectrum of the supernatant, which showed no sign of the drug. In contrast, the resuspended bacteria pellet showed strong drug signals [4].

Further investigations demonstrated that after cell lysis, almost all of the drug was still associated with the cell debris, suggesting that phenoxybenzamine is mainly associated with the bacterial membrane. The high local concentration of phenoxy-benzamine near the bacterial membrane is most likely responsible for the observed weak interaction between the calcium-free calmodulin and the drug. While the example reported here—calmodulin expressed in *E. coli*—is not a biologically relevant system, it demonstrates the advantages of in-cell NMR experiments, which are able to detect the protein resonances as well as the drug resonances.

As already mentioned, in-cell NMR experiments are also capable of determining whether the ion-binding sites of a protein are occupied in a certain cellular environment. The example of the interaction between phenoxybenzamine and calmodulin also demonstrates that the ion-binding state of a protein can be important for its potential interaction with a drug. During their in-cell NMR drug-binding study

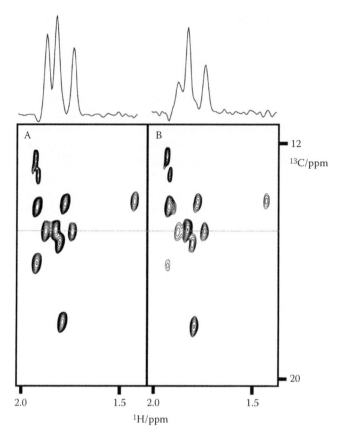

FIGURE 15.2 Comparison of two in-cell [^{13}C, ^{1}H]-HSQC spectra of methionine methyl group labeled calmodulin. The sample in B contained 40 mg/L of phenoxybenzamine hydrochloride while the sample in A did not contain any additives. The spectra were measured on a 500-MHz NMR spectrometer equipped with a cryogenic probe. (Reprinted with permission from Serber, Z. et al., *J. Am. Chem. Soc.*, 126, 7119, 2004.)

between CheY and BRL-16492PA, Hubbard et al. also investigated the ion-binding status of the protein [7]. Through comparison of the in-cell spectrum with *in vitro* spectra of the protein complexed with different ions, they could show that CheY preferentially binds Mg^{2+} ions in the *E. coli* cytoplasm. Small additional changes in the chemical shifts might indicate further interactions with other components; however, this could not be conclusively interpreted so far.

15.3 TECHNICAL ASPECTS

Classical *in vivo* NMR spectroscopy is a well established technique and has been used for many years for the observation of metabolites and the investigation of metabolic fluxes in systems ranging from suspensions of bacteria and other cells to entire perfused organs [26–32]. So far, investigations of macromolecules in living cells have been possible only in a few very special cases [33–36]. Recently, however, we and others have demonstrated that magnetic resonance spectroscopy can be used to study proteins and other biological macromolecules inside living cells. The main difference between these in-cell NMR techniques to study macromolecules and the classical *in vivo* NMR spectroscopy to investigate metabolites and other small molecules is in how one distinguishes the resonance lines of the molecule of interest from the background of all other molecules present in a cell.

In the case of small molecule *in vivo* NMR this distinction is often achieved because the molecule of interest is highly abundant and has the most prominent resonances of the entire spectrum. Alternatively, the resonances can be made visible by adding molecules labeled with NMR-active isotopes (mainly ^{13}C) to the cell suspension, where they diffuse through the cellular membrane or are actively transported into the interior of the cells. Due to the small number of resonance lines of the labeled molecules, one-dimensional spectra are normally sufficient to monitor changes in their intracellular concentration and investigate their interactions with other cellular components. In contrast, in-cell NMR spectroscopy of macromolecules relies on expressing these molecules and labeling them with NMR-active isotopes directly in the cells.

This requirement for labeling the protein inside the cells that are also used during the experiments makes efficient labeling strategies a very important technical aspect of in-cell NMR spectroscopy. In most cases, the protein or other macromolecule of interest will be labeled with the stable and NMR active isotopes ^{15}N and/or ^{13}C. In some cases ^{19}F labeling has also been employed [37,38]. The advantage of ^{19}F labeling is the virtually zero background due to the low abundance of fluorine compounds in cells. For example, fluorine labeling with 5-fluorotryptophan of phospho-glycerate kinase in yeast was used to investigate the binding of nucleotides to this protein *in vivo* [38]. An advantage of fluorine NMR is the high sensitivity that almost reaches the sensitivity obtained in ^1H NMR and enables direct detection schemes. On the other hand, a strong disadvantage of fluorine-based in-cell NMR spectroscopy is its requirement for chemical modification of amino acids by replacing hydrogen with fluorine. This modification changes the chemical properties of the amino acid and can potentially lead to a different behavior of the protein. In addition, fluorine-labeled amino acids are toxic for certain cell types [38].

Observation of macromolecules without chemical modification requires labeling either with ^{15}N or with ^{13}C. The specific labeling scheme will depend on the kind of scientific question that should be investigated by the in-cell NMR experiment. ^{15}N labeling allows the observation of the amide protons of a macromolecule, which mainly provides information about the backbone of the molecule. Exceptions are the amide groups in the side chains of glutamine and asparagine, the indole proton of tryptophan, the $\delta1$ (or $\epsilon2$) proton of histidine, and the ϵ-proton of arginine. All other nitrogen-bound protons in amino acids (the guanidinium group of arginine and the amino group of lysine) are usually not observable (or only as very broad peaks) in NMR experiments.

While these nitrogen-bound side chain protons present important functional groups that are involved in many catalytic reactions, obtaining information on side chains—for example, during binding studies—requires labeling with ^{13}C. In addition, the particular labeling scheme to be used depends on the types of cells selected for the experiment. In some cases (e.g., xenopus oocytes), it is possible to microinject

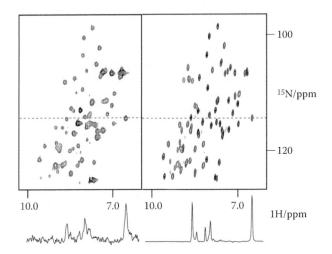

FIGURE 15.3 Comparison of an in-cell [^{15}N, ^{1}H] spectrum of NmerA (left) with an *in vitro* spectrum of the same protein (right). In these spectra each amide proton is represented by one peak. With the exception of the larger line width observed in the in-cell spectrum, both spectra are very similar, demonstrating the low background level observed in amide proton-detected in-cell NMR experiments. One-dimensional slices are shown at the bottom of each spectrum taken at the indicated position. Each spectrum was measured on a 500-MHz NMR spectrometer equipped with a cryogenic probe and four scans in less than 10 minutes.

a labeled protein directly into the cell. Injection has the advantage that the only background signals are the signals produced by the natural abundance of ^{15}N and ^{13}C. Of these, only the natural abundance of ^{13}C (1.1%) plays a significant role. In most cases, however, the macromolecule of interest will be expressed directly inside the cells, which, depending on the labeling scheme, can cause severe problems with background signals due to labeling of other cellular components.

Experience with ^{15}N labeling in *E. coli* so far has shown that only a minimum background level is produced, which does not interfere with most protein resonances [1–3] (fig. 15.3). The only requirement for the observation of a protein's backbone resonances in the in-cell NMR experiment is that the protein be expressed above a certain threshold [2]. This threshold is approximately 1–2% of the entire soluble protein content of a cell or roughly 200–300 μM intracellular concentration. This minimal background level can be further suppressed to virtually zero by using amino acid type selective labeling schemes [2].

Unfortunately, not all amino acids can be used for selective labeling procedures in standard *E. coli* strains such as BL21. Good candidates for labeling are lysine, arginine, and histidine, which are at the end of a biosynthetic pathway and do not serve as precursors for other amino acids [39]. Labeling with other amino acid types has to rely on special media [40] that contain not only the labeled amino acid but also unlabeled amino acids that suppress cross-labeling. Alternatively, specialized auxotrophic bacterial strains can be used that have been created for all 20 amino acids [39–41]. In addition, yeast geneticists have engineered many different auxotrophic yeast strains, albeit not for NMR labeling purposes. One drawback, at least of the auxotrophic bacterial strains, is that they often show a reduced expression level, which reduces the quality of the in-cell NMR spectra.

As mentioned previously, the observation of most side chain resonances requires carbon-based labeling schemes. ^{13}C-based in-cell NMR experiments provide several advantages over ^{15}N-based experiments. First, the sensitivity of detecting methylene and methyl groups is higher due to the larger number of protons directly attached to the heteronucleus as compared to the single amide proton. Second, carbon-bound protons do not chemically exchange with protons of the bulk water as amide protons do. This exchange can significantly reduce the signal intensity and even broaden the resonance beyond detection. Finally, methyl groups belong to the slowest relaxing spins, based on their fast internal rotation, which further increases the sensitivity of methyl group detection [17,18]. Since methyl groups also show the best proton-to-heteronucleus ratio, they are the most attractive side chain probes for in-cell NMR experiments; we used ^{13}C methyl group-labeled methionine to observe FKBP. Analysis of spectra of doubly ^{15}N and ^{13}C labeled in-cell NMR samples has revealed that the sensitivity of methyl group-detected experiments is indeed approximately three times higher than the sensitivity of amide group-detected spectra [4] (fig. 15.4).

Unfortunately, ^{13}C-based labeling schemes also suffer from disadvantages. Full carbon labeling with ^{13}C-labeled glucose produces such a high background level that, with the exception of a few high field-shifted methyl groups of calmodulin, no other protein resonances could be unambiguously identified [4]. The greater abundance of carbon than nitrogen in small molecules is the most likely reason for the high ^{13}C background level as compared to the minimal background observed with ^{15}N labeling. In addition, many nitrogen-bound protons in these small molecules will exchange very fast with the bulk water, thus effectively broadening their resonances beyond detection. More selective labeling schemes—for example, based on using ^{13}C-labeled pyruvate [42]—improve the situation but still produce a high level of background signals [4]. The lowest level of background signals can be achieved by amino acid type selective labeling. As mentioned earlier, labeling of methyl groups is, with regard to the sensitivity of the experiments, the most attractive labeling option. However, similar to the ^{15}N-based amino acid type selective labeling, not all amino acids can be used equally well. While methyl group-labeled methionine produces virtually zero background, the high background level caused by methyl group-labeled alanine has to be suppressed by using special media composition [4,40].

[13]C-based labeling schemes also offer the possibility of observing other biologically important macromolecules besides proteins. The research group of Guy Lippens has used ^{13}C labeling to observe cyclic osmoregulated periplasmic glucan in *Ralstonia solanacearum* [5]. Similar to experiments with proteins, they observed a high background level when using fully ^{13}C-labeled glucose but could reduce it by using glucose that was selectively ^{13}C labeled on the C_1 position.

15.4 TUMBLING RATE OF PROTEINS INSIDE CELLS

A prerequisite for the observation of macromolecules by liquid state NMR spectroscopy is that they tumble freely in solution with a rotational correlation time that is not longer than a couple of tens of nanoseconds. Slow tumbling of a molecule causes

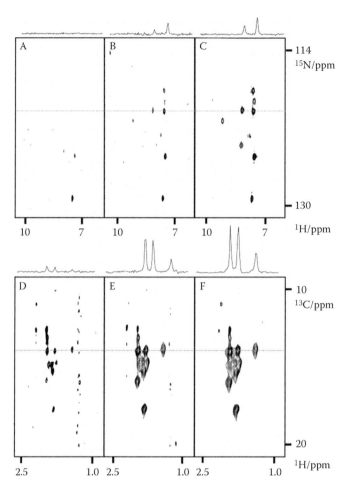

FIGURE 15.4 Comparison of in-cell [^{15}N, ^{1}H] HSQC spectra (A-C) with in-cell [^{13}C, ^{1}H]-HSQC spectra (D–F) of calmodulin selectively labeled with ^{15}N lysine and with ^{13}C methyl group methionine. The spectra A and D were measured simultaneously 25 minutes after induction, the spectra B and E 50 minutes after induction, and the spectra C and F 75 minutes after induction. One-dimensional slices taken at the indicated position are shown on top of each spectrum. These spectra demonstrate that detection of methyl groups is approximately three times more sensitive than the detection of amide protons in in-cell NMR experiments. The strong peak at 129 ppm in the [^{15}N, ^{1}H] HSQC spectrum is not a calmodulin resonance. Spectra were measured on a 500-MHz NMR spectrometer equipped with a cryogenic probe in an interleaved mode. (Reprinted with permission from Serber, Z. et al., *J. Am. Chem. Soc.*, 126, 7119, 2004.)

broadening of the resonance lines, which can lead to their complete disappearance—for example, as observed with the amide proton resonances of FKBP. For in-cell NMR experiments, the intracellular viscosity, which determines the tumbling rate of a macromolecule inside a cell, becomes an important parameter. Several different techniques, including ^{19}F- and ^{13}C-based relaxation measurements, fluorescence polarization experiments, and investigations of the NMR line width of protons and electron spin experiments have been used to study the intracellular viscosity in a wide variety of cell types [43–50]. All these experiments have shown that the intracellular viscosity is at most twice as high as the viscosity of water.

These measurements are in agreement with our investigations. Measurements of the longitudinal relaxation time of the backbone nitrogen spins of NmerA in the bacterial cytoplasm and in an *in vitro* sample showed an increase in the rotational correlation time by a factor of two for the *in vivo* sample. It is important to distinguish this rotational correlation time from the translational diffusion, which can be dramatically reduced in the intracellular environment—for example, due to the presence of a cytoskeleton. For NMR investigations, however, only the rotational correlation time is relevant. Based on the linear relationship among the viscosity, rotational correlation time, and molecular mass of a protein, this twofold increase in the viscosity leads to a twofold increase in the apparent molecular mass of a macromolecule. Fortunately, the molecular weight limit of macromolecules that can be studied by NMR spectroscopy has been significantly extended by the introduction of relaxation optimized pulse sequences such as TROSY (transverse relaxation optimized spectroscopy) [51] and similar techniques [52,53]. With these methods, NMR investigations of even the 900-kDa complex consisting of GroEL and GroES were possible [54].

The availability of these techniques in combination with the rather modest increase in the rotational correlation time for proteins inside the cellular environment predicts that the intracellular viscosity will not be a limiting factor for in-cell NMR experiments. However, binding of a protein to other cellular components can significantly increase the rotational correlation time and can lead to the disappearance of its resonances, as described for FKBP. In these cases, the detection of methyl groups might still be possible due to their very fast internal rotation [4]. For even larger complexes with slower correlation times, solid-state NMR experiments might be the method of choice.

15.5 CELLULAR SURVIVAL DURING NMR EXPERIMENTS

Another parameter that strongly influences the applicability of in-cell NMR experiments is the survival rate of the cells in the NMR tube. In particular, the high cell density can lead to oxygen starvation and lack of nutrients. If the sensitivity of the selected system (mainly the overexpression level) is high enough, the NMR spectra can be measured relatively fast (less than an hour). During longer experiments or series of experiments such as relaxation studies, however, significant changes in the cellular status can occur. These changes can range from shifts of the intracellular pH in *E. coli* cells to cell death, which is observed, for example, with insect cell samples.

On the other hand, in classical *in vivo* NMR experiments cell cultures have been kept alive for long time periods [55–57]. Cell survival can be increased by using modified NMR sample tubes that enable a continuous exchange of the media, thus providing the cells with nutrients and oxygen. In order to ensure that the continuous flow of media through the NMR tube does not remove the cells, several different designs have been tested. While solutions with semipermeable fibers and microcarriers have been used, the easiest method for keeping nonattached cells in an NMR tube is to encapsulate them—for example, with low melting agarose [55,58]. Previous *in vivo* NMR experiments have shown that bacteria can be kept alive in these gels for long periods of time. In addition, our experiments with *E. coli* bacteria expressing ^{15}N-labeled NmerA have demonstrated that the spectral quality of the in-cell NMR spectra is not affected by encapsulation with low melting agarose [59].

15.6 FUTURE PROSPECTS

In-cell NMR experiments are aimed at investigating the behavior of a protein or other macromolecule in its natural environment—for example, the cellular cytoplasm. Results from several research groups have demonstrated that biological macromolecules can indeed be observed in cells ranging from bacteria and yeast to insect cells. The inherent insensitivity of the NMR method, however, requires a high intracellular concentration of the macromolecule of interest. Of course, overexpression of a protein can have a significant impact on its behavior. In order to investigate the true natural behavior of a macromolecule, it is necessary to keep its concentration as close to its natural level as possible. The current detection limit for proteins in in-cell NMR experiments is 200 μM for amide proton-based and 70 μM for methyl group-based experiments.

Fortunately, the introduction of cryoprobes has dramatically increased the sensitivity of NMR instruments over the past years and further increase in the sensitivity are expected. However, while cryogenic probes are very useful for increasing the sensitivity, they are not strictly a prerequisite for the observation of proteins in in-cell NMR experiments with bacteria. In particular, for in-cell NMR experiments with eukaryotic cells, which usually do not reach the same protein expression levels as bacteria, further enhancement of the sensitivity will be of paramount importance. Additional improvements—again, in particular for eukaryotic cells—can be expected from the use of modified NMR tubes that enable a continuous exchange of media. With these devices, longer experiments will become possible and will therefore extend the scope of in-cell NMR spectroscopy to more detailed studies of the behavior of biological macromolecules inside living cells.

REFERENCES

1. Serber, Z. et al., High-resolution macromolecular NMR spectroscopy inside living cells, *J. Am. Chem. Soc.*, 123, 2446, 2001.
2. Serber, Z., et al., Evaluation of parameters critical to observing proteins inside living *Escherichia coli* by in-cell NMR spectroscopy, *J. Am. Chem. Soc.*, 123, 8895, 2001.
3. Serber, Z. and Dötsch, V., In-cell NMR spectroscopy, *Biochemistry*, 40, 14317, 2001.
4. Serber, Z., et al., Methyl groups as probes for proteins and complexes in in-cell NMR experiments, *J. Am. Chem. Soc.*, 126, 7119, 2004.
5. Wieruszeski, J.-M. et al., *In vivo* detection of the cyclic osmoregulated oeriplasmic glucan of *Ralstonia solanacearum* by high-resolution magic angle spinning NMR, *J. Magn. Reson.*, 151, 118, 2001.

6. Dedmon, M. M. et al., FlgM gains structure in living cells, *Proc. Natl. Acad. Sci. USA*, 99, 12681, 2002.

7. Hubbard, J. A. et al., Nuclear magnetic resonance spectroscopy reveals the functional state of the signaling protein CheY *in vivo* in *Escherichia coli*, *Mol. Microbiol.*, 49, 1191, 2003.

8. Shimba, N. et al., Quantitative identification of the protonation state of histidines *in vitro* and *in vivo*, *Biochemistry*, 42, 9227, 2003.

9. Hajduk, P. J. et al., NMR-based screening of proteins containing 13C-labeled methyl groups, *J. Am. Chem. Soc.*, 122, 7898, 2000.

10. Fesik, S. W., NMR structure-based drug design, *J. Biomol. NMR*, 3, 261, 1993.

11. Shuker, S. B. et al., Discovering high-affinity ligands for proteins: SAR by NMR, *Science*, 274, 1531, 1996.

12. Minton, A. P., Implication of macromolecular crowding for protein assembly, *Curr. Opin. Struct. Biol.*, 10, 34, 2000.

13. Berg, B. V. D. et al., Macromolecular crowding perturbs protein refolding kinetics, *EMBO J.*, 19, 3870, 2000.

14. Minton, A. P., The influence of macromolecular crowding and macromolecular confinement on biochemical reactions in physiological media, *J. Biol. Chem.*, 276, 10577, 2001.

15. Michnick, S. W. et al., Solution structure of FKBP, a rotamase enzyme and receptor for FK506 and rapamycin, *Science*, 252, 836, 1991.

16. Moore, J. M. et al., Solution structure of the major binding protein for the immuno-suppressant FK506, *Nature*, 351, 248, 1991.

17. Kay, L. E. and Torchia, D. A., The effect of dipolar cross correlation on 13C methyl-carbon T1, T2 and NOE measurements in macromolecules, *J. Magn. Reson.*, 95, 536, 1991.

18. Pellecchia, M. et al., NMR-based structural characterization of large protein–ligand interactions, *J. Biomol. NMR*, 22, 165, 2002.

19. Kreishman-Deitrick, M. et al., NMR analysis of methyl groups at 100–500 kDa: Model systems and Arp2/3 complex, *Biochemistry*, 42, 8579, 2003.

20. Palmer, A. G., Kroenke, C. D., and Loria, J. P., Nuclear magnetic resonance methods for quantifying microsecond-to-millisecond motions in biological macromolecules. In *Methods in Enzymology*, ed. James, T. L., Dötsch, V., and Schmitz, U., Academic Press, San Diego, 339, 204, 2001.

21. Miller, S. M., Bacterial detoxification of Hg(II) and organomercurials, *Essays Biochem.*, 34, 17, 1999.

22. Fox, B. and Walsh, C. T., Mercuric reductase. Purification and characterization of a transposon-encoded flavoprotein containing an oxidation-reduction-active disulfide, *J. Biol. Chem.*, 257, 2498, 1982.

23. Misra, T. K. et al., Mercuric reductase structural genes from plasmid R100 and transposon Tn501: Functional domains of the enzyme, *Gene*, 34, 253, 1985.

24. Stockman, B. J. and Dalvit, C., NMR screening techniques in drug discovery and drug design, *Prog. Nuc. Magn. Reson. Spectrom.*, 41, 187, 2002.

25. Gnegy, M. E., Calmodulin: Effects of cell stimuli and drugs on cellular activation, *Prog. Drug Res.*, 45, 33, 1995.

26. Bachert, P., Pharmacokinetics using fluorine NMR *in vivo*, *Prog. Nuc. Magn. Reson. Spectrom.*, 33, 1, 1998.

27. Cohen, J. S., Lyon, R. C., and Daly, P. F., Monitoring intracellular metabolism by nuclear magnetic resonance. In *Methods in Enzymology*, ed. Oppenheimer, N. J. and James, T. J., Academic Press, San Diego, 177, 435, 1989.

28. Gillies, R. J. In *NMR in Physiology and Biomedicine*, ed. Gillies, R. J., Academic Press, San Diego, 1994.

29. Krishnan, P., Kruger, N. J., and Ratcliffe, R. G., Metabolite fingerprinting and profiling in plants using NMR, *J. Exp. Bot.*, 56, 255, 2005.

30. Golder, W., Magnetic resonance spectroscopy in clinical oncology, *Onkologie*, 27, 304, 2004.

31. de Graaf, R. A. et al., *In vivo* 1H-[13C]-NMR spectroscopy of cerebral metabolism, *NMR Biomed.*, 16, 339, 2003.

32. Zwingmann, C. and Leibfritz, D., Regulation of glial metabolism studied by 13C-NMR, *NMR Biomed.*, 16, 370, 2003.

33. Brown, F. F. et al., Human erythrocyte metabolism studies by 1H spin echo NMR, *FEBS Lett.*, 82, 12, 1977.

34. Jue, T. and Anderson, S., 1H NMR observation of tissue myoglobin: An indicator of cellular oxygenation *in vivo*, *Magn. Reson. Med.*, 13, 524, 1990.

35. Kreutzer, U., Wang, D. S., and Jue, T., Observing the 1H NMR signal of the myoglobin Val-E11 in myocardium: an index of cellular oxygenation, *Proc. Natl. Acad. Sci. USA*, 89, 4731, 1992.

36. Tran, T. K., Kreutzer, U., and Jue, T., Observing the deoxy myoglobin and hemoglobin signals from rat myocardium *in situ*, *FEBS Lett.*, 434, 309, 1998.

37. Brindle, K. M., Williams, S. P., and Boulton, M., 19F NMR detection of a fluorine-labeled enzyme *in vivo*, *FEBS Lett.*, 255, 121, 1989.

38. Brindle, K. M., Fulton, A. M., and Williams, S. P., Combined NMR and molecular genetics approach to studying enzymes *in vivo*. In *NMR in Physiology and Biomedicine*, ed. Gillies, R. J., Academic Press, San Diego, 237, 1994.

39. Waugh, D. S., Genetic tools for selective labeling of proteins with a-15N-aminoacids, *J. Biomol. NMR*, 8, 184, 1996.

40. Cheng, H. et al., Protein expression, selective isotope labeling, and analysis of hyperfine-shifted NMR signals of Anabaena 7120 vegetative [2Fe-2S]ferredoxin, *Arch. Biochem. Biophys.*, 316, 619, 1995.

41. McIntosh, L. P. and Dahlquist, F. W., Biosynthetic incorporation of 15N and 13C for assignment and interpretation of nuclear magnetic resonance spectra of proteins, *Q. Rev. Biophys.*, 23, 1, 1990.

42. Rosen, M. K. et al., Selective methyl group protonation of perdeuterated proteins, *J. Mol. Biol.*, 263, 627, 1996.

43. Williams, S. P., Haggle, P. M., and Brindle, K. M., F-19 NMR measurements of the rotational mobility of proteins *in vivo*, *Biophys. J.*, 72, 490, 1997.

44. Fushimi, K. and Verkman, A. S., Low viscosity in the aqueous domain of cell cytoplsam measured by picosecond polarization microfluorimetry, *J. Cell Biol.*, 112, 719, 1991.

45. Bicknese, S. et al., Cytoplasmic viscosity near the cell plasma membrane: Measurement by evanescent field frequency-domain microfluorimetry, *Biophys. J.*, 65, 1272, 1993.

46. Dayel, M. J., Hom, E. F., and Verkman, A. S., Diffusion of green fluorescent protein in the aqueous-phase lumen of endoplasmic reticulum, *Biophys. J.*, 76, 2843, 1999.

47. Luby-Phelps, K. et al., A novel fluorescence ratiometric method confirms the low solvent viscosity of the cytoplasm, *Biophys. J.*, 65, 236, 1993.

48. Endre, Z. H., Chapman, B. E., and Kuchel, P. W., Intra-erythrocyte microviscosity and diffusion of specifically labeled [glycyl-alpha-13C]glutathione by using 13C NMR, *Biochem. J.*, 216, 655, 1983.

49. Livingston, D. J., La Mar, G. N., and Brown, W. D., Myoglobin diffusion in bovine heart muscle, *Science*, 220, 71, 1983.

50. Mastro, A. M. et al., Diffusion of a small molecule in the cytoplasm of mammalian cells, *Proc. Natl. Acad. Sci. USA*, 81, 3414, 1984.

51. Pervushin, K. et al., Attenuated T2 relaxation by mutual cancellation of dipole–dipole coupling and chemical shift anisotropy indicates an avenue to NMR structures of very large biological macromolecules in solution, *Proc. Natl. Acad. Sci. USA*, 94, 12366, 1997.

52. Riek, R. et al., Polarization transfer by cross-correlated relaxation in solution NMR with very large molecules, *Proc. Natl. Acad. Sci. USA*, 96, 4918, 1999.

53. Riek, R. et al., Solution NMR techniques for large molecular and supramolecular structures, *J. Am. Chem. Soc.*, 124, 12144, 2002.

54. Fiaux, J. et al., NMR analysis of a 900K GroEL GroES complex, *Nature*, 418, 207, 2002.

55. McGovern, K. A., Bioreactors. In *NMR in Physiology and Biomedicine*, ed. Gillies, R. J., Academic Press, San Diego, 279, 1994.

56. Egan, W. M., The use of perfusion systems for nuclear magnetic resonance studies in cells. In *NMR Spectroscopy of Cells and Organs*, ed. Gupta, R. K., CRC Press, Boca Raton, FL, 1, 1987.

57. Szwergold, B. S., NMR spectroscopy of cells, *Annu. Rev. Phys.*, 54, 775, 1992.

58. Egan, W., Barile, M., and Rottem, S., 31P-NMR studies of *Mycoplasma gallisepticum* cells using a continuous perfusion technique, *FEBS Lett.*, 204, 373, 1986.

59. Reckel, S., Löhr, F., and Dötsch, V., In-cell NMR spectroscopy, *ChemBioChem*, 6, 1601, 2005.

16 An Overview of Metabonomics Techniques and Applications

John C. Lindon, Elaine Holmes, and Jeremy K. Nicholson

CONTENTS

16.1 INTRODUCTION

The determination of the human genome sequence has spurred a great deal of interest in using changes in levels of gene expression in individuals to discover the basis of disease and to identify new drug targets. While this process has been successful and there are indeed specific gene changes in certain diseases, it has also recently been shown that variation in life span and longevity in identical and nonidentical twins is mainly explained by environmental effects such as smoking and diet [1]. This new approach has been incorporated into drug discovery programs in pharmaceutical companies and although some significant advances have been made—notably in some aspects of understanding cancer susceptibility [2,3]—it often remains difficult to relate any changes seen to conventional end points used in disease diagnosis and to optimize efficacy and minimize toxicity in pharmaceutical development. For example, in toxicogenomics studies, candidate drugs can give rise to many gene expression changes that have no actual pathological consequences.

The simultaneous measurement of many gene expression changes is termed transcriptomics and is mainly carried out in an automatic fashion using a technology known as gene microarrays (gene chips) based on fluorescence changes when extracted RNA binds to specific nucleotide sequences on the chip [4]. Since the identified genes from this process can code for protein synthesis, much effort has subsequently been focused on the consequent protein level changes in tissues, cells, and biofluids (known as proteomics), and these matters form the subject of this book [5]. Since the techniques and applications of proteomics are covered extensively in this volume and the advantages and disadvantages are well discussed, it is not necessary to elaborate on this subject.

However, another approach provides data sets of similar high information content. This is metabonomics, the systemic profiling of metabolites and metabolic pathways in whole organisms through the study of biofluids and tissues [6]. A parallel approach has led to the term metabolomics also [7], but the two disciplines and their definitions are highly convergent. This approach holds out the promise of a means by which real disease and drug effect end points can be obtained. In this chapter, the main technologies used in metabonomics are summarized, brief details of the types of samples used are given, and the applications of metabonomics are described.

In complex organisms, the three levels of biomolecular organization and control (transcriptomic, proteomic, and metabonomic) are highly interdependent, but all of them can have very variable time scales of change. This makes it difficult to correlate, mathematically, time courses that can be very rapid (e.g., some gene switching events), require much longer time scales (protein synthesis), or encompass enormous ranges of time scales (metabolite levels). Additionally, such biochemical changes do not always occur in the order—transcriptomic, proteomic, metabolic—because, for example, pharmacological or toxicological effects at the metabolic level can induce subsequent adaptation effects at the proteomic or transcriptomic levels.

In addition, as captured in figure 16.1, environmental and lifestyle effects have a large effect on gene and protein expression and metabolite levels, and these have to be considered as part of intersample and interindividual variation. Interpretation of genomic data, in terms of real biological end-points, is a major challenge because of the conditional interactions of specific gene expression levels with environmental factors that nonlinearly change disease risks. The measurement and modeling of such diverse information sets poses significant challenges at the analytical and bioinformatic modeling levels. Highly complex animals such as man can be considered as "superorganisms" with an internal ecosystem of diverse symbiotic microbiota and parasites that have interactive metabolic processes and for which, in many cases, the genome is not known. The many levels of complexity of the mammalian system and the diverse features that need to be measured to allow "-omic" data to be utilized fully have been reviewed recently [8]. Novel approaches continue to be required to measure and model metabolic compartments in different interacting cell types and genomes connected by cometabolic processes in the global mammalian system [9].

All metabonomics studies, which rely on analytical chemistry methods, result in complex multivariate data sets that require a variety of chemometric and bioinformatic tools for effective interpretation. The aim of these procedures is to produce biochemically based fingerprints that are of diagnostic or other classification value.

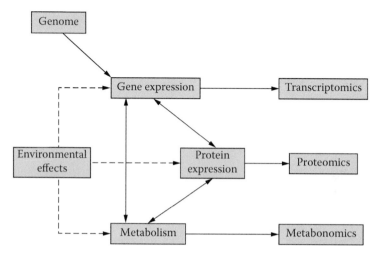

FIGURE 16.1 The relationships among the genome, gene expression levels, protein expression levels, and metabolite concentrations. The influence of environmental effects such as food, gut microflora, and drugs are included.

A second stage, crucial in such studies, is to identify the substances causing the diagnosis or classification; these become biomarkers that reflect actual biological events. Thus, metabonomics allows real-world or biomedical end-point observations to be related to the measurements provided by all of the -omic technologies. To carry out metabonomics studies, it is not necessary to have the genome sequence of all of the organisms involved.

The main areas where metabonomics has an impact include:

- validation of animal models of disease, including genetically modified animals
- preclinical evaluation of drug safety studies, allowing ranking of candidate compounds
- assessment of drug side effects in humans in clinical trials and for marketed drugs
- quantitation, or ranking, of the beneficial effects of pharmaceuticals, in development and clinically
- improved understanding of the causes of highly sporadic idiosyncratic toxicity of marketed drugs
- improved, differential diagnosis and prognosis of human diseases, particularly for chronic and degenerative diseases and for diseases caused by genetic effects
- better understanding of large-scale human population differences through epidemiological studies
- patient stratification for clinical trials and drug treatment (pharmaco-metabonomics)
- sports medicine and lifestyle studies, including the effects of diet, exercise, and stress

- evaluation of the effects of interactions between drugs, and between drugs and diet
- identification of new drug targets
- metabolic profiles of plants, including genetically modified species and quality-control studies (essential equivalence)
- microbiological characterization
- forensic science, including the biochemical effects of drug overdose
- environmental effects of pollutants monitored using marker terrestrial and aquatic species

Importantly, because many metabonomics studies can use biofluids that can be collected noninvasively or minimally so, metabonomics also allows time-dependent patterns of change in response to disease, drug effects, or other stimuli to be measured. In complex multicellular organisms, there will be a wide variety of time scales of change, not always predictable even in order, varying according to the involved genes, proteins, pathways, and tissues. One important potential role for metabonomics, therefore, is to direct the timing of proteomic and genomic analyses in order to maximize the probability of observing -omic biological changes relevant to functional outcomes.

Metabonomics has been formally defined in a biological context as the quantitative measurement of the dynamic multiparametric metabolic response of living systems to pathophysiological stimuli or genetic modification [6]. It provides an approach that is complementary to other -omic measurements, and in general it gives real-world end points since biochemical changes at the metabolite level (usually measured in a univariate manner in the past) can be related to health and disease and are changeable by therapeutic intervention.

A number of reviews of metabonomics have described the various techniques used and summarized the main areas of application. These provide more detail than can be given here and also serve to act as pointers to the original literature studies [10–15].

16.2   METABONOMICS ANALYTICAL TECHNOLOGIES

The two principal methods that can produce metabolic profiles of biomaterials comprise ^{1}H nuclear magnetic resonance (NMR) spectroscopy [10] and mass spectrometry (MS) [16], the latter usually including a separation stage such as gas chromatography (GC) or high-performance liquid chromatography (HPLC). Both of these disciplines are also used heavily in proteomics; MS is one of the mainstay techniques for protein identification [17] and NMR spectroscopy is widely used for determination of the three-dimensional structures of proteins [18].

NMR spectroscopy is a nondestructive technique that provides detailed information on molecular structure for pure compounds and in complex mixtures. In addition and uniquely, NMR spectroscopic methods can also be used to probe metabolite molecular dynamics and mobility as well as substance concentrations through the interpretation of NMR spin relaxation times and by the determination of molecular diffusion coefficients. MS is considerably more sensitive than NMR spectroscopy, but in complex mixtures of very variable composition such as biofluids, it is generally necessary to employ different separation techniques for different classes of substances. MS is also a mainstay technique for molecular identification purposes, especially through the use of MS^n methods for fragment ion studies. Analyte quantitation by MS in complex mixtures of highly variable composition can also be impaired by variable ionization and ion suppression effects, although improved chromatographic resolution such as that offered by UPLC can reduce this problem considerably. Chemical derivatization might be necessary to ensure volatility and analytical reproducibility in complex mixtures such as biofluids. Most published metabonomics studies on mammalian biological systems have used NMR spectroscopy, but HPLC-MS techniques are increasing in use, particularly using electrospray ionization.

Typically, metabonomics is carried out on biofluids, since these can provide an integrated systems biology view. The biochemical profiles of the main diagnostic fluids—blood plasma, cerebrospinal fluid (CSF), and urine—reflect normal variation and the impact of drug toxicity or disease on single- or multiple-organ systems [19]. Urine and plasma are obtained essentially noninvasively and hence are most appropriate for monitoring clinical trials and diagnosing disease. However, a wide range of fluids can be and has been studied, including seminal fluids, amniotic fluid, cerebrospinal fluid, synovial fluid, digestive fluids, blister and cyst fluids, lung aspirates, and dialysis fluids [19].

A standard ^1H NMR spectrum of urine, as seen in figure 16.2, typically contains thousands of sharp lines from hundreds of predominantly low molecular weight metabolites. The position of each spectral band (its chemical shift) gives information on molecular group identity (e.g., methyl group, olefinic hydrogen, phenyl ring, aldehyde, etc.). The splitting pattern on each band (J-coupling) provides information about nearby protons, their through-bond connectivities, and molecular conformations. The band areas relate directly to the number of protons giving rise to the peak and hence to the relative concentrations of the substances in the sample. Absolute concentrations can be obtained if an internal standard of known concentration is added to the sample [20].

Plasma and serum separated from whole blood contain low and high molecular weight components, and these give a wide range of signal line widths. Broad bands from protein and lipoprotein signals contribute strongly to the ^1H NMR spectra, with sharp peaks from small molecules superimposed on them. Standard NMR pulse sequences, where the observed peak intensities are edited on the basis of molecular diffusion coefficients or on NMR relaxation times (T_1, $T_{1\rho}$, T_2), can be used to select only the contributions from proteins and other macromolecules and micelles or, alternatively, to select only the signals from the small molecule metabolites, respectively. Typical ^1H NMR spectra from human blood serum are shown in figure 16.3, showing a complete profile and a series of edited spectra. In some cases, it is also possible to investigate molecular mobility and to study intermolecular interactions such as the reversible binding between small molecules and proteins [21,22].

Standard NMR spectra typically take only a few minutes to acquire using robotic flow-injection methods, with automatic sample preparation involving buffering and addition of D_2O as a magnetic field lock signal for the spectrometer. For large-scale

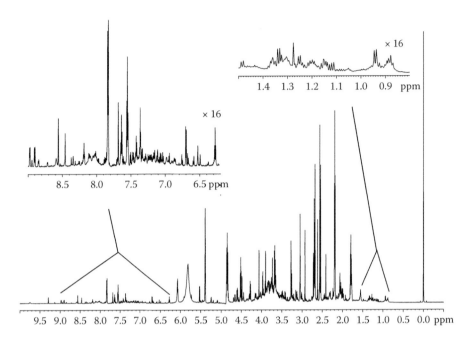

FIGURE 16.2 800-MHz ^{1}H NMR spectrum of control human urine with a series of expansions to show the spectral complexity. The peaks arise from different chemical types of hydrogen in the biochemicals present. The peak areas are related to molar concentrations and the peak positions and splittings allow information to be obtained, after expert interpretation, on the molecules responsible for the peaks. Many of these peaks have been assigned to specific metabolites and a list of these can be found in the literature [10,19]. The signal from water has been suppressed by an NMR procedure in order to avoid problems of dynamic range in the detection process.

studies, bar-coded vials containing the biofluid are used and the contents of these can be transferred and prepared for analysis using robotic liquid-handling technology into 96-well plates with the whole process under LIMS system control. The large interfering NMR signal arising from water in all biofluids is easily eliminated by use of standard NMR solvent suppression methods. Using NMR flow probes, the capacity for NMR analysis has increased enormously recently, and around 200 samples per day can be measured on one spectrometer.

A major improvement in the scope of metabonomics has been made possible by the commercialization of miniaturized NMR probes. Now it is possible to study metabolic profiles by NMR using as little as 2–20 µL of sample and examples have been published using CSF and blood plasma [23,24].

Cryogenic NMR probe technology where the detector coil and preamplifier are cooled to around 20 K, provides an improvement in spectral signal–noise ratios of up to four to five times by reducing the thermal noise in the circuits. This allows use of less material or, conversely, shorter data acquisition times by up to a factor of 20–25 for the same amount of sample. This technology has also permitted routine use of natural abundance ^{13}C NMR spectroscopy of biofluids, such as urine or plasma,

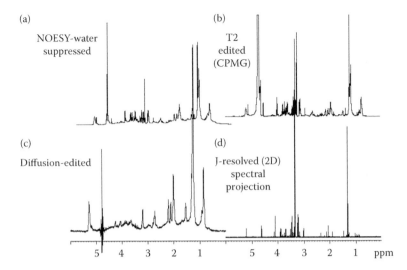

FIGURE 16.3 ^{1}H NMR spectra of rat serum. (a) Standard water-suppressed spectrum, showing all metabolites; (b) CPMG spin–echo spectrum, with attenuation of peaks from fast relaxing components such as macromolecules and lipoproteins; (c) diffusion-edited spectrum, with attenuation of peaks from fast diffusing components such as small molecules; and (d) a projection of a two-dimensional J-resolved spectrum onto the chemical shift axis, showing removal of all spin–spin coupling and peaks from fast relaxing species. This illustrates the various NMR responses possible through the use of different pulse sequences, which edit the spectral intensities.

with acquisition times that enable a high throughput of samples. Information-rich ^{13}C NMR spectra of urine can be obtained using appropriately short acquisition times suitable for biochemical samples when using a cryogenic probe [25].

More recently, the use of mass spectrometry has enjoyed a resurgence in metabonomics with the commercialization of directly coupled HPLC-MS [26], although the first use of pattern recognition for interpreting mass spectra was recognized many years earlier [27]. Thus, for metabonomics applications on biofluids such as urine, an HPLC chromatogram is generated with MS detection, nowadays using electrospray ionization, and usually both positive and negative ion chromatograms are measured. At each sampling point in the chromatogram, there is a full mass spectrum, so the data are three-dimensional in nature (i.e., retention time, mass, and intensity). Given this very high resolution, it is possible to cut out any mass peaks from drug metabolites without compromising the data set. The data can then be used as input to pattern recognition studies, as for NMR spectra [26].

Within the last few years, the development of a technique called high-resolution ^{1}H magic angle spinning (MAS) NMR spectroscopy has made feasible the acquisition of high-resolution NMR data on small pieces of intact tissues with no pretreatment [28–30]. Rapid spinning of the sample (typically at ~4–6 kHz) at an angle of 54.7° relative to the applied magnetic field serves to reduce the line-broadening effects seen in heterogeneous, nonliquid samples such as tissues. These are caused by sample heterogeneity and residual anisotropic NMR parameters normally averaged out in

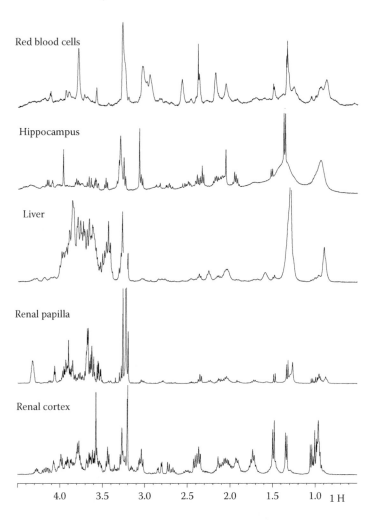

FIGURE 16.4 High-resolution 400-MHz ¹H MAS NMR spectra of various tissues, with sample spinning at 4.2 kHz.

free solution where molecules can tumble isotropically and rapidly. NMR spectroscopy on a tissue matrix in an MAS experiment is the same as solution-state NMR, and all common pulse techniques can be employed in order to study metabolic changes and to perform molecular structure elucidation. Typical ¹H NMR spectra from a range of tissue types are shown in figure 16.4. In most cases, a standard set of one-dimensional sequences is used to describe the biochemical changes in the pool of low molecular weight metabolites and lipids. The different spin properties of macromolecules and small molecules can also be exploited using spectral editing techniques to filter out certain subgroups of peaks, as in solution-state NMR spectroscopy.

MAS NMR spectroscopy has straightforward sample preparation, although this still has to be carried out manually. Samples of as little as ~10 mg can be prepared; initially, snap-frozen tissue samples, which have been stored at –80°C, are defrosted and cut in order to select the region of interest. The original samples should be frozen rapidly using small specimens in liquid nitrogen to avoid microcrystallization of the water in the cells and consequent cell damage. Then, the tissue is rinsed with about 50 µL of 0.9% D_2O/saline in order to wash off remaining blood, and the specimen is transferred into a 4-mm diameter zirconia rotor. A Teflon spacer is used in order to restrict the sample volume, eliminate trapped air bubbles, and increase sample homogeneity.

As soon as the sample has been transferred into the rotor, a trace volume of 0.9% D_2O/saline is added to provide a field-frequency lock for the ^1H MAS NMR data acquisitions. In cases where the chemical shifts of signals have not been identified previously, it is necessary to add a reference standard such as 3-(trimethylsilyl) [2,2,3,3-$^2H_4$] propionic acid, sodium salt (TSP) in the saline solution, although sometimes it is possible to use a well characterized peak of an easily assigned compound, such as the ^1H resonance of α-glucose, as a secondary chemical shift standard.

A number of studies have shown that diseased or toxin-affected tissues have characteristically different metabolic profiles relative to those taken from healthy organs. In addition, ^1H MAS NMR spectroscopy can be used to access information regarding the restriction of metabolites in different compartments within cellular environments. NMR-based metabonomics can also be applied to *in vitro* systems such as tissue extracts [29,30], Caco-2 cells [31], or spheroids [32]. A combined metabonomic analysis of biofluids such as urine and plasma, tissue extracts, and intact tissues is possible, thus providing a comprehensive view of the biochemical responses to a pathological situation, an approach termed "integrated metabonomics" [29,30].

Identification of the biomarkers of a pharmacological, toxicological, or disease situation detected in a biofluid can involve the application of a range of techniques including two-dimensional NMR experiments [20]. Although all of the armory of the usual analytical physical chemistry can and should be used, including mass spectrometry, ^1H NMR spectra of urine and other biofluids, even though they are very complex, allow many resonances to be assigned directly based on their chemical shifts, signal multiplicities, and by adding authentic material. Further information can be obtained by using spectral editing techniques, as described previously.

Two-dimensional NMR spectroscopy can be useful for increasing signal dispersion and for elucidating the connectivities between signals, thereby enhancing the information content and helping to identify biochemical substances [20]. These include the ^1H–^1H 2D J-resolved experiment, which attenuates the peaks from macromolecules and yields information on the multiplicity and coupling patterns of resonances, a good aid to molecule identification. Other 2D experiments known as COSY and TOCSY provide ^1H–^1H spin–spin coupling connectivities, thus giving information about which hydrogens in a molecule are close in chemical bond terms.

Use of other types of nuclei such as naturally abundant ^{13}C or ^{15}N or, where present, ^{31}P can be important to help assign NMR peaks. In this case, heteronuclear correlation NMR experiments, usually ^1H–^{13}C, can also be obtained by use of appropriate NMR pulse sequences [20]. These benefit by the use of so-called inverse detection, where the lower sensitivity or less abundant nucleus NMR spectrum (such as ^{13}C) is detected indirectly using the more sensitive/abundant nucleus (^1H) by making use of spin–spin interactions such as the one-bond ^{13}C–^1H spin–spin coupling. These yield ^1H and ^{13}C NMR chemical shifts of CH, CH_2, and CH_3 groups, useful again for identification purposes. There is also a sequence that allows correlation of protons to quaternary carbons based on long-range $^{13}C-^{1}H$ spin–spin coupling between the nuclei. These NMR approaches are well known also in the field of protein structure determination, where the situation is usually eased by ensuring that the protein under study has been fully enriched with ^{13}C and ^{15}N by production of the protein from a genetically modified bacterium, such as *Escherichia coli* grown in medium enriched in isotopically labeled precursors.

It is also possible to separate substances of interest from a complex biofluid sample using techniques such as solid-phase extraction or HPLC. For metabolite identification, directly coupled chromatography-NMR spectroscopy methods can be used. The most powerful of these "hyphenated" approaches is HPLC-NMR-MS [33] and this can provide the full array of NMR- and MS-based molecular identification tools. These include tandem MS (MS-MS) for identification of fragment ions and Fourier transform (FT)-MS or time of flight (TOF)-MS for accurate mass measurement and hence derivation of empirical molecular formulae.

Some systematic biochemical changes in ^{1}H NMR spectroscopic profiles of biofluids can be obscured in pattern recognition analyses by interfering factors such as variations in pH, which can cause changes in NMR chemical shifts because of differences in degree of protonation of some molecules. In NMR spectroscopy of urine, one means of limiting the effects of pH on the chemical shift of sensitive moieties is to add a standard amount of buffer to the sample prior to NMR spectroscopic analysis. Another widely used procedure is the segmentation of the NMR spectra into regions of equal chemical shift ranges (typically, each is 0.01–0.04 ppm wide) followed by signal integration within those ranges [34].

Automatic data reduction of 2D NMR spectra can be performed using a procedure similar to that for 1D spectra, in which the spectrum is divided by a grid containing squares or rectangles of equal size and the spectral integral in each volume element is calculated. The intensity in some spectral regions, such as those containing water or urea, is highly variable due to water NMR peak suppression artifacts. In addition, many drug compounds or their metabolites are excreted in biofluids and these can obscure significant changes in the concentration of endogenous components. Therefore, it is usual to remove these redundant spectral regions prior to chemometric analysis.

Other approaches are possible and have been used, including shifting peak positions using mathematical methods to take into account small pH-dependent variations in chemical shift, in which case the full NMR spectrum can be used for PR (pattern recognition) [35,36]. More recently, in studies where large numbers of samples are analyzed, the variation in peak positions can be modeled explicitly, especially when the whole spectra are used without any segmentation of the NMR signal intensity into "bins" [37].

16.3 DATA INTERPRETATION USING CHEMOMETRICS

In chemistry, the term chemometrics is generally applied to describe the use of PR and related multivariate statistical approaches to chemical numerical data. The general aim of PR is to classify an object or to predict the origin of an object based on identification of inherent patterns in a set of experimental measurements or descriptors. PR can also be used for reducing the dimensionality of complex data sets—for example, by 2D or 3D mapping procedures—to enable easy visualization of any clustering or similarity of the various samples. Alternatively, multiparametric data can be modeled using PR techniques so that the class of separate samples (a "validation set") can be predicted based on a series of mathematical models derived from the original data or "training set" [13].

The complex data that arise from, for example, an NMR spectrum of a sample can be thought of as an object with a multidimensional set of metabolic coordinates, the values of which are the spectral intensities at each data point. Thus, each spectrum becomes a point in a multidimensional metabolic hyperspace. Similarity or differences between samples can then be evaluated using multivariate statistical methods or other pattern recognition (PR) approaches, such as those based on interpoint distances. In general, the methods allow the quantitative description of the multivariate boundaries that characterize and separate each class of sample in terms of their metabolic profiles [13]; some provide a level of probability for the classification.

Principal components analysis (PCA), one of the simplest techniques used extensively in metabonomics, allows for the expression of most of the variance within a data set using a smaller number of factors or principal components (PCs). Properties of PCs are such that each PC is a linear combination of the original data parameters whereby each successive PC explains the maximum amount of variance possible, not accounted for by the previous PCs. Each PC is orthogonal and therefore independent of the other PCs. Thus, the variation in the spectral set is usually described by many fewer PCs than the original data point values because the less important PCs simply describe the noise variation in the spectra.

Conversion of the data matrix to PCs results in two matrices known as scores and loadings. Scores are the coordinates for the samples in the established model and may be regarded as the new variables; these are linear combinations of the original variables. In a scores plot, each point represents a single NMR spectrum. The PC loadings define the way in which the old variables are linearly combined to form the new variables. The loadings define the orientation of the computed PC plane with respect to the original variables and indicate which variables carry the greatest weight in transforming the position of the original samples from the data matrix into their new position in the scores matrix. In the loadings plot, each point represents a different NMR spectral region. Thus, the cause of any spectral clustering observed in a PC scores plot is interpreted by examination of the loadings that leverage the cluster separation.

Unsupervised methods such as PCA are useful for comparing pathological samples with control samples, but supervised analyses that model each class individually are preferred where the number of classes is large. Supervised methods are used also in regression analysis where a quantitative figure can be provided for a biological endpoint, and this can be a univariate end point or a multivariate end point matrix.

A widely used supervised method is partial least squares (PLS). This is a method that relates a data matrix containing independent variables from samples (an X matrix) to a matrix containing dependent variables (or measurements of response) for those samples (a Y matrix). PLS can also be used to examine the influence of time on a data set, which is particularly useful for biofluid NMR data collected from samples taken over a time course for the progression of a pathological effect. PLS can also be combined with discriminant analysis (DA) to establish the optimal position in which to place a discriminant surface that best separates classes. Finally, because biofluid samples are often collected over a period of time from the same animals as subjects, the analysis can take this into account by utilizing the principles of batch processing [38].

Apart from the methods described here that use linear combinations of parameters for dimension reduction or classification, other methods exist that are not limited in this way. For example, neural networks comprise a widely used nonlinear approach for modeling data. A training set of data is used to develop algorithms that "learn" the structure of the data and can cope with complex functions. The basic software network consists of three or more layers, including an input level of neurons (spectral descriptors or other variables); one or more hidden layers of neurons that adjust the weighting functions for each variable; and an output layer, which designates the class of the object or sample. Recently, probabilistic neural networks, which represent an extension to this approach, have shown promise for metabonomics applications in toxicity [39]. Other approaches currently being tested include genetic algorithms, machine learning, and Bayesian modeling [13].

Biochemical changes, whether naturally induced by diet or lifestyle or artificially by beneficial or adverse effects of drugs, develop and recover over time and, as a consequence, there can be complex, time-related changes in metabolic profiles as detected by NMR spectroscopy of biofluids. Hence, in order to develop automatic classification methods, it has proved efficient to use supervised chemometrics approaches that explicitly take time into account [40,41].

It is possible to use such supervised models to provide classification probabilities or even quantitative response factors for a wide range of sample types. However, given the strong possibility of chance correlations when the number of descriptors is large, it is important to build and test such chemometric models using independent training data and validation data sets.

An initiative has been under way to investigate the reporting needs and to consider recommendations for standardizing reporting arrangements for metabonomics studies, and to this end a standard metabolic reporting structures (SMRS) group has been formed (see www.smrsgroup.org). This has produced a draft policy document that covers all of those aspects of a metabolic study recommended for recording, from the origin of a biological sample, the analysis of material from that sample, and chemometric and statistical approaches to retrieve information from the sample data. The various levels and consequent detail for reporting needs, including journal submissions, public databases, and regulatory submissions, have also been addressed.

16.4 SELECTED APPLICATIONS OF METABONOMICS

16.4.1 Genetic Differences and Other Physiological Effects

In order to determine therapeutic or toxic effects or to understand the biochemical alterations caused by disease, it is necessary first to understand any underlying physiological sources of variation. To this end, metabonomics has been used to separate classes of experimental animals such as mice and rats according to a number of inherent and external factors based on the endogenous metabolite patterns in their biofluids [42]. Such differences may help explain differential toxicity of drugs between strains and interanimal variation within a study.

For example, the importance of gut microfloral populations on urine composition has been highlighted by a study in which axenic (germ-free) rats were allowed to acclimatize in normal laboratory conditions and their urine biochemical makeup was monitored for 21 days [43]. Recently, the influence of parasitic infections on urinary metabolite profiles has been elucidated [44]. Many other effects can be distinguished using metabonomics, including male/female differences, age-related changes, estrus cycle effects in females, diet, diurnal effects, and interspecies differences and similarities [42].

Metabonomics has also been used for the phenotyping of mutant or transgenic animals and investigation of the consequences of transgenesis such as the transfection process [45]. Genetic modifications used in the development of genetically engineered animal models of disease are often made using transfection procedures and it is important to differentiate often seen unintended consequences of this process from the intended result. Metabonomic approaches can give insight into the metabolic similarities or differences between mutant or transgenic animals and the human disease processes that they are intended to simulate and hence their appropriateness for monitoring the efficacy of novel therapeutic agents. This suggests that the method may be appropriate for following treatment regimes such as gene therapy.

16.4.2 PRECLINICAL DRUG CANDIDATE SAFETY ASSESSMENT

Despite the huge investment by pharmaceutical companies in safety testing in animals, unexpected results are sometimes observed in the clinic and drugs occasionally still have to be withdrawn from the market place. This is an indication that drug safety assessment approaches used in the pharmaceutical industry can still fail, and hence there is a need for methodologies that can detect potential problems earlier, faster, more cheaply, and more reliably. A recent survey of market withdrawals during the period 1960–1999 identified the most common reason for withdrawal as hepatotoxicity [46]. The selection of robust candidate drugs for development based upon minimization of the occurrence of adverse drug effects is therefore one of the most important aims of pharmaceutical R&D, and the pharmaceutical industry is now embracing metabonomics for evaluating the adverse effects of candidate drugs. The National Center for Toxicological Research, a part of the U.S. Food and Drug Administration, is also investigating the usefulness of the approach.

In this application, NMR-based metabonomics can be used for (1) definition of the metabolic hyperspace occupied by normal samples; (2) the consequential rapid classification of a biofluid sample as normal or abnormal (this enables NMR spectrometer automation for data acquisition); (3) if abnormal, classification of the target organ or region of toxicity; (4) biochemical mechanism of that toxin; (5) identification of combination biomarkers of toxic effects; and (6) evaluation of the time course of the effect (e.g., the onset, evolution and regression of toxicity). An example of a

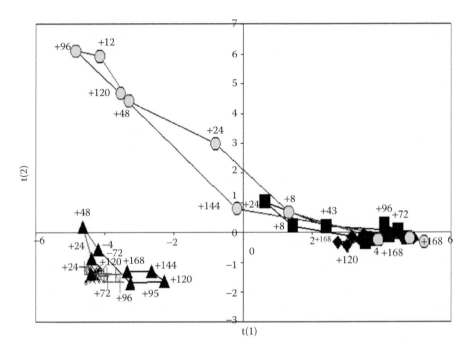

FIGURE 16.5 Principal components scores trajectories (PC1 vs. PC2) for 600-MHz ^{1}H NMR spectra of urine from hydrazine-treated rats and mice illustrating the differences in metabolic starting positions between the rat and the mouse and different magnitudes of toxic response to hydrazine dosing. (Bollard, M.E. et al. *Toxicol. Appl. Pharmacol.*, 204, 135–151, 2005.) Key: × = control mice; + = low-dose mice (100 mg/kg); ▲ = high-dose mice (250 mg/kg); 0 = control rats; ■ = low-dose rats (30 mg/kg); ⬭ = high-dose rats (90 mg/kg).

metabolic trajectory is shown in figure 16.5 where each spectrum, following a single dose of a toxin, is plotted as a single point in a principal component scores plot [47]. There have been many studies using ^{1}H NMR spectroscopy of biofluids to characterize drug toxicity. The role of metabonomics in particular and magnetic resonance in general in toxicological evaluation of drugs has been comprehensively reviewed recently [15].

Metabonomics has already been applied in fields outside human and other mammalian systems. For example, studies in the environmental pollution field have highlighted the potential benefits of this approach by studies of caterpillar hemolymph [48] and earthworm biochemical changes as a result of soil pollution by model toxic substances [49]. In addition, a study of heavy metal toxicity (As^{3+} and Cd^{2+}) in wild rodents living on polluted sites has been concluded successfully [50]. Finally, in terms of monitoring water quality, one study has evaluated adverse effects in abalone using NMR-based metabonomics [51].

The usefulness of metabonomics for the evaluation of xenobiotic toxicity effects has recently been comprehensively and successfully explored by the Consortium for Metabonomic Toxicology (COMET). This was formed between five pharmaceutical companies and Imperial College, London [52], with the aim of developing

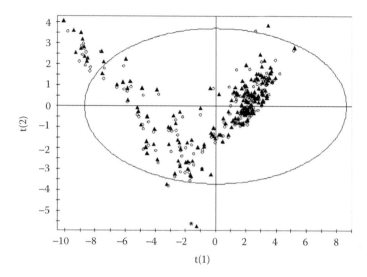

FIGURE 16.6 Principal components scores plot where each point represents a ^{1}H NMR spectrum of rat urine taken at various times as shown after a single acute dose of hydrazine, with each urine sample split and measurements made at 600 MHz at Imperial College, London, and at 500 MHz at Roche, Basel, Switzerland. (Keun H.C. et al. *Chem. Res. Toxicol.*, 15, 1380–1386, 2002.) This is a plot of the first and second PC scores for all NMR spectra. Triangles and circles represent NMR spectra measured at IC and Roche, respectively. The ellipse denotes the 95% significance limit. Samples appear as pairs showing similarity of biochemical profiles. The asterisk denotes a sample that gave an anomalous spectrum when measured automatically at IC, due to an instrumental artifact. The multivariate coefficient of variation for 250 paired samples is 1.6%.

methodologies for the generation and evaluation of metabonomic data generated using ^{1}H NMR spectroscopy of urine and blood serum for preclinical toxicological screening of candidate drugs. The successful outcome is evidenced by the databases of spectral and conventional results for a wide range of model toxins (147 in total) that serve as the raw material for computer-based expert systems for toxicity prediction. The project goals for the generation of comprehensive metabonomic databases (now around 35,000 NMR spectra) and multivariate statistical models (expert systems) for prediction of toxicity—initially for liver and kidney toxicity in the rat and mouse—have now been achieved, and the predictive systems and databases have been transferred to the sponsoring companies.

A feasibility study was carried out at the start of the project, using the same detailed protocol and the same model toxin, over seven sites in the companies and their appointed contract research organizations. This was used to evaluate the levels of analytical and biological variation that could arise through the use of metabonomics on a multisite basis. The intersite NMR analytical reproducibility revealed the high degree of robustness expected for this technique when the same samples were analyzed at Imperial College and at various company sites. This gave a multivariate coefficient of regression between paired samples of only about 1.6%. This is illustrated in figure 16.6, which shows a principal components scores plot for

^{1}H NMR spectra from split urine samples from a toxicity study following a single dose of hydrazine, a model liver toxin. Each point represents a single NMR spectrum with the triangular and circular points measured at different sites, one in Switzerland and the other in the United Kingdom.

The biological variability was evaluated by a detailed comparison of the ability of the six companies to provide consistent urine and serum samples for an in-life study of the same toxin, with all samples measured at Imperial College. There was a high degree of consistency between samples from the various companies and dose-related effects could be distinguished from intersite variation [53].

As a precursor to developing the final predictive expert systems, metabonomic models were constructed for urine from control rats and mice, enabling identification of outlier samples and the metabolic reasons for the deviation. To achieve the project goals, new methodologies for analyzing and classifying the complex data sets were developed. For example, since the expert system takes into account the metabolic trajectory over time, a new way of comparing and scaling these multivariate trajectories was required and has been developed [37]. Additionally, a novel classification method for identifying the class of toxicity based on all of the NMR data for a given study has been generated. This has been termed "classification of unknowns by density superposition (CLOUDS)" and is a novel non-neural implementation of a classification technique developed from probabilistic neural networks [54]. Modeling the urinary NMR data according to organ of effect (control, liver, kidney, or other organ), using a model training set of 50% of the samples and predicting the other 50%, over 90% of the test samples were classified as belonging to the correct group with only a 2% misclassification rate between these classes. This work showed that it is possible to construct predictive and informative models of metabonomic data delineating the whole time course of toxicity—the ultimate goal of the COMET project.

The value of obtaining multiple NMR data sets from various biofluid samples and tissues of the same animals collected at different time points has been demonstrated. This procedure has been termed "integrated metabonomics" [11] and can be used to describe the changes in metabolic chemistry in different body compartments affected by exposure to toxic drugs [29,30]. An illustration of the types of information that can be obtained from a study of the acute toxicity of α-naphthylisothiocyanate, a model liver toxin, is shown in figure 16.7. Such timed profiles in multiple compartments are characteristic of particular types and mechanisms of pathology and can be used to give a more complete description of the biochemical consequences than can be obtained from one fluid or tissue alone.

Integration of data and findings from other multivariate techniques in molecular biology, such as from gene array experiments, is also feasible. Thus, it has been possible to integrate data from transcriptomics and metabonomics to find common metabolic pathways implicated by both gene expression changes and changes in metabolism after acetaminophen administration to mice [55].

16.4.3 DISEASE DIAGNOSIS AND THERAPEUTIC EFFICACY

Many examples exist in the literature of the use of NMR-based metabolic profiling to aid human disease diagnosis, including the use of plasma to study diabetes; CSF

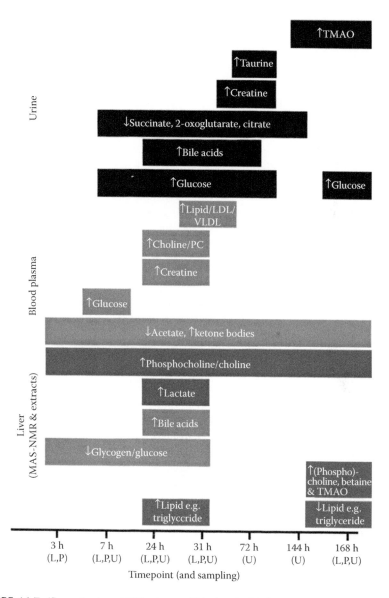

FIGURE 16.7 (See color insert.) The types of biochemical information that can be obtained from an integrated metabonomics study of the acute toxicity of the liver toxin α-naphthyl-isothiocyanate. The figure illustrates the major fluctuations in metabolite profiles observed by ¹H NMR spectroscopy over the experimental time course in the liver, blood plasma and urine. Key: ↑, above control levels; ↓, below control levels; LDL, very low-density lipoprotein. Metabolite changes in purple were observed in both extract and MAS NMR spectra. L, P, and U refer to sampling of liver, blood plasma and urine, respectively [29].

for investigating Alzheimer's disease; synovial fluid for osteoarthritis; seminal fluid for male infertility; and urine in the investigation of drug overdose, renal transplantation, and various renal diseases. Most of the earlier studies using NMR spectroscopy have been reviewed [19]. For example, a promising use of NMR spectroscopy of urine and plasma, as evidenced by the number of publications on the subject, is in the diagnosis of inborn errors of metabolism in children caused by single gene defects [56] with an example shown in Figure 16.8.

More recently, some studies have been undertaken in the area of cancer diagnosis using perchloric acid extracts of various types of human brain tumor tissue [57] and the spectra could be classified using neural network software giving ~85% correct classification. Tissues can be studied by metabonomics through the magic-angle-spinning technique and published examples include prostate cancer [58,59], renal cell carcinoma [60], breast cancer [61,62], and various brain tumors [63,64]. Other recent studies include an NMR-based urinary metabonomic study of multiple sclerosis in humans and nonhuman primates [65].

Currently, the "gold standard" diagnostic method for coronary heart disease (CHD) requires the injection of x-ray opaque dye into the blood stream and visualization of the coronary arteries using x-ray angiography. This is both expensive and invasive with an associated 0.1% mortality and 1–3% of patients experiencing adverse effects. Recently, metabonomics has been applied to provide a method for diagnosis of CHD noninvasively through analysis of a blood serum sample using NMR spectroscopy [66]. Patients were classified, based on angiography, into two groups: those with normal coronary arteries and those with triple coronary vessel disease. Around 80% of the NMR spectra were used as a training set to provide a two-class model after appropriate data filtering techniques had been applied and the samples from the two classes were easily distinguished. The remaining 20% of the samples were used as a test set and their class was then predicted based on the derived model with a sensitivity of 92% and a specificity of 93% calculated according to standard methods, based on a 99% confidence limit for class membership. Sensitivity is defined as the number of true positives, divided by the sum of the numbers of true positives and false negatives, while specificity is defined as the number of true negatives divided by the sum of the numbers of true negatives and false positives.

It was also possible to diagnose the severity of the CHD that was present by employing serum samples from patients with stenosis of one, two, or three of the coronary arteries. Although this is a simplistic indicator of disease severity, separation of the three sample classes was evident even though none of the wide range of conventional clinical risk factors that had been measured was significantly different between the classes. The visualization of the separation of patients with no, one, two, or three coronary arteries occluded is shown in figure 16.9.

One of the long-term goals of using pharmacogenomic approaches is to understand the genetic makeup of different individuals (their genetic polymorphism) and their varying abilities to handle pharmaceuticals for their beneficial effects and for identifying adverse effects. If personalized health care is to become a reality, an individual's drug treatments must be balanced so as to achieve maximal efficacy and

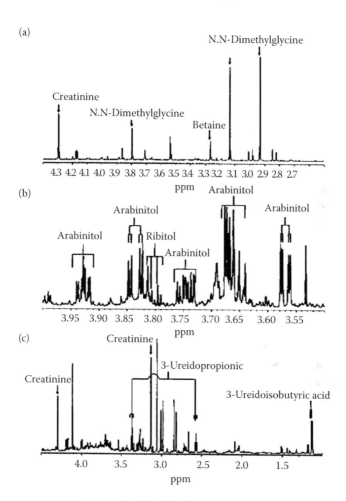

FIGURE 16.8 ¹H NMR spectra showing the abnormal metabolic profile observed from patients with inborn errors of metabolism, (a) dimethylglycine dehydrogenase deficiency, (b) a polyol disease involving arabitol and ribitol, and (c) ureidopropionase deficiency (c). These diseases were first recognized as novel inborn errors of metabolism using NMR spectroscopy of body fluids [56].

avoid adverse drug reactions. An alternative approach to understanding intersubject variability in response to drug treatment has been proposed that uses a combination of system multivariate metabolic profiling and chemometrics (metabonomics) to predict the metabolism, disposition, efficacy, toxicity, and other effects of a dosed substance, based solely on the analysis and modeling of a predose metabolic profile [67]. This new approach, which has been termed "pharmacometabonomics," should be sensitive to genetic and modifying environmental influences, such as the presence of different types of gut microflora [43] that determine the basal metabolic "starting position" of an individual and that will in turn determine the outcome of a chemical intervention. This approach has recently ben exemplified [68].

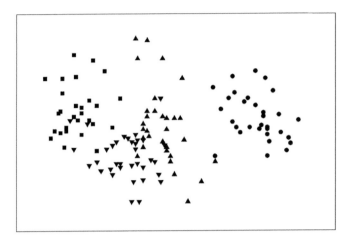

FIGURE 16.9 Partial least squares discriminant analysis model for the classification of blood plasma samples in terms of coronary artery disease, based on their ^{1}H NMR spectra with visualization of the degree of coronary artery occlusion. Each point is based on data from a ^{1}H NMR spectrum of human blood plasma from subjects with different degrees of coronary artery occlusion. Key: circles, no stenosis; triangles, stenosis of one artery; inverted triangles, stenosis of two arteries; squares, stenosis of three arteries. (Brindle, J.T. et al. *Nat. Med.*, 8, 1439–1445, 2002.)

16.5   FUTURE PROSPECTS

Although there continues to be a need for advances in metabonomic analytical technologies in NMR and MS, NMR is likely to remain the method of choice for a broad impartial survey of metabolic profiles, especially given recent gains in sensitivity through the use of cryoprobe detectors. MS coupled to a separation stage is always likely to yield better detection limits for specific classes of metabolites.

NMR-based metabonomics is now recognized as an independent and widely used technique for evaluating the toxicity of drug candidate compounds, and it has been adopted by a number of pharmaceutical companies in their drug development protocols. For drug safety studies, it is possible to identify the target organ of toxicity, derive the biochemical mechanism of the toxicity, and determine the combination of biochemical biomarkers for the onset, progression and regression of the lesion. Additionally, the technique has been shown to be able to provide a metabolic fingerprint for an organism ("metabotyping") as an adjunct to functional genomics, and hence has applications in design of drug clinical trials and for evaluation of genetically modified animals as disease models.

Using metabonomics, it has proved possible to derive new biochemically based assays for disease diagnosis and to identify combination biomarkers for disease that can then be used to monitor the efficacy of drugs in clinical trials. Thus, based on differences observed in metabonomic databases from control animals and from animal models of disease, diagnostic methods and biomarker combinations might be derivable in a preclinical setting. Similarly, the use of databases to derive predictive expert systems for human disease diagnosis and the effects of therapy requires compilations from normal human populations and patients before, during, and after therapy.

The ultimate goal of systems biology must be the integration of data acquired from living organisms at the genomic, protein, and metabolite levels. In this respect, transcriptomics, proteomics, and metabonomics will play an important role. Through the combination of these and related approaches will come an improved understanding of an organism's total biology and, with this, better understanding of the causes and progression of human diseases and, given the 21st century goal of personalized health care, the improved design and development of new and better targeted pharmaceuticals.

REFERENCES

1. Zaretsky, M.D. Communication between identical twins: Health behavior and social factors are associated with longevity that is greater among identical than fraternal US World War II veteran twins. *J. Gerontol. Ser A–Biol. Sci. Med. Sci.*, 58, 566–572, 2003.
2. Antoniou, A.C., Pharoah, P.D.P., McMullan, G., Day, N.E., Stratton, M.R., Peto, J., Ponder, B.J. A comprehensive model for familial breast cancer incorporating BRCA1, BRCA2 and other genes. *Br. J. Cancer*, 86, 76–83, 2002.
3. Armstrong, K., Weiner, J., Weber, B., Asch, D.A. Early adoption of BRCA1/2 testing: Who and why. *Genet Med.*, 5, 92–98, 2003.
4. Copland, J.A., Davies, P.J., Shipley, G.L., Wood, C.G., Luxon, B.A., Urban, R.J. The use of DNA microarrays to assess clinical samples: the transition from bedside to bench to bedside. *Rec. Prog. Hormone Res.*, 58, 25–53, 2003.
5. Kenyon, G.L., DeMarini, D.M., Fuchs, E., Galas, D.J., Kirsch, J.F., Leyh, T.S., Moos, W.H., Petsko, G.A., Ringe, D., Rubin, G.M., Sheahan, L.C. Defining the mandate of proteomics in the post-genomics era: Workshop report. *Mol. Cell Proteomics*, 1, 763–780, 2002.
6. Nicholson, J.K., Lindon, J.C., Holmes, E. "Metabonomics": Understanding the metabolic responses of living systems to pathophysiological stimuli via multivariate statistical analysis of biological NMR spectroscopic data. *Xenobiotica*, 29, 1181–1189, 1999.
7. Fiehn, O. Metabolomics—The link between genotypes and phenotypes. *Plant Molecular Biol.*, 48, 155–171, 2002.
8. Nicholson, J.K., Wilson, I.D. Understanding "global" systems biology: Metabonomics and the continuum of metabolism. *Nat. Rev. Drug Discovery*, 2, 668–676, 2003.
9. Nicholson, J.K., Holmes, E., Lindon, J.C., Wilson, I.D. The challenges of modeling mammalian biocomplexity. *Nat. Biotech.*, 22, 1268–1274, 2004.
10. Lindon, J.C., Nicholson, J.K., Holmes, E., Everett J.R. Metabonomics: Metabolic processes studied by NMR spectroscopy of biofluids. *Concepts Magn. Reson.*, 12, 289–320, 2000.
11. Nicholson, J.K., Connelly, J., Lindon, J.C., Holmes, E. Metabonomics: A platform for studying drug toxicity and gene function. *Nat. Rev. Drug Discovery*, 1, 153–162, 2002.
12. Lindon, J.C., Holmes, E., Nicholson, J.K. So what's the deal with metabonomics? *Anal. Chem.*, 75, 384A–391A, 2003.

13. Lindon, J.C., Holmes, E., Nicholson, J.K. Pattern recognition methods and applications in biomedical magnetic resonance. *Prog. NMR Spectrosc.*, 39, 1–40, 2001.

14. Lindon, J.C., Holmes, E., Bollard, M.E., Stanley, E.G., Nicholson, J.K. Metabonomics technologies and their applications in physiological monitoring, drug safety assessment and disease diagnosis. *Biomarkers*, 9, 1–31, 2004.

15. Lindon, J.C., Holmes, E., Nicholson, J.K. Toxicological applications of magnetic resonance. *Prog. NMR Spectrosc.*, 45, 109–143, 2004.

16. Plumb, R.S., Stumpf, C.L., Gorenstein, M.V., Castro-Perez, J.M., Dear, G.J., Anthony, M.L., Sweatman, B.C., Connor, S.C., Haselden, J.N. Metabonomics: The use of electrospray mass spectrometry coupled to reversed-phase liquid chromatography shows potential for the screening of rat urine in drug development. *Rapid Commun. Mass Spectrom.*, 16, 1991–1996, 2002.

17. Carr, S., Aebersold, R., Baldwin M., Burlingame, A., Clauser, K., Nesvizhskii, A. The need for guidelines in publication of peptide and protein identification data— Working group on publication guidelines for peptide and protein identification data. *Mol. Cell. Proteomics*, 3, 531–533, 2004.

18. Kanelis, V., Forman-Kay, J.D., Kay, L.E. Multidimensional NMR methods for protein structure determination. *IUBMB Life*, 52, 291–302, 2001.

19. Lindon, J.C., Nicholson, J.K., Everett, J.R. NMR spectroscopy of biofluids. In *Annual Reports on NMR Spectrometry*, ed. Webb G.A., Academic Press, Oxford, U.K., 38, 1–88, 1999.

20. Claridge, T.D.W. *High-Resolution NMR Techniques in Organic Chemistry.* Elsevier Science, Oxford, U.K. 384. ISBN: 0080427995.

21. Liu, M., Nicholson, J.K., Lindon, J.C. High resolution diffusion and relaxation edited one- and two-dimensional ^{1}H NMR spectroscopy of biological fluids. *Anal. Chem.*, 68, 3370–3376, 1996.

22. Price, W.S. Recent advances in NMR diffusion techniques for studying drug binding. *Aust. J. Chem.*, 56, 855–860, 2003.

23. Khandelwal, P., Beyer, C.E., Lin, Q., McGonigle, P., Schechter, L.E., Bach, C. Nanoprobe NMR spectroscopy and *in vivo* microdialysis: New analytical methods to study brain neurochemistry. *J. Neurosci. Methods*, 133, 181–189, 2004.

24. Griffin, J.L., Nicholls, A.W., Keun, H.C., Mortishire-Smith, R.J., Nicholson, J.K., Kuehn, T. Metabolic profiling of rodent biological fluids via ^{1}H NMR spectroscopy using a 1-mm microlitre probe. *Analyst*, 127, 582–584, 2002.

25. Keun, H.C., Beckonert, O., Griffin, J.L., Richter, C., Moskau, D., Lindon, J.C., Nicholson, J.K. Cryogenic probe ^{13}C NMR spectroscopy of urine for metabonomic studies. *Anal. Chem.*, 74, 4588–4593, 2002.

26. Plumb, R., Granger, J., Stumpf, C., Wilson, I.D., Evans, J.A., Lenz, E.M. Metabonomic analysis of mouse urine by liquid-chromatography/time of flight mass spectrometry (LC-TOFMS): Detection of strain, diurnal and gender differences. *Analyst*, 128, 819–823, 2003.

27. van der Greef, J., Tas, A.C., Bouwman, J., Ten Noever de Brauw, M.C. Evaluation of field-desorption and fast atom-bombardment mass spectrometric profiles by pattern recognition techniques. *Anal. Clin. Acta*, 150, 45–52, 1983.

28. Garrod, S.L., Humpfer, E., Spraul, M., Connor, S.C., Polley, S., Connelly, J., Lindon, J.C., Nicholson, J.K., Holmes, E. High-resolution magic angle spinning ^{1}H NMR spectroscopic studies on intact rat renal cortex and medulla. *Magn. Reson. Med.*, 41, 1108–1118, 1999.

29. Waters, N.J., Holmes, E., Williams, A., Waterfield, C.J., Farrant, R.D., Nicholson, J.K. NMR and pattern recognition studies on the time-related metabolic effects of α-naphthylisothiocyanate on liver, urine, and plasma in the rat: An integrative metabonomic approach. *Chem. Res. Toxicol.*, 14, 1401–1412, 2001.
30. Coen, M., Lenz, E.M., Nicholson, J.K., Wilson, I.D., Pognan, F., Lindon, J.C. An integrated metabonomic investigation of acetaminophen toxicity in the mouse using NMR spectroscopy. *Chem. Res. Tox.*, 16, 295–303, 2003.
31. Lamers, R.J.A.N., Wessels, E.C.H.H., van der Sandt, J.J.M., Venema, K., Schaafsma, G., van der Greef, J., van Nesselrooij, J.H.J. A pilot study to investigate effects of inulin on Caco-2 cells through *in vitro* metabolic fingerprinting. *J. Nutrition*, 133, 3080–3084, 2003.
32. Bollard, M.E., Xu, J.S., Purcell, W., Griffin, J.L., Quirk, C., Holmes, E., Nicholson, J.K. Metabolic profiling of the effects of D-galactosamine in liver spheroids using ^{1}H NMR and MAS NMR spectroscopy. *Chem. Res. Toxicol.*, 15, 1351–1359, 2002.
33. Lindon, J.C., Nicholson, J.K., Wilson, I.D. Directly coupled HPLC-NMR and HPLC-NMR-MS in pharmaceutical research and development. *J. Chromatogr. B.*, 748, 233–258, 2000.
34. Farrant, R.D., Lindon, J.C., Rahr, E., Sweatman, B.C. An automatic data reduction and transfer method to aid pattern-recognition analysis and classification of NMR spectra. *J. Pharmaceut. Biomed. Anal.*, 10, 141–144, 1992.
35. Brown, T.R., Stoyanova, R.J. NMR spectral quantitation by principal component analysis. 2. Determination of frequency and phase shifts. *J. Magn. Reson.*, 112, 32–43, 1996.
36. Stoyanova, R., Nicholls, A.W., Nicholson, J.K., Lindon, J.C., Brown, T.R. Automatic alignment of individual peaks in large high-resolution spectral data sets. *J. Magn. Reson.*, 170, 329–335, 2004.
37. Cloarec, O., Dumas, M.E., Trygg, J., Craig, A., Barton, R.H., Lindon, J.C., Nicholson, J.K., Holmes, E. Evaluation of the O-PLS model limitations caused by chemical shift variability and improved visualization of biomarker changes in ^{1}H NMR spectroscopic metabonomic studies. *Anal. Chem.*, 77, 517–526, 2005.
38. Antti, H., Bollard, M.E., Ebbels, T., Keun, H., Lindon, J.C., Nicholson, J.K., Holmes, E. Batch statistical processing of H-1 NMR-derived urinary spectral data. *J. Chemometrics*, 16, 461–468, 2002.
39. Holmes, E., Nicholson, J.K., Tranter, G. Metabonomic characterization of genetic variations in toxicological and metabolic responses using probabilistic neural networks. *Chem. Res. Toxicol.*, 14, 182–191, 2001.
40. Holmes, E., Bonner, F.W., Sweatman, B.C., Lindon, J.C., Beddell, C.R., Rahr, E., Nicholson, J.K. Nuclear magnetic resonance spectroscopy and pattern recognition analysis of the biochemical processes associated with the progression of and recovery from nephrotoxic lesions in the rat induced by mercury(II) chloride and 2-bromo-ethanamine. *Mol. Pharmacol.*, 42, 922–930, 1992.
41. Keun, H.C., Ebbels, T.M.D., Bollard, M.E., Beckonert, O., Antti, H., Holmes, E., Lindon, J.C., Nicholson, J.K. Geometric trajectory analysis of metabolic responses to toxicity can define treatment specific profiles. *Chem. Res. Tox.*, 17, 579–587, 2004.
42. Bollard, M.E., Stanley, E.G., Lindon, J.C., Nicholson, J.K. Homes, E. NMR-based metabonomics approaches for evaluating physiological influences on biofluid composition *NMR. Biomedicine*, 18, 143–162, 2005.

43. Nicholls, A.W., Mortishire-Smith, R.J., Nicholson, J.K. NMR spectroscopic based metabonomic studies of urinary metabolite variation in acclimatizing germ-free rats. *Chem. Res. Toxicol.*, 16, 1395–1404, 2003.

44. Wang, Y., Holmes, E., Nicholson, J.K., Cloarec, O., Chollet, J., Tanner, M., Singer, B.H., Utzinger, J. Metabonomic investigations in mice infected with *Schistosoma mansoni*: An approach for biomarker identification. *Proc. Nat. Acad. Sci. USA*, 101, 12676–12681, 2004.

45. Griffin, J.L., Sang, E., Evens, T., Davies, K., Clarke, K. Metabolic profiles of dystrophin and utrophin expression in mouse models of Duchenne muscular dystrophy. *FEBS Lett.*, 530, 109–116, 2002.

46. Fung, M., Thornton, A., Mybeck, K., Wu, J.H.-H., Hornbuckle, K., Muniz, E. Evaluation of the characteristics of safety withdrawal of prescription drugs from worldwide pharmaceutical markets—1960 to 1999. *Drug Inf. J.*, 35, 293–317, 2001.

47. Bollard, M.E., Keun, H.C., Beckonert, O., Ebbels, T.M.D., Antti, H., Nicholls, A.W., Shockcor, J.P., Cantor, G.H., Stevens, G., Lindon, J.C., Holmes, E., Nicholson, J.K. Comparative metabonomics of differential hydrazine toxicity in the rat and mouse. *Toxicol. Appl. Pharmacol.*, 204, 135–151, 2005.

48. Phalaraksh, C., Lenz, E.M., Nicholson, J.K., Lindon, J.C., Farrant, R.D., Reynolds, S.E., Wilson, I.D., Osborn, D., Weeks, J. NMR spectroscopic studies on the hemolymph of the tobacco hornworm, *manduca sexta*: Assignment of ^{1}H and ^{13}C NMR spectra. *Insect Biochem. Mol. Biol.*, 29, 795–805, 1999.

49. Bundy, J.G., Lenz, E.M., Bailey, N.J., Gavaghan, C.L., Svendsen, C., Spurgeon, D., Hanjard, P.K., Osborn, D., Weeks, J.M., Teauger, S.A., Speir, P., Sanders, I., Lindon, J.C., Nicholson, J.K., Tang, H. Metabonomic investigation into the toxicity of 4-fluoroaniline, 3,5-difluoroaniline and 2-fluoro-4-methylaniline to the earthworm *Eisenia veneta* (Rosa): Identification of novel endogenous biomarkers. *Environ. Toxicol. Chem.*, 21, 1966–1972, 2002.

50. Griffin, J.L., Walker, L.A., Shore, R.F., Nicholson, J.K. High-resolution magic angle spinning ^{1}H NMR spectroscopy studies on the renal biochemistry in the bank vole (*Clethrionomys glareolus*) and the effects of arsenic (As^{3+}) toxicity. *Xenobiotica*, 31, 377–385, 2001.

51. Viant, M.R., Rosenblum, E.S., Tjeerdema, R.S. NMR based metabolomics: A powerful approach for characterizing the effects of environmental stressors on organism health. *Environ. Sci. Technol.*, 37, 4982–4989, 2003.

52. Lindon, J.C., Nicholson, J.K., Holmes, E., Antti, H., Bollard, M.E., Keun, H., Beckonert, O., Ebbels, T.M., Reily, M.D., Robertson, D., Stevens, G.J., Luke, P., Breau, A.P., Cantor, G.H., Bible, R.H., Niederhauser, U., Senn, H., Schlotterbeck, G., Sidelmann, U.G., Laursen, S.M., Tymiak, A., Car, B.D., Lehman-McKeeman, L., Colet, J.M., Loukaci, A., Thomas, C. Contemporary issues in toxicology: The role of metabonomics in toxicology and its evaluation by the COMET project. *Toxicol. Appl. Pharmacol.*, 187, 137–146, 2003.

53. Keun H.C., Ebbels, T.M.D., Antti, H., Bollard, M., Beckonert, O., Schlotterbeck, G., Senn, H., Niederhauser, U., Holmes, E., Lindon, J.C., Nicholson, J.K. Analytical reproducibility in ^{1}H NMR-based metabonomic urinalysis. *Chem. Res. Toxicol.*, 15, 1380–1386, 2002.

54. Ebbels, T., Keun, H., Beckonert, O., Antti, H., Bollard, M., Holmes, E., Lindon, J., Nicholson J. Toxicity classification from metabonomic data using a density superposition approach: "CLOUDS." *Anal. Chim. Acta*, 490, 109–122, 2003.

55. Coen, M., Ruepp, S.U., Lindon, J.C., Nicholson, J.K., Pognan, F., Lenz, E.M., Wilson, I.D. Integrated application of transcriptomics and metabonomics yields new insight into the toxicity due to paracetamol in the mouse. *J. Pharm. Biomed. Anal.*, 35, 93–105, 2004.

56. Moolenaar, S.H., Engelke, U.F.H., Wevers, R.A. Proton nuclear magnetic resonance spectroscopy of body fluids in the field of inborn errors of metabolism. *Ann. Clin. Biochem.*, 40, 16–24, 2003.

57. Maxwell, R.J., Martinez-Perez, I., Cerdan, S., Cabanas, M.E., Arus, C., Moreno, A., Capdevila, A., Ferrer, E., Bartomeus, F., Aparicio, A., Conesa, G., Roda, J.M., Carcellar, F., Pascual, J.M., Howells, S.L., Mazucco, R., Griffiths, J.R. Pattern recognition analysis of ^{1}H NMR spectra from perchloric acid extracts of human brain tumor biopsies. *Magn. Res. Med.*, 39, 869–877, 1998.

58. Tomlins, A., Foxall, P.J.D., Lindon, J.C., Lynch, M.J., Spraul, M., Everett, J.R., Nicholson, J.K. High-resolution magic angle spinning ^{1}H nuclear magnetic resonance analysis of intact prostatic hyperplastic and tumor tissues. *Anal. Commun.*, 35, 113–115, 1998.

59. Swanson, M.G., Vigneron, D.B., Tabatabai, Z.L., Males, R.G., Schmitt, L., Carroll, P.R., James, J.K., Hurd, R.E., Kurhanewicz, J. Proton HR-MAS spectroscopy and quantitative pathologic analysis of MRI/3D-MRSI-targeted postsurgical prostate tissues. *Magn. Reson. Med.*, 50, 944–954, 2003.

60. Moka, D., Vorreuther, R., Schicha, H., Spraul, M., Humpfer, E., Lipinski, M., Foxall, P.J.D., Nicholson, J.K., Lindon, J.C. Biochemical classification of kidney carcinoma biopsy samples using magic-angle-spinning ^{1}H nuclear magnetic resonance spectroscopy. *J. Pharmaceut. Biomed. Anal.*, 17, 125–132, 1998.

61. Cheng, L.L., Chang, I.W., Smith, B.L., Gonzalez, R.G. Evaluating human breast ductal carcinomas with high-resolution magic-angle spinning proton magnetic resonance spectroscopy. *J. Magn. Reson.*, 135, 194–202, 1998.

62. Sitter, B., Sonnewald, U., Spraul, M., Fjosne, H.E., Gribbestad, I.S. High-resolution magic angle spinning MRS of breast cancer tissue. *NMR Biomed.*, 15, 327–337, 2002.

63. Cheng, L.L., Chang, I.W., Louis, D.N., Gonzalez, R.G. Correlation of high-resolution magic angle spinning proton magnetic resonance spectroscopy with histopathology of intact human brain tumor specimens. *Cancer Res.*, 58, 1825–1832, 1998.

64. Barton, S.J., Howe, F.A., Tomlins, A.M., Cudlip, S.A., Nicholson, J.K., Bell, B.A., Griffiths, J.R. Comparison of *in vivo* ^{1}H MRS of human brain tumors with ^{1}H HR-MAS spectroscopy of intact biopsy samples *in vitro*. *MAGMA*, 8, 121–128, 1999.

65. 't Hart, B.A., Vogels, J.T.W.E., Spijksma, G., Brok, H.P.M., Polman, C., van der Greef, J. NMR spectroscopy combined with pattern recognition analysis reveals characteristic chemical patterns in urines of MS patients and nonhuman primates with MS-like disease. *J. Neuro. Sci.*, 212, 21–30, 2003.

66. Brindle, J.T., Antti, H., Holmes, E., Tranter, G., Nicholson, J.K., Bethell, H.W.L., Clarke, S., Schofield, P.M., McKilligin, E., Mosedale, D.E., Grainger, D.J. Rapid and noninvasive diagnosis of the presence and severity of coronary heart disease using H-1 NMR-based metabonomics. *Nat. Med.*, 8, 1439–1445, 2002.

67. Lindon, J.C., Holmes, E., Nicholson, J.K. Metabonomics: Systems biology in pharmaceutical research and development. *Curr. Op. Med. Therap.*, 6, 265–272, 2004.

68. Clayton, T.A., Lindon, J.C., Cloarec, O., Antti, H., Chareul, C., Hanton, G., Provost, J.P., Le Net, J.L., Baker, D., Walley, R.J., Everett, J.R., Nicholson, J.K. Pharmacometabonomic phenotyping and personalized drug treatment. *Nature* 440, 1073–1077, 2006.

Part V

*Structural Proteomics:
Parallel Studies of Proteins*

17 NMR-Based Structural Proteomics

John L. Markley

CONTENTS

17.1 INTRODUCTION

Structural proteomics (often called structural genomics) is the systematic investigation of the three-dimensional structures of the protein products of genes. Because gene sequencing has become automated and inexpensive, our knowledge of predicted sequences of proteins far overshadows what we know of their three-dimensional structures and functions. Owing to the enormity of the challenge of determining structures of large numbers of proteins, the field of structural proteomics is driving the development of economic, high-throughput methodology. Although the major sites involved are large centers, much of the new technology they are developing is applicable to the general field of protein biochemistry.

17.1.1 STRUCTURAL PROTEOMICS CENTERS, WEB SITES, AND INFORMATION

A list of structural proteomics centers along with their goals and technologies is maintained on the "structural genomics" web pages at the Protein Data Bank (http://sg.pdb.org). Of the 24 centers listed as contributing to the TargetDB (November, 2006), 15 are cited as using NMR spectroscopy in addition to x-ray crystallography as a method for structure determination. Several structural proteomics centers focus exclusively on x-ray crystallography, primarily because it is a more mature and high-throughput approach. However, centers that are using both x-ray and NMR platforms are finding that the two methods are highly complementary. Many proteins that can be prepared in soluble form in adequate quantities for structural analysis fail to crystallize; however, some of these prove to be suitable for NMR structural analysis. Conversely, some proteins that fail as NMR structural targets can be crystallized to yield x-ray structures.

A recent volume of *Methods in Enzymology* (vol. 394) devoted to "nuclear magnetic resonance of biological macromolecules" contains a number of useful chapters relevant to NMR-based structural proteomics [1–4]. The several structural proteomics centers that make extensive use of NMR are making important technological contributions. This chapter provides a snapshot of the advantages of, current state of, and future potential for NMR-based structural proteomics. While focusing primarily on the approaches being used at the Center for Eukaryotic Structural Genomics (CESG), this chapter attempts a general survey of the field. In addition to the publications cited, each of the CESG protocols referred to here is supported by a detailed written description available from the CESG Web site (http://uwstructuralgenomics.org).

17.1.2 ADVANTAGES OF NMR-BASED STRUCTURAL PROTEOMICS

As documented in greater detail, the major roles that NMR spectroscopy is playing in structural proteomics are (1) to determine structures of smaller proteins that fail to form crystals suitable for structure determination by x-ray crystallography; (2) to screen structural candidates (full-length proteins, protein domains, or re-engineered proteins) for folding and aggregation state as a function of solution conditions; and (3) to screen proteins for binding of metal ions, cofactors, or other small molecules.

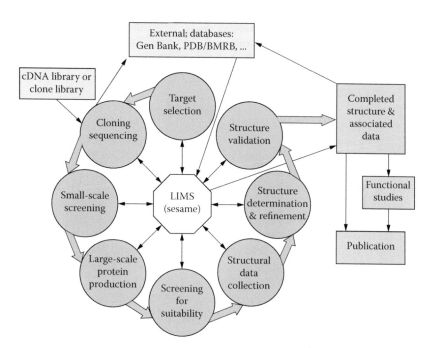

FIGURE 17.1 Schematic view of the pipeline in use at the Center for Eukaryotic Structural Genomics. A specialized laboratory information management system (LIMS), "Sesame" [5], developed at CESG is used to track operations at the center. Thick arrows indicate flow of materials and products; thin arrows indicate information flow.

Unlike x-ray crystallography where months may be spent in testing various crystallization approaches, screening for the suitability of a protein for structure determination takes at most a week. NMR screening includes solvent optimization and testing for long-term stability of the protein under the conditions proposed for the structural study.

17.2 STEPS IN NMR-BASED STRUCTURAL PROTEOMICS

Structural proteomics centers have developed pipelines (fig. 17.1) leading from target selection to structure determination and data deposition.

17.2.1 INFORMATION MANAGEMENT

The pipeline typically covers many steps and involves different investigators performing separate operations at different sites. In the interest of quality control and capturing information that can be used to improve procedures, it is essential to record information about decisions, protocols, and intermediate as well as final results. Given the magnitude of these projects, the information must be in digital form. Many groups have developed their laboratory information management systems independently, with site-specific formats and different granularity of the data collected. With support and encouragement from the NIGMS, the public databases (PDB and

BMRB) have participated in workshops designed to develop standards for data representation and interchange for structural proteomics. These efforts have resulted in the published standards used by two Web sites developed by the U.S. National Institute of General Medical Sciences (National Institutes of Health) and maintained at the Protein Data Bank (PDB).

One of these, TargetDB (http://targetdb.pdb.org/) is a target registration database that provides the status of and tracking information on the progress of the protein production and structure determinations. Many of the worldwide structural proteomics centers provide information to this site, which is aimed at avoiding duplication of effort in structure determinations. The second Web site, PEPCdb (http://pepcdb.pdb.org/), has been created initially for use by all of the NIH-supported centers under phase 2 of the Protein Structure Initiative. PEPCdb aims to capture the protocols used by these centers for a given target, along with information about the failures or successes of these protocols when applied to the target. In addition, the Research Collaboratory for Structural Biology (RCSB), which includes the PDB and BMRB, has independently been working to extend its data models and data dictionaries with the goal of capturing all information that normally would go into a journal article describing a structure determination. Although provision of this level of detail will be optional, it soon will be possible to archive information on this scale as part of the deposition of an x-ray or NMR structure. The structural proteomics centers are developing the capability of depositing information at increasing levels of detail from their laboratory information management systems (LIMS).

The LIMS developed at CESG, named "Sesame" [5], is an extensive suite of applications that supports various phases of structural proteomics and center activities. The software has been adopted by several other centers and is being made available in "open source" format. Sesame is easily configured by the user and can be used to support information management needs of a single laboratory, a center, or a far-flung consortium. Information on Sesame is available from its Web site (http://www.sesame.wisc.edu/).

Sesame is designed to organize and record data relevant to complex scientific projects, to launch computer-controlled processes, and to help decide about subsequent steps on the basis of information available. The Sesame system, which is based on the multitier paradigm, consists of a framework and application modules that carry out specific tasks and can support high-throughput centers and small labs (down to individual users). Users interact with Sesame through a series of Web-based Java applet applications designed to organize data generated by projects in structural genomics, structural biology, and shared laboratory resources. Sesame allows collaborators on a given project to enter, process, view, and extract relevant data, regardless of location, so long as Web access is available. Sesame serves as a digital laboratory notebook and allows users to attach numerous files and images. It can launch computations that utilize local computers or distributed computer clusters. The system has the capability of printing and reading barcodes relevant to various parts of the pipeline. Sesame can create reports and output data as XML files, including those used in generating CESG's weekly contributions to the TargetDB and PEPCdb.

17.2.2 TARGET SELECTION

The criteria for selecting proteins whose structures are to be determined differs among structural proteomics centers. Focus areas of established centers include determination of structures of proteins that (1) enlarge our knowledge of sequence-fold space (generally proteins with ≤30% sequence identity with any protein with a structure in the PDB); (2) represent the set of proteins from a given organism; (3) are related to human health and disease; (4) are produced by particular pathogenic organisms; (5) are involved in specific functions, such as signaling or gene expression; (6) are targets for drug discovery; and (7) correspond to more difficult protein targets, such as membrane proteins or proteins from humans or other eukaryotes.

In practice, the rate of success from a chosen target to a sequence can be very low—as low as 2–3% for eukaryotic proteins with sequence identity <30% to a known structure in the Protein Data Bank and with unknown function. Thus, one of the strategies in target selection is to improve the success rate by enriching the target list with proteins more likely to generate structures [6]. CESG, along with other structural proteomics centers that work with unknown gene products, uses bioinformatics tools to remove targets that have low complexity, are predicted to have large regions of dynamic disorder, or are predicted to be membrane proteins. Genes coding for proteins with leader sequences are redesigned so as to remove these. Generally, gene targets destined for NMR spectroscopy are limited to those coding for proteins of 20 kDa or less. Larger genes are broken into domains for NMR structural analysis. As discussed, as new NMR methods for dealing with larger proteins are developed, this limit may be relaxed.

17.2.3 CHOICE OF PROTEIN PRODUCTION PLATFORMS

Protein NMR spectroscopy requires ~1- to 5-mg samples of purified protein, and the protein must be labeled with stable isotopes (uniformly with nitrogen-15, [U-^{15}N]; and uniformly with carbon-13, [U-^{13}C]) to enable multinuclear spectroscopy. Most of the major structural proteomics centers currently rely nearly exclusively on protein production from *Escherichia coli* cells [1]. An exception is the RIKEN Structural Genomics Initiative, which produces most of its proteins with an *E. coli* cell-free platform [7,8].

CESG uses two platforms for production of its eukaryotic gene products for NMR analysis (fig. 17.2): heterologous expression in *E. coli* cells [9] and *in vitro* enzymatic transcription followed by translation by wheat-germ cell-free lysates [10]. These platforms were chosen on the basis of comparisons with other options, including protein production from *Pichia pastoris*, insect cells, and *E. coli* cell-free approaches. The two platforms used by CESG have proved to be complementary [11]. Overall, the wheat-germ cell-free system provides greater coverage in terms of successful expression of soluble proteins than the cell-based method. Although few targets are found to be produced in *E. coli* cells that are not expressed in the cell-free system, the *E. coli* cells approach can provide significantly higher protein yields at a given cost. Currently, CESG's wheat-germ cell-free protocol for small-scale screening for protein production and solubility is more highly automated and more economical than the corresponding screening with *E. coli* cells.

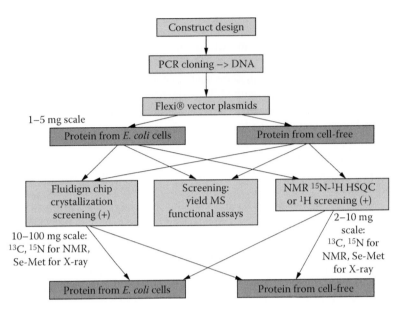

FIGURE 17.2 (See color insert.) Platforms for protein production for structure determination at the Center for Eukaryotic Structural Genomics. The red arrows represent flow of proteins produced by cell-free and cell-based platform. This scheme emphasizes the complementarity of the cell-based and cell-free platforms and the ways in which NMR spectroscopy and x-ray crystallography can be used to attempt structures of proteins that fail with the other approach.

CESG has concluded from this experience that cell-free screening should be used first to test constructs for expression and solubility. Then, constructs that pass these tests should be evaluated by small-scale screening in *E. coli* cells. The comparative outcome then is used to decide which approach to use for large-scale protein production (fig. 17.2). Constructs that fail the initial cell-free screen can be redesigned with the goal of rectifying problems with expression or solubility.

Cell-free platforms have certain advantages for NMR spectroscopy, as described in recent reviews of an *E. coli* cell-free platform [12] and a wheat-germ cell-free platform [13]. Although cell-free protein production required labeled amino acids rather than the less expensive precursors utilized by *E. coli* cell-based platforms ([U-^{13}C]-glucose and ^{15}NH$_4$Cl), the level of incorporation of the amino acids is relatively high (~10%). Because isotopic exchange is minimal, cell-free platforms support selective labeling. The volumes involved in cell-free protein production tend to be small. Sufficient protein for a structure determination can be prepared from a 4- to 12-mL reaction volume. The preparation of labeled protein is fast and efficient and can be automated.

17.2.4 CLONING AND CONSTRUCT PRODUCTION

Structural proteomics centers have three options for obtaining DNA coding for protein targets; the gene of interest can be (1) cloned from a cDNA library (prepared in house or purchased); (2) obtained from another laboratory or gene repository; or (3) synthesized on the basis of the sequence in the genome model. CESG developed its own *Arabidopsis thaliana* cDNA library and found that roughly 80% of selected genes could be obtained by PCR from that library. Interestingly, about 25% of the resulting genes, when sequenced, revealed differences from the published gene models. Most of the differences concerned exon/intron boundaries, but others included termination sites and amino acid substitutions. CESG has demonstrated that the odds for successful cloning can be improved by using results from gene chip analysis of the cDNA library to remove targets unlikely to be present at levels required for PCR amplification [14].

Gene repositories or other sources of cloned and sequenced DNA offer very convenient sources for target genes. Although currently expensive, DNA synthesis provides a way to produce any gene from its sequence. As the costs go down, DNA synthesis may reach a level where it offers a very attractive alternative for structural studies of domains or full-length proteins. Aside from permitting the production of proteins not represented in cDNA or gene libraries, gene synthesis allows the sequence of the synthesized DNA to be engineered to maximize protein yields by minimizing mRNA secondary structure and/or by incorporating optimal codon frequencies.

A powerful approach to the production of constructs is to insert the gene of interest into an entry clone, which then can be easily transferred to multiple destination clones so as to support protein production from multiple platforms (e.g., cell-free, *E. coli* cells, *Pichia pastoris*, insect cells) and multiple constructs (e.g., N-terminal (His)$_x$ tag, N-terminal GST tag, N-terminal MBP tag). CESG initially used the Gateway® system from Invitrogen for this purpose [15]; however, it was limited in that a suitable Gateway destination construct was unavailable for wheat-germ cell-free system. Recently, CESG, in collaboration with Promega, has found through extensive tests that Promega's Flexi®Vector system will support both of CESG's protein production platforms (B. G. Fox et al., personal communication). This approach enables a single cloning step to support both approaches. As a quality-control step, all entry clones must be sequenced.

17.2.5 SMALL-SCALE SCREENING FOR PROTEIN PRODUCTION LEVEL AND SOLUBILITY

To achieve economical, high-throughput results, it is necessary to identify targets in the pipeline that ultimately will fail to yield structures as soon and as accurately as possible. One key step is, therefore, the screening of each target for protein expression, cleavage (if the construct produces a fusion protein), and solubility of the protein product. Failed expression, low-level production, and insolubility each indicate that the protein likely will not yield a structure. These tests should be rapid and inexpensive so as to conserve resources, yet need to reflect accurately what will happen during large-scale protein production and labeling.

CESG routinely uses the wheat-germ cell-free system on a 50-μL level to screen NMR target constructs [10]. The transcription/translation product is analyzed by SDS PAGE for total protein, soluble protein (supernatant following centrifugation),

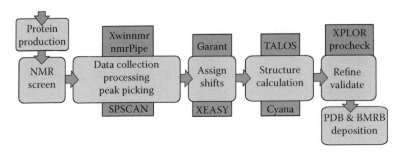

FIGURE 17.3 NMR steps leading to a protein solution structure. Also shown are some of the software packages available to support individual steps in the pipeline.

and insoluble protein. If the protein is being made as a cleavable fusion, a cleavage step is also inserted to determine its success rate. In our experience, this test correlates extremely well with large-scale (4–12 mL) cell-free protein production runs. The default procedure for cell-free production of proteins for NMR spectroscopy is to incorporate a noncleavable, N-terminal $(His)_6$ tag. If this fails, we have the option of trying a cleavable N-terminal His tag or a cleavable GST tag (3C protease). Only proteins that are produced in high yield (~2 mg/mL) and are more than 75% soluble are carried forward.

Small-scale trials are carried out to test the *E. coli* cell-based approach. Here our default construct is an N-terminal maltose binding protein (MBP) with a cleavable linker (tobacco etch virus protease). This has been chosen following extensive trials with a variety of alternative tags and cleavage options. Protein targets are scored as high (H), medium (M), and low (L) on the basis of cleavability, yield of cleaved protein, and solubility of cleaved protein. The goal of small-scale screening is to identify proteins that have at least an 80% probability of success in a subsequent large-scale production run. Any protein with one or more "L" values is excluded.

17.2.6 SCREENING OF [U-^{15}N]-PROTEIN FOR SUITABILITY AS AN NMR TARGET

Screening is the first in a number of steps leading to an NMR structure (fig. 17.3). Although ^{1}H NMR screening has been advocated as an approach for screening targets in NMR-based structural proteomics [16,17], in our experience ^{15}N-^{1}H HSQC analysis provides the most reliable predictor of success. The screening is carried out more economically with a ^{15}N-labeled protein than with a more expensive double-labeled protein. CESG's cell-free and cell-based platforms for producing ^{15}N-labeled proteins have been described in detail in the literature, and current protocols are available from the CESG Web site. A 0.5-mL sample is prepared in a concentration of about 1 m*M*. The volume can be reduced to 0.3 mL if susceptibility-matched NMR cells are used or to 0.15 mL if a 3-mm NMR tube is used. It is important to screen at the concentration that will be used for subsequent data collection for the structure determination so as to detect possible solubility or stability problems. A 2D ^{15}N-^{1}H HSQC spectrum of the protein is acquired, along with a 1D ^{1}H NMR spectrum. These spectra provide a good diagnostic for whether the protein, under the conditions tested, will be a suitable target for an NMR structural investigation. Only targets that pass the screens are prepared as more expensive $^{13}C/^{15}N$-labeled protein.

What one looks for (fig. 17.4) are (1) good dispersion of the NMR chemical shifts, particularly in the ^1H NMR dimension; (2) a peak count that matches that expected from the protein sequence (fewer peaks may indicate partial disorder and more peaks may indicate multiple conformational states or covalent structural heterogeneity); (3) uniformity of HSQC peak intensities; (4) lack of regions containing broad overlapped signals); and (5) presence of high-field methyl signals in the ^1H NMR spectrum. In addition, the protein must exhibit stability over a period (generally 7–10 days) comparable to that required for the full collection of NMR spectra.

Targets that do not pass these tests may be salvageable. It may be possible to improve the solubility, the appearance of the HSQC fingerprint, or the stability of the protein by varying the solution conditions (buffer, pH, salt concentration, temperature, addition of a reductant such as DTT or a protease inhibitor) or by adding a metal ion (Zn^{2+}, Ca^{2+}). A microdrop analysis approach [18] makes it possible to screen a large set of solution conditions for protein solubility and stability with very little protein. One or more of the successful conditions can then be evaluated by $^{15}N-^1H$ HSQC screening. If the solvent optimization approach fails, the protein product can be altered by removing or changing a tag, by trimming residues off one or both ends (fig. 17.4D), or by mutating one or more residues.

17.2.7 PRODUCTION OF [U-^{13}C, U-^{15}N]-PROTEIN

Double-labeled protein is produced by the same platform used for [^{15}N] protein and is transferred to the optimal solution conditions previously determined. If a protein has been shown to have limited stability, it may be advantageous to prepare enough protein for multiple NMR samples.

17.2.8 NMR DATA COLLECTION

The best strategy is to collect all data required for a structure at one time on one sample and one NMR spectrometer. This minimizes difficulties associated with differences between samples and calibration differences between spectrometers; the strategy presupposes that the sample is stable over the data collection period. A standard protocol developed for CESG by Brian Volkman's group at the Medical College of Wisconsin is available from the CESG Web site. Under favorable conditions, all data for a structure can be collected in one week on a 600-MHz NMR spectrometer with a cryogenic probe. All data sets are acquired on a [U-^{15}N, U-^{13}C] protein (≥ 0.5 mM) in H_2O solution: 2D $^{15}N-^1H$ HSQC, 3D $^{15}N-^1H$ NOESY-HSQC, 2D $^1H-^{13}C$ HSQC (aromatic), 2D $^1H-^{13}C$ CT-HSQC (aliphatic), 3D HNCO, 3D HNCA, 3D HNCOCA, 3D HNCACO, 3D CCONH, 3D HCCONH, 3D HNCACB, 3D HBHACONH, 3D HCCH-TOCSY, 3D $^1H-^{13}C$ NOESY-HSQC (aliphatic), and 3D $^1H-^{13}C$ NOESY-HSQC (aromatic).

The National Magnetic Resonance Facility at Madison (NMRFAM) in collaboration with CESG is developing a new generation of tools to support a comprehensive probabilistic approach to protein NMR data collection and analysis. On the data collection side, an adaptive reduced-dimensionality approach to fast data collection

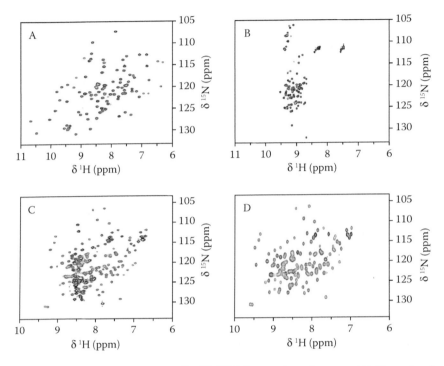

FIGURE 17.4 Sample results from ^{15}N-^{1}H HSQC screening of targets. (A) Example of a target judged to be suitable for high-throughput structure determination (HSQC+). This protein is from *Arabidopsis thaliana* (gene At1g77540) with a predicted pI of 8.2. Solution conditions were 10 m*M* KH$_2$PO$_4$, 50 m*M* KCl, 2 m*M* DTT, and pH = 6. The protein contains 103 residues (7 Pro, 0 Gln, 4 Asn, 2 Trp). Thus, the theoretical number of backbone resonances is 96. After subtracting side chain resonances from the observed number of cross-peaks, the observed backbone peak number is 89. Thus, 89/96 = ~92% of the expected peaks are resolved. (B) Example of a target judged to be unsuitable for structure determination (HSQC−). The target is from the mouse genome (gene Mm295552) and has a predicted pI of 5.6. The solution conditions were 10 m*M* Bis Tris,100 m*M* NaCl, 2 m*M* DTT, and pH = 7. The protein contains 110 residues (4 Pro, 4 Gln, 3 Asn, and 0 Trp). The ratio of observed/theoretical backbone resonances is 73/106 = 68%. The poor dispersion indicates that the protein is dynamically disordered. (C) Example of a target with problems that may be overcome by changing the solvent or redesigning the protein construct (HSQC+/−). The protein is from zebrafish (gene Dr15775). The solution conditions were 10 m*M* Bis Tris, 100 m*M* NaCl, 2 m*M* DTT, and pH = 6. The protein has a predicted pI of 5.9 and contains 176 residues (4 Pro, 8 Gln, 5 Asn, 0 Trp). The ratio of observed/theoretical backbone resonances is 150/172 = 87%. The presence of very intense peaks suggested that the protein contains unstructured regions. (D) The target shown in C following truncation to improve its degree of foldedness. After removal of the first 17 and last 29 residues, the ratio of observed/theoretical backbone resonances is 103/114 = 90%. The modified protein is an excellent candidate for structure determination.

(high-resolution iterative frequency identification [HIFI]-NMR) promises to reduce data collection time requirements by a factor of 5 or more [46]. The approach used by HIFI is to collect 3D spectral information in the form of tilted planes. Peaks are identified in each plane collected; information from a spectra model is used to determine whether an additional experimental plane would improve that model and, if so, the optimal tilt angle for the next plane.

The mathematical and computational tools used in HIFI-NMR are sufficiently general to allow the combination of information from various sources into a single model. This will make it possible to improve the overall process through the integration of data from multiple experiments. For example, by combining data from 3D experiments with common data planes (for example, CBCA(CO)NH, HNCACB, and HNCO), it should be possible to improve the discrimination between signal and noise. Most importantly, the direct output from HIFI-NMR is a probabilistic peak list that can be used as input to automated assignment software packages.

17.2.9 BACKBONE AND SIDE-CHAIN ASSIGNMENTS

CESG initially chose Garant [19,20] as the software tool for semiautomated assignments. The first step is to generate backbone resonance assignments by Garant, using peak lists generated from NMR spectra HNCA, HN(CO)CA, CCONH, HNCACB, HNCO, HN(CA)CO, and ^{15}N-^{1}H-HSQC as input. The assigned peak lists are then inspected, corrected, and amended using the XEASY or Sparky programs. Upon completion of the backbone assignments, the side chain ^{1}H and ^{13}C resonances are assigned by a second run of Garant that adds peak lists generated from H(CCO)NH, HBHA(CO)NH, and HCCH-TOCSY data. Manual intervention after this run may be needed to resolve ambiguities. In order to achieve the combination of manual and automated assignment, simple software tools have been built for interconverting data (peak lists and resonance list) formats between Sparky (sparky@cgl.ucsf.edu) and XEASY [21].

A new software package was developed at NMRFAM with the goal of achieving fully automated probabilistic assignment of protein NMR data [19]. PISTACHIO (probabilistic identification of spin systems and their assignments including coil-helix inference as output) uses as input peak lists from 2D and 3D NMR experiments and the sequence of the protein. Currently, 15 different experiments are supported, and the data can be used in any combination or order. PISTACHIO recasts the NMR assignment problem, which normally has been posed as a constrained weighted bipartite graph-matching problem, as an energetic model. PISTACHIO provides assignments in terms of probabilities. Since its run time is short compared to data collection times, one can analyze the results as they come out and decide when sufficient data are on hand for the assignments. PISTACHIO screens input data and provides a rapid assessment of the level of assignment supported by the peak lists provided. PISTACHIO, along with other NMRFAM tools discussed later, is available from the Web site http://bija.nmrfam.wisc.edu. With typical data sets, PISTACHIO yields backbone assignments at greater than 90% and side chain assignments at greater than 75%.

As a second step in the probabilistic analysis of NMR data, the assigned data are run through the LACS (linear analysis of chemical shifts) software, which tests

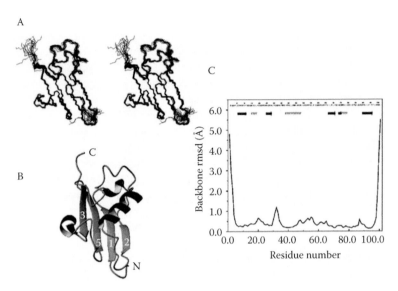

FIGURE 17.5 (See color insert.) NMR solution structure of AAH26994.1. (Singh, S. et al. *Protein Sci.*, 14, 2095–2102, 2005.) (A) Stereo view of the ensemble of 20 conformers that represent the structure (backbone of residues 1–101). Coloring scheme: β-strand residues, blue; α-helical residues, red; other residues, green. (B) Ribbon view of the representative AAH26994.1 structure (that closest to the average) showing residues1–101. Numbers represent the order of β-strands, and the N- and C-termini of the protein are indicated. The molecular graphics program MOLMOL [76] was used in generating these views of the structure. (C) Backbone (N, C^α, C') atomic root mean square deviation plotted as a function of residue number. In solving this structure, the NMR data sets were processed with NMRPipe [74] and peak picked using NMRView [75].

for possible referencing problems and assignment outliers [20]. Following corrections, the data are rerun through PISTACHIO to refine the assignments.

The third step is to use the assigned chemical shifts along with the protein sequence to determine the secondary structure of the protein. A new tool for accomplishing this is PECAN (protein energetic conformational analysis) [21], which has been shown to outperform other available approaches. The level of correctness and the small variance of predictions make PECAN a particularly reliable tool. Once the secondary structure has been determined, another round of PISTACHIO analysis can be used to refine the assignments further.

The most widely used software tool for determining dihedral angle restraints from assigned NMR chemical shifts is the TALOS package developed at the NIH [22]. TALOS generates constraints representing ranges of ϕ and ψ dihedral angles from $^1H^\alpha$, $^{13}C^\alpha$, $^{13}C^\beta$, $^{13}C^\gamma$, and ^{15}N secondary shifts. As a next-generation tool, NMRFAM and CESG are developing an algorithm that uses the results from PECAN to provide robust probabilistic dihedral angle restraints to be used as input for NMR structure determinations.

A recent structure determination from CESG (fig. 17.5) made use of this novel technology. Peak lists from CBCA(CO)NH, HNCACB, HNCO, and ^{15}N-^1H-HSQC experiments were provided as input to PISTACHIO [19] for the initial assignment step. The assigned peak lists were verified and corrected by manual intervention by use of NMRView software. Upon completion of the backbone assignments, the side chain ^{1}H and ^{13}C resonances were assigned by a second run of PISTACHIO, with peak lists generated from H(CCO)NH, CCONH, ^{15}N-^{1}H-HSQC, and HBHACONH as input. Again, ambiguities were resolved by manual intervention. Assignments were verified and corrected with a CCH-TOCSY data set. TALOS constraints [22] and 3D ^{15}N-edited and ^{13}C-edited NOESY spectra with ~100-ms mixing time for NOE constraints were used for structure calculations. The nuclear Overhauser enhancement (NOE) peak lists were then submitted to ARIA [23] for automated assignment and for structure calculation [24]. This structure determination took a total of two weeks from the start of data collection to the completion of PDB-ready structures.

17.2.10 NOE ASSIGNMENTS AND DISTANCE CONSTRAINTS

Proton–proton nuclear Overhauser enhancements (NOEs) are the most widely used source of distance constraints used in protein structure determinations. Typically, NOE distance constraints are generated from 3D ^{15}N-edited NOESY and 3D ^{13}C-edited NOESY spectra collected with mixing times of ~125 ms.

The most widely used NMR structure determination packages are CYANA [25], ARIA [23], and X-PLOR-NIH [26]. The recent versions of these contain algorithms for dealing with ambiguously assigned NOE data. They use as input experimental NOESY results plus peak assignments. Tolerance levels are assigned for the ^{1}H, ^{13}C, and ^{15}N chemical shifts. Assignments are refined iteratively along with the structure refinement.

17.2.11 ADDITIONAL NMR DATA COLLECTION

Residual dipolar couplings (RDCs) provide another valuable source of information for structure determination [3,27]. Although structures can be determined with RDC data alone (without NOEs), our experience has been that they are more useful for refining structures and for validating structures determined with NOE constraints. It is useful to collect relaxation data and/or NMR diffusion data to determine whether a protein target is multimeric. It is fairly common for structural proteomics proteins to be homodimers. If the protein is dimeric, it may be useful to prepare a sample containing a 1:1 mixture of unlabeled protein and [U-^{13}C,^{15}N] protein. The protein needs to be incubated under conditions where equilibration of the subunits can take place. Then intra- and intersubunit NOEs can be distinguished from the analysis of ^{13}C-filtered/edited and ^{15}N-filtered/edited NOESY data sets.

Additional information may be required to complete the structure of a homo-dimeric or homo-oligomeric protein. Generally, evidence for a complex comes from gel filtration chromatography during protein purification, from a low-angle light scattering assay or from NMR diffusion or relaxation measurements. Software packages, such as ARIA and CYANA, provide assistance with determining structures of homo–subunit complexes. Usually, the surest method is to prepare a complex by mixing unlabeled protein with double-labeled protein and to use NOE filtering/editing [28] to distinguish between intra- and intersubunit NOEs.

17.2.12 APPROACHES TO STREAMLINING THE STRUCTURE DETERMINATION PROCESS

CESG has explored various approaches to NMR structure determination based on different combinations of available software packages. Algorithms and software are evolving rapidly, so we have maintained an interest in comparing different approaches. The NESGC has developed their platform [1] around the AutoAssign and AutoStructure packages developed in the Montelione laboratory. AutoStructure makes use of XPLOR/CNS or DYANA as its structure calculation module.

The CESG-MCW group has tested and applied a comprehensive method that makes use of SPSCAN (www.molebio.uni-jena.de/~rwg/spscan/), Garant [29], CANDID/CYANA [30], and XPLOR-NIH [31] to perform peak picking, chemical shift assignment, and structure refinement and validation in a series of fully automated steps (fig. 17.3). This approach has been described in recent publications [10,32] and is documented in the NMR structure determination protocol deposited in the CESG protocol library.

17.2.13 VALIDATION OF NMR STRUCTURES

A final water refinement [33,34] generally is carried out prior to the validation steps. PROCHECK-NMR [35], WhatCheck [36,37], and Wattos [38,39] are useful programs for validating coordinates based on prior knowledge. PROCHECK-NMR and Wattos look at the agreement between the coordinates and distance and dihedral angle restraints. A recent analysis has shown that factors such as root mean square deviations (RMSDs), agreement of restraints with structures, NOE completeness, and number of restraints are not statistically valid predictors of the quality of NMR structures [34]. A better approach is to compare back-calculated chemical shifts as done by SHIFTS [40], SHIFTX [41], PROSHIFT [42], and a module in NMRPipe (Frank Delaglio, personal communication) versus the experimentally measured values. The chemical shifts of the backbone atoms in proteins are commonly used to define the two backbone dihedral angles ϕ and ψ. The chemical shifts of side-chain nuclei (e.g., the methyl protons and carbons), however, normally are not used in structure calculation; they may be used as truly independent validation criteria and can be back-calculated with some of the previously mentioned tools.

17.2.14 DATA DEPOSITION

The Worldwide Protein Data Bank (wwPDB) recently established a single site for the deposition of all data relevant to a biomolecular NMR structure. This obviates the need for separately depositing the coordinates and restraints at PDB and the assigned chemical shifts and other NMR parameters at BMRB and streamlines the process considerably. At this new site, it is possible to deposit assignments first and then a structure later without having to reenter information common to both. It is possible to bypass the normal data deposition process by preparing files compliant with the formats used by PDB and BMRB. The Northeast Structural Genomics Consortium first demonstrated their capability to do this for NMR data in 2005. The approach is to create the deposition from the contents of the center's information management system. Files constructed in this way still require validation by the wwPDB but avoid much manual work. BMRB is accepting primary data (free induction decay data sets) associated with structure depositions. These data sets provide the means for recalculating structures as methods improve, and they also are useful for software developers and persons learning to determine structures.

17.2.15 FUNCTIONAL STUDIES

One of the motivations for structural proteomics is to use structure determination as a shortcut for developing hypotheses about the function of a target. In some cases, the structure may reveal a bound metal ion or cofactor that gives a clue to the function. In other cases, a search for structural homologues by submitting the coordinates to the VAST (www.ncbi.nlm.nih.gov/Structure/VAST/vast.shtml) and DALI (www.ebi.ac.uk/dali/) servers may reveal a match that may lead to a functional hypothesis. One of the advantages of carrying out an NMR structure determination is that NMR screening can be used to rapidly test hypotheses about binding of substrate-like molecules or suspected cofactors.

An example is the putative protein from *Arabidopsis thaliana* (At2g24940.1), whose structure [43] unexpectedly revealed that it has a cytochrome b5 fold. Although the biochemical function of At2g24940.1 was unknown, it has a sequence similarity (~40% sequence identity) with mammalian MAPR. This suggested that At2g24940.1 may act as a steroid-binding protein. At present, no structure of an MAPR is available from the Protein Data Bank. Thus, we suggested that the structure of At2g24940.1 may provide clues to the function of a class of steroid-binding proteins in plants. To test the hypothesis that this protein binds steroids, the HSQC spectrum of ^{15}N-labeled protein was monitored on the addition of progesterone. The results (fig. 17.6) clearly show binding and identify the predicted sterol binding site [43] as correct.

17.2.16 PUBLICATION

The major goal of most structural proteomics centers is to solve and deposit structures. Such centers may not have the time or the means to prepare lengthy articles describing the structures. Although many proteins are represented only by wwPDB depositions, it is useful to produce at least a structure note (for example, for *Protein Science*, *Proteins*, or the *Journal of Biomolecular NMR*) so that the structure will show up in a literature search. Structure notes contain information about the biological relevance, how the sample was prepared, and how the data were collected and analyzed.

17.3 FUTURE PROSPECTS

Structural proteomics has demonstrated the ability to develop high-throughput pipelines capable of lowering the costs of protein structure determinations. These

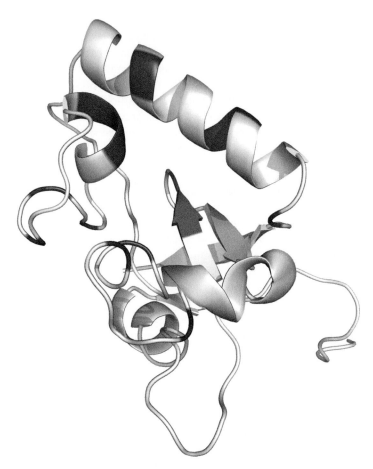

FIGURE 17.6 (See color insert.) Ribbon diagram (red) showing the three-dimensional structure of *Arabidopsis thaliana* putative protein At2g24940.1. (Song, J. et al. *J. Biomol. NMR* 30, 215–218, 2004.) Sites exhibiting the largest ^{15}N-^{1}H chemical shift perturbations upon progesterone binding ($\Delta_{HN} > 0.15$ ppm, where $\Delta_{HN} = \{((\Delta_H)^2 + (\Delta_N/5)^2)/2\}^{1/2}$) are highlighted in blue. (J. Song et al., unpublished.) This identifies the sterol binding pocket.

projects have greatly expanded our knowledge of sequence/3D structure relationships and have enabled the prediction of structures for many more protein families. The pilot projects have also demonstrated how structures can provide valuable insights into function. Structures can provide starting points for more interesting investigations of mechanisms of action and inhibition. Several structural proteomics groups, including CESG, incorporate outreach efforts aimed at putting new structures into the hands of biochemists who may be interested in following up on the initial steps taken.

The major challenges for the future are to decrease the costs of protein production and structure determination still further and to enlarge the scope of proteins and protein domains amenable to a high-throughput approach. Costs will be lowered as processes become more standardized and automated. Reduced dimensionality NMR data collection approaches offer ways of speeding up one of the most serious bottlenecks [44,45]. The basic approach is to collect lower dimensionality data for use in reconstructing higher dimensionality spectra. A promising alternative is to determine peak positions as the two-dimensional planes are collected and to use a dynamic model of the peaks in the higher dimensional spectrum to choose the optimal next plane to be collected; data collection is terminated when the addition of another two-dimensional plane is not predicted to improve the model peak list [46].

The more challenging targets include protein domains, protein:ligand complexes, protein:protein complexes, and membrane proteins. Success in high-throughput approaches to these more difficult targets hinges on the development of efficient small-scale screening appropriate to the task.

17.3.1 PROTEIN DOMAINS

Many proteins are unsuitable for NMR structural analysis in their intact, full-length forms because they are too large or contain elements (such as transmembrane domains or highly degenerate regions) that reduce solubility or are unstructured. When independent domains can be separated from disordered or dynamic regions, they often become highly desirable targets for NMR structural studies. Thus, there is significant opportunity for developing methods to identify and produce protein domains. Two possible approaches are *de novo* prediction of the boundaries of domains and experimental definition of domains from intact, full-length targets that do not exhibit suitable properties for structure determination.

17.3.2 PROTEIN–LIGAND COMPLEXES

It will be important to the field of structural proteomics to develop routine and reliable methods for screening all targets against a panel of common cofactors and metal ions. High-throughput methods are needed to detect binding and/or increased folding or stability of the protein. Hypotheses developed in this way can be tested by HSQC screening of [U-^{15}N] protein upon addition of the ligand. The simple screen usually identifies the ligand binding site and provides a measure of the magnitude of any accompanying conformational change. Depending on these results, it may be of interest to use NMR spectroscopy to determine the structure of the complex.

17.3.3 PROTEIN–PROTEIN COMPLEXES

Past structural proteomics efforts have focused on preparing proteins one by one. Hence, as noted above, homo-oligomers are generally the only protein:protein complexes discovered. For NMR, these are most often homodimers, given the molecular weight limitations. As bioinformatics tools for predicting protein:protein interactions improve, it should be possible to develop high-throughput coproduction of proteins likely to interact with one another and to assay for their interaction by a gel filtration step followed by mass spectrometric analysis of the proteins. Structures of AB protein complexes are most efficiently determined by making two differentially labeled complexes: [U-^{13}C, U-^{15}N]-A:[NA]-B and [NA]-B:[U-^{13}C, U-^{15}N]-B where NA stands for natural abundance. The first is used to determine the conformation of A and the second is used to determine the conformation of B in the complex. Then filtered/edited experiments [28] and docking methods [47] are used to determine and refine the structure of the complex.

17.3.4 LARGER PROTEINS

The NMR structural biology field has recently developed a number of approaches for dealing with larger proteins [48]. These include data collection approaches and specialized labeling patterns. In collecting data with larger proteins, lines can be narrowed by pulse sequences, such as TROSY (transverse relaxation optimized spectroscopy) and CRINEPT (cross relaxation-enhanced polarization transfer) that engineer the destructive interference of different relaxation pathways [49]. Alternatively, data can be read out on carbon signals, which tend to be sharper in larger proteins than the much more commonly used proton signals [50,51]. In addition, spectra can be simplified and lines can be narrowed by residue selective labeling [52], perdeuteration [53], perdeuteration with methyl labeling [54], segmental labeling [55], or stereo array isotope labeling [56]. It will be of interest to see which approach or combination of these approaches may be adaptable to high-throughput structural proteomics applications.

17.3.5 MEMBRANE PROTEINS

NMR methods for determining structures of membrane proteins are evolving steadily [57]. The most general NMR approach is to solubilize labeled proteins in detergent or lipid micelles as described in recent publications. Initial NMR structures were of β-sheet membrane proteins, but more recent success has been achieved with α-helical membrane proteins [58].

17.3.6 PROTEIN DISORDER–ORDER

One of the dividends of NMR-based structural proteomics is the capture of experimental information on sequences that are dynamically disordered or exist in two or more interconverting conformations. Keith Dunker has set up a *Database of Protein Disorder* (DisProt) (www.disprot.org/), which is designed to capture relevant experimental data on protein disorder. This has the potential of organizing useful information largely outside the purview of the Protein Data Bank. DisProt has the potential of becoming a valuable resource for research on the connection between disorder and physical properties, such as NMR chemical shifts and relaxation parameters. It will also be a place where scientists can obtain information to refine the already useful predictors of protein disorder [59–67].

Information coming from structural proteomics should catalyze experiments designed to investigate the biology of why proteins are disordered or contain disordered domains [68,69]. Documented cases exist for proteins that fold only when presented with a metal ion [70], nucleic acid [71], or second peptide chain. This is a challenging area of structural biology, but some fundamental principles are emerging. For example, a peptide chain may fail to fold because its sequence does not contain the hydrophobic amino acids needed to form a hydrophobic core; however, when it is complemented by a second chain containing the needed residues, the two may form a stable fold [72].

ACKNOWLEDGMENTS

I thank the many CESG and NMRFAM members who have contributed to work described here and the NIH Protein Structure Initiative (www.nigms.nih.gov/psi/) of the National Institutes for General Biomedical Sciences for funding (1 P50 GM64598 and 1 U54 GM074901). Jikui Song, Robert C. Tyler, and Brian F. Volkman contributed to the figures.

REFERENCES

1. Huang, Y.J., Moseley, H.N., Baran, M.C., Arrowsmith, C., Powers, R., Tejero, R., Szyperski, T., and Montelione,G.T. An integrated platform for automated analysis of protein NMR structures. *Methods Enzymol.*, 394, 111–141, 2005.

2. Atreya, H.S. and Szyperski, T. Rapid NMR data collection. *Nucl. Magn. Resonance Biological Macromolec., Part C*, 394, 78–108, 2005.

3. Prestegard, J.H., Mayer, K.L., Valafar, H., and Benison, G.C. Determination of protein backbone structures from residual dipolar couplings. *Nucl. Magn. Resonance Biological Macromolec., Part C*, 394, 175, 2005.

4. Acton, T.B., Gunsalus, K.C., Xiao, R., Ma, L.C., Aramini, J., Baran, M.C., Chiang, Y.W., Climent, T., Cooper, B., Denissova, N.G., Douglas, S.M., Everett, J.K., Ho, C.K., Macapagal, D., Rajan, P.K., Shastry, R., Shih, L.Y., Swapna, G.V.T., Wilson, M., Wu, M., Gerstein, M., Inouye, M., Hunt, J.F., and Montelione, G.T. Robotic cloning and protein production platform of the Northeast Structural Genomics Consortium. *Nucl. Magn. Resonance Biological Macromolec., Part C*, 394, 210, 2005.

5. Zolnai, Z., Lee, P.T., Li, J., Chapman, M.R., Newman, C.S., Phillips, G.N., Jr., Rayment, I., Ulrich, E.L., Volkman, B.F., and Markley, J.L. Project management system for structural and functional proteomics: Sesame. *J. Struct. Funct. Genomics*, 4, 11–23, 2003).

6. Oldfield, C.J., Ulrich, E.L., Cheng, Y., Dunker, A.K., and Markley, J.L. Addressing the intrinsic disorder bottleneck in structural proteomics. *Proteins*, 59, 444–453, 2005.

7. Kigawa, T., Yabuki, T., Yoshida, Y., Tsutsui, M., Ito, Y., Shibata, T., and Yokoyama, S. Cell-free production and stable-isotope labeling of milligram quantities of proteins. *FEBS Lett.*, 442, 15–19, 1999.

8. Yokoyama, S. Protein expression systems for structural genomics and proteomics. *Curr. Opin. Chem Biol.*, 7, 39–43, 2003.

9. Tyler, R.C., Sreenath, H.K., Singh, S., Aceti, D.J., Bingman, C.A., Markley, J.L., and Fox, B.G. Auto-induction medium for the production of [U-15N]- and [U-13C, U-15N]-labeled proteins for NMR screening and structure determination. *Protein Expr. Purif.*, 40, 268–278, 2005.

10. Vinarov, D.A., Lytle, B.L., Peterson, F.C., Tyler, E.M., Volkman, B.F., and Markley, J.L. Cell-free protein production and labeling protocol for NMR-based structural proteomics. *Nat. Methods*, 1, 149–153, 2004.

11. Tyler, R.C., Aceti, D.J., Bingman, C.A., Cornilescu, C.C., Fox, B.G., Frederick, R.O., Jeon, W.B., Lee, M.S., Newman, C.S., Peterson, F.C., Phillips, G.N., Jr., Shahan, M.N., Singh, S., Song, J., Sreenath, H.K., Tyler, E.M., Ulrich, E.L., Vinarov, D.A., Vojtik, F.C., Volkman, B.F., Wrobel, R.L., Zhao,Q., and Markley, J.L. Comparison of cell-based and cell-free protocols for producing target proteins from the *Arabidopsis thaliana* genome for structural studies. *Proteins*, 59, 633–643, 2005.

12. Torizawa, T., Shimizu, M., Taoka, M., Miyano, H., and Kainosho, M. Efficient production of isotopically labeled proteins by cell-free synthesis: a practical protocol. *J. Biomol. NMR*, 30, 311–325, 2004.

13. Vinarov, D.A. and Markley, J.L. High-throughput automated platform for nuclear magnetic resonance-based structural proteomics. *Exp. Rev. Proteomics*, 2, 49–55, 2005.

14. Stolc, V., Samanta, M.P., Tongprasit, W., Sethi, H., Liang, S., Nelson, D.C., Hegeman, A., Nelson, C., Rancour, D., Bednarek, S., Ulrich, E.L., Zhao, Q., Wrobel, R.L., Newman, C.S., Fox, B.G., Phillips, G.N., Jr., Markley, J.L., and Sussman, M.R. Identification of transcribed sequences in *Arabidopsis thaliana* by using high-resolution genome tiling arrays. *Proc. Natl. Acad. Sci. U.S.A.*, 102, 4453–4458, 2005.

15. Thao, S., Zhao, Q., Kimball, T., Steffen, E., Blommel, P.G., Riters, M., Newman, C.S., Fox, B.G., and Wrobel, R.L. Results from high-throughput DNA cloning of *Arabidopsis thaliana* target genes using site-specific recombination. *J. Struct. Funct. Genomics*, 5, 267–276, 2004.

16. Hoffmann, B., Eichmuller, C., Steinhauser, O., and Konrat, R. Rapid assessment of protein structural stability and fold validation via NMR. *Methods Enzymol.*, 394, 142–175, 2005.

17. Peti, W., Etezady-Esfarjani, T., Herrmann, T., Klock, H.E., Lesley, S.A., and Wuthrich, K. NMR for structural proteomics of *Thermotoga maritima*: Screening and structure determination. *J Struct. Funct. Genomics*, 5, 205–215, 2004.

18. Lepre, C.A. and Moore, J.M. Microdrop screening: A rapid method to optimize solvent conditions for NMR spectroscopy of proteins. *J. Biomol. NMR*, 12, 493–499, 1998.

19. Eghbalnia, H.R., Bahrami, A., Wang, L., Assadi, A., and Markley, J.L. Probabilistic identification of spin systems and their assignments including coil–helix inference as output (PISTACHIO). *J. Biomol. NMR*, 32, 219–233, 2005.

20. Wang, L., Eghbalnia, H.R., Bahrami, A., and Markley, J.L. Linear analysis of carbon-13 chemical shift differences and its application to the detection and correction of errors in referencing and spin system identifications. *J. Biomol. NMR*, 32, 13–22, 2005.

21. Eghbalnia, H.R., Wang, L.Y., Bahrami, A., Assadi, A., and Markley, J.L. Protein energetic conformational analysis from NMR chemical shifts (PECAN) and its use in determining secondary structural elements. *J. Biomol. NMR*, 32, 71–81, 2005.

22. Cornilescu, G., Delaglio, F., and Bax, A. Protein backbone angle restraints from searching a database for chemical shift and sequence homology. *J. Biomol. NMR*, 13, 289–302, 1999.

23. Linge, J.P., O'Donoghue, S.I., and Nilges, M. Automated assignment of ambiguous nuclear Overhauser effects with ARIA. *Methods Enzymol.*, 339, 71–90, 2001.

24. Brunger, A.T., Adams, P.D., Clore, G.M., Delano, W.L., Gros, P., Grossekunstleve, R.W., Jiang, J.S., Kuszewski, J., Nilges, M., Pannu, N.S., Read, R.J., Rice, L.M., Simonson, T., and Warren, G.L. Crystallography and NMR system—A new software suite for macro-molecular structure determination. *Acta Crystallographica Section D-Biological Crystallography*, 54, 905–921, 1998.

25. Guntert, P. Automated NMR structure calculation with CYANA. *Methods Mol. Biol.*, 278, 353–378, 2004.

26. Kuszewski, J., Schwieters, C.D., Garrett, D.S., Byrd, R.A., Tjandra, N., and Clore, G.M. Completely automated, highly error-tolerant macromolecular structure determination from multidimensional nuclear Overhauser enhancement spectra and chemical shift assignments. *J. Am. Chem. Soc.*, 126, 6258–6273, 2004.

27. Tjandra, N., Omichinski, J.G., Gronenborn, A.M., Clore, G.M., and Bax, A. Use of dipolar ^{1}H-^{15}N and ^{1}H-^{13}C couplings in the structure determination of magnetically oriented macromolecules in solution. *Nat. Struct. Biol.*, 4, 732–738, 1997.

28. Zwahlen, C., Legault, P., Vincent, S.J.F., Greenblatt, J., Konrat, R., and Kay, L.E. Methods for measurement of intermolecular NOEs by multinuclear NMR spectroscopy: Application to a bacteriophage L N-peptide/boxB RNA complex. *J. Am. Chem. Soc.*, 119, 6711–6721, 1997.

29. Bartels, C., Güntert, P., Billeter, M., and Wüthrich, K. GARANT—A general algorithm for resonance assignment of multidimensional nuclear magnetic resonance spectra. *J. Computer Chem.*, 18, 139–149, 1997.

30. Herrmann, T., Guntert, P., and Wuthrich, K. Protein NMR structure determination with automated NOE assignment using the new software CANDID and the torsion angle dynamics algorithm DYANA. *J. Mol. Biol.*, 319, 209–227, 2002.

31. Schwieters, C.D., Kuszewski, J.J., Tjandra, N., and Marius, C.G. The Xplor-NIH NMR molecular structure determination package. *J. Magn. Resonance*, 160, 65–73, 2003.

32. Peterson, F.C., Lytle, B.L., Sampath, S., Vinarov, D., Tyler, E., Shahan, M., Markley, J.L., and Volkman, B.F. Solution structure of thioredoxin h1 from *Arabidopsis thaliana*. *Protein Sci.*, 14, 2195–2200, 2005.

33. Linge, J.P., Williams, M.A., Spronk, C.A., Bonvin, A.M., and Nilges, M. Refinement of protein structures in explicit solvent. *Proteins*, 50, 496–506, 2003.

34. Nederveen, A.J., Doreleijers, J.F., Vranken, W., Miller, Z., Spronk, C.A., Nabuurs, S.B., Guntert, P., Livny, M., Markley, J.L., Nilges, M., Ulrich, E.L., Kaptein, R., and Bonvin, A.M. RECOORD: A recalculated coordinate database of 500+ proteins from the PDB using restraints from the BioMagResBank. *Proteins*, 59, 662–672, 2005.

35. Laskowski, R.A., Rullmann, J.A.C., MacArthur, M.W., Kaptein, R., and Thornton, J.M. AQUA and PROCHECK-NMR: Programs for checking the quality of protein structures solved by NMR. *J. Biomol. NMR*, 8, 477–486, 1996.

36. Doreleijers, J.F. Validation of biomolecular structures solved by NMR. Ph.D. thesis, University of Utrecht, the Netherlands, 1999.

37. Hooft, R.W.W., Vriend, G., Sander, C., and Abola, E.E. Errors in protein structures. *Nature*, 381, 272, 1996.

38. Doreleijers, J.F., Mading, S., Maziuk, D., Sojourner, K., Yin, L., Zhu, J., Markley, J.L., and Ulrich, E.L. BioMagResBank database with sets of experimental NMR constraints corresponding to the structures of over 1400 biomolecules deposited in the Protein Data Bank. *J. Biomol. NMR*, 26, 139–146, 2003.

39. Doreleijers, J.F., Nederveen, A.J., Vranken, W., Lin, J., Bonvin, A.M., Kaptein, R., Markley, J.L., and Ulrich, E.L. BioMagResBank databases DOCR and FRED containing converted and filtered sets of experimental NMR restraints and coordinates from over 500 protein PDB structures. *J. Biomol. NMR*, 32, 1–12, 2005.

40. Xu, X.P. and Case, D.A. Automated prediction of N-15, C-13(alpha), C-13(beta) and C-13′ chemical shifts in proteins using a density functional database. *J. Biomol. NMR*, 21, 321–333, 2001.

41. Neal, S., Nip, A.M., Zhang, H., and Wishart, D.S. Rapid and accurate calculation of protein 1H, 13C and 15N chemical shifts. *J Biomol. NMR*, 26, 215–240, 2003.

42. Meiler, J. PROSHIFT: Protein chemical shift prediction using artificial neural networks. *J. Biomol. NMR*, 26, 25–37, 2003.

43. Song, J., Vinarov, D., Tyler, E.M., Shahan, M.N., Tyler, R.C., and Markley, J.L. Hypothetical protein At2g24940.1 from *Arabidopsis thaliana* has a cytochrome b5-like fold. *J. Biomol. NMR*, 30, 215–218, 2004.

44. Shen, Y., Atreya, H.S., Liu, G., and Szyperski, T. G-matrix Fourier transform NOESY-based protocol for high-quality protein structure determination. *J. Am. Chem. Soc.*, 127, 9085–9099, 2005.

45. Kupce, E. and Freeman, R. Projection-reconstruction technique for speeding up multidimensional NMR spectroscopy. *J. Am. Chem. Soc.*, 126, 6429–6440, 2004.

46. Eghbalnia, H.R., Bahrami, A., Tonelli, M., Hallenga, K., and Markley, J.L. High-resolution iterative frequency identification for NMR as a general strategy for multi-dimensional data collection. *J. Am. Chem. Soc.*, 127, 12528–12536, 2005.

47. Dominguez, C., Boelens, R., and Bonvin, A.M. HADDOCK: A protein–protein docking approach based on biochemical or biophysical information. *J. Am. Chem. Soc.*, 125, 1731–1737, 2003.

48. Wider, G. NMR techniques used with very large biological macromolecules in solution. *Nucl. Magn. Resonance Biological Macromolec., Part C*, 394, 382–398, 2005.

49. Riek, R., Pervushin, K., and Wuthrich, K. TROSY and CRINEPT: NMR with large molecular and supramolecular structures in solution. *Trends Biochem. Sci.*, 25, 462–8, 2000.

50. Stockman, B.J., Reily, M.D., Westler, W.M., Ulrich, E.L., and Markley, J.L. Concerted two-dimensional NMR approaches to hydrogen-1, carbon-13, and nitrogen-15 resonance assignments in proteins. *Biochemistry*, 28, 230–236, 1989.

51. Bermel, W., Bertini, I., Duma, L., Felli, I.C., Emsley, L., Pierattelli, R., and Vasos, P.R. Complete assignment of heteronuclear protein resonances by protonless NMR spectroscopy. *Angewandte Chemie-Int. Ed.*, 44, 3089–3092, 2005.

52. Shi, J., Pelton, J.G., Cho, H.S., and Wemmer, D.E. Protein signal assignments using specific labeling and cell-free synthesis. *J. Biomol. NMR*, 28, 235–247, 2004.

53. Venters, R.A., Farmer, B.T., Fierke, C.A., and Spicer, L.D. Characterizing the use of perdeuteration in NMR studies of large proteins: ^{13}C, ^{15}N and ^{1}H assignments of human carbonic anhydrase II. *J. Mol. Biol.*, 264, 1101–1116, 1996.

54. Rosen, M.K., Gardner, K.H., Willis, R.C., Parris, W.E., Pawson, T., and Kay, L.E. Selective methyl group protonation of perdeuterated proteins. *J. Mol. Biol.*, 263, 627–636, 1996.

55. Yamazaki, T., Otomo, T., Oda, N., Kyogoku, Y., Uegaki, K., Ito, N., Ishino, Y., and Nakamura, H. Segmental isotope labeling for protein NMR using peptide splicing. *J. Am. Chem. Soc.*, 120, 5591–5592, 1998.

56. Kainosho, M., Torizawa, T., Iwashita, Y., Terauchi, T., Mei Ono, A., and Guntert, P. Optimal isotope labeling for NMR protein structure determinations. *Nature* 440, 52–57, 2006.

57. Tian, C., Karra, M.D., Ellis, C.D., Jacob, J., Oxenoid, K., Sonnichsen, F., and Sanders, C.R. Membrane protein preparation for TROSY NMR screening. *Methods Enzymol.*, 394, 321–334, 2005.

58. Tian, C., Breyer, R.M., Kim, H.J., Karra, M.D., Friedman, D.B., Karpay, A., and Sanders, C.R. Solution NMR spectroscopy of the human vasopressin V2 receptor, a G protein-coupled receptor. *J. Am. Chem. Soc.*, 127, 8010–8011, 2005.

59. Obradovic, Z., Peng, K., Vucetic, S., Radivojac, P., Brown, C.J., and Dunker, A.K. Predicting intrinsic disorder from amino acid sequence. *Proteins*, 53 Suppl 6, 566–572, 2003.

60. Romero, P., Obradovic, Z., Li, X., Garner, E.C., Brown, C.J., and Dunker, A.K. Sequence complexity of disordered protein. *Proteins*, 42, 38–48, 2001.

61. Vucetic, S., Brown, C.J., Dunker, A.K., and Obradovic, Z. Flavors of protein disorder. *Proteins*, 52, 573–584, 2003.

62. Linding, R., Jensen, L.J., Diella, F., Bork, P., Gibson, T.J., and Russell, R.B. Protein disorder prediction: Implications for structural proteomics. *Structure* (Camb.), 11, 1453–1459, 2003.

63. Ward, J.J., Sodhi, J.S., McGuffin, L.J., Buxton, B.F., and Jones, D.T. Prediction and functional analysis of native disorder in proteins from the three kingdoms of life. *J. Mol. Biol.*, 337, 635–645, 2004.

64. Linding, R., Russell, R.B., Neduva, V., and Gibson, T.J. GlobPlot: Exploring protein sequences for globularity and disorder. *Nucleic Acids Res.*, 31, 3701–3708, 2003.

65. Dosztanyi, Z., Csizmok, V., Tompa, P., and Simon, I. The pairwise energy content estimated from amino acid composition discriminates between folded and intrinsically unstructured proteins. *J. Mol. Biol.*, 347, 827–839, 2005.

66. Coeytaux, K. and Poupon, A. Prediction of unfolded segments in a protein sequence based on amino acid composition. *Bioinformatics*, 21, 1891–1900, 2005.

67. Thomson, R., Hodgman, T.C., Yang, Z.R., and Doyle, A.K. Characterizing proteolytic cleavage site activity using bio-basis function neural networks. *Bioinformatics*, 19, 1741–1747, 2003.

68. Dyson, H.J. and Wright, P.E. Unfolded proteins and protein folding studied by NMR. *Chem Rev.*, 104, 3607–3622, 2004.

69. Wright, P.E. and Dyson, H.J. Intrinsically unstructured proteins: re-assessing the protein structure–function paradigm. *J. Mol. Biol.*, 293, 321–331, 1999.

70. Lytle, B.L., Volkman, B.F., Westler, W.M., and Wu, J.H. Secondary structure and calcium-induced folding of the *Clostridium thermocellum* dockerin domain determined by NMR spectroscopy. *Arch. Biochem. Biophys.*, 379, 237–244, 2000.

71. Spolar, R.S. and Record, M.T., Jr. Coupling of local folding to site-specific binding of proteins to DNA. *Science*, 263, 777–784, 1994.

72. Dyson, H.J. and Wright, P.E. Intrinsically unstructured proteins and their functions. *Nat. Rev. Mol. Cell Biol.*, 6, 197–208, 2005.

73. Singh, S., Tonelli, M., Tyler, R.C., Bahrami, A., Lee, M.S., and Markley, J.L. Three-dimensional structure of the AAH26994.1 protein from *Mus musculus*, a putative eukaryotic Urm 1. *Protein Sci.*, 14, 2095–2102, 2005.

74. Delaglio, F., Grzesiek, S., Vuister, G.W., Zhu, G., Pfeifer, J., and Bax, A. NMRPIPE— A multidimensional spectral processing system based on UNIX pipes. *J. Biomol. NMR*, 6, 277–293, 1995.

75. Johnson, B.A. and Blevins, R.A. NMR view—A computer program for the visualization and analysis of NMR data. *J. Biomol. NMR*, 4, 603–614, 1994.

18 Leveraging X-Ray Structural Information in Gene Family-Based Drug Discovery: Application to Protein Kinases

Marc Jacobs, Harmon Zuccola, Brian Hare, Alex Aronov, Al Pierce, and Guy Bemis

CONTENTS

18.1 INTRODUCTION

In structure-based drug design (SBDD), x-ray structures of small-molecule inhibitors bound to protein targets are used to guide the synthesis of increasingly potent and selective inhibitors in an iterative process. Traditionally, this iterative design method is applied to one target at a time. If the target of interest belongs to a large family of proteins with related sequences and structures, new opportunities and challenges arise. Lessons learned from one protein may be used to speed progress on others.

Specifically, similar structural features of related proteins may be exploited to under-stand and predict ligand binding across targets within the gene family, while subtle differences may be used to design compounds that are selective. Here, we present a gene-family approach for the design of new inhibitors of protein kinases, with emphasis on computational methods that maximize the ability to leverage structural data from one protein-inhibitor structure across related proteins in the family.

18.2 X-RAY CRYSTALLOGRAPHY—ADVANTAGES OF A GENE-FAMILY APPROACH

The steps necessary to produce the initial x-ray structure of a protein in a gene-family approach are similar to those in a single-target approach: design of the expression construct, cloning, protein expression and purification, crystallization, x-ray data collection, and structure determination. While high-throughput and parallel methods have improved the efficiency of all these steps, only *expression construct design* and *structure determination* become dramatically more efficient when working with sub-sequent targets in a gene family. For instance, high-throughput parallel cloning, protein expression, and protein purification methods have greatly reduced the cost and time required to produce protein suitable for crystallization [1–4]. Likewise, automation in crystallization screening has reduced the quantity of protein needed to produce diffraction quality crystals [5–7]. Such automation, however, is not unique to any given protein or protein family and does not generally rely on reuse of information. For example, proteins with similar tertiary structure do not necessarily crystallize under similar conditions. This observation is not surprising since the surface residues that form protein crystal lattice contacts are not as highly conserved as buried residues in the hydrophobic core of the protein. Efforts are under way, however, to better predict the outcome of crystallization experiments based upon prior data [8,9].

In contrast, the design of protein expression constructs is directly aided by prior data. Gene-family sequence homologies and existing crystal structures may be used to guide the identification of the catalytic domain and flanking sequences desired for crystallization trials. Typically, multiple expression constructs are designed, differing in their N- and C-termini with the goal of producing a protein that is folded and soluble and lacks disordered regions [10].

Beyond construct design, the process of crystallographic structure determination is also accelerated when structures of related proteins are known. This productivity enhancement results from the ability to employ molecular replacement methods routinely for phasing of crystallographic data. When protein crystal x-ray diffraction data are collected, only part of the information needed to solve the structure is observable. One part of the x-ray data, the phase of the x-rays, cannot be directly observed; instead, it must be inferred from additional experiments or calculated using prior knowledge of the protein structure. The phases can be determined experimen-tally by collecting x-ray data using crystals modified with heavy metals, halides, or incorporation of selenomethionine in place of methionine during protein expression.

However, if an approximate model of the protein structure exists, molecular replacement can be used [11]. Molecular replacement is a computational method that is significantly less cumbersome and consequently much faster than heavy atom methods, since it does not require chemical modification of crystals or additional data collection. The protein model is usually based upon the structure of a homologous protein such as another member of the same gene family. Here, rotation and translation searches are performed to determine whether a particular placement (orientation and position) of the model is consistent with the observed x-ray data. The success of molecular replacement depends upon the structural similarity between the protein in the crystal and the model.

When working within a gene family where a large number of related structures have been determined, many different starting models are available. Typically, the structure of the protein with the most similar sequence to the target of interest is evaluated first as a molecular replacement model. Also, many models can be evaluated in parallel or can be combined into a single model where a larger emphasis is placed upon regions of the protein with the highest degree of similarity. A number of computer programs that perform molecular replacement are available. One such program, PHASER, uses an ensemble of models and iterates through the models to determine the best solution [12]. As the number of structures in a gene family increases, the ability to use known structures to solve new structures is enhanced, greatly reducing the amount of time required for structure determination.

18.3 PROTEIN KINASES

18.3.1 KINASES AS THERAPEUTIC TARGETS

Protein phosphorylation is one of the major post-translational modifications required for regulation of cellular activities, and kinases that catalyze these reactions comprise a large family of enzymes in eukaryotic cells. Protein kinases mediate diverse cellular processes such as signal transduction, metabolism, cell-cycle control, apoptosis, and cytoskeletal rearrangement. Disruption of a kinase activity through mutation or deregulation may have profound consequences on the functioning of a cell, leading to a wide variety of disease states. Further, since kinases play a regulatory role in many pathways, modulation of kinase function presents opportunities for therapeutic intervention. It follows that significant resources in the pharmaceutical industry have been focused on the discovery and development of small-molecule kinase inhibitors.

The human genome encodes more than 500 protein kinases, which constitute 1.7% of all human genes [13]. It has been estimated that within this set of proteins, there are 130–300 therapeutically relevant targets [14,15]. Some of these are well validated as pharmacological targets for which small-molecule inhibitors have demonstrated clinical efficacy, such as p38 for rheumatoid arthritis and Abl (Abelson tyrosine kinase) in chronic myeloid leukemia [16]. Most kinases, however, are not so well established as therapeutic targets and instead may be implicated to be relevant for a particular disease through indirect evidence, such as regulation of a related signaling pathway.

18.3.2 COMMON STRUCTURAL FEATURES OF KINASES

The catalytic domains of protein kinases share a common fold: two lobes joined by a hinge region (fig. 18.1). The N-terminal lobe contains a twisted five-stranded

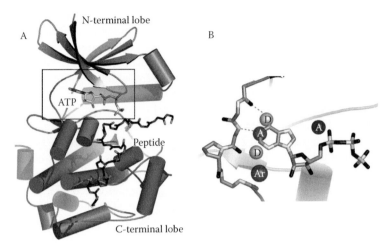

FIGURE 18.1 A: Overall structure of a protein kinase catalytic domain. The structure of c-AMP dependent kinase (PKA) bound to ATP and a peptide substrate is drawn with β-sheets as arrows and α-helices as cylinders (Protein Data Bank accession number: 1ATP). N- and C-terminal residues outside the limits of the kinase domain have been omitted for clarity. The ATP binding site is indicated by a box. Within this region, the hinge between the N- and C-terminal lobes is on the left. ATP and the main chain of the peptide substrate are drawn with bonds represented as sticks. The two β-strands connected by a short loop, directly above the ATP, comprise the glycine-rich loop. This figure and subsequent molecular graphics figures were prepared using Pymol. (DeLano, W.L. The Pymol molecular graphics system. DeLano Scientific, San Carlos, CA, 2002.) B: Kinase frequent-hitter pharmacophore. The ATP binding site is enlarged with the hinge drawn as sticks (left). Hydrogen bonds between the hinge and the ATP are shown as dotted lines. The frequent-hitter pharmacophore elements are shown as spheres, labeled A for hydrogen bond acceptor, D for hydrogen bond donor, and Ar for aromatic.

β-sheet and an α-helix (referred to as the C-helix). The C-terminal lobe is largely α-helical. The cleft between the two lobes forms the ATP binding site, which includes the hinge region, two β-strands in the N-terminal lobe called the glycine-rich loop, and an extended strand in the C-terminal domain referred to as the activation loop. The conformation of the activation loop varies widely among kinase structures. In some inactive kinase conformations, the activation loop blocks the active site and interferes with peptide and ATP substrate binding, while in active kinases this loop adopts a conformation in which the active site is accessible [17,18].

Since protein kinases share structural and sequence similarities in their active sites as well as a common catalytic mechanism, one of the largest challenges in kinase inhibitor design is achieving selectivity [19]. The main obstacle to selectivity prediction appears to be the highly flexible nature of the ATP binding site, which includes a wide range of side chain motions within the active site and the presence of flexible loops such as the activation loop and the glycine-rich loop, as well as domain motions [20]. Indeed, the extent of structural variation between different kinases is comparable to the variation found in different structures of a single kinase [21]. Only when the sequence identity is 60% or more can the binding site variations of related kinases be distinguished from variations observed for the binding sites of the same kinase, thus making structure-based kinase selectivity prediction extremely challenging [22].

Pursuing ligand potency for a particular kinase target often leads to the synthesis of compounds that exhibit promiscuous behavior across the gene family. Staurosporine is perhaps the archetypal example of a "kinase frequent hitter"—a compound that exhibits potent ATP-competitive inhibition across much of the kinase gene family. Elucidation of structural features characteristic of this class of compounds constitutes an alternative approach to kinase selectivity prediction: the prediction of ligand "unselectivity." Utilizing x-ray structural information for promiscuous kinase inhibitors, a five-point pharmacophore was proposed for kinase frequent hitters that includes two hydrogen bond donors, two hydrogen bond acceptors, and an aromatic ring feature (fig. 18.1) [23]. This pharmacophore is able to discriminate between frequent hitters and selective ligands and captures the molecular features that coincide with a propensity for nonselective inhibition of multiple kinase targets. Chemical space information about the kinase frequent hitters can then be related to the biological space (i.e., contacts with conserved residues within the active site that predispose a kinase ligand to promiscuous behavior). Knowledge of the kinase frequent-hitter pharmacophore should enable medicinal chemists to make more informed decisions in the context of predicting cross-kinase reactivity.

18.4 APPLICATION OF STRUCTURAL INFORMATION—MOLECULAR MODELING

In any SBDD project, it is obviously inefficient and unnecessary to determine an x-ray structure for every synthesized compound bound to the protein target. Rather, molecular modeling methods are used to analyze one structure to draw conclusions for similar compounds and to guide decisions for the synthesis of subsequent molecules. The information from a particular structure can be used more efficiently if it can be applied broadly to understand more varied inhibitors or related protein targets. When working simultaneously on multiple targets within a gene family, there are opportunities to reuse information from one protein–ligand structure and apply it to the design of inhibitors for another protein. The extent to which protein–ligand structures can be used to predict the binding mode of similar ligands depends upon how often similar ligands bind in similar ways. For kinase inhibitors, one ligand structure can often be used reliably to predict the conformation of related ligands (see following discussion) [24,25]. Further, compounds can be selected for crystallographic studies when their experimentally measured activity deviates from predictions ("outliers" in terms of their inhibition potency), thus increasing the likelihood of obtaining novel structural information.

Reuse of structural information in an SBDD project can be applied in two ways: "target hopping" and "scaffold morphing" (fig. 18.2). We define target hopping as the use of information from one protein–ligand structure to modify a compound, creating or enhancing inhibition of a related protein. Target hopping often involves changes to the periphery of the ligand and is typically used when a new target is pursued. This is conceptually similar to the concept of "latent hits"—inactive

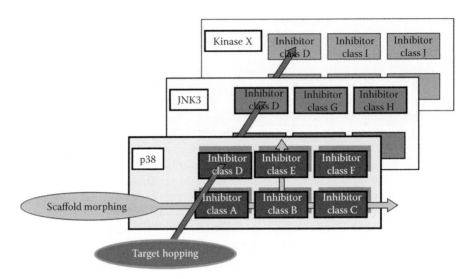

FIGURE 18.2 Within a gene family, structural data can be used to model and design new ligands for the same target (scaffold morphing) and to understand similar ligands in related protein targets (target hopping).

compounds that could potentially become inhibitors after simple chemical modifications are applied [26]. Scaffold morphing is a process of molecular recombination by which several protein–ligand structures are used to generate a novel inhibitor for the same target. Scaffold morphing usually utilizes changes to the core structure of the ligand and is used when new scaffolds with similar binding but different physicochemical properties are desired.

Modeling the conformation of inhibitors in protein complexes is typically performed using computational docking methods [27,28]. The three-dimensional structure of a protein–ligand complex is predicted by optimizing the ligand in the protein active site using all of the conformational, translational, and rotational degrees of freedom available to the ligand. These methods use fairly general criteria to select the most likely model from all of the possible orientations. However, if the orientation of a similar inhibitor is already known from an x-ray structure, that knowledge can be applied to the ligand of interest instead of evaluating all possibilities. In the three ligand modeling methods described next, existing structural data are directly utilized to predict new ligand conformations and to design novel inhibitors rapidly. CORES (complexes restricted by experimental structures) is a method used to model compounds accurately based upon similar x-ray structures [24]. The program BREED is a procedure by which three-dimensional fragments from existing inhibitor structures are combined to create novel compounds [29]. The program COREGEN builds upon this idea by altering the scaffold to create and predict new molecules [30].

18.4.1 CORES

To determine whether one protein–ligand structure could be used directly as a template for similar ligands, published and proprietary structures were analyzed to

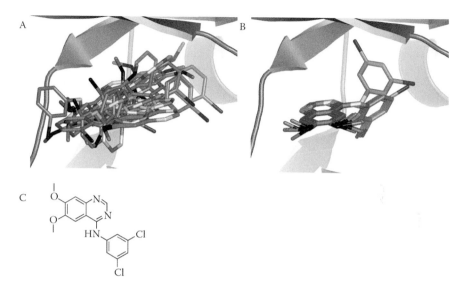

FIGURE 18.3 Comparison of the conformations sampled by traditional docking and CORES. The inhibitor SU-5271 was modeled in the active site of EGFR (PDB code: 1M17). The resulting predicted ligand structures obtained with molecular docking (GLIDE 2.5, Schrodinger, Inc., NY) are shown in panel A. Molecular modeling of the same compound using CORES is shown in panel B.

quantitate the frequency with which related ligands bind in the same orientation in the kinase active site. For 74 different inhibitor scaffolds, we have determined multiple protein–ligand x-ray structures per scaffold. For some of these scaffolds, ligand structures were determined in complex with more than one protein kinase. We found that most compounds containing a common scaffold bind to protein kinases in the same orientation. In fact, greater than 75% of the scaffolds are always observed to bind in a single orientation. Larger scaffolds bind in a single orientation more frequently than smaller scaffolds. For instance, an analysis of public protein kinase x-ray structures revealed that more than 80% of scaffolds containing three or more rings bind in a single orientation [24]. In contrast, only 20% of scaffolds containing fewer than three rings bind in a single orientation.

The propensity of similar ligands to bind in similar ways enables the use of three-dimensional structures of protein–ligand complexes for model building related complexes using a procedure called CORES (complexes restricted by experimental structures). In the CORES method, ligands are first reduced to frameworks that are used as the modeling templates. Ligand frameworks are a representation of ligand atoms consisting only of rings and the linkers between the rings; side chains are ignored [31]. The ligand templates are then used to build models that constrain the ligand conformation in the active site of the target protein (fig. 18.3).

The CORES method is accurate and widely applicable. The accuracy of the method was shown in a test case where 15 out of 19 CDK2-ligand complexes from the Protein Data Bank built using this method deviated from the x-ray structure by less than 2 Å root mean square (RMS) [24]. The method can be used to build models for over 70% of small-molecule protein kinase inhibitors published in the *Journal of Medicinal Chemistry* since 1993 using templates extracted from kinase structures in the Protein Data Bank. Templates containing at least three rings are available in the protein databank for 87% of the small-molecule inhibitors in the *Journal of Medicinal Chemistry* database that can be modeled using the CORES method. Also, the CORES method can be used to model about 80% of kinase inhibitors in the Vertex corporate database.

To reuse structural information in gene families other than protein kinases, we may be able to use natural ligands for predicting the size of molecular templates that will likely adopt unique binding orientations in a protein binding pocket. For instance, molecules containing the ATP framework (e.g., ATP analogs and adenosine) bind in the same orientation in kinase complex structures. Endogenous cofactors and substrates may have to bind in a single orientation since nonproductive orientations could inhibit biological pathways.

18.4.2 BREED

As advances in technology have facilitated the more routine determination of protein–ligand crystal structures, the amount of structural information used for ligand design has grown rapidly. Unfortunately, as this information has become more abundant, it has become increasingly difficult to utilize fully. While it may be relatively easy to draw conclusions from examination of a single protein–ligand structure, extracting all relevant data from a pair of structures is more difficult. With larger numbers of structures, simultaneous comparison becomes virtually impossible. However, in a gene-family SBDD approach, many protein–ligand complexes may be available for analysis for a particular project. For instance, approximately 300 ligand-bound structures of protein kinases are publicly available [32]. There is a clear need for methods to handle this quantity of information efficiently, extracting as much guidance as possible for the design of novel ligands.

One simple method for taking advantage of structural information is to combine fragments from two known ligands to generate a novel inhibitor. This is a common medicinal chemistry strategy where the hope is that this new ligand will be structurally complementary to the ligand binding site of the enzyme so that binding affinity can be maintained. While recombining fragments of ligands can be based upon two-dimensional chemical structures, presumably the results will be improved when ligands are compared in three dimensions in the context of the protein active site. In this manner, the relative positions of substituents on the two initial ligands can be accurately compared to determine which substituents can be transferred among scaffolds while still maintaining a good fit within the active site. Typically, recombining fragments in this manner requires that each pair of ligands be manually inspected for all possible recombinations—a process that becomes difficult and error prone as the number of input ligand structures increases.

BREED was developed to automate this process, comparing three-dimensionally superimposed structures and making all structurally appropriate recombinations among an arbitrarily large collection of starting ligands [29]. The bond-matching and fragment-swapping algorithm used by BREED is similar to one developed earlier by Ho and Marshall [33] and provides additional enhancements in order to maximize the creation of novel structures. Because the fragment recombination is automated and the BREED-produced ligands are still in the protein active site reference frame, the process can be applied recursively. The "offspring" of the original ligands can be added into the pool of input compounds for further recombination into a third generation of ligands. In this manner, a small number of input structures can be processed, combining scaffold and side chain elements from multiple lead compounds, to generate a very large set of novel inhibitors, many of which bear little resemblance to any of the starting ligands.

In this respect, the method has some capabilities of *de novo* design techniques, though obviously it can only be applied in cases where several structurally characterized ligands exist. Though this is a disadvantage, the method is applicable in cases where initial lead classes prove unusable or where a second-generation compound is sought. The value to such projects more likely will apply to problems other than binding affinity (e.g., insolubility, cell permeability, metabolism, toxicity, etc.) that can be addressed by modifying the ligand. Gene family-based projects should also be well served by BREED, where structures from prior targets can be used to design new inhibitors. Further, methods such as BREED may be applied to ligand design where structures are available from proteins sharing topologically similar active sites that do not have similar sequences [34].

BREED has been applied to a single target (HIV protease) and to a family of targets (protein kinases). The HIV protease results demonstrated that novel scaffolds could be generated when there are a large number of high-quality x-ray structures available from the Protein Data Bank [32]. The kinase gene-family example showed that by branching out into closely related enzyme structures, this method is still able to produce structurally relevant ligands. The output from both studies produced several useful categories of molecules: entirely novel scaffolds, known scaffolds with novel substituents, and known scaffolds that were not included in the BREED input set. These results confirmed that BREED is able to use the wealth of structural information available to contemporary medicinal chemists to design compounds of high relevance to a structure-based project automatically.

18.4.3   COREGEN

Most pharmaceutically interesting ligands can be represented in terms of the ring-linker frameworks that comprise them [31]. Recent analysis of 119 published kinase inhibitors from at least 18 different targets illustrated that a basis set of four rings and eight linkers is sufficient to describe about 90% of ring and linker occurrences, respectively. A similar result was derived from a larger set of approximately 40,000 kinase inhibitors from curated patents [30]. Tools that combine elements of *de novo* molecular design with experimental structural information are becoming increasingly important in processing the avalanche of structural data.

COREGEN is a novel method for stepwise ring linker-based scaffold assembly, which integrates fragment-by-fragment ligand design approaches with high-throughput virtual library screening [30]. It utilizes experimental structural information for weak ligands and the framework-based fragment libraries to guide the optimization process from the small fragments to drug-like molecules. The goal of COREGEN is not to generate completely new scaffolds *de novo*, but rather to derive novel scaffold classes using the experimental information about the orientation of known anchor fragments within the site of interest. By starting from a known binding mode of an anchor fragment, speed and accuracy of COREGEN benefit from the constrained docking approach to searching over the available chemical space. The utility of this approach was demonstrated on a set of reported inhibitors of the kinases Bcr-Abl, CDK2, and Src [30]. COREGEN reproduced predominant structural features of a number of known kinase inhibitors for these targets, as well as proposed a number of novel ligand structures. It is expected that design such as this will benefit from a gene family-specific approach to fragment selection, making scaffold libraries built from kinase inhibitor fragments more kinase inhibitor-like and libraries assembled from protease inhibitor fragments more protease inhibitor-like, etc.

18.4.4 PROTEIN HOMOLOGY MODELING

In the absence of an x-ray structure for a particular protein, structures of related proteins can be used to make a homology model. The kinase catalytic domains are suitable subjects for comparative modeling since they all adopt similar folds. Comparative modeling uses a three-dimensional structure of a protein with the same fold as a modeling template. The backbone conformation of the template structure is preserved, and side chains of the target protein are built onto the template. Side chain rotamers are selected that provide good packing and minimal clashes with other side chains and with the backbone. Typically, the model is refined using molecular dynamics to minimize bad contacts.

The abundance of protein kinase x-ray structures creates the opportunity to select among multiple modeling templates. A typical approach is to choose as a template the structure with highest sequence similarity to the target sequence. The distribution of similarities between all human protein kinase sequences and the most closely related protein kinase with an x-ray structure is shown in figure 18.4. For most kinases, however, the most closely related structure has less than 50% sequence identity, and frequently multiple structures with nearly the same sequence similarity are available. In these cases, using ligand binding data may aid the selection of a suitable modeling template.

Automated approaches for incorporating ligand binding data into homology modeling have been described previously [35]. Rather than using automated approaches, more often ligand information is incorporated manually by identifying compounds that bind the target kinase and utilize pockets unique to particular conformations. For example, shifts in the C-helix [36] and activation loop [37] expose unique spaces that are not present in active kinase conformations. Knowledge that a ligand binds to a target kinase and utilizes a unique space in the active site implies that the kinase can adopt a conformation that makes the unique space accessible. A structure with the appropriate conformation can then be selected as a modeling template.

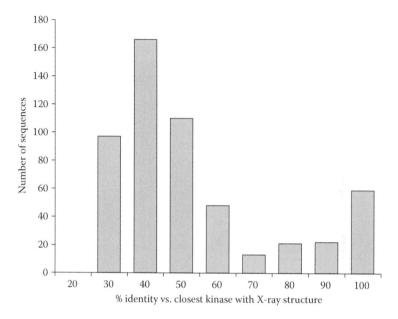

FIGURE 18.4 The distribution of amino acid sequence identity between each human kinase domain and the most closely related kinase domain with an x-ray structure. The sequence of each human kinase domain was compared to the sequences of the subset of human kinase domains with x-ray structures using BLAST. (Altschul, S.F. et al., *J Molecular Biol* 215(3), 403–410, 1990.) The most closely related sequence was identified as the highest scoring BLAST hit, and the percent sequence identity was calculated using the BLAST alignment. The right-most bar in the histogram includes kinase sequences with a known x-ray structure.

18.5 DESIGNING LIGAND SELECTIVITY

To apply and reuse protein and ligand structural data from one target to another, features common to gene family members can be exploited to make predictions applicable to multiple members of the family. However, to design a selective inhibitor, sequence and structural differences among the gene family members must be utilized. Traditionally, the design of kinase inhibitors has targeted the ATP binding site of the kinase. Such inhibitors often contain an anchor element that binds to the kinase via hydrogen bonds to the main chain atoms of the hinge region. While the ATP binding sites of protein kinases are generally conserved, there are several ways to design potent and selective kinase inhibitors. To make a selective inhibitor, one can utilize contacts with uncommon residues in the ATP site or contact residues outside the ATP binding site. Also, selectivity can be achieved when a ligand is made that binds to a protein conformation that only a small subset of kinases can adopt (depicted schematically in fig. 18.5).

Nonconserved residues and uncommon features in the ATP binding site have been targeted to design selective inhibitors. For example, nonconserved residues in the ATP binding site have recently been exploited in the design of inhibitors of p90 ribosomal protein S6 kinase (RSK) [38]. A structural bioinformatics approach was

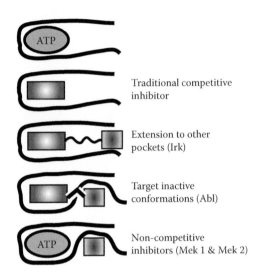

FIGURE 18.5 Schematic of strategies for selective kinase inhibitor design. The kinase ATP side is depicted as a curved line. ATP is shown as an oval while inhibitors are represented as rectangles.

taken to identify two "selectivity filters." One filter was a threonine residue located at the entrance to a lipophilic pocket adjacent to the ATP binding pocket. Typically, a residue with a larger side chain occupies this position [39]. The smaller amino acid side chain, as the gatekeeper to this hydrophobic pocket, allows the placement of bulky ligand side chains. The second selectivity filter was that of a rarely occurring cysteine residue in the glycine-rich loop of the kinase. Using the two selectivity filters reduced the number of kinases with these features from 491 to 3, all members of the RSK family of kinases. Compounds were made that would have an electrophilic group designed to interact with the cysteine side chain and a hydrophobic group to occupy the lipophilic pocket. The resulting inhibitor had an IC_{50} of 15 nM and appeared to be selective for RSK1 and RSK2.

The idea of using nonconserved residues and uncommon structural features was also exploited in the design of inhibitors of p38 MAP kinase [40]. The quinazolinone and pyridol-pyrimidine classes of p38 MAP kinase inhibitors have a high degree of selectivity for p38 over other MAP kinases. These compounds bind to a p38 protein conformation that has undergone a main chain peptide flip between two residues in the hinge, M109 and G110 (fig. 18.6). In other MAP kinases, including the δ isoform of p38, bulkier side chains occupying the position corresponding to G110 prevent the peptide backbone from adopting a similar conformation. The local conformational change provides an avenue to pursue selectivity using an additional hydrogen bond donor not present in other kinases.

Another selectivity design strategy is to contact amino acid positions distal from the ATP binding site. It is expected that the amino acid sequence becomes more diverse once residues not required for catalysis are examined. This is exemplified in the structure of a bisubstrate inhibitor bound to insulin receptor kinase (Irk) (fig. 18.7) [41]. The inhibitor was synthesized by linking ATPγS with a peptide

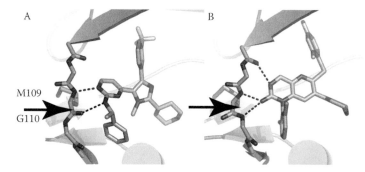

FIGURE 18.6 p38 ligand-induced hinge conformations. In panels A and B, the hinge region is shown on the left as sticks. Hydrogen bonds between the hinge and the ligand are shown as dotted lines. The main chain conformational change is highlighted with an arrow. The structure of p38 bound to a pyridinylimidazole inhibitor (PDB code: 1OUK) is shown in panel A, and a dihydropyrido-pyrimidine inhibitor (PDB code: 1OUY) is shown in panel B.

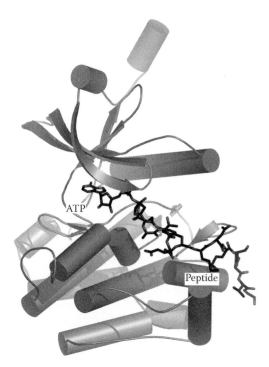

FIGURE 18.7 Insulin receptor kinase (Irk) bound to a bisubstrate inhibitor. The structure of Irk bound to a bisubstrate inhibitor (PDB code: 1GAG) is drawn in an orientation similar to that of figure 18.1. The inhibitor is drawn as sticks and the ATP and peptide binding sites are labeled.

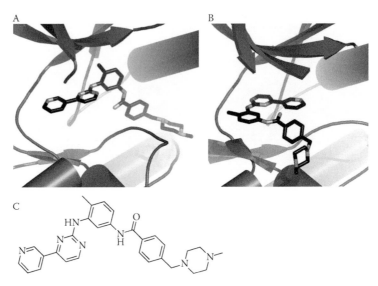

FIGURE 18.8 Gleevec binding orientations. The ATP binding site of the Abl/Gleevec complex is shown in panel A (PDB code: 1IEP). Gleevec contacts the hinge (left) and binds between the C-helix (upper right) and the activation loop (lower right). Gleevec bound to Syk is shown in panel B (PDB code: 1XBB). The chemical structure of Gleevec is drawn in panel C.

substrate analog via a 2-carbon spacer. The design of this inhibitor was aided by the high-resolution structure of the enzyme bound to an ATP analog and a peptide substrate [42]. The resulting compound binds the protein more tightly than either of the two substrates; the binding energy of the bisubstrate inhibitor is 8.9 kcal mol^{-1}, while the binding energies for the peptide and for ATPγS are 4.1 and 5.1 kcal mol^{-1}, respectively. Although the low bioavailability of peptides limits the use of such inhibitors in the clinic, it provides a template for the design of peptidomimetic compounds, a method used successfully in the design of protease inhibitors [43].

While active kinase conformations share a common structure since they must all catalyze the same reaction, a large variety of inactive conformations has been observed. Often the activation loop adopts a conformation that blocks the peptide and ATP binding sites. One inhibitor, Gleevec, appears to gain selectivity by binding to an inactive kinase conformation. Gleevec is a small-molecule inhibitor of a number of kinases including the Abl kinase. It is a potent inhibitor of the unphosphorylated form of the enzyme, but binds rather poorly to the active, phosphorylated form of Abl [37,44,45]. The crystal structure of the Abl kinase domain in complex with Gleevec revealed that the pyrimidine and the pyridine rings of the drug overlap with the ATP binding site (fig. 18.8). However, the rest of the molecule is wedged between the activation loop and the C-helix, locking the kinase in an inactive conformation [44].

Recently the structure of Gleevec bound to the activated form of spleen tyrosine kinase, Syk, has been determined [46]. In this structure, Gleevec binds in the active site adopting a *cis-* rather than *trans*-conformation (as observed in Abl) and occupies space similar to traditional ATP site inhibitors (fig. 18.8) [44,47]. The molecule is a poor inhibitor of this form of Syk, which has a structure similar to that of activated Abl.

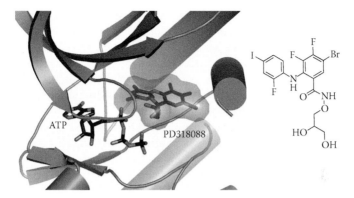

FIGURE 18.9 Mek1/ATP/PD318088 complex. The structure of Mek1 bound to ATP is drawn as sticks, and the noncompetitive inhibitor PD318088 is shown as sticks highlighted with a surface (PDB code: 1S9J). The chemical structure of PD318088 is shown on the right.

Finally, there has been increased interest in the discovery and design of non-competitive ATP inhibitors. PD184352 was identified as an inhibitor of the kinases MEK1 and MEK2. This ligand is noncompetitive for ATP and the protein substrate MAP kinase [48]. The resulting crystal structure of MEK1 bound to ATP and PD318008 (an analog of PD184352) provides a starting point for the design of future MEK inhibitors. These structures revealed a unique binding pocket adjacent to the ATP binding site (fig. 18.9). The binding of the ligand induces a conformational change in the protein that locks the protein into a closed, catalytically inactive state, with ATP bound in the active site.

18.6 FUTURE PROSPECTS AND CONCLUSIONS

While the overall structure of protein kinases is generally conserved among members of the family, structural variations exist that allow us to design selective inhibitors. Several selective ligands have been shown to distinguish between different conformations of the same kinase. This kind of selectivity seems to have often been discovered by chance, most likely by pursuing novel experimental observations inconsistent with structure-based predictions. By extension of this idea, one could attempt to systematically force conformational changes where none have been previously observed (reviewed by Teague [49]).

Many variations on this theme are possible. For kinases, one could start with a tight binding ligand and systematically alter positions on the periphery of the inhibitor with the aim of disrupting the protein structure without abolishing inhibitor binding. Optimization of "hits" from this approach would then follow traditional medicinal synthetic chemistry approaches. Alternatively, one could take the "slinky" approach, where a readily derivatizable side chain is attached to the inhibitor in a position structurally consistent with ligand binding. A large variety of substituents could then be attached with the hope of drastically enhancing binding. A variation of this approach is the *in situ* "click" chemistry of Sharpless and coworkers used to prepare acetylcholinesterase inhibitors [50,51]. In these experiments, two libraries of compounds with complementary reactive functional groups are combined with the protein target. When two compounds bind in the protein active site simultaneously, they can react with one another to form a single molecule. These reaction products can then be identified by mass spectroscopy. Here, the protein active site serves as a template, selecting compounds from the two libraries for chemical addition. Once tight binding to the target has been established, the large and unwieldy inhibitors produced by this method could be trimmed back. Ultimately, novel inhibitors with divergent binding characteristics could be produced by this approach.

In SBDD, protein–ligand complex structures of one ligand are used in the design of subsequent compounds. While many improvements in efficiency have come from automation and parallel approaches and the rate at which data can be generated has increased, the nature of the drug discovery process has not changed. In a gene family-based approach, however, structural data from one target can be used to predict ligand binding orientations for other related targets, and the structure of one protein can be used to model the active site of other proteins. Further, inhibitors from multiple protein–ligand structures can be recombined to form novel molecules. Thus, when working on targets within a gene family, the application rather than the generation of structural data is transformed.

ACKNOWLEDGMENTS

We thank Jonathan Moore for helpful discussions and critical comments on this manuscript. Portions of this work were conducted within the protein kinase collaboration between Vertex Pharmaceuticals and Novartis Pharma AG.

REFERENCES

1. Chambers, S.P. et al., High-throughput screening for soluble recombinant expressed kinases in *Escherichia coli* and insect cells, *Protein Expression Purification* 36(1), 40–47, 2004.
2. Dieckman, L. et al., High throughput methods for gene cloning and expression, *Protein Expression Purification* 25(1), 1–7, 2002.
3. Ding, H.T. et al., Parallel cloning, expression, purification and crystallization of human proteins for structural genomics, *Acta Crystallogr D Biol Crystallogr* 58(Pt 12), 2102–2108, 2002.
4. Park, J. et al., Building a human kinase gene repository: Bioinformatics, molecular cloning, and functional validation, *Proc Natl Acad Sci USA* 102(23), 8114–8119, 2005.
5. Stewart, L., Clark, R., and Behnke, C., High-throughput crystallization and structure determination in drug discovery, *Drug Discovery Today* 7(3), 187–196, 2002.
6. Stevens, R.C., Design of high-throughput methods of protein production for structural biology, *Struct Fold Design* 8(9), R177–185, 2000.
7. Stevens, R.C., High-throughput protein crystallization, *Curr Opin Struct Biol* 10(5), 558–563, 2000.
8. Jurisica, I. et al., Intelligent decision support for protein crystal growth, *IBM Syst J* 40(2), 394–409, 2001.

9. Luft, J.R. et al., A deliberate approach to screening for initial crystallization conditions of biological macromolecules, *J Struct Biol* 142(1), 170–179, 2003.

10. Dale, G.E., Oefner, C., and D'Arcy, A., The protein as a variable in protein crystallization, *J Struct Biol* 142(1), 88–97, 2003.

11. Rossmann, M.G., Molecular replacement—Historical background, *Acta Crystallogr D Biol Crystallogr* 57(Pt 10), 1360–1366, 2001.

12. McCoy, A.J. et al., Likelihood-enhanced fast translation functions, *Acta Crystallogr D Biol Crystallogr* 61(Pt 4), 458–464, 2005.

13. Manning, G. et al., The protein kinase complement of the human genome, *Science* 298(5600), 1912–1934, 2002.

14. Hopkins, A.L. and Groom, C.R., The druggable genome, *Nat Rev Drug Discovery* 1(9), 727–730, 2002.

15. ter Haar, E. et al., Kinase chemogenomics: Targeting the human kinome for target validation and drug discovery, *Mini Rev Med Chem* 4(3), 235–253, 2004.

16. Cohen, P., Protein kinases—The major drug targets of the twenty-first century? *Nat Rev Drug Discovery* 1(4), 309–315, 2002.

17. Huse, M. and Kuriyan, J., The conformational plasticity of protein kinases, *Cell* 109(3), 275–282, 2002.

18. Nolen, B., Taylor, S., and Ghosh, G., Regulation of protein kinases; controlling activity through activation segment conformation, *Mol Cell* 15(5), 661–675, 2004.

19. Dancey, J. and Sausville, E.A., Issues and progress with protein kinase inhibitors for cancer treatment, *Nat Rev Drug Discovery* 2(4), 296–313, 2003.

20. Diller, D.J. and Li, R., Kinases, homology models, and high throughput docking, *J Med Chem* 46(22), 4638–4647, 2003.

21. Lowrie, J.F. et al., The different strategies for designing GPCR and kinase targeted libraries, *Comb Chem High Throughput Screen* 7(5), 495–510, 2004.

22. Vieth, M. et al., Kinomics: Characterizing the therapeutically validated kinase space, *Drug Discovery Today* 10(12), 839–846, 2005.

23. Aronov, A.M. and Murcko, M.A., Toward a pharmacophore for kinase frequent hitters, *J Med Chem* 47(23), 5616–5619, 2004.

24. Hare, B.J. et al., CORES: An automated method for generating three-dimensional models of protein/ligand complexes, *J Med Chem* 47(19), 4731–4740, 2004.

25. Rockey, W.M. and Elcock, A.H., Rapid computational identification of the targets of protein kinase inhibitors, *J Med Chem* 48(12), 4138–4152, 2005.

26. Mestres, J. and Veeneman, G.H., Identification of "latent hits" in compound screening collections, *J Med Chem* 46(16), 3441–3444, 2003.

27. Krumrine, J. et al., Principles and methods of docking and ligand design, *Methods Biochem Anal* 44, 443–476, 2003.

28. Brooijmans, N. and Kuntz, I.D., Molecular recognition and docking algorithms, *Annu Rev Biophys Biomol Struct* 32, 335–373, 2003.

29. Pierce, A.C., Rao, G., and Bemis, G.W., BREED: Generating novel inhibitors through hybridization of known ligands. Application to CDK2, p38, and HIV protease, *J Med Chem* 47(11), 2768–2775, 2004.

30. Aronov, A.M. and Bemis, G.W., A minimalist approach to fragment-based ligand design using common rings and linkers: Application to kinase inhibitors, *Proteins* 57(1), 36–50, 2004.

31. Bemis, G.W. and Murcko, M.A., The properties of known drugs. 1. Molecular frameworks, *J Med Chem* 39(15), 2887–2893, 1996.

32. Berman, H.M. et al., The Protein Data Bank, *Nucleic Acids Res* 28(1), 235–242, 2000.

33. Ho, C.M.W. and Marshall, G.R., Splice: A program to assemble partial query solutions from three-dimensional database searches into novel ligands, *J Computer-Aided Molecular Design* 323, 387–406, 1993.
34. Schmitt, S., Kuhn, D., and Klebe, G., A new method to detect related function among proteins independent of sequence and fold homology, *J Molecular Biol* 323(2), 387–406, 2002.
35. Evers, A., Gohlke, H., and Klebe, G., Ligand-supported homology modeling of protein binding-sites using knowledge-based potentials, *J Molecular Biol* 334(2), 327–45, 2003.
36. Wood, E.R. et al., A unique structure for epidermal growth factor receptor bound to GW572016 (Lapatinib): Relationships among protein conformation, inhibitor off-rate, and receptor activity in tumor cells, *Cancer Res* 64(18), 6652–6659, 2004.
37. Nagar, B. et al., Crystal structures of the kinase domain of c-Abl in complex with the small molecule inhibitors PD173955 and Imatinib (STI-571), *Cancer Res* 62(15), 4236–4243, 2002.
38. Cohen, M.S. et al., Structural bioinformatics-based design of selective, irreversible kinase inhibitors, *Science* 308(5726), 1318–1321, 2005.
39. Liu, Y. et al., Structural basis for selective inhibition of Src family kinases by pp1, *Chem Biol* 6(9), 671–678, 1999.
40. Fitzgerald, C.E. et al., Structural basis for p38alpha map kinase quinazolinone and pyridol–pyrimidine inhibitor specificity, *Nat Struct Biol* 10(9), 764–769, 2003.
41. Parang, K. et al., Mechanism-based design of a protein kinase inhibitor, *Nat Struct Biol* 8(1), 37–41, 2001.
42. Hubbard, S.R., Crystal structure of the activated insulin receptor tyrosine kinase in complex with peptide substrate and ATP analog, *Embo J* 16(18), 5572–5581, 1997.
43. Wlodawer, A. and Vondrasek, J., Inhibitors of HIV-1 protease: A major success of structure-assisted drug design, *Annu Rev Biophys Biomol Struct* 27, 249–284, 1998.
44. Schindler, T. et al., Structural mechanism for STI-571 inhibition of Abelson tyrosine kinase, *Science* 289(5486), 1938–1942, 2000.
45. Zimmermann, J. et al., Potent and selective inhibitors of the Abl-kinase: Phenyl-lamino–pyrimidine (PAP) derivatives, *Bioorganic Medicinal Chem Lett* 7(2), 187–192, 1997.
46. Atwell, S. et al., A novel mode of gleevec binding is revealed by the structure of spleen tyrosine kinase, *J Biol Chem* 279(53), 55827–55832, 2004.
47. Mol, C.D. et al., Structural basis for the autoinhibition and STI-571 inhibition of c-Kit tyrosine kinase, *J Biol Chem*, 2004.
48. Ohren, J.F. et al., Structures of human map kinase kinase 1 (MEK1) and MEK2 describe novel noncompetitive kinase inhibition, *Nat Struct Molecular Biol* 11(12), 1192–1197, 2004.
49. Teague, S.J., Implications of protein flexibility for drug discovery, *Nat Rev Drug Discovery* 2(7), 527–541, 2003.
50. Krier, M. et al., Design of small-sized libraries by combinatorial assembly of linkers and functional groups to a given scaffold: Application to the structure-based optimization of a phosphodiesterase 4 inhibitor, *J Med Chem* 48(11), 3816–3822, 2005.
51. Krasinski, A. et al., *In situ* selection of lead compounds by click chemistry: Target-guided optimization of acetylcholinesterase inhibitors, *J Am Chem Soc* 127(18), 6686–6692, 2005.
52. DeLano, W.L., The Pymol molecular graphics system, DeLano Scientific, San Carlos, CA, 2002.
53. Altschul, S.F. et al., Basic local alignment search tool, *J Molecular Biol* 215(3), 403–410, 1990.

19 EPR Spectroscopy in Genome-Wide Expression Studies

Richard Cammack

CONTENTS

19.1 INTRODUCTION

19.1.1 THE SCOPE OF EPR SPECTROSCOPY

Electron paramagnetic resonance (EPR), also known as electron spin resonance or electronic magnetic resonance spectroscopy, is a means of investigating paramagnetic molecules (i.e., those containing unpaired electrons) [1]. The majority of molecules are diamagnetic, having all their electrons paired in atomic or molecular orbitals. As a result, EPR is a selective technique capable of yielding specific molecular information. EPR spectroscopy has many different applications in biochemistry and molecular biology. These include the observation of transient free radicals by spin trapping and the observation of molecular motion and environment by spin labeling—methods that involve the introduction of a paramagnet to the biological system. However, this chapter will focus principally on the application of EPR spectroscopy to observe naturally occurring stable paramagnetic complexes of transition metals, such as iron, copper, molybdenum, or nickel. Examples of these are found in the active sites of metalloenzymes and metalloproteins [2–5].

EPR spectroscopy is a quantitative spectroscopic technique. If measured under the appropriate conditions, the integrated EPR signal of a species with spins of $S = 1/2$ is proportional to concentration, independent of the type of paramagnet. Unlike optical spectroscopy, quantitative measurements do not require prior knowledge of a proportionality constant or extinction coefficient. Thus, EPR can be used to estimate the concentration of spins, even for unidentified species [6].

The sample may be a homogeneous solution or a cell suspension and may include membranes and precipitates such as inclusion bodies. Therefore, EPR may be used to monitor the purification of paramagnetic molecules. The method is nondestructive, so materials can be recovered after measurement and used for other purposes.

EPR is not just a method for measuring the metal content of protein preparations; it can provide detailed information about the metal's state of coordination. There are a considerable number of conserved metal-binding motifs [7], and each tends to produce a characteristic EPR signature.

19.1.2 PRINCIPLES OF EPR SPECTROSCOPY

Electron paramagnetic resonance is the resonant absorption of microwave radiation of frequency ν by a paramagnetic material in an applied magnetic field, B_0 (fig. 19.1a). The condition for resonance is given by

$$h\nu = g\mu_B B_0 \tag{19.1}$$

where h is Planck's constant and μ_B is the Bohr magneton, two physical constants; g is a parameter known as the g-factor or g-value, which is a characteristic of the paramagnetic species. The g-factor for the unpaired electron is approximately 2.0023, but in transition ions this can be very different.

For instrumental reasons, the EPR spectrum is recorded by following the absorption of microwave energy as a function of magnetic field B_0, as illustrated in figure 19.1b. The g-factor is inversely proportional to the magnetic field where resonance is observed. A characteristic feature of the conventional EPR spectrum is that it appears as the first derivative of microwave absorption with respect to B_0 (dP_{abs}/dB_0). This is because it is acquired by modulation of B_0 and phase-sensitive detection. The conventional absorption spectrum may be obtained by integration of the spectrum, but this is usually only done when quantifying the spin concentration.

The strength of the EPR signal depends on the population difference between the $m_S = +1/2$ and $m_S = -1/2$ electron energy levels, which is determined by the Boltzmann distribution,

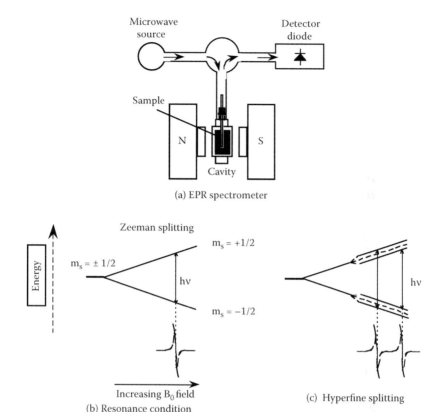

(a) EPR spectrometer

(b) Resonance condition

(c) Hyperfine splitting

FIGURE 19.1 Principles of EPR spectroscopy. (Cammack, R. In *Encyclopedia of Spectroscopy and Spectrometry*, ed. Tranter, G. and Holmes, J. London: Academic Press, 1999, 457–470.)

$$\frac{N_+}{N_-} = e^{-\frac{g\mu_B B_0}{kT}}$$

(19.2)

and maintained by electron spin-lattice relaxation. The signal decreases if the populations become equalized, which tends to happen if too much microwave power is applied, a phenomenon known as *power saturation*.

EPR spectroscopy is analogous to the more familiar nuclear magnetic resonance (NMR) method used in chemistry of organic compounds. However, there are some significant differences from NMR. Unpaired electrons are generally much more rare than nuclear spins, so fewer species are present in the spectrum. Also, the magnetic moment of the electron is much greater than that of the proton, so the spectra are more intense. Moreover, EPR spectra of proteins show considerable broadening due to anisotropy of the hyperfine parameters.

The electron-spin relaxation of unpaired electrons is generally more rapid than for protons. If relaxation is too fast, the spectra become too broad for detection. In

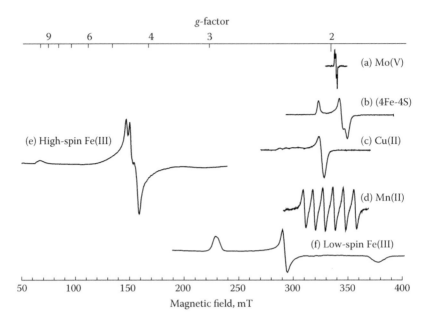

FIGURE 19.2 EPR spectra of transition metal ions. Mn^{II}: a 5-μM solution of MnCl$_2$, measured in a flat cell at room temperature. Spectra of various transition metals in proteins were recorded in the frozen state at the temperatures indicated: high-spin Fe^{III} in transferrin (12 K); low-spin Fe^{III} in myoglobin azide (30 K); Mo^V in xanthine oxidase (150 K); Cu^{II} in plastocyanin (60 K); [4Fe-4S] cluster in *Desulfovibrio africanus* ferredoxin, reduced with dithionite (17 K). Calculated *g*-factors are presented for comparison at the top of the figure, but this is for guidance only; the abscissa of EPR spectra should be magnetic field.

order to record spectra for fast relaxing paramagnets, including most transition metal ions, it is usually necessary to cool the sample to cryogenic temperatures. As a result, EPR spectroscopy is often applied to materials in the frozen state. This inconvenience has the compensating advantage that the sensitivity is enhanced by the Boltzmann distribution factor (equation 19.2).

19.1.3 SPECTRA OF TRANSITION METAL IONS AND COMPLEXES

Figure 19.2 illustrates some EPR spectra of transition metals and metal clusters in biological systems. Most transition metal ions are paramagnetic in certain oxidation states. Examples are given in table 19.2. Although this paramagnetism is stable, in most cases the sample has to be cooled to low temperatures, using liquid nitrogen or liquid helium, for the EPR signal to be observed. There are a few exceptions, such as Mn^{2+}, which may be observed in solution at physiological temperatures (fig. 19.2d).

Organic free radicals can be produced by one-electron oxidation or reduction reactions, by irradiation processes, or by homolytic cleavage of a chemical bond. Organic radicals in solution give complex EPR spectra and tend to be unstable, decaying rapidly by recombination processes. However, some radicals in proteins are relatively stable; some are stabilized by the presence of metal ions. A notable example

TABLE 19.1
Elements with Nuclear Spins and the Expected Splittings in EPR Spectra

Isotope	% Natural abundance	Nuclear spin I	No. hyperfine lines
^1H	99.985	1/2	2
^2H	0.015	1	3
^{13}C	1.11	1/2	2
^{14}N	99.63	1	3
^{15}N	0.37	1/2	2
^{17}O	0.037	5/2	6
^{19}F	100	1/2	2
^{31}P	100	1/2	2
^{33}S	0.76	3/2	4
^{35}Cl	75.53	3/2	4
^{37}Cl	24.47	3/2	4
^{51}V	99.76	7/2	8
^{53}Cr	9.55	3/2	4
^{55}Mn	100	5/2	6
^{57}Fe	2.19	1/2	2
^{59}Co	100	7/2	8
^{61}Ni	1.134	3/2	4
^{77}Se	7.58	1/2	2
^{63}Cu	69.09	3/2	4
^{65}Cu	30.91	3/2	4
^{95}Mo	15.72	5/2	6
^{97}Mo	9.46	5/2	6

is the dinuclear iron-containing ribonucleotide reductase, which contains a tyrosyl cation radical and can readily be examined by conventional EPR spectroscopy.

Just as in NMR, EPR spectra are strongly influenced by the magnetic interaction of the unpaired electron with nuclear spins, which is known as the hyperfine interaction (fig. 19.1c). Examples of nuclei with nuclear spins are shown in table 19.1. The natural abundance of the isotope represents the proportion of nuclei of the element having that particular nuclear spin, which determines the proportion of the paramagnetic material undergoing that splitting. The signal is split into a number of lines, depending on the nuclear spin, I; a nucleus with spin I will split the spectrum into $[2I+1]$ lines. If the transition metal ion has a nuclear spin, a large and measurable splitting will usually result.

A clear example is shown in figure 19.2d, where the EPR spectrum of manganese is split by the interaction with the naturally abundant ^{55}Mn ($I = 5/2$) nucleus. Similarly, the EPR spectrum of copper (fig. 19.2c) is split by nuclear hyperfine interactions with the naturally abundant $^{63}Cu + ^{65}Cu$ (both $I = 3/2$). In this case the magnitude of the splitting is *anisotropic*; it depends on the orientation of the molecule relative to the B_0 field, so in a randomly distributed sample, some parts of the spectrum are split more than others. This strong splitting can be seen on the left-hand side of the spectrum in figure 19.2c. Hyperfine interactions may also be observed with nuclei of ligands to the metal ion, such as ^{15}N; this is sometimes referred to as *superhyperfine* coupling. Hyperfine interactions are very useful for identifying paramagnetic species.

EPR spectra tend to be spread over a much wider bandwidth than in NMR. Instead of parts per million, the signal may cover most of the spectrum. This is particularly the case for paramagnets with spins greater than $S = 1/2$, such as high-spin Fe^{III} (fig. 19.2e), where internal electrostatic fields cause the so-called zero-field splitting. When spectra are measured in frozen samples, further broadening is caused by the anisotropy of the g-factor and hyperfine interactions. This is due partly to the random distribution of molecules with different orientations relative to the B_0 field, which gives a characteristic rhombic or axial line shape to the spectrum. Such anisotropy is not usually observed in NMR, where rapid motion leads to motional narrowing of the spectra. In EPR, motional narrowing of spectra in solution occurs only in small molecules (less than 1 kDa). Another cause of line broadening is a random structural perturbation of the molecules due to freezing of the sample, a phenomenon known as *strain broadening*.

In summary, EPR can be used to identify paramagnetic materials and to measure the amount present. Many complex factors lead to the shape of the EPR spectrum of a particular paramagnet, but what is important for biological EPR is that the resulting spectrum often behaves as a characteristic fingerprint for a particular paramagnet. Moreover this fingerprint is often remarkably conserved in evolution, so spectra of the same protein from widely differing species are the same. If distortions of the paramagnetic center occur, they will usually reveal themselves as changes in the EPR spectrum. We will now briefly examine the type of instrument in which the spectra are measured.

19.1.4 EQUIPMENT NEEDED

The type of commercial EPR spectrometer most often used is the continuous-wave spectrometer operating at X-band microwave frequency, approximately 10 GHz. A helium cryostat is employed to control the temperature. For a sample with a g-factor near two, the required magnetic field as calculated from equation 19.1 is of the order of 0.35 T (3500 G). This is conveniently obtained with an electromagnet. Note that, unlike in optical spectroscopy, one can choose the frequency one uses by varying the magnetic field. The choice of frequency is a compromise: Higher frequencies give a larger Boltzmann distribution and larger transition probability, but a smaller cavity and sample volume; as a result, fewer spins are present and the signal is smaller. This is in contrast to NMR, where increased magnetic fields result in considerable improvements in sensitivity and resolution.

There is another contrast with the NMR spectrometer, in that the application of pulsed radiation and Fourier transformation offers no improvement in sensitivity in EPR spectroscopy. There are two fundamental reasons for this. First, the relaxation time of the electron is much shorter than that of the proton, so the spectrum can only be observed very transiently (less than a microsecond). Second, in contrast to NMR, a microwave pulse cannot excite the broad bandwidth of the typical EPR spectrum. But, pulsed EPR spectrometers certainly do exist [8]. They have many specialty applications, such as the resolution of weak hyperfine interactions with quadrupolar nuclei such as ^{14}N, but they require very concentrated samples and are not used for screening biological materials.

19.2 APPLICATIONS IN PROTEOMICS

In view of its application to a wide range of paramagnetic centers, EPR provides a potential screening method for metal-containing centers in proteins of unknown function. It is able to provide some structural information about the proteins through the analysis of spin–spin interactions. As in many aspects of proteomics, the proper interpretation of spectra ultimately requires specialized expertise.

Interpretation of genomes usually relies on comparison of predicted gene sequences with those of proteins of known function. In order to check whether the protein function is as predicted, the usual procedure is to express the gene, often in a heterologous system, with a suitable tag for easy purification. Inherent in these techniques are a number of assumptions: that the protein has been expressed completely and correctly; that it is correctly folded; and that any necessary post-translational processing has been correctly performed. When a native protein containing a cofactor or metal ion is heterologously expressed, spectroscopic methods are an invaluable aid to the characterization of the expressed product.

The same considerations apply in structural genomics when the structure of a protein is to be determined [9]. Metalloproteins containing complex metal centers or clusters are quite common, but they are probably underrepresented in the Protein Databank. This is because the incorrect or incomplete assembly of metal centers makes them more difficult to crystallize [10,11]. EPR provides a means of checking samples of metalloproteins to verify that the paramagnetic cofactors are correctly assembled, prior to crystallization.

19.2.1 SPECTROSCOPIC EXAMINATION OF PROTEIN COFACTORS

Often a combination of techniques is advantageous. For example, optical (UV/visible absorption) spectroscopy is useful where the protein is expected to contain a chromophoric prosthetic group. EPR spectroscopy is particularly useful to establish the correct insertion of transition metal ions and/or stable free radicals [12].

Identification of cofactors. As already noted, the EPR spectrum of a particular paramagnetic species represents the combined effects of g-factors, hyperfine splittings, and zero-field splitting. These effects are also influenced by the distortions in the active site due to strains in the protein. For metal centers in proteins, the shapes of the spectra are often highly characteristic; moreover, they are often conserved over evolution. They can be used to identify the molecular species by comparison with other known metalloproteins. Where a new type of paramagnetic species is encountered, it may be identified from hyperfine interactions to the metal center or to its ligands. Substitution with stable isotopes that have different nuclear spins is an unambiguous way of identifying the species involved. This can be done by expressing the protein in growth medium enriched with the particular isotope.

Correct folding and assembly of proteins. The EPR spectra of a metal center in a protein are often indicative of a native conformational state. Any denaturation may lead to a distortion of the signal, loss of the signal, or induction of new signals that are not characteristic of the native state. EPR may be used to estimate the proportion of protein molecules that contain the paramagnetic cofactor and those that are distorted.

19.2.2 PARAMETERS THAT AFFECT DETECTION OF AN EPR SPECTRUM

Before carrying out a series of measurements on a particular compound in a batch of samples, it is necessary to find the optimum conditions for high sensitivity for the spectrum, without loss of spectral detail. This requires some experimentation. Once found, the optimum conditions provide further characteristics that identify the species being studied. When presenting EPR spectra of metalloproteins, these parameters should be specified.

Sample temperature. The optimum measurement temperature depends on the type of paramagnetic material. The Boltzmann distribution, upon which the signal amplitude depends, increases as the inverse of temperature in Kelvin (equation 19.2). In addition, if the temperature is too high, the spectrum may broaden out due to fast relaxation; if is too low, the signal may become saturated with microwave power. Therefore, it is necessary to explore variations in temperature and microwave power.

Microwave power. The amplitude of the signal should be proportional to the square root of applied microwave power, until at higher power the signal becomes saturated.

Microwave frequency. Normally, the resonant frequency of the EPR cavity determines the frequency at which the EPR spectrum is taken. Spectrometers can operate at higher frequencies such as Q-band (35 GHz) or W-band (95 GHz). These can resolve features such as *g*-factors that are hidden under the line width at X-band [13].

Field modulation amplitude and *field modulation frequency*. These parameters result from the use of field modulation and phase-sensitive detection in EPR, which gives rise to the first-derivative line shape. Too low a modulation amplitude will produce a weak signal and too high will broaden and distort the spectrum. For modulation frequency, the higher the better, until electron-spin relaxation becomes limiting and the spectra are distorted. Generally, the modulation amplitude should be less than 1/3 of the line width, and a modulation frequency of 100 kHz is standard.

Quantification. EPR spectroscopy is a quantitative technique; the amplitude of the signal is proportional to the number of spins present in the sample. Usually this is determined by comparison with a standard sample of known concentration. It is generally possible to determine the spin concentration of a sample from the double integral of a first-derivative spectrum, comparing the integral with that of a standard sample such as a copper standard or a stable radical compound such as a nitroxide. Quantitative estimates are typically accurate to about 10%, and to obtain this, careful attention must be paid to consistency of sample geometry and position, temperature, and other measurement conditions.

Chemical state. For radicals, the type of radical can often be determined. For transition-metal ions, EPR provides information about the coordination geometry

TABLE 19.2
Examples of Paramagnets Detectable in Their Intermediate Oxidation States

Center	Oxidation states	Paramagnetic state
Flavin	FAD $\leftrightarrow$ FADH· $\leftrightarrow$ FADH$_2$	FADH·
Manganese	Mn^{II} $\leftrightarrow$ Mn^{III} $\leftrightarrow$ Mn^{IV}	Mn^{II}
Cobalt	Co^{I} $\leftrightarrow$ Co^{II} $\leftrightarrow$ Co^{III}	Co^{II}
Heme- or nonheme iron	Fe^{II} $\leftrightarrow$ Fe^{III} $\leftrightarrow$ Fe^{IV}=O	Fe^{III}
Nickel	Ni^{I} $\leftrightarrow$ Ni^{II} $\leftrightarrow$ Ni^{III}	Ni^{I}, Ni^{III}
Copper	Cu^{I} $\leftrightarrow$ Cu^{II}	Cu^{II}
Molybdenum	Mo^{VI} $\leftrightarrow$ Mo^{V} $\leftrightarrow$ Mo^{IV}	Mo^{V}
Oxo-bridged dinuclear iron clusters	Fe^{III}-Fe^{III} $\leftrightarrow$ Fe^{III}-Fe^{II} $\leftrightarrow$ Fe^{II}-Fe^{II}	Fe^{III}-Fe^{II}
[4Fe-4S] Clusters	[4Fe-4S]$^{+}$ $\leftrightarrow$ [4Fe-4S]$^{2+}$ $\leftrightarrow$ [4Fe-4S]$^{3+}$	[4Fe-4S]$^{+}$, [4Fe-4S]$^{3+}$
Cupredoxin Cu$_2$ clusters	Cu^{I}-Cu^{I} $\leftrightarrow$ Cu^{II}-Cu^{I}	Cu^{II}-Cu^{I}

and types of ligands. For example, figure 19.2e shows the spectrum of a protein, oxidized transferrin, containing Fe^{III} (S = 5/2). Many different types of nonheme iron proteins give signals of this type. The exact shape of such spectra depends on the coordination geometry. It may be noted that a signal in this region of the spectrum is observed in most biological samples, due partly to contaminant iron and partly to iron in particulate matter or glassware. The signal is particularly prominent because the paramagnet is high spin and partly because, statistically, there is a higher probability in a powder sample that there will be a prominent feature around g = 4.27. Therefore, a small quantity of this material will give a strong signal.

19.2.3 OXIDATION–REDUCTION STATE

Most proteins involved in oxidation-reduction reactions can be studied by EPR in their reduced or oxidized state. By following the magnitude of the EPR signal as a function of applied reduction potential, it is possible to estimate the redox potentials of the individual redox centers in a protein [14]. In many cases, detection by EPR is a good indication of the oxidation state of the species observed—for example, quinone radicals in their semireduced state and metal ions in their paramagnetic oxidation states. A case that is sometimes difficult to resolve is nickel, where Ni^{I} and Ni^{III} can show similar types of spectra.

Some redox centers can undergo two one-electron reduction steps and are only detectable in their intermediate oxidation states. Examples are shown in table 19.2. In these cases, it is not possible in general to obtain the center quantitatively in the intermediate state. The state may be obtained by stoichiometric titration with an oxidant or reductant, monitored by spectroscopy, or by careful poising of the redox potential [15].

Spin–spin interactions. EPR spectra are also sensitive to magnetic interactions between an electron spin and a nucleus or between electron spins; they can be interpreted to give a measure of the distance between the spins. This information is provided by the dipolar component of the spin–spin interaction. At short distances, the interactions are observed as changes in the shape of the EPR signal. At longer distances, electron spin–spin interactions are detected by changes in microwave power saturation.

EPR is particularly useful for resolving the spectra of samples containing multiple iron-sulfur clusters, as well as copper and molybdenum, which do not give sharp optical absorption peaks.

19.2.4 LIMITATIONS OF THE METHOD

EPR does not detect diamagnetic metal ions such as zinc or calcium; there has to be paramagnetism. Moreover, if the paramagnet has an even spin, such as $S = 1, 2, 3\ldots$, the detection of a spectrum is not guaranteed. This applies to ions such as high-spin ferrous Fe^{II} ($S = 2$). The same applies to paramagnetic systems that are strongly spin coupled, such as dimeric Cu^{II}–Cu^{II} centers or oxidized [2Fe-2S] clusters. The paramagnet may usually be converted to a state with odd spin, such as $S = 1/2, 3/2, 5/2$, by one-electron oxidation or reduction—for example, oxidation of high-spin ferrous (Fe^{II}) to high-spin ferric (Fe^{III}) or reduction of the iron-sulfur cluster. Fe^{II} can also be rendered EPR detectable by nitric oxide, a stable free radical that forms paramagnetic nitrosyl iron complexes.

19.3 SAMPLE PREPARATION

Liquid water absorbs at microwave frequencies, so aqueous solutions have to be contained in very thin flat cells or capillaries. An advantage of using frozen samples in EPR spectroscopy is that they do not suffer from this problem. Samples are typically of 100- to 200-µL volume, with a concentration of the paramagnet of $1-100 \, \mu M$. They are prepared in pure quartz sample tubes of 3-mm diameter and can be stored frozen in liquid nitrogen. Most of them will have optical absorption in the visible region, so the density of color gives an indication of the signal strength expected.

Samples may be prepared in the EPR-detectable redox state by addition of substrates, oxidizing agents such as $K_3Fe(CN)_6$, or reducing agents such as sodium dithionite. In the latter case, the protein should be maintained under an inert atmosphere such as argon or nitrogen prior to freezing. Some proteins from anaerobes, such as certain iron-sulfur proteins, are sensitive to oxygen and require isolation under strictly anaerobic conditions. This requires the use of anaerobic glove-boxes and so far has not been widely used in high-throughput investigations.

Once conditions have been optimized for measurement of the signal of interest, a spectrum typically requires a few minutes to record. Preliminary screening—for example, by UV/visible absorption spectroscopy—is required to select the samples likely to give EPR signals. Cell preparations sometimes contain significant amounts of free Mn^{2+}. The signal can be alleviated by addition of EDTA, which forms a complex with a broad EPR signal.

When a sample contains a mixture of paramagnetic species, the spectra may be difficult to resolve. This may occur when a sample is of a protein with multiple

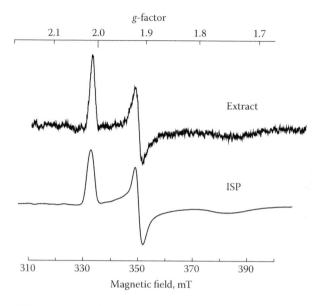

FIGURE 19.3 EPR spectrum of the Rieske iron-sulfur cluster in benzene dioxygenase, expressed in extract from *E. coli* cells, expressing the α-subunit of ISP, purified catalytic iron-sulfur protein. Samples were reduced with dithionite before freezing. Conditions of measurement: temperature, 15 K; microwave frequency, 9.35 GHz; modulation amplitude, 1 mT.

centers or a mixture of EPR-active proteins. When the overlapping spectra are resolved, it is sometimes helpful to add substrates, which should reduce the relevant enzymes specifically, or to control the redox potential of the sample to select for individual centers. It is possible to use spectral subtraction to extract the desired spectrum from composite spectra recorded on samples poised at different redox potentials [14,15].

19.4 WHOLE-CELL STUDIES

When proteins are highly expressed in host cells (e.g., *Escherichia coli*), it is of great assistance to be able to measure their EPR spectra without the need for purification. This provides the means of screening for the presence of metal centers in expressed proteins. An example is the expression of the iron-containing protein, benzene dioxygenase from *Pseudomonas putida*, encoded on a plasmid [16]. This protein contains a Rieske iron-sulfur cluster, which is a [2Fe-2S] cluster that has histidine and cysteine coordination. The protein could be expressed to represent more than 40% of the cellular protein, making it possible to observe the spectra of the unstable iron-sulfur clusters without the need for purification (fig. 19.3).

As can be envisaged from figure 19.2, the presence of several paramagnetic centers in a sample leads to overlapping spectra. Various methods can be used to resolve them. The first is to try to render some of the centers EPR silent by adjustment of the redox potential. It is also possible to exploit the different temperature dependence of the signals. Variation of other measurement conditions, such as microwave power, may also be used.

19.5 INTERPRETATION OF RESULTS

It is important to be aware that paramagnetism may be associated with denatured proteins. Occasionally, sophisticated EPR studies have been carried out on essentially dead enzyme systems; the active systems are EPR silent under the conditions used. It is also important to select the appropriate host organism for expression of gene products. If it does not have the necessary machinery, the metal center or cofactor will not be inserted or the wrong one will be inserted.

19.5.1 NO INSERTION OF METAL CENTER

An example is the active site or H-cluster of the iron-containing hydrogenases. When the hydrogenases from *Desulfovibrio vulgaris* were expressed in *E. coli*, EPR spectroscopy showed that the [4Fe-4S] clusters were assembled correctly, but not the complex H-cluster in the active site [17,18].

19.5.2 INSERTION OF INAPPROPRIATE METAL IONS

Another possible outcome of heterologous cloning studies is the insertion of inappropriate metals into expressed gene products. This may take place, for example, when individual subunits of a multisubunit protein are expressed, exposing potential ligands that would normally be protected from metal binding by the other subunits. This can lead to heterogeneity, making protein crystallization very difficult. EPR spectroscopy provides a means to observe when this happens. An example is the observation of Cu^{II} in proteins from the hyperthermophilic archeaon *Pyrococcus furiosus*, when expressed in *E. coli* (fig. 19.4). *P. furiosus* is an anaerobe and is not known to contain any Cu^{II} proteins, suggesting that *E. coli* may mistake a feature of the foreign protein as a copper-binding site. A possible location for the metal-binding site is the polyhistidine tag on the protein, introduced for ease of purification. Such a situation appears to be relatively rare, however: EPR studies have been carried out on many other proteins containing His tags, but copper binding has not been reported.

Iron-sulfur proteins are commonly found in electron-transfer proteins, as well as other proteins such as hydrolases. There are numerous different types of iron-sulfur cluster in these proteins. Depending on the host organism, they are assembled in one or more of three different generic systems: nif, isc, and suf [19]. It has been found that under some circumstances, these systems may operate inappropriately on heterologously expressed proteins. The cysteine-rich zinc-finger region of a Lim transcription factor from *Caenorhabditis elegans*, when expressed in *E. coli*, was found in inclusion bodies containing iron-sulfur clusters [20]. Inclusion bodies formed from cysteine-rich proteins tend to have a brown color and it may be a common occurrence that iron-sulfur clusters are inserted inappropriately. The soluble form of the same protein, when cloned in *E. coli*, contained only zinc [21].

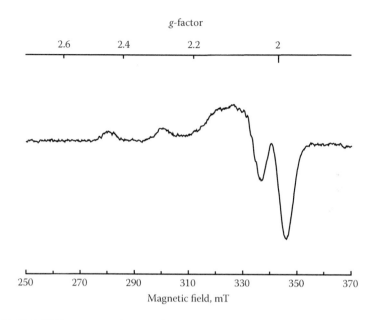

FIGURE 19.4 EPR spectrum for a polypeptide from *Pyrococcus furiosus*, expressed in *E. coli.* Conditions of measurement: temperature, 10 K; microwave power, 20 mW; frequency, 9.603 GHz; modulation amplitude, 1 mT; frequency, 100 kHz.

19.6 FUTURE PROSPECTS

Spectroscopic methods are an invaluable aid for the characterization of expressed proteins. Application of EPR spectroscopy to the large-scale screening of proteomics samples is possible and likely to produce unique information. It has not been carried out so far, but EPR has a special part to play in studies of proteins, such as iron-sulfur proteins, that are difficult to investigate by other methods. The method can be scaled up to genome-wide studies, provided that appropriate selection of samples is first carried out. The EPR spectrum will monitor the correct insertion of the metal cofactor and indirectly reports on the correct folding and assembly of the protein. It is to be expected that EPR will play a role in the functional characterization of metalloproteins identified from the genome data.

The insertion of inappropriate metal ions into proteins that are expressed in a foreign host is an intriguing observation. Such investigations may help to elucidate the signals that are employed by the host organism to direct the appropriate metals into proteins.

ACKNOWLEDGMENTS

I thank Dr. C. Bagneris and Prof. M. W. W. Adams for permission to use spectra in figures 19.3 and 19.4, and Prof. M. K. Johnson and Prof. Adams for providing facilities for the studies of expressed *P. furiosus* proteins. This work was supported by grants from the U.K. Biotechnology and Biological Sciences Research Council.

REFERENCES

1. Weil, J. A., Bolton, J. R., and Wertz, J. E., *Electron Paramagnetic Resonance: Elementary Theory and Practical Applications.* Wiley, New York, 1994.
2. Riordan, J. F. and Vallee, B. L., eds., *Methods in Enzymology,* vol. 227 *Metallobiochemistry, Parts C & D: Spectroscopic Methods for Probing Metal Ion Environments in Metalloproteins.* Academic Press, New York, 1993.
3. Brudvig, G. W., Electron paramagnetic resonance spectroscopy, *Biochemical Spectroscopy: Methods in Enzymology* 246, 536–554, 1995.
4. Cammack, R. and Shergill, J. K., EPR spectroscopy. In *Enzymology Labfax,* ed. Engel, P. C. Bios Scientific Publishers, Oxford, 1995, 249–268.
5. Cammack, R. and Shergill, J. K., EPR spectroscopy. In *Proteins Labfax,* ed. Price, N. C. Bios Scientific Publishers, Oxford, 1996, 217–231.
6. Randolph, M. L., Quantitative considerations in electron spin resonance studies of biological materials. In *Biological Applications of Electron Spin Resonance,* ed. Swartz, H. M., Bolton, J. R., and Borg, D. C. Wiley-Interscience, New York, 1972, 119–153.
7. Degtyarenko, K., Bioinorganic motifs: Towards functional classification of metalloproteins, *Bioinformatics* 16, 851–864, 2000.
8. Schweiger, A. and Jeschke, G., *Principles of Pulse Electron Paramagnetic Resonance Spectroscopy.* Oxford University Press, Oxford, 2001.
9. Chance, M. R., Fiser, A., Sali, A., Pieper, U., Eswar, N., Xu, G. P., Fajardo, J. E., Radhakannan, T., and Marinkovic, N., High-throughput computational and experimental techniques in structural genomics, *Genome Research* 14, 2145–2154, 2004.
10. Jenney, F. E., Brereton, P. S., Izumi, M., Poole, F. L., Shah, C., Sugar, F. J., Lee, H. S., and Adams, M. W. W., High-throughput production of *Pyrococcus furiosus* proteins: Considerations for metalloproteins, *Journal of Synchrotron Radiation* 12, 8–12, 2005.
11. Scott, R. A., Shokes, J. E., Cosper, N. J., Jenney, F. E., and Adams, M. W. W., Bottlenecks and roadblocks in high-throughput XAS for structural genomics, *Journal of Synchrotron Radiation* 12, 19–22, 2005.
12. Cammack, R., Magnetic resonance: EPR methods. In *Encyclopedia of Spectroscopy and Spectrometry,* ed. Tranter, G. and Holmes, J. Academic Press, London, 1999, 457–470.
13. Möbius, K., High-field EPR and ENDOR in bioorganic systems. In *EMR of Paramagnetic Molecules,* ed. Berliner, L. J. and Reuben, J. Plenum, New York, 1993, 253–274.
14. Dutton, P. L., REDOX potentiometry: Determination of midpoint potentials of oxidation-reduction components of biological electron-transfer systems, *Methods in Enzymology: Biomembranes* 54, 411–434, 1978.
15. Cammack, R., REDOX states and potentials. In *Bioenergetics: A Practical Approach,* ed. Brown, G. C. and Cooper, C. E. IRL Press, Oxford, 1995, 85–109.
16. Mason, J. R., Butler, C. S., Cammack, R., and Shergill, J. K., Structural studies on the catalytic component of benzene dioxygenase from *Pseudomonas putida, Biochemical Society Transactions* 25, 90–95, 1997.
17. Voordouw, G., Hagen, W. R., Krusewolters, K. M., Van Berkel-Arts, A., and Veeger, C., Purification and characterization of *Desulfovibrio vulgaris* (Hildenborough) hydrogenase expressed in *Escherichia coli, European Journal of Biochemistry* 162, 31–36, 1987.
18. Nicolet, Y., Piras, C., Legrand, P., Hatchikian, E. C., and Fontecilla-Camps, J. C., *Desulfovibrio desulfuricans* iron hydrogenase: The structure shows unusual coordination to an active-site Fe binuclear center, *Structure* 7, 13–23, 1999.

19. Frazzon, J. and Dean, D. R., Formation of iron-sulfur clusters in bacteria: An emerging field in bioinorganic chemistry, *Current Opinion in Chemical Biology* 7(2), 166–173, 2003.

20. Li, P. M., Reichert, J., Freyd, G., Horvitz, H. R., and Walsh, C. T., The Lim region of a presumptive *Caenorhabditis elegans* transcription factor is an iron sulfur-containing and zinc-containing metallodomain, *Proceedings of the National Academy of Sciences of the United States of America* 88, 9210–9213, 1991.

21. Archer, V. E. V., Breton, J., Sanchez-Garcia, I., Osada, H., Forster, A., Thomson, A. J., and Rabbitts, T. H., Cysteine-Rich LIM domains of LIM-homeodomain and LIM-only proteins contain zinc but not iron, *Proceedings of the National Academy of Sciences of the United States of America* 91, 316–320, 1994.

Part VI

Summary

20 Summary of Chapters and Future Prospects for Spectral Techniques in Proteomics

Daniel S. Sem

CONTENTS

20.1 PART I. THE SCOPE OF PROTEOMIC AND CHEMICAL PROTEOMIC STUDIES

Proteomic studies strive to characterize groups of proteins in a systematic manner. This can involve studies of pools of proteins isolated from tissue extracts or subsets (subproteomes) of these protein mixtures (part II). Studies can also be of individual purified proteins, but carried out in a highly parallel way, as in structural proteomics. In both cases, studies are "systems based" (chapter 1) because they focus on groups of proteins that are related by the networks of interactions they participate in or by similarities in their binding sites.

20.1.1 GROUPING PROTEINS BY BINDING SITE SIMILARITIES

In order to simplify the complexity of a proteome, it is often useful to group proteins into families based on similarities in their binding sites. Once this is done, many of the spectral techniques presented in this book (reviewed in chapter 3) can be used to characterize protein–protein (part III) or protein–ligand (part IV) interactions, one protein family at a time. To achieve the full benefit of this approach to proteomic studies, Villar et al. (chapter 2) noted the need for more sophisticated tools to extract information on protein interactions and to "identify the relationships across datasets and disciplines." They note the importance of striving towards this goal, since it is crucial to understanding the selectivity issues associated with "drug–drug interactions, environmental challenges, and toxicological risk assessment." The interface of chemistry and the proteome is defining the field of chemoproteomics (chemical proteomics) and is an underlying theme in many chapters of this book. The fact that small molecules can be used to classify proteomes (and vice versa) has led to the affinity-based separation methods (e.g., surface-enhanced laser desorption/ionization [SELDI], discussed in chapters 7 and 8) that are being integrated with spectral-analysis methods.

20.2 PART II. MASS SPECTRAL STUDIES OF PROTEOME AND SUBPROTEOME MIXTURES

20.2.1 CHAPTER 4: CAPILLARY ELECTROPHORESIS-MASS SPECTROMETRY (CE-MS) FOR CHARACTERIZATION OF PEPTIDES AND PROTEINS

Efficient analysis of protein mixtures often requires in-line separation technology before the mass spectral analysis. In this regard, capillary electrophoresis-mass spectrometry (CE-MS) is complementary to liquid chromatography (LC)-MS. It offers an orthogonal separation strategy, usually coupled to electrospray ionization (ESI). It also offers advantages in situations where sample is limited or when fast separation is needed. Different variations are possible, but capillary zone electrophoresis (CZE) is presented in greatest detail by Neusüß and Pelzing. CZE-MS offers very good ionization in micro- and nanospray, with sharp peaks and attomolar sensitivity. In comparison, capillary isoelectric focusing (CIEF) offers extremely high separation capability that, when combined with the high resolution of Fourier transform ion cyclotron resonance (FTICR), can produce powerful separations in a two-dimensional (pI vs. mass) experiment analogous to 2D electrophoresis.

Furthermore, 2D separations such as CIEF-LC(-MS), CIEF-CZE(-MS), reverse phase liquid chromatography (RPLC)-CZE(-MS) and others are possible. Neusüß and Pelzing note the important role CZE-MS will continue to play in 2D separations, coupled with in-line MS analysis. They also note that CE separation techniques may play a dominant role in interfacing with the microfluidics associated with chip-MS coupling and with "lab-on-a-chip" approaches as that technology develops. CE is well suited to meet the future demands of "miniaturization and multidimensional analysis." Finally, although CE-MALDI-MS is also possible, this off-line approach loses the benefit of the fast CE separation, which is realized with the more commonly implemented CE-ESI-MS approaches. Still, there are benefits to coupling CE to matrix-assisted laser desorption/ionization (MALDI), enabled by recent advances in MALDI MS/MS automation (chapter 5).

20.2.2 CHAPTER 5: PROTEIN AND PEPTIDE ANALYSIS BY MATRIX-ASSISTED LASER DESORPTION/IONIZATION TANDEM MASS SPECTROMETRY (MALDI MS/MS)

The primary methods of ionization used in protein/peptide mass spectrometry are ESI and MALDI. Sachon and Jenson briefly compare these methods and then provide a thorough overview of mass spectrometers in general, followed by a detailed explanation of MALDI technology along with various approaches for preparing samples. One of the most exciting applications of MALDI in proteomics is protein sequencing with tandem MS, which is described in detail. Sachon and Jensen note advances on the horizon relating to continued improvements in sample preparation strategies and MALDI hardware, along with bioinformatics and data analysis tools for sequence analysis. Studies of increasingly complex protein mixtures are also being enabled by the coupling of CE and LC (e.g., chapter 4), along with automated MALDI MS/MS.

20.2.3   CHAPTER 6: CHARACTERIZATION OF GLYCOSYLATED PROTEINS BY MASS SPECTROMETRY USING MICROCOLUMNS AND ENZYMATIC DIGESTION

Glycosylation is one of the most prevalent post-translational modifications to proteins, with importance in immunity, cell signaling, and cell adhesion. Hägglund and Larsen present the application of MALDI and ESI MS methods to the analysis of glycosylated proteins, especially for identifying sites of glycosylation, and for defining the composition and structure. They emphasize the use of tandem MS-based fragmentation and enzymatic cleavage coupled to MS analysis and separation techniques. It is anticipated that new sensitive hybrid MS instruments, performing fragmentation with electron capture dissociation (peptide sequencing) and infrared multiphoton photodissociation, will contribute significantly to the field of glycomics. The authors note the importance of quantifying the glycosylation state of a proteome and that this can only be accomplished with proteomic methods since post-translational modifications cannot be detected by RNA/array-based technology.

20.2.4   CHAPTER 7: SURFACE-ENHANCED LASER DESORPTION/IONIZATION (SELDI) PROTEIN BIOCHIP TECHNOLOGY FOR PROTEOMICS RESEARCH AND ASSAY DEVELOPMENT

Although many MS techniques require additional in-line separation techniques, such as CE or LC, SELDI is unique in that separation is integral to the chip and the method. Surface arrays are derivatized with commonly used chromatographic separation groups, including reverse phase, ion exchange, immobilized metal capture, or normal phase supports. They can even be tagged with a ligand of interest for chemical proteomic applications. Weinberger et al. present a thorough description of SELDI technology, with applications in biomarker discovery and assay creation for clinical studies, drug discovery, and basic research. Analysis can be of crude samples (mixtures) because separation occurs on the chip. The authors expect future advances in surface chemistry, protein purification technology, bioinformatics, and laser desorption-based MS to further improve the SELDI technique. One exciting advance on the horizon is the use of "microscale multidimensional fractionation schemes" based on solid-phase combinatorial chemistry libraries used to affinity purify subproteomes in a way that favors less abundant proteins, thereby addressing the dynamic range problems that plague proteomic studies.

20.2.5   CHAPTER 8: AN APPROACH TO THE REPRODUCIBILITY OF SELDI PROFILING

Given the potentially important role of SELDI profiles in identifying and applying biomarkers in a clinical setting, there is a need to develop statistical measures of reproducibility. To this end, Liggett et al. analyzed 88 SELDI mass spectra, replicate measurements of a human serum standard, and assessed scores of variation in the measurement system. Their approach was to find long-distance correlations between peaks of very different mass-to-charge ratios (m/z) using a functional canonical correlation analysis (CCA). Although they did succeed in identifying a source of variation relating to the sample preparation step, they note that there is a need to expand the technique more broadly, as well as to address key questions such as (1) What are the sources of variation in SELDI results between labs? and (2) Is a SELDI profile sufficiently reproducible to serve as a biomarker?

20.3   PART III. PROTEIN–PROTEIN (OR PEPTIDE) INTERACTIONS: STUDIES IN PARALLEL AND WITH MIXTURES

20.3.1   CHAPTER 9: MASS SPECTROMETRIC APPLICATIONS IN IMMUNOPROTEOMICS

Purcell et al. present a clear and thorough background for the field of immunoproteomics, along with some of the significant challenges it faces. For example, since MHC classes I and II can present as many as 10,000–100,000 peptides on the surface of antigen presenting cells, the diversity needing analysis is immense. This collection of peptides is called the immunoproteome. Various methods of separating and analyzing the immunoproteome are reviewed, with in-depth discussion of multidimensional chromatography (capillary reverse phase high-performance liquid chromatography [RP-HPLC]; immunoaffinity) followed by MALDI-time-of-flight (TOF) MS/MS.

Many of the challenges faced in immunoproteomics are similar to those faced in the broader field of proteomics: quantitation, data management/mining, sensitivity, and dealing with complex mixtures. The authors feel that although there will continue to be hardware advances providing improved sensitivity and resolution, the real power of MS in immunoproteomics can only be realized with better approaches to sample preparation. To this end, they feel that immunochemical reagents (antibodies, major histocompatibility complex [MHC] tetramers) may actually drive future immunoproteomics research. Besides permitting the purification of peptide subproteomes, such reagents would permit the isolation of homogeneous cells from tissues. Also important in this regard will be isolation and fractionation techniques like cell sorting, flow cytometry, and laser capture microscopy.

20.3.2   CHAPTER 10: NEAR-INFRARED FLUORESCENCE DETECTION OF ANTIGEN–ANTIBODY INTERACTIONS ON MICROARRAYS

Protein antigens or antibodies can be immobilized on solid supports, permitting the microarray-based studies of antigen–antibody interactions. Vehary and Garabet present the use of near-infrared (NIR) fluorescence, using labeled probes to detect such interactions in direct binding and competition assays. Especially exciting applications include the monitoring of changes in phosphorylation state in breast cancer cell lines, using immobilized antibodies raised against phosphopeptides. A similar strategy to detect the phosphyorylation of 62 signaling components was performed, this time with immobilized cell extracts in a "reverse phase protein array." Antigen arrays are also used to detect viral infections (HIV, hepatitis) as well as cancer (prostate, lung).

Methods for addressing specificity issues and the value of doing subproteome analysis are also discussed. These examples demonstrate that NIR fluorescence-based detections will likely play an increasingly important role in disease diagnosis and in the identification of new biomarkers. But an important challenge to be addressed is the simplification of protein samples, since concentrations can vary by up to 12 orders of magnitude, which is clearly a dynamic range problem. Another challenge noted by Vehary and Garabet is the organization and statistical analysis of the large volume of data generated from antibody and antigen microarrays. Finally, they note the importance of the complementary use of other multiplexed methods and bio-informatic tools, in order to make sense of the complexity of biological systems.

20.3.3   CHAPTER 11: APPLICATION OF SHOTGUN PROTEOMICS TO TRANSCRIPTIONAL REGULATORY PATHWAYS

Shotgun proteomics, often referred to as MudPIT (multidimensional protein identification technology), permits the analysis of crude protein mixtures, such as cell lysates or multiprotein complexes. It typically involves proteolysis of the protein mixture (e.g., with trypsin), then separation on a microcolumn packed with reverse phase and strong cation exchange resins, with various configurations possible. Proteins are then identified by sequencing with tandem MS. Mosley and Washburn provide an overview of shotgun proteomics, with applications to proteins in the nucleus, especially those involved in transcriptional regulation. These studies present a significant challenge for proteomics because these proteins are often present at extremely low levels. Indeed, one study showed that, of proteins expressed at levels of <1,000 molecules per cell, only 15% were identified by MudPIT. Mosley and Washburn discuss efforts to address this dynamic range challenge, which is thought to be most significant in eukaryotic cells.

One of the most effective strategies is to purify organelles (e.g., endoplasmic reticulum, Golgi apparatus, peroxisomes, mitchondria, lysosomes, and the nucleus) to enrich in a subproteome of interest and to eliminate highly abundant proteins that can mask the less abundant proteins of interest. This fractionation can be carried even further, to components within an organelle such as structures within the nucleus: nuclear pore complex, nucleolus, interchromatic granules, clastosome, promyelocytic leukemia bodies, and Cajal bodies. Various MudPIT studies are presented, including the identification of 337 proteins unique to the nuclear envelope. Methods such as tandem affinity purification (TAP)/MS to characterize multiprotein complexes are also presented. Thus, the dynamic range challenge is largely addressed by enriching proteins of interest by purifying organelles or multiprotein complexes. Future improvements in MS hardware may also help to address the dynamic range challenge. The long-term goal is to obtain increasingly detailed and sensitive dynamic snapshots of the cell in terms of protein quantities and post-translational modification state, under different cellular conditions.

20.3.4   CHAPTER 12: ELECTROPHORETIC NMR OF PROTEIN MIXTURES AND ITS PROTEOMICS APPLICATIONS

Electrophoretic NMR (ENMR) permits the study of protein mixtures, without the need for physical separation, by resolving proteins *in situ* based on electrophoretic mobility.

He et al. present recent advances in the technique, including new hardware, pulse sequences, signal processing strategies, and applications. The technique has now matured, with recent innovations such as development of a microsolenoidal-coil ENMR probe, development of capillary array ENMR, and use of the maximum entropy method (MEM) for processing. Especially exciting are 3D ENMR experiments, with correlation spectroscopy (COSY) and heteronuclear single quantum coherence (HSQC) 2D planes resolved in a third dimension of electrophoretic mobility.

The truly unique aspect of this method is that proteins can be visualized separately, based on electrophoretic mobility, but without actually separating the proteins. In that sense, it is a powerful complement to traditional proteomic studies in gels because ENMR permits the study of protein–ligand interactions in solution. Applications are presented for proteins ranging in size from 14 kDa (lysozyme) up to 480 kDa (urease). Promising applications on the horizon include studies of protein–ligand and protein–protein interactions, especially towards the goal of characterizing signaling cascades and for discovery of cancer biomarkers.

20.4 PART IV. CHEMICAL PROTEOMICS: STUDIES OF PROTEIN–LIGAND INTERACTIONS IN POOLS AND PATHWAYS

20.4.1 CHAPTER 13: CHARACTERIZING PROTEINS AND PROTEOMES USING ISOTOPE-CODED MASS SPECTROMETRY

Central to the field of chemical proteomics is the use of chemical probes to label categories of proteins specifically. To this end, Aebersold introduced the ICAT (isotope coded affinity tag) approach to isotopically encode two separate proteome samples that are later mixed and relative abundances of proteins determined. The ICAT technique has since been extended significantly to include labeling of cysteine as well as lysine and tryptophan residues and N- and C-termini. Kota and Goshe describe these advances, along with various labeling schemes ($^{1}H/^{2}H$, $^{12}C/^{13}C$, $^{14}N/^{15}N$, and $^{16}O/^{18}O$) using chemical, metabolic, and enzymatic approaches for labeling. Especially exciting applications are in monitoring changes in phosphorylation or glycosylation state across a proteome. They cite data analysis as a key future challenge and note the importance of the "transproteomic pipeline," which will permit "a uniform analysis of MS/MS spectra" from various sources and instruments. This will enable more efficient data mining and data validation.

20.4.2 CHAPTER 14: SURFACE PLASMON RESONANCE BIOSENSORS' CONTRIBUTIONS TO PROTEOME MAPPING

The complete systems-based view of a proteome would include a description of all protein–protein, protein–DNA, protein–ligand, and enzyme–substrate interactions. Rich and Myszka provide a description of how advances in SPR technology are permitting increasingly high throughput characterization of these interactions in terms of kinetics and thermodynamics. This is moving us closer to the goal of obtaining a comprehensive and dynamic picture of the proteome. Recent coupling of SPR to MS has allowed downstream identification of many proteins involved in these interactions. Microfluidics coupled to multiplexed/arrayed chips permit analysis of up to 400 ligands at a time, spanning a broad range of affinities. Future improvements in sampling rate, detection limits, chip surfaces, and an expanded range of applications can be expected. It is hoped that it provides a more comprehensive and well-defined (systems-based) picture of the proteome and its dynamic state.

20.4.3 CHAPTER 15: APPLICATION OF IN-CELL NMR SPECTROSCOPY TO INVESTIGATION OF PROTEIN BEHAVIOR AND LIGAND–PROTEIN INTERACTION INSIDE LIVING CELLS

Since chemical proteomics is devoted to the study of protein–ligand interactions, it requires techniques that can describe them with as much structural detail as possible. While only NMR and x-ray crystallography can provide high-resolution structural characterizations, NMR is unique in its ability to probe structural differences *in vitro* and inside cells. This is important because it gives additional biological relevance to structural characterizations. Such studies are the focus of the in-cell NMR technique developed and reviewed here by Volker Dötsch. Different labeling schemes are presented (^{15}N, ^{13}C, ^{19}F) and the sensitivity benefits of labeling the methyl groups of specific amino acids (especially Met) are presented. Examples of structural changes seen inside cells are presented, including the important observation that some proteins, which appear unfolded *in vitro*, may actually adopt a folded state inside cells, due to molecular crowding. In-cell NMR has been applied in bacteria, yeast, and insect cells, with a detection limit of 70 μM. While NMR cryoprobes have improved sensitivity to this level, additional NMR hardware improvements, combined with developments in NMR tubes capable of maintaining viable cells for longer periods, may increase sensitivity further. This would permit more studies in eukaryotic cells, where it is difficult to obtain high protein concentrations.

20.4.4 CHAPTER 16: AN OVERVIEW OF METABONOMICS TECHNIQUES AND APPLICATIONS

A complete systems biology view of an organism would include quantification of all protein (proteomics), mRNA (transcriptomics), and metabolite (metabonomics) levels. Although this book is focused on proteomics, all metabolites are made by the proteome and bind the enzymes that make or metabolize them, so a discussion of metabonomics is fitting. In fact, one could view the complete description of how the metabolome relates to the proteome as being squarely in the realm of chemical proteomics, defining all the protein–ligand (enzyme metabolite) interactions in a proteome. Lindon et al. define metabonomics as "the systematic profiling of metabolites and metabolic pathways in whole organisms through the study of biofluids." The methods used include LC-MS, magic angle spinning NMR of tissue samples, and solution NMR—usually, automated/flow NMR of biofluids. Various pattern recognition/chemometric strategies have been used to analyze biofluid data to diagnose coronary heart disease, various cancers, inborn metabolic disorders, and many other pathologies. As a result of numerous studies validating it as a diagnostic tool in clinical and research settings, it is expected that metabonomics will play an increasingly prominent role in:

assessing drug toxicity
choosing proper animal models for clinical studies
diagnosing disease/biomarkers
assessing drug efficacy
personalized medicine—treating based on metabolic fingerprint

From a basic research perspective, a complete metabonomic profile provides a more comprehensive systems-based view of an organism when integrated with proteomic and other data. A significant step in this direction has been made by the Consortium of Metabonomic Toxicology (five pharmaceutical companies and Imperial College, London), which has worked to generate a metabonomic database of ~35,000 NMR spectra to define the effects of 147 model toxins. Continued growth and integration of metabonomic databases in a consistent format is being facilitated by the "standard metabolic reporting structures" (SMRS) group.

20.5 PART V: STRUCTURAL PROTEOMICS: PARALLEL STUDIES OF PROTEINS

20.5.1 Chapter 17: NMR-Based Structural Proteomics

Protein function can include binding to other proteins, ligands, and DNA, as well as catalysis of reactions, signal transduction, and mechanical motion, among other possibilities. But, function cannot be fully understood in the absence of high-resolution three-dimensional structures. For this reason, much emphasis has recently been placed on structural proteomics efforts using x-ray crystallography or NMR, with 30 centers involved. John Markley provides an overview of NMR-based structural proteomics, with emphasis on future potential and technological advances, as well as tools used at the Center for Eukaryotic Structural Genomics (CESG).

NMR is the method of choice for small proteins (<20 kDa) that are not easily crystallized and serves as a complementary technique to x-ray crystallography. Advantages of NMR include the ability to rapidly screen for protein folding and aggregation state, as well as ligand binding. Many methods and software tools are presented for automating all aspects of structure determination so that, in ideal cases, it is possible to go from protein expression to deposited structure in only two weeks. Emphasis of structural proteomics projects is usually on proteins with ≤30% sequence identity to other proteins already in the PDB and on proteins involved in human disease.

Advances on the horizon include better tools to predict and prepare protein domains for NMR studies and the use of rapid screening against ligands and cofactors to profile proteins for functional characterizations, as with chemical proteomics. A reduced dimensionality approach to data collection (high-resolution iterative frequency identification [HIFI] NMR) is able to accelerate data collection fivefold or more. NMR approaches to studying protein–protein interactions will help to identify systems-based relationships, and the challenges of characterizing larger and membrane-bound proteins are being increasingly addressed, albeit slowly. Finally, the recent creation of the Database of Protein Disorder (DisProt) will help to increase our understanding of why proteins are disordered.

20.5.2   CHAPTER 18: LEVERAGING X-RAY STRUCTURAL INFORMATION IN GENE FAMILY-BASED DRUG DISCOVERY: APPLICATION TO PROTEIN KINASES

An efficient and systems-based way to approach the structural characterization of proteins is by focusing on families with related structures and binding sites (chapter 2). Protein kinases share the same backbone fold and have similar binding sites, since they all bind the same ligand, ATP. They are a large gene family, with ~500 protein kinases (1.7% of all genes) in the human genome. Already, over 300 structures of complexes comprising 74 ligand scaffolds have been obtained with x-ray crystallography. Jacobs et al. present a systems-based structural characterization of protein kinases from the perspective of inhibitors they bind (~40,000 inhibitors are known from the patent literature). Although their focus is on drug design, the approaches that are described apply equally well to the design of chemical proteomic and chemical genetic probes tailored to protein kinases.

Advantages of a family-centered approach to x-ray crystallography include more efficient design of optimized protein expression constructs for making protein and more efficient structure determination using molecular replacement. Further efficiency improvements, common to other structural proteomic approaches like NMR, come from dramatic improvements in the high-throughput and parallel approaches to protein expression and crystallization due to technological advances in automation and liquid handling. Jacobs et al. note that the complexity of characterizing protein kinase binding sites is confounded by the conformational variability within a given kinase. For example, when sequence identity is >60%, there is often more conformational variability in a given kinase than there is between different kinases.

For this reason, their approach to categorizing binding sites focuses on categorizing ligand complexes with computational analysis to find common pharmacophores, often performing fragment swapping of known complexes to design new inhibitors. Inhibitor design efforts can be for more cross-reactivity, targeting common binding site features and conformations, or for more specificity, targeting binding site differences. The ongoing innovation in this field pertains not so much to the increasing speed of structure determination with x-ray crystallography via automation, but rather to the approaches and computational tools to analyze the abundance of data on gene family-related protein–ligand complexes.

20.5.3   CHAPTER 19: EPR SPECTROSCOPY IN GENOME-WIDE EXPRESSION STUDIES

EPR is a powerful spectroscopic tool for characterizing paramagnetic species (especially metal ions and cofactors) bound to purified proteins. Richard Cammack provides a cogent description of EPR methodology, theory, hardware, and sample preparation considerations. The scope of EPR applications in protein studies and their limitations are also presented. Relevant for proteomics is the ability to study protein mixtures with EPR and to assess proper folding and assembly of metallo-proteins expressed in *Escherichia coli*. The latter is especially important as functional and structural proteomics projects ramp up the parallel expression and production of many new genomics-derived proteins, often of unknown function. Given the precedent for misincorporation of metal ions into proteins expressed in *E. coli*, EPR-based characterization of potential metalloproteins may become an important complementary step in x-ray crystallography and NMR-based structural genomics efforts. This may be especially true for the iron-sulfur proteins.

Index

421